Elementary Algebra

Denny Burzynski

Wade Ellis, Jr.

West Valley College
Saratoga, California

 SAUNDERS COLLEGE PUBLISHING

Philadelphia   New York   Chicago
San Francisco   Montreal   Toronto
London   Sydney   Tokyo

Requests for permission to make copies of any part of the work should be mailed to: Permissions, Holt, Rinehart and Winston, 111 Fifth Avenue, New York, New York 10003

Text Typeface: Century Schoolbook
Compositor: Progressive Typographers, Inc.
Acquisitions Editor: Robert B. Stern
Developmental Editor: Ellen Newman
Project Editors: Margaret Mary Anderson, Janet B. Nuciforo
Copy Editor: Charlotte Nelson
Art Director: Carol C. Bleistine
Art Assistant: Doris Bruey
Text Designer: Emily Harste
Cover Designer: Lawrence R. Didona
Text Artwork: ANCO/Boston, Inc. and Grafacon, Inc.
Layout Artist: I. Raymond Chronister, Jr.
Managing Editor: Carol Field
Production Manager: Merry Post
Production Coordinator: Joanne Cassetti

Cover Credit: Coonley Playhouse Triptych (center panel) stained glass window © 1962 The Frank Lloyd Wright Foundation. Oakbrook-Esser Studios, exclusive Frank Lloyd Wright art glass licensee.

Printed in the United States of America

ELEMENTARY ALGEBRA

0-03-063906-9

Library of Congress Catalog Card Number: 88-26517

901   071   987654321

Preface

Elementary Algebra is a work text that covers the traditional topics studied in a modern elementary algebra course. It is intended for students who (1) have no exposure to elementary algebra, (2) have had a previously unpleasant experience with elementary algebra, or (3) need to review algebraic concepts and techniques.

Use of this book will help the student develop the insight and intuition necessary to master algebraic techniques and manipulative skills. The text is written to promote problem-solving ability so that the student has the maximum opportunity to see that the concepts and techniques are logically based and to be comfortable enough with these concepts to know when and how to use them in subsequent sections, courses, and non-classroom situations. Intuition and understanding are some of the keys to creativity; we believe that the material presented will help make these keys available to the student.

This text can be used in standard lecture or self-paced classes. To help students meet these objectives and to make the study of algebra a pleasant and rewarding experience, *Elementary Algebra* is organized as follows.

❑ PEDAGOGICAL FEATURES

The work text format gives the student space to practice algebraic skills with ready reference to sample problems. The chapters are divided into sections, and each section is a complete treatment of a particular topic, which includes the following features: Section Overview, Sample Sets, Practice Sets, Section Exercises, Exercises for Review, and Answers to Practice Sets. The chapters begin with Objectives and end with a Summary of Key Concepts, an Exercise Supplement, and a Proficiency Exam.

Objectives

Each chapter begins with a set of objectives identifying the material to be covered. Each section begins with an overview that repeats the objectives for that particular section. Sections are divided into subsections that correspond to the section objectives, which makes for easier reading.

Sample Sets

Elementary Algebra contains examples that are set off in boxes for easy reference. The examples are referred to as Sample Sets for two reasons: (1) They serve as a representation to be imitated, which we believe will foster understanding of algebraic concepts and provide experience with algebraic techniques. (2) Sample Sets also serve as a preliminary representation of problem-solving techniques that may be used to solve more general and more complicated problems. The examples have been carefully chosen to illustrate and develop concepts and techniques in the most instructive, easily remembered way. Concepts and techniques preceding the examples are introduced at a level below that normally used in similar texts and are thoroughly explained, assuming little previous knowledge.

Practice Sets

A parallel Practice Set follows each Sample Set, which reinforces the concepts just learned. There is adequate space for the student to work each problem directly on the page.

Answers to Practice Sets

The Answers to all Practice Sets are given at the end of each section and can be located easily by referring to the page number, which appears after the last Practice Set in each section.

Section Exercises

The exercises at the end of each section are graded in terms of difficulty, although they are not grouped into categories. There is an ample number of problems; after working through the exercises, the student will be capable of solving a variety of challenging problems.

The problems are paired so that the odd-numbered problems are equivalent in kind and difficulty to the even-numbered problems. Answers to the odd-numbered problems are provided at the back of the book.

Exercises for Review

This section consists of problems that form a cumulative review of the material covered in the preceding sections of the text and is not limited to material in that chapter. The exercises are keyed by section for easy reference. Since these exercises are intended for review only, no work space is provided.

Summary of Key Concepts

A summary of the important ideas and formulas used throughout the chapter is included at the end of each chapter. More than just a list of terms, the summary is a valuable tool that reinforces concepts in preparation for the Proficiency Exam at the end of the chapter, as well as future exams. The summary keys each item to the section of the text where it is discussed.

Exercise Supplement

In addition to numerous section exercises, each chapter includes approximately 100 supplemental problems, which are referenced by section. Answers to the odd-numbered problems are included in the back of the book.

Proficiency Exam

Each chapter ends with a Proficiency Exam that can serve as a chapter review or a chapter evaluation. The Proficiency Exam is keyed to sections, which enables the student to refer back to the text for assistance. Answers to all Proficiency Exam problems are included in the back of the book.

❑ CONTENT

The writing style is informal and friendly, offering a no-nonsense, straightforward approach to algebra. We have made a deliberate effort not to write another text that minimizes the use of words because we believe that students can best study algebraic concepts and understand algebraic techniques by using words *and* symbols rather than symbols alone. It has been our experience that students at the elementary level are not experienced enough with mathematics to understand symbolic explanations alone; they also need to read the explanation.

We have taken great care to present concepts and techniques so they are understandable and easily remembered. After concepts have been developed, students are warned about common pitfalls.

Chapter R — Arithmetic Review This chapter contains many examples of arithmetic techniques that are used directly or indirectly in algebra. Since the chapter is intended as a review, the problem-solving techniques are presented without being developed. Therefore, no work space is provided, nor does the chapter contain all of the pedagogical features of the text. As a review, this chapter can be assigned at the discretion of the instructor and can also be a valuable reference tool for the student.

Chapter 1 — Basic Properties of Real Numbers The symbols, notations, and properties of numbers that form the basis of algebra, as well as exponents and the rules of exponents, are introduced in Chapter 1. Each property of real numbers and the rules of exponents are expressed both symbolically and literally. Literal explanations are included because symbolic explanations alone may be difficult for a student to interpret.

Chapter 2 — Basic Operations with Real Numbers The basic operations with real numbers are presented in this chapter. The concept of absolute value is discussed both geometrically and symbolically. The geometric presentation offers a visual understanding of the meaning of $|x|$. The symbolic presentation includes a literal explanation of how to use the definition. Negative exponents are developed, using reciprocals and the rules of exponents the student has already learned. Scientific notation is also included, using unique and real-life examples.

Chapter 3 — Algebraic Expressions and Equations   Operations with algebraic expressions and numerical evaluations are introduced in Chapter 3. Coefficients are described rather than merely defined. Special binomial products have both literal and symbolic explanations and since they occur so frequently in mathematics, we have been careful to help the student remember them. In each example problem, the student is "talked" through the symbolic form.

Chapter 4 — Solving Linear Equations and Inequalities   In this chapter, the emphasis is on the mechanics of equation solving, which clearly explains how to isolate a variable. The goal is to help the student feel more comfortable with solving applied problems. Ample opportunity is provided for the student to practice translating words to symbols, which is an important part of the "Five-Step Method" of solving applied problems (discussed in Sections 4.5 and 4.6).

Chapter 5 — Factoring Polynomials   Factoring is an essential skill for success in algebra and higher level mathematics courses. Therefore, we have taken great care in developing the student's understanding of the factorization process. The technique is consistently illustrated by displaying an empty set of parentheses and describing the thought process used to discover the terms that are to be placed inside the parentheses.

The factoring scheme for special products is presented with both verbal and symbolic descriptions, since not all students can interpret symbolic descriptions alone. Two techniques, the standard "trial and error" method, and the "collect and discard" method (a method similar to the "ac" method), are presented for factoring trinomials with leading coefficients different from 1.

Chapter 6 — Graphing Linear Equations and Inequalities in One and Two Variables   In this chapter the student is shown how graphs provide information that is not always evident from the equation alone. The chapter begins by establishing the relationship between the variables in an equation, the number of coordinate axes necessary to construct its graph, and the spatial dimension of both the coordinate system and the graph. Interpretation of graphs is also emphasized throughout the chapter, beginning with the plotting of points. The slope formula is fully developed, progressing from verbal phrases to mathematical expressions. The expressions are then formed into an equation by explicitly stating that a ratio is a comparison of two quantities of the same type (e.g., distance, weight, or money). This approach benefits students who take future courses that use graphs to display information.

The student is shown how to graph lines using the intercept method, the table method, and the slope-intercept method, as well as how to distinguish, by inspection, oblique and horizontal/vertical lines.

Chapter 7 — Rational Expressions   A detailed study of arithmetic operations with rational expressions is presented in this chapter, beginning with the definition of a rational expression and then proceeding immediately to a discussion of the domain. The process of reducing a rational expression and illustrations of multiplying, dividing, adding, and subtracting rational expressions are also included. Since the operations of addition and subtraction can cause the most difficulty, they are given particular attention. We have tried to make the written explanation of the examples clearer by using a "freeze frame" approach. For example, the operation $\dfrac{3x}{x-2} + \dfrac{5}{x+3}$ is treated as follows:

$$\frac{3x}{x-2} + \frac{5}{x+3}$$

The LCD is $(x-2)(x+3)$. Rewrite each fraction with the LCD as the denominator.

We now try the "freeze frame" approach by writing each step.

$$\frac{}{(x-2)(x+3)} + \frac{}{(x-2)(x+3)}$$

The first denominator has been multiplied by $x + 3$. Multiply the numerator by $x + 3$.

$$3x(x+3) = 3x^2 + 9x$$

$$\frac{3x(x+3)}{(x-2)(x+3)} + \frac{}{(x-2)(x+3)}$$

The second denominator has been multiplied by $x - 2$. Multiply the numerator by $x - 2$.

$$5(x-2) = 5x - 10$$

$$\frac{3x(x+3)}{(x-2)(x+3)} + \frac{5(x-2)}{(x-2)(x+3)}$$

Now, add the fractions.

$$\frac{3x^2 + 9x + 5x + 10}{(x-2)(x+3)} + \frac{3x^2 + 14x + 10}{(x-2)(x+3)}$$

The five-step method of solving applied problems is included in this chapter to show the problem-solving approach to number problems, work problems, and geometry problems. The chapter also illustrates simplification of complex rational expressions, using the combine-divide method and the LCD-multiply-divide method.

Chapter 8 — Roots, Radicals, and Square Root Equations The distinction between the principal square root of the number x, $\sqrt{x}$, and the secondary square root of the number x, $\sqrt{x}$, is made by explanation and by example. The simplification of radical expressions that both involve and do not involve fractions is shown in many detailed examples; this is followed by an explanation of how and why radicals are eliminated from the denominator of a radical expression. Real-life applications of radical equations have been included, such as problems involving daily output, daily sales, electronic resonance frequency, and kinetic energy.

Chapter 9 — Quadratic Equations Methods of solving quadratic equations as well as the logic underlying each method are discussed. Factoring, extraction of roots, completing the square, and the quadratic formula are carefully developed. The zero-factor property of real numbers is reintroduced. The chapter also includes graphs of quadratic equations based on the standard parabola, $y = x^2$, and applied problems from the areas of manufacturing, population, physics, geometry, mathematics (numbers and volumes), and astronomy, which are solved using the five-step method.

Chapter 10 — Systems of Linear Equations Beginning with the graphical solution of systems, this chapter includes an interpretation of independent, inconsistent, and dependent systems and examples to illustrate the applications for these systems. The substitution method and the addition method of solving a system by elimination are explained, noting when to use each method. The five-step method is again used to illustrate the solutions of value and rate problems (coin and mixture problems), using drawings that correspond to the actual situation.

❑ ANCILLARY PACKAGE

Users of *Elementary Algebra* will receive an extensive set of ancillary items that can substantially assist the instructor in the presentation of the course as well as motivate the student. This package includes the following:

Math Review Pak   With every copy of the text, students will receive a package of index cards containing important ideas and formulas from the Summary of Key Concepts at the end of each chapter. Each index card lists a key word on one side with the definition or explanation on the reverse side; the key word is referenced to the section of the text where it is discussed. This is a useful device to help students remember the concepts learned in the text and to study for exams.

Instructor's Manual   The printed manual contains chapter-by-chapter objectives, lecture suggestions, and recommendations for using the entire ancillary package, as well as:

- Answers to both even- and odd-numbered exercises
- Pretests and answers for each chapter of the text

Prepared Tests   This set of tests contains six written tests for each of the chapters in the book. Half of the tests have open-ended questions and the other half have multiple-choice questions. Thus, instructors are provided with more flexibility in testing. The answers to all of the questions are included. There is also a Diagnostic Test (with answers included) that can be used for placing students in the proper course.

Computerized Test Bank   Available for the Apple II and the IBM PC, this bank contains over 2,000 questions, both open-ended and multiple choice, enabling the instructor to create many unique tests. The instructor is able to edit the questions as well as add new questions. The software solves each problem and prints the answer in a grading key on a separate sheet.

Test Bank   This printed bank contains five tests for each chapter of the book that were generated from the Computerized Test Bank and bound as another source of tests. Answers to the tests are included on separate grading keys.

MAXIS Interactive Software   This program disk contains practice problems from each chapter of the text. Use of the software will provide the student with an alternative way to learn the material and, at the same time, receive individualized attention. The program automatically advances to the next level of difficulty once the student has successfully solved a few problems; the student may also ask to see the solution to check his or her understanding of the process used. Once the student has completed a section, a printout is available showing how many problems were answered correctly. The software is keyed to the text and refers the student to the appropriate section of the text if an incorrect answer is input. A useful tool to check skills and to identify and correct any difficulties in finding solutions, this software is available for the Apple II and IBM microcomputers.

Videotapes   A complete set (15 hours) of videotapes is free to adopters and will give added assistance or serve as a quick review of the book. Keyed to the text, the videotapes "walk" the student through each section of the book and provide another approach toward mastery of the given topic. Prepared by the authors of the text, the videotapes often include a short summary of the theory used in solving the problems before working some of the odd-numbered problems in the text.

Student Solutions Manual and Study Guide   This guide contains step-by-step solutions to one fourth of the problems in the exercise sets (every other odd-numbered problem) in addition to providing the student with a short summary of each chapter. This will help the student practice the techniques used for solving problems.

Acknowledgments

Many extraordinarily talented people are responsible for helping to create this text. We wish to acknowledge the efforts and skills of the following mathematicians. Their contributions have been invaluable.

Jerald T. Ball, Chabot College
Ron L. Bohuslov, College of Alameda
Anita Buker, Miami-Dade Community College
Ann Bretscher, University of Georgia
Loren Gaither, Paul D. Camp Community College
John Gordon, Georgia State University
Patricia Hauss, Arapahoe Community College
Jean Holton, Tidewater Community College
Katherine Huppler, St. Cloud State University
Bruce Jacobs, Laney College
Donald R. Johnson, Scottsdale Community College
John Lenhert, Long Beach Community College
Roland E. Lentz, Mankato State University
Jean Moran, Donnelley College
Patricia Morgan, San Diego State University
Charles Peselnick, Devry Institute of Technology
Mazina S. Porter, Paul D. Camp Community College
David Price, Tarrant County Junior College
Harvey Reynolds, Golden West College
J. Doug Richey, Northeast Texas Community College
Joyce L. Riseberg, Montgomery College
Mark Saks, Community College of Philadelphia
Nancy Wadlington Spears, Everett Community College
Molly Sumner, Pikes Peak Community College
Ian Walton, Mission College
Elizabeth M. Wayt, Tennessee State University
John Whitcomb, University of North Dakota

Special thanks to the following individuals for their careful accuracy reviews of manuscript, galleys, and page proofs: Steve Blasberg, West Valley College; Wade Ellis, Sr., University of Michigan; John R. Martin, Tarrant County Junior College; Jane Ellis, Amy Miller, and Guy Sanders, Branham High School for their help.

Our sincere thanks to Debbie Wiedemann for her encouragement, suggestions concerning psychobiological examples, proofreading much of the manuscript, and typing many of the section exercises; Sandi Wiedemann for collating the annotated reviews, counting the examples and exercises, and her untiring use of "white-out"; and Jane Ellis for solving and typing all the exercise solutions.

We thank the following people for their excellent work on the various ancillary items that accompany *Elementary Algebra:* Jane Ellis (Instructor's Manual); John R. Martin, Tarrant County Junior College (Student Solutions Manual and Study Guide); Virginia Hamilton, Shawnee State University (Computerized Test Bank); Patricia Morgan, San Diego State University (Prepared Tests); and George W. Bergeman, Northern Virginia Community College (MAXIS Interactive Software).

We also wish to thank the talented people at Saunders College Publishing whose efforts made this text run smoothly and less painfully than we had imagined. Our particular thanks to Bob Stern, Mathematics Editor; Ellen Newman, Developmental Editor; and Janet B. Nuciforo, Project Editor. Their guidance, suggestions, open minds to our suggestions and concerns, and encouragement have been extraordinarily helpful. Although there were times we thought we might be permanently damaged from rereading and rewriting, their efforts have improved this text immensely. It is a pleasure to work with such high-quality professionals.

San Jose, California

Denny Burzynski
Wade Ellis, Jr.

I would like to thank Doug Campbell, Ed Lodi, and Guy Sanders for listening to my frustrations and encouraging me on. Thanks also go to my cousin, David Raffety, who long ago in Sequoia National Forest told me what a differential equation is.

Particular thanks go to each of my colleagues at West Valley College. Our everyday conversations regarding mathematics instruction have been of the utmost importance to the development of this text and to my teaching career.

D.B.

À Sandi

C'est pour toi, l'étoile au centre de mon univers.

Contents

R ☐ Arithmetic Review R-1

R.1 Factors, Products, and Exponents R-2
R.2 Prime Factorization R-4
R.3 The Least Common Multiple R-6
R.4 Equivalent Fractions R-9
R.5 Operations with Fractions R-12
R.6 Decimal Fractions R-17
R.7 Percent R-23

1 ☐ Basic Properties of Real Numbers 1

1.1 Symbols and Notations 2
1.2 The Real Number Line and the Real Numbers 8
1.3 Properties of the Real Numbers 13
1.4 Exponents 19
1.5 Rules of Exponents 25
1.6 The Power Rules for Exponents 31
SUMMARY OF KEY CONCEPTS 36
EXERCISE SUPPLEMENT 38
PROFICIENCY EXAM 41

2 ☐ Basic Operations with Real Numbers 43

2.1 Signed Numbers 44
2.2 Absolute Value 47
2.3 Addition of Signed Numbers 50
2.4 Subtraction of Signed Numbers 56
2.5 Multiplication and Division of Signed Numbers 60
2.6 Negative Exponents 67
2.7 Scientific Notation 75
SUMMARY OF KEY CONCEPTS 82
EXERCISE SUPPLEMENT 83
PROFICIENCY EXAM 85

3 ☐ Algebraic Expressions and Equations 87

3.1 Algebraic Expressions 88
3.2 Equations 94
3.3 Classification of Expressions and Equations 100
3.4 Combining Polynomials Using Addition and Subtraction 104
3.5 Combining Polynomials Using Multiplication 110
3.6 Special Binomial Products 119
3.7 Terminology Associated with Equations 124
SUMMARY OF KEY CONCEPTS 126
EXERCISE SUPPLEMENT 128
PROFICIENCY EXAM 131

4 ☐ Solving Linear Equations and Inequalities 133

4.1 Solving Equations 134
4.2 Solving Equations of the Form $ax = b$ and $x/a = b$ 139
4.3 Further Techniques in Equation Solving 145
4.4 Applications I – Translating from Verbal to Mathematical Expressions 155
4.5 Applications II – Solving Problems 160

4.6 Linear Inequalities in One Variable   171
4.7 Linear Equations in Two Variables   180
SUMMARY OF KEY CONCEPTS   187
EXERCISE SUPPLEMENT   188
PROFICIENCY EXAM   191

5 ❑ Factoring Polynomials   193

5.1 Finding the Factors of a Monomial   194
5.2 Factoring a Monomial from a Polynomial   197
5.3 The Greatest Common Factor   202
5.4 Factoring by Grouping   207
5.5 Factoring Two Special Products   210
5.6 Factoring Trinomials with Leading Coefficient 1   217
5.7 Factoring Trinomials with Leading Coefficient Other Than 1   222
SUMMARY OF KEY CONCEPTS   233
EXERCISE SUPPLEMENT   234
PROFICIENCY EXAM   237

6 ❑ Graphing Linear Equations and Inequalitie, in One and Two Variables   239

6.1 Graphing Linear Equations and Inequalities in One Variable   240
6.2 Plotting Points in the Plane   245
6.3 Graphing Linear Equations in Two Variables   251
6.4 The Slope Intercept Form of a Line   264
6.5 Graphing Equations in Slope-Intercept Form   277
6.6 Finding the Equation of a Line   284
6.7 Graphing Linear Inequalities in Two Variables   292
SUMMARY OF KEY CONCEPTS   300
EXERCISE SUPPLEMENT   302
PROFICIENCY EXAM   305

7 ❑ Rational Expressions   309

7.1 Rational Expressions   310
7.2 Reducing Rational Expressions   317
7.3 Multiplying and Dividing Rational Expressions   323
7.4 Building Rational Expressions and the LCD   333
7.5 Adding and Subtracting Rational Expressions   347
7.6 Rational Equations   360
7.7 Applications   370
7.8 Complex Rational Expressions   383
7.9 Dividing Polynomials   390
SUMMARY OF KEY CONCEPTS   401
EXERCISE SUPPLEMENT   403
PROFICIENCY EXAM   407

8 ❑ Roots, Radicals, and Square Root Equations   409

8.1 Square Root Expressions   410
8.2 Simplifying Square Root Expressions   416
8.3 Multiplication of Square Root Expressions   422
8.4 Division of Square Root Expressions   427
8.5 Addition and Subtraction of Square Root Expressions   433
8.6 Square Root Equations with Applications   441
SUMMARY OF KEY CONCEPTS   446
EXERCISE SUPPLEMENT   448
PROFICIENCY EXAM   451

9 ❑ Quadratic Equations — 453

9.1	Solving Quadratic Equations	454
9.2	Solving Quadratic Equations by Factoring	458
9.3	Solving Quadratic Equations Using the Method of Extraction of Roots	465
9.4	Solving Quadratic Equations Using the Method of Completing the Square	471
9.5	Solving Quadratic Equations Using the Quadratic Formula	478
9.6	Applications	486
9.7	Graphing Quadratic Equations	498
	SUMMARY OF KEY CONCEPTS	504
	EXERCISE SUPPLEMENT	505
	PROFICIENCY EXAM	507

10 ❑ Systems of Linear Equations — 511

10.1	Solutions by Graphing	512
10.2	Elimination by Substitution	519
10.3	Elimination by Addition	526
10.4	Applications	533
	SUMMARY OF KEY CONCEPTS	541
	EXERCISE SUPPLEMENT	542
	PROFICIENCY EXAM	543

Answers to Selected Exercises	A-1
Index	I-1

R

Arithmetic Review

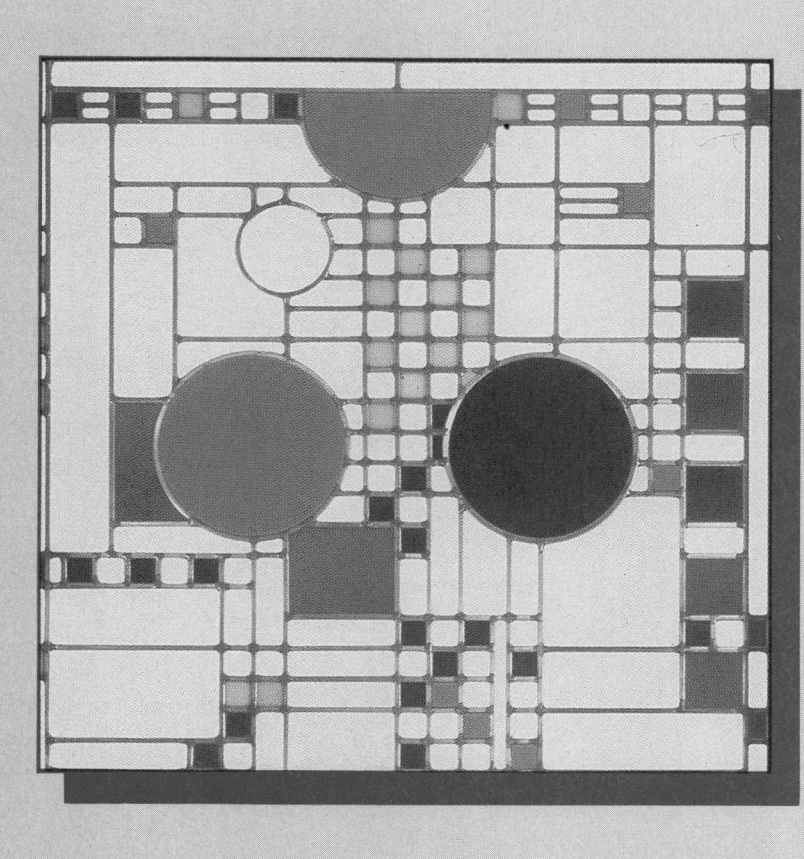

This chapter contains many examples of arithmetic techniques that are used directly or indirectly in algebra. Since the chapter is intended as a review, the problem-solving techniques are presented without being developed. If you would like a quick review of arithmetic before attempting the study of algebra, this chapter is recommended reading. If you feel your arithmetic skills are pretty good, then move on to Chapter 1. However you feel, do not hesitate to use this chapter as a *quick reference of arithmetic techniques.*

Chapters 1 through 10 include Practice Sets paired with Sample Sets with sufficient space for the student to work out the problems. In addition, these chapters include a Summary of Key Concepts, Exercise Supplements, and Proficiency Exams.

R.1 Factors, Products, and Exponents

Section Overview

☐ **FACTORS**
☐ **EXPONENTIAL NOTATION**

☐ FACTORS

Let's begin our review of arithmetic by recalling the meaning of multiplication for whole numbers (the counting numbers and zero).

Multiplication

Multiplication is a description of repeated addition.

In the addition

$7 + 7 + 7 + 7$

the number 7 is repeated as an *addend** 4 *times*. Therefore, we say we have *four times seven* and describe it by writing

$4 \cdot 7$

The raised dot between the numbers 4 and 7 indicates multiplication. The dot directs us to multiply the two numbers that it separates. In algebra, the dot is preferred over the symbol $\times$ to denote multiplication because the letter x is often used to represent a number. Thus,

$4 \cdot 7 = 7 + 7 + 7 + 7$

Factors and Products

In a multiplication, the numbers being multiplied are called **factors.** The result of a multiplication is called the **product.** For example, in the multiplication

$4 \cdot 7 = 28$

the numbers 4 and 7 are factors, and the number 28 is the product. We say that 4 and 7 are factors of 28. (They are not the only factors of 28. Can you think of others?)

Now we know that

$(\text{factor}) \cdot (\text{factor}) = \text{product}$

This indicates that a first number is a factor of a second number if the first number divides into the second number with no remainder. For example, since

$4 \cdot 7 = 28$

both 4 and 7 are factors of 28 since both 4 and 7 divide into 28 with no remainder.

☐ EXPONENTIAL NOTATION

Quite often, a particular number will be repeated as a factor in a multiplication. For example, in the multiplication

$7 \cdot 7 \cdot 7 \cdot 7$

the number 7 is repeated as a factor 4 times. We describe this by writing 7^4. Thus,

$7 \cdot 7 \cdot 7 \cdot 7 = 7^4$

---

* An addend is a number that is to be added to another number.

The repeated factor is the lower number (the base), and the number recording how many times the factor is repeated is the higher number (the superscript). The superscript number is called an *exponent*.

Exponent

> An **exponent** is a number that records how many times the number to which it is attached occurs as a factor in a multiplication.

☆ SAMPLE SET A

For Examples 1, 2, and 3, express each product using exponents.

1. $3 \cdot 3 \cdot 3 \cdot 3 \cdot 3 \cdot 3$.   Since 3 occurs as a factor 6 times,

   $3 \cdot 3 \cdot 3 \cdot 3 \cdot 3 \cdot 3 = 3^6$

2. $8 \cdot 8$.   Since 8 occurs as a factor 2 times,

   $8 \cdot 8 = 8^2$

3. $5 \cdot 5 \cdot 5 \cdot 9 \cdot 9$.   Since 5 occurs as a factor 3 times, we have 5^3. Since 9 occurs as a factor 2 times, we have 9^2. We should see the following replacements.

   $$\underbrace{5 \cdot 5 \cdot 5}_{5^3} \cdot \underbrace{9 \cdot 9}_{9^2}$$

   Then we have

   $5 \cdot 5 \cdot 5 \cdot 9 \cdot 9 = 5^3 \cdot 9^2$

4. Expand 3^5.   The base is 3 so it is the repeated factor. The exponent is 5 and it records the number of times the base 3 is repeated. Thus, 3 is to be repeated as a factor 5 times.

   $3^5 = 3 \cdot 3 \cdot 3 \cdot 3 \cdot 3$

5. Expand $6^2 \cdot 10^4$.   The notation $6^2 \cdot 10^4$ records the following two facts: 6 is to be repeated as a factor 2 times and 10 is to be repeated as a factor 4 times. Thus,

   $6^2 \cdot 10^4 = 6 \cdot 6 \cdot 10 \cdot 10 \cdot 10 \cdot 10$

Section R.1 EXERCISES

For problems 1–8, express each product using exponents.

1. $8 \cdot 8 \cdot 8$
2. $12 \cdot 12 \cdot 12 \cdot 12 \cdot 12$
3. $5 \cdot 5 \cdot 5 \cdot 5 \cdot 5 \cdot 5 \cdot 5$
4. $1 \cdot 1$
5. $3 \cdot 3 \cdot 3 \cdot 3 \cdot 3 \cdot 4 \cdot 4$
6. $8 \cdot 8 \cdot 8 \cdot 15 \cdot 15 \cdot 15 \cdot 15$
7. $2 \cdot 2 \cdot 2 \cdot 9 \cdot 9 \cdot 9 \cdot 9 \cdot 9 \cdot 9 \cdot 9 \cdot 9$
8. $3 \cdot 3 \cdot 10 \cdot 10 \cdot 10 \cdot 10$
9. Suppose that the letters x and y are each used to represent numbers. Use exponents to express the following product.

   $x \cdot x \cdot x \cdot y \cdot y$

10. Suppose that the letters x and y are each used to represent numbers. Use exponents to express the following product.

    $x \cdot x \cdot x \cdot x \cdot x \cdot y \cdot y \cdot y$

For problems 11–18, expand each product (do not compute the actual value).

11. 3^4
12. 4^3
13. 2^5
14. 9^6
15. $5^3 \cdot 6^2$
16. $2^7 \cdot 3^4$
17. $x^4 \cdot y^4$
18. $x^6 \cdot y^2$

For problems 19–28, specify all the whole number factors of each number. For example, the complete set of whole number factors of 6 is 1, 2, 3, 6.

19. 20
20. 14
21. 12
22. 30

23. 21
24. 45
25. 11
26. 17
27. 19
28. 2

R.2 Prime Factorization

Section Overview

❑ PRIME AND COMPOSITE NUMBERS
❑ THE FUNDAMENTAL PRINCIPLE OF ARITHMETIC
❑ THE PRIME FACTORIZATION OF A WHOLE NUMBER

❑ PRIME AND COMPOSITE NUMBERS

Notice that the only factors of 7 are 1 and 7 itself, and that the only factors of 23 are 1 and 23 itself.

Prime Number

A whole number greater than 1 whose only whole number factors are itself and 1 is called a **prime number.**

The first seven prime numbers are

2, 3, 5, 7, 11, 13, and 17

The number 1 is not considered to be a prime number, and the number 2 is the first and only even prime number.

Many numbers have factors other than themselves and 1. For example, the factors of 28 are 1, 2, 4, 7, 14, and 28 (since each of these whole numbers and only these whole numbers divide into 28 without a remainder).

Composite Numbers

A whole number that is composed of factors other than itself and 1 is called a **composite number.** Composite numbers are not prime numbers.

Some composite numbers are 4, 6, 8, 10, 12, and 15.

❑ THE FUNDAMENTAL PRINCIPLE OF ARITHMETIC

Prime numbers are very important in the study of mathematics. We will use them soon in our study of fractions. We will now, however, be introduced to an important mathematical principle.

The Fundamental Principle of Arithmetic

Except for the order of the factors, every whole number, other than 1, can be factored in one and only one way as a product of prime numbers.

Prime Factorization

When a number is factored so that all its factors are prime numbers, the factorization is called the **prime factorization** of the number.

☆ SAMPLE SET A

1. Find the prime factorization of 10.

$$10 = 2 \cdot 5$$

Both 2 and 5 are prime numbers. Thus, $2 \cdot 5$ is the prime factorization of 10.

2. Find the prime factorization of 60.

$$60 = 2 \cdot 30 \qquad \text{30 is not prime. } 30 = 2 \cdot 15$$
$$ = 2 \cdot 2 \cdot 15 \qquad \text{15 is not prime. } 15 = 3 \cdot 5$$
$$ = 2 \cdot 2 \cdot 3 \cdot 5 \qquad \text{We'll use exponents. } 2 \cdot 2 = 2^2$$
$$ = 2^2 \cdot 3 \cdot 5$$

The numbers 2, 3, and 5 are all primes. Thus, $2^2 \cdot 3 \cdot 5$ is the prime factorization of 60.

3. Find the prime factorization of 11.

11 is a prime number. Prime factorization applies only to composite numbers.

❑ THE PRIME FACTORIZATION OF A WHOLE NUMBER

The following method provides a way of finding the prime factorization of a whole number. The examples that follow will use the method and make it more clear.

1. Divide the number repeatedly by the smallest prime number that will divide into the number without a remainder.
2. When the prime number used in step 1 no longer divides into the given number without a remainder, repeat the process with the next largest prime number.
3. Continue this process until the quotient is 1.
4. The prime factorization of the given number is the product of all these prime divisors.

☆ SAMPLE SET B

1. Find the prime factorization of 60.

Since 60 is an even number, it is divisible by 2. We will repeatedly divide by 2 until we no longer can (when we start getting a remainder). We shall divide in the following way.

2⟌60	
2⟌30	30 is divisible by 2 again.
3⟌15	15 is not divisible by 2, but is divisible by 3, the next largest prime.
5⟌5	5 is not divisible by 3, but is divisible by 5, the next largest prime.
1	The quotient is 1 so we stop the division process.

The prime factorization of 60 is the product of all these divisors.

$$60 = 2 \cdot 2 \cdot 3 \cdot 5 \qquad \text{We will use exponents when possible.}$$
$$60 = 2^2 \cdot 3 \cdot 5$$

Continued

2. Find the prime factorization of 441.

Since 441 is an odd number, it is not divisible by 2. We'll try 3, the next largest prime.

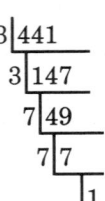

$3\lfloor 441$	
$3\lfloor 147$	**147** is divisible by 3.
$7\lfloor 49$	**49** is not divisible by 3 nor by 5, but by 7.
$7\lfloor 7$	**7** is divisible by 7.
1	The quotient is 1 so we stop the division process.

The prime factorization of 441 is the product of all the divisors.

$441 = 3 \cdot 3 \cdot 7 \cdot 7$ **We will use exponents when possible.**

$441 = 3^2 \cdot 7^2$

Section R.2 EXERCISES

For problems 1–15, determine which whole numbers are prime and which are composite.

1. 23

3. 27

5. 3

7. 7

9. 11

11. 55

13. 1044

15. 209

2. 25

4. 2

6. 5

8. 9

10. 34

12. 63

14. 339

For problems 16–25, find the prime factorization of each whole number. Use exponents on repeated factors.

16. 26

18. 54

20. 56

22. 480

24. 2025

17. 38

19. 62

21. 176

23. 819

25. 148,225

R.3 The Least Common Multiple

Section Overview

☐ **MULTIPLES**
☐ **COMMON MULTIPLES**
☐ **THE LEAST COMMON MULTIPLE (LCM)**
☐ **FINDING THE LEAST COMMON MULTIPLE**

☐ MULTIPLES

Multiples

When a whole number is multiplied by other whole numbers, with the exception of zero, the resulting products are called **multiples** of the given whole number.

Multiples of 2	Multiples of 3	Multiples of 8	Multiples of 10
$2 \times 1 = 2$	$3 \times 1 = 3$	$8 \times 1 = 8$	$10 \times 1 = 10$
$2 \times 2 = 4$	$3 \times 2 = 6$	$8 \times 2 = 16$	$10 \times 2 = 20$
$2 \times 3 = 6$	$3 \times 3 = 9$	$8 \times 3 = 24$	$10 \times 3 = 30$
$2 \times 4 = 8$	$3 \times 4 = 12$	$8 \times 4 = 32$	$10 \times 4 = 40$
$2 \times 5 = 10$	$3 \times 5 = 15$	$8 \times 5 = 40$	$10 \times 5 = 50$
$\vdots$	$\vdots$	$\vdots$	$\vdots$

❏ COMMON MULTIPLES

There will be times when we are given two or more whole numbers and we will need to know if there are any multiples that are common to each of them. If there are, we will need to know what they are. For example, some of the multiples that are common to 2 and 3 are 6, 12, and 18.

☆ SAMPLE SET A

We can visualize common multiples using the number line.

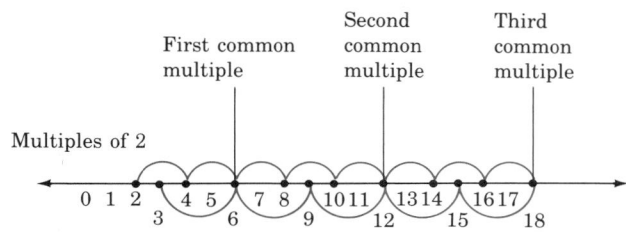

Notice that the common multiples can be divided by both whole numbers.

❏ THE LEAST COMMON MULTIPLE (LCM)

Notice that in our number line visualization of common multiples (above) the first common multiple is also the smallest, or **least common multiple,** abbreviated by **LCM.**

Least Common Multiple

> The **least common multiple, LCM,** of two or more whole numbers is the smallest whole number that each of the given numbers will divide into without a remainder.

❏ FINDING THE LEAST COMMON MULTIPLE

Finding the LCM

> To find the LCM of two or more numbers,
> 1. Write the prime factorization of each number, using exponents on repeated factors.
> 2. Write each base that appears in each of the prime factorizations.
> 3. To each base, attach the largest exponent that appears on it in the prime factorizations.
> 4. The LCM is the product of the numbers found in step 3.

☆ SAMPLE SET B

Find the LCM of the following numbers.

1. 9 and 12

    1. $9 = 3 \cdot 3 = 3^2$
       $12 = 2 \cdot 6 = 2 \cdot 2 \cdot 3 = 2^2 \cdot 3$
    2. The bases that appear in the prime factorizations are 2 and 3.

Continued

3. The largest exponents appearing on 2 and 3 in the prime factorizations are, respectively, 2 and 2 (or 2^2 from 12, and 3^2 from 9).
4. The LCM is the product of these numbers.

$$LCM = 2^2 \cdot 3^2 = 4 \cdot 9 = 36$$

Thus, 36 is the smallest number that both 9 and 12 divide into without remainders.

2. 90 and 630

1. $90 = 2 \cdot 45 = 2 \cdot 3 \cdot 15 = 2 \cdot 3 \cdot 3 \cdot 5 = 2 \cdot 3^2 \cdot 5$
   $630 = 2 \cdot 315 = 2 \cdot 3 \cdot 105 = 2 \cdot 3 \cdot 3 \cdot 35 = 2 \cdot 3 \cdot 3 \cdot 5 \cdot 7$
   $$= 2 \cdot 3^2 \cdot 5 \cdot 7$$
2. The bases that appear in the prime factorizations are 2, 3, 5, and 7.
3. The largest exponents that appear on 2, 3, 5, and 7 are, respectively, 1, 2, 1, and 1.

   2^1 from either 90 or 630
   3^2 from either 90 or 630
   5^1 from either 90 or 630
   7^1 from 630

4. The LCM is the product of these numbers.

   $$LCM = 2 \cdot 3^2 \cdot 5 \cdot 7 = 2 \cdot 9 \cdot 5 \cdot 7 = 630$$

Thus, 630 is the smallest number that both 90 and 630 divide into with no remainders.

3. 33, 110, and 484

1. $33 = 3 \cdot 11$
   $110 = 2 \cdot 55 = 2 \cdot 5 \cdot 11$
   $484 = 2 \cdot 242 = 2 \cdot 2 \cdot 121 = 2 \cdot 2 \cdot 11 \cdot 11 = 2^2 \cdot 11^2$
2. The bases that appear in the prime factorizations are 2, 3, 5, and 11.
3. The largest exponents that appear on 2, 3, 5, and 11 are, respectively, 2, 1, 1, and 2.

   2^2 from 484
   3^1 from 33
   5^1 from 110
   11^2 from 484

4. The LCM is the product of these numbers.

   $$LCM = 2^2 \cdot 3 \cdot 5 \cdot 11^2$$
   $$= 4 \cdot 3 \cdot 5 \cdot 121$$
   $$= 7260$$

Thus, 7260 is the smallest number that 33, 110, and 484 divide into without remainders.

Section R.3 EXERCISES

For problems 1–25, find the least common multiple of given numbers.

1. 8, 12

2. 8, 10

3. 6, 12

4. 9, 18

5. 5, 6

6. 7, 9

7. 28, 36

8. 24, 36

9. 28, 42

10. 20, 24

11. 25, 30

12. 24, 54

13. 16, 24

14. 36, 48

15. 15, 21

16. 7, 11, 33

17. 8, 10, 15
18. 4, 5, 21
19. 45, 63, 98
20. 15, 25, 40
21. 12, 16, 20

22. 12, 16, 24
23. 12, 16, 24, 36
24. 6, 9, 12, 18
25. 8, 14, 28, 32

R.4 Equivalent Fractions

**Section
Overview**

❑ **EQUIVALENT FRACTIONS**
❑ **REDUCING FRACTIONS TO LOWEST TERMS**
❑ **RAISING FRACTIONS TO HIGHER TERMS**

❑ EQUIVALENT FRACTIONS

Equivalent Fractions

Fractions that have the same value are called **equivalent fractions.**

For example, $\dfrac{2}{3}$ and $\dfrac{4}{6}$ represent the same part of a whole quantity and are therefore equivalent. Several more collections of equivalent fractions are listed below.

1. $\dfrac{15}{25}, \dfrac{12}{20}, \dfrac{3}{5}$

2. $\dfrac{1}{3}, \dfrac{2}{6}, \dfrac{3}{9}, \dfrac{4}{12}$

3. $\dfrac{7}{6}, \dfrac{14}{12}, \dfrac{21}{18}, \dfrac{28}{24}, \dfrac{35}{30}$

❑ REDUCING FRACTIONS TO LOWEST TERMS

It is often useful to convert one fraction to an equivalent fraction that has reduced values in the numerator and denominator. When a fraction is converted to an equivalent fraction that has the smallest numerator and denominator in the collection of equivalent fractions, it is said to be **reduced to lowest terms.** The conversion process is called **reducing a fraction.**

Reduced to Lowest Terms

We can reduce a fraction to lowest terms by

1. Expressing the numerator and denominator as a product of prime numbers. (Find the prime factorization of the numerator and denominator. See Section R.2 for this technique.)
2. Divide the numerator and denominator by all common factors. (This technique is commonly called "cancelling.")

☆ **SAMPLE SET A**

Reduce each fraction to lowest terms.

1. $\dfrac{6}{18} = \dfrac{2 \cdot 3}{2 \cdot 3 \cdot 3}$

$= \dfrac{\cancel{2} \cdot \cancel{3}}{\cancel{2} \cdot \cancel{3} \cdot 3}$ **2 and 3 are common factors.**

$= \dfrac{1}{3}$

Continued

2. $\dfrac{16}{20} = \dfrac{2 \cdot 2 \cdot 2 \cdot 2}{2 \cdot 2 \cdot 5}$

$= \dfrac{\not2 \cdot \not2 \cdot 2 \cdot 2}{\not2 \cdot \not2 \cdot 5}$ **2 is the only common factor.**

$= \dfrac{4}{5}$

3. $\dfrac{56}{70} = \dfrac{2 \cdot 4 \cdot 7}{2 \cdot 5 \cdot 7}$

$= \dfrac{\not2 \cdot 4 \cdot \not7}{\not2 \cdot 5 \cdot \not7}$ **2 and 7 are common factors.**

$= \dfrac{4}{5}$

4. $\dfrac{8}{15} = \dfrac{2 \cdot 2 \cdot 2}{3 \cdot 5}$ **There are no common factors.**

Thus, $\dfrac{8}{15}$ is reduced to lowest terms.

❏ RAISING A FRACTION TO HIGHER TERMS

Equally important as reducing fractions is **raising fractions to higher terms.** Raising a fraction to higher terms is the process of constructing an equivalent fraction that has higher values in the numerator and denominator. The higher, equivalent fraction is constructed by multiplying the original fraction by 1.

Notice that $\dfrac{3}{5}$ and $\dfrac{9}{15}$ are equivalent, that is $\dfrac{3}{5} = \dfrac{9}{15}$. Also,

$$\dfrac{3}{5} \cdot 1 = \dfrac{3}{5} \cdot \dfrac{3}{3} = \dfrac{3 \cdot 3}{5 \cdot 3} = \dfrac{9}{15}$$

$$1 = \dfrac{3}{3}$$

This observation helps us suggest the following method for raising a fraction to higher terms.

Raising a Fraction to Higher Terms

> A fraction can be raised to higher terms by multiplying both the numerator and denominator by the same nonzero number.

For example, $\dfrac{3}{4}$ can be raised to $\dfrac{24}{32}$ by multiplying both the numerator and denominator by 8, that is, multiplying by 1 in the form $\dfrac{8}{8}$.

$$\dfrac{3}{4} = \dfrac{3 \cdot 8}{4 \cdot 8} = \dfrac{24}{32}$$

How did we know to choose 8 as the proper factor? Since we wish to convert 4 to 32 by multiplying it by some number, we know that 4 must be a factor of 32. This means that 4 divides into 32. In fact, $32 \div 4 = 8$. We divided the original denominator into the new, specified denominator to obtain the proper factor for the multiplication.

⭐ **SAMPLE SET B**

Determine the missing numerator or denominator.

1. $\frac{3}{7} = \frac{?}{35}$. Divide the original denominator, 7, into the new denominator, 35. $35 \div 7 = 5$. Multiply the original numerator by 5.

$$\frac{3}{7} = \frac{3 \cdot 5}{7 \cdot 5} = \frac{15}{35}$$

2. $\frac{5}{6} = \frac{45}{?}$. Divide the original numerator, 5, into the new numerator, 45. $45 \div 5 = 9$. Multiply the original denominator by 9.

$$\frac{5}{6} = \frac{5 \cdot 9}{6 \cdot 9} = \frac{45}{54}$$

Section R.4 EXERCISES

For problems 1–25, reduce, if possible, each fraction to lowest terms.

1. $\frac{6}{8}$

2. $\frac{5}{10}$

3. $\frac{6}{14}$

4. $\frac{4}{14}$

5. $\frac{18}{12}$

6. $\frac{20}{8}$

7. $\frac{10}{6}$

8. $\frac{14}{4}$

9. $\frac{10}{12}$

10. $\frac{32}{28}$

11. $\frac{36}{10}$

12. $\frac{26}{60}$

13. $\frac{12}{18}$

14. $\frac{18}{27}$

15. $\frac{18}{24}$

16. $\frac{32}{40}$

17. $\frac{11}{22}$

18. $\frac{17}{51}$

19. $\frac{27}{81}$

20. $\frac{16}{42}$

21. $\frac{39}{13}$

22. $\frac{44}{11}$

23. $\frac{121}{132}$

24. $\frac{30}{105}$

25. $\frac{108}{76}$

For problems 26–35, determine the missing numerator or denominator.

26. $\frac{1}{3} = \frac{?}{12}$

27. $\frac{1}{5} = \frac{?}{30}$

28. $\frac{3}{3} = \frac{?}{9}$

29. $\frac{3}{4} = \frac{?}{16}$

30. $\frac{5}{6} = \frac{?}{18}$

31. $\frac{4}{5} = \frac{?}{25}$

32. $\frac{1}{2} = \frac{4}{?}$

33. $\frac{9}{25} = \frac{27}{?}$

34. $\frac{3}{2} = \frac{18}{?}$

35. $\frac{5}{3} = \frac{80}{?}$

R.5 Operations with Fractions

☐ **MULTIPLICATION OF FRACTIONS**
☐ **DIVISION OF FRACTIONS**
☐ **ADDITION AND SUBTRACTION OF FRACTIONS**

☐ MULTIPLICATION OF FRACTIONS

Multiplication of Fractions

> To multiply two fractions, multiply the numerators together and multiply the denominators together. Reduce to lowest terms if possible.

For example, multiply $\frac{3}{4} \cdot \frac{1}{6}$.

$$\frac{3}{4} \cdot \frac{1}{6} = \frac{3 \cdot 1}{4 \cdot 6}$$

$$= \frac{3}{24} \qquad \text{Now reduce.}$$

$$= \frac{3 \cdot 1}{2 \cdot 2 \cdot 2 \cdot 3}$$

$$= \frac{\cancel{3} \cdot 1}{2 \cdot 2 \cdot 2 \cdot \cancel{3}} \qquad \text{3 is the only common factor.}$$

$$= \frac{1}{8}$$

Notice that we since had to reduce, we nearly started over again with the original two fractions. If we factor first, then cancel, then multiply, we will save time and energy and still obtain the correct product.

☆ **SAMPLE SET A**

Perform the following multiplications.

1. $\dfrac{1}{4} \cdot \dfrac{8}{9} = \dfrac{1}{2 \cdot 2} \cdot \dfrac{2 \cdot 2 \cdot 2}{3 \cdot 3}$

$\qquad = \dfrac{1}{\cancel{2} \cdot \cancel{2}} \cdot \dfrac{\cancel{2} \cdot \cancel{2} \cdot 2}{3 \cdot 3} \qquad$ 2 is a common factor.

$\qquad = \dfrac{1}{1} \cdot \dfrac{2}{3 \cdot 3}$

$\qquad = \dfrac{1 \cdot 2}{1 \cdot 3 \cdot 3}$

$\qquad = \dfrac{2}{9}$

2. $\dfrac{3}{4} \cdot \dfrac{8}{9} \cdot \dfrac{5}{12} = \dfrac{3}{2 \cdot 2} \cdot \dfrac{2 \cdot 2 \cdot 2}{3 \cdot 3} \cdot \dfrac{5}{2 \cdot 2 \cdot 3}$

$\qquad = \dfrac{\cancel{3}}{\cancel{2} \cdot \cancel{2}} \cdot \dfrac{\cancel{2} \cdot \cancel{2} \cdot \cancel{2}}{\cancel{3} \cdot 3} \cdot \dfrac{5}{\cancel{2} \cdot 2 \cdot 3} \qquad$ 2 and 3 are common factors.

$\qquad = \dfrac{1 \cdot 1 \cdot 5}{3 \cdot 2 \cdot 3}$

$\qquad = \dfrac{5}{18}$

☐ DIVISION OF FRACTIONS

Reciprocals

Two numbers whose product is 1 are **reciprocals** of each other. For example, since $\frac{4}{5} \cdot \frac{5}{4} = 1$, $\frac{4}{5}$ and $\frac{5}{4}$ are reciprocals of each other. Some other pairs of reciprocals are listed below.

$$\frac{2}{7}, \frac{7}{2} \qquad \frac{3}{4}, \frac{4}{3} \qquad \frac{6}{1}, \frac{1}{6}$$

Reciprocals are used in division of fractions.

Division of Fractions

> To divide a first fraction by a second fraction, multiply the first fraction by the reciprocal of the second fraction. Reduce if possible.

This method is sometimes called the "invert and multiply" method.

☆ SAMPLE SET B

Perform the following divisions.

1. $\frac{1}{3} \div \frac{3}{4}$.   The divisor is $\frac{3}{4}$. Its reciprocal is $\frac{4}{3}$.

$$\frac{1}{3} \div \frac{3}{4} = \frac{1}{3} \cdot \frac{4}{3}$$

$$= \frac{1 \cdot 4}{3 \cdot 3}$$

$$= \frac{4}{9}$$

2. $\frac{3}{8} \div \frac{5}{4}$.   The divisor is $\frac{5}{4}$. Its reciprocal is $\frac{4}{5}$.

$$\frac{3}{8} \div \frac{5}{4} = \frac{3}{8} \cdot \frac{4}{5}$$

$$= \frac{3}{2 \cdot 2 \cdot 2} \cdot \frac{2 \cdot 2}{5}$$

$$= \frac{3}{\cancel{2} \cdot \cancel{2} \cdot 2} \cdot \frac{\cancel{2} \cdot \cancel{2}}{5} \qquad \text{2 is a common factor.}$$

$$= \frac{3 \cdot 1}{2 \cdot 5}$$

$$= \frac{3}{10}$$

Continued

3. $\dfrac{5}{6} \div \dfrac{5}{12}$. The divisor is $\dfrac{5}{12}$. Its reciprocal is $\dfrac{12}{5}$.

$$\dfrac{5}{6} \div \dfrac{5}{12} = \dfrac{5}{6} \cdot \dfrac{12}{5}$$

$$= \dfrac{5}{2 \cdot 3} \cdot \dfrac{2 \cdot 2 \cdot 3}{5}$$

$$= \dfrac{\cancel{5}}{\cancel{2} \cdot \cancel{3}} \cdot \dfrac{\cancel{2} \cdot 2 \cdot \cancel{3}}{\cancel{5}}$$

$$= \dfrac{1 \cdot 2}{1}$$

$$= 2$$

☐ ADDITION AND SUBTRACTION OF FRACTIONS

Fractions with Like Denominators

To add (or subtract) two or more fractions that have the same denominators, add (or subtract) the numerators and place the resulting sum over the common denominator. Reduce if possible.

CAUTION

Add or subtract only the numerators. *Do not* add or subtract the denominators!

☆ SAMPLE SET C

Find the following sums.

1. $\dfrac{3}{7} + \dfrac{2}{7}$. The denominators are the same. Add the numerators and place the sum over 7.

$$\dfrac{3}{7} + \dfrac{2}{7} = \dfrac{3+2}{7} = \dfrac{5}{7}$$

2. $\dfrac{7}{9} - \dfrac{4}{9}$. The denominators are the same. Subtract 4 from 7 and place the difference over 9.

$$\dfrac{7}{9} - \dfrac{4}{9} = \dfrac{7-4}{9} = \dfrac{3}{9} = \dfrac{1}{3}$$

Fractions can only be added or subtracted conveniently if they have like denominators.

Fractions with Unlike Denominators

To add or subtract fractions having unlike denominators, convert each fraction to an equivalent fraction having as the denominator the least common multiple of the original denominators.

The least common multiple of the original denominators is commonly referred to as the **least common denominator** (LCD). See Section R.3 for the technique of finding the least common multiple of several numbers.

☆ **SAMPLE SET D**

Find each sum or difference.

1. $\frac{1}{6} + \frac{3}{4}$.   The denominators are not alike. Find the LCD of 6 and 4.

$\begin{cases} 6 = 2 \cdot 3 \\ 4 = 2^2 \end{cases}$ The LCD is $2^2 \cdot 3 = 4 \cdot 3 = 12$.

Convert each of the original fractions to equivalent fractions having the common denominator 12.

$$\frac{1}{6} = \frac{1 \cdot 2}{6 \cdot 2} = \frac{2}{12} \qquad \frac{3}{4} = \frac{3 \cdot 3}{4 \cdot 3} = \frac{9}{12}$$

Now we can proceed with the addition.

$$\frac{1}{6} + \frac{3}{4} = \frac{2}{12} + \frac{9}{12}$$
$$= \frac{2 + 9}{12}$$
$$= \frac{11}{12}$$

2. $\frac{5}{9} - \frac{5}{12}$.   The denominators are not alike. Find the LCD of 9 and 12.

$\begin{cases} 9 = 3^2 \\ 12 = 2^2 \cdot 3 \end{cases}$ The LCD is $2^2 \cdot 3^2 = 4 \cdot 9 = 36$.

Convert each of the original fractions to equivalent fractions having the common denominator 36.

$$\frac{5}{9} = \frac{5 \cdot 4}{9 \cdot 4} = \frac{20}{36} \qquad \frac{5}{12} = \frac{5 \cdot 3}{12 \cdot 3} = \frac{15}{36}$$

Now we can proceed with the subtraction.

$$\frac{5}{9} - \frac{5}{12} = \frac{20}{36} - \frac{15}{36}$$
$$= \frac{20 - 15}{36}$$
$$= \frac{5}{36}$$

Section R.5 EXERCISES

For problems 1–40, perform each indicated operation.

1. $\frac{1}{3} \cdot \frac{4}{3}$

2. $\frac{1}{3} \cdot \frac{2}{3}$

3. $\frac{2}{5} \cdot \frac{5}{6}$

4. $\frac{5}{6} \cdot \frac{14}{15}$

5. $\frac{9}{16} \cdot \frac{20}{27}$

6. $\frac{35}{36} \cdot \frac{48}{55}$

7. $\dfrac{21}{25} \cdot \dfrac{15}{14}$

8. $\dfrac{76}{99} \cdot \dfrac{66}{38}$

9. $\dfrac{3}{7} \cdot \dfrac{14}{18} \cdot \dfrac{6}{2}$

10. $\dfrac{14}{15} \cdot \dfrac{21}{28} \cdot \dfrac{45}{7}$

11. $\dfrac{5}{9} \div \dfrac{5}{6}$

12. $\dfrac{9}{16} \div \dfrac{15}{8}$

13. $\dfrac{4}{9} \div \dfrac{6}{15}$

14. $\dfrac{25}{49} \div \dfrac{4}{9}$

15. $\dfrac{15}{4} \div \dfrac{27}{8}$

16. $\dfrac{24}{75} \div \dfrac{8}{15}$

17. $\dfrac{57}{8} \div \dfrac{7}{8}$

18. $\dfrac{7}{10} \div \dfrac{10}{7}$

19. $\dfrac{3}{8} + \dfrac{2}{8}$

20. $\dfrac{3}{11} + \dfrac{4}{11}$

21. $\dfrac{5}{12} + \dfrac{7}{12}$

22. $\dfrac{11}{16} - \dfrac{2}{16}$

23. $\dfrac{15}{23} - \dfrac{2}{23}$

24. $\dfrac{3}{11} + \dfrac{1}{11} + \dfrac{5}{11}$

25. $\dfrac{16}{20} + \dfrac{1}{20} + \dfrac{2}{20}$

26. $\dfrac{3}{8} + \dfrac{2}{8} - \dfrac{1}{8}$

27. $\dfrac{11}{16} + \dfrac{9}{16} - \dfrac{5}{16}$

28. $\dfrac{1}{2} + \dfrac{1}{6}$

29. $\dfrac{1}{8} + \dfrac{1}{2}$

30. $\dfrac{3}{4} + \dfrac{1}{3}$

31. $\dfrac{5}{8} + \dfrac{2}{3}$

32. $\dfrac{6}{7} - \dfrac{1}{4}$

33. $\dfrac{8}{15} - \dfrac{3}{10}$

34. $\dfrac{1}{15} + \dfrac{5}{12}$

35. $\dfrac{25}{36} - \dfrac{7}{10}$

36. $\dfrac{9}{28} - \dfrac{4}{45}$

37. $\dfrac{8}{15} - \dfrac{3}{10}$

38. $\dfrac{1}{16} + \dfrac{3}{4} - \dfrac{3}{8}$

39. $\dfrac{8}{3} - \dfrac{1}{4} + \dfrac{7}{36}$

40. $\dfrac{3}{4} - \dfrac{3}{22} + \dfrac{5}{24}$

R.6 Decimal Fractions

Section Overview

☐ **DECIMAL FRACTIONS**
☐ **ADDING AND SUBTRACTING DECIMAL FRACTIONS**
☐ **MULTIPLYING DECIMAL FRACTIONS**
☐ **DIVIDING DECIMAL FRACTIONS**
☐ **CONVERTING DECIMAL FRACTIONS TO FRACTIONS**
☐ **CONVERTING FRACTIONS TO DECIMAL FRACTIONS**

☐ DECIMAL FRACTIONS

Fractions are one way we can represent parts of whole numbers. Decimal fractions are another way of representing parts of whole numbers.

Decimal Fractions

A **decimal fraction** is a fraction in which the denominator is a power of 10.

A decimal fraction uses a **decimal point** to separate whole parts and fractional parts. Whole parts are written to the *left* of the decimal point and fractional parts are written to the *right* of the decimal point. Just as each digit in a whole number has a particular value, so do the digits in decimal positions.

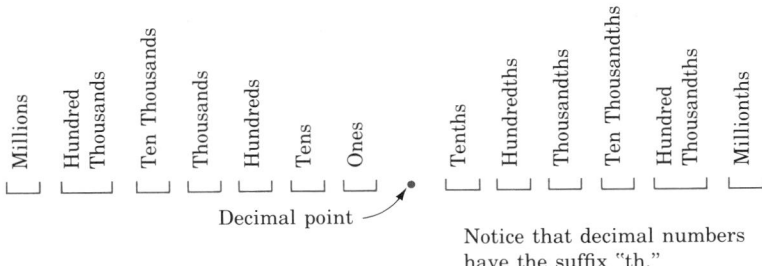

Decimal point

Notice that decimal numbers have the suffix "th."

☆ **SAMPLE SET A**

The following numbers are decimal fractions.

1. 57.9

The 9 is in the *tenths* position. $57.9 = 57\frac{9}{10}$.

2. 6.8014

The 8 is in the *tenths* position.
The 0 is in the *hundredths* position.
The 1 is in the *thousandths* position.
The 4 is in the *ten thousandths* position.

$6.8014 = 6\frac{8014}{10000}$.

❏ ADDING AND SUBTRACTING DECIMAL FRACTIONS

Adding/Subtracting Decimal Fractions

To add or subtract decimal fractions,

1. Align the numbers vertically so that the decimal points line up under each other and corresponding decimal positions are in the same column. Add zeros if necessary.
2. Add or subtract the numbers as if they were whole numbers.
3. Place a decimal point in the resulting sum or difference directly under the other decimal points.

☆ SAMPLE SET B

Find each sum or difference.

1. $9.183 + 2.140$

$$
\begin{array}{r}
\downarrow \\
9.183 \\
+\ 2.140 \\
\hline
11.323
\end{array}
$$

The decimal points are aligned in the same column.

2. $841.0056 + 47.016 + 19.058$

$$
\begin{array}{r}
\downarrow \\
841.0056 \\
47.016 \\
+\ 19.058 \\
\end{array}
$$

The decimal points are aligned in the same column.

Place a 0 into the thousandths position.
Place a 0 into the thousandths position.

$$
\begin{array}{r}
\downarrow \\
841.0056 \\
47.0160 \\
+\ 19.0580 \\
\hline
907.0796
\end{array}
$$

The decimal points are aligned in the same column.

3. $16.01 - 7.053$

$$
\begin{array}{r}
\downarrow \\
16.01 \\
-\ 7.053 \\
\end{array}
$$

The decimal points are aligned in the same column.
Place a 0 into the thousandths position.

$$
\begin{array}{r}
\downarrow \\
16.010 \\
-\ 7.053 \\
\hline
8.957
\end{array}
$$

The decimal points are aligned in the same column.

❏ MULTIPLYING DECIMAL FRACTIONS

Multiplying Decimal Fractions

To multiply decimals,

1. Multiply the numbers as if they were whole numbers.
2. Find the sum of the number of decimal places in the factors.
3. The number of decimal places in the product is the sum found in step 2.

Find the following products.

1. 6.5×4.3

$$
\begin{array}{r}
6.5 \\
\underline{\times\,4.3} \\
195 \\
\underline{260} \\
27.95
\end{array}
$$

← 1 decimal place ⎫ $1 + 1 = 2$ decimal places in the
← 1 decimal place ⎬ product.

← 2 decimal places

$6.5 \times 4.3 = 27.95$

2. 23.4×1.96

$$
\begin{array}{r}
23.4 \\
\underline{\times\,1.96} \\
1404 \\
2106 \\
\underline{234} \\
45.864
\end{array}
$$

← 1 decimal place ⎫ $1 + 2 = 3$ decimal places in the
← 2 decimal places ⎬ product.

← 3 decimal places

$23.4 \times 1.96 = 45.864$

☐ DIVIDING DECIMAL FRACTIONS

Dividing Decimal Fractions

To divide a decimal by a nonzero decimal,

1. Convert the divisor to a whole number by moving the decimal point to the position immediately to the right of the divisor's last digit.
2. Move the decimal point of the dividend to the right the same number of digits it was moved in the divisor.
3. Set the decimal point in the quotient by placing a decimal point directly above the decimal point in the dividend.
4. Divide as usual.

Find the following quotients.

1. $32.66 \div 7.1$

$7.1\overline{)32.66}$ The divisor has one decimal place.

$$
\begin{array}{r}
4.6 \\
71.\overline{)326.6} \\
\underline{284} \\
42.6 \\
\underline{42.6} \\
0
\end{array}
$$

Move the decimal point of both the divisor and the dividend 1 place to the right.
Set the decimal point.
Divide as usual.

$32.66 \div 7.1 = 4.6$

Continued

Check: $32.66 \div 7.1 = 4.6$ if $4.6 \times 7.1 = 32.66$

$$
\begin{array}{r}
4.6 \\
\underline{7.1} \\
46 \\
\underline{322} \\
32.66
\end{array}
$$
   True

2. $1.0773 \div 0.513$

$$
\begin{array}{r}
2.1 \\
.513\overline{)1.077\,3} \\
\underline{1\,026} \\
51\,3
\end{array}
$$

The divisor has 3 decimal places.

Move the decimal point of both the divisor and the dividend 3 places to the right.
Set the decimal place and divide.

$1.0773 \div 0.513 = 2.1$

Check by multiplying 2.1 and 0.513. This will show that we have obtained the correct result.

3. $12 \div 0.00032$

$$0.00032\overline{)12}$$

$$0.00032\,\overline{)12.00000}$$

The divisor has 5 decimal places.

Move the decimal point of both the divisor and the dividend 5 places to the right. We will need to add 5 zeros to 12.

$$
\begin{array}{r}
37500. \\
32\overline{)1200000.} \\
\underline{96} \\
240 \\
\underline{224} \\
160 \\
\underline{160} \\
000
\end{array}
$$

Set the decimal place and divide.

☐ CONVERTING DECIMAL FRACTIONS TO FRACTIONS

We can convert a decimal fraction to a fraction by reading it and then writing the phrase we have just read. As we read the decimal fraction, we note the place value farthest to the right. We may have to reduce the fraction.

☆ **SAMPLE SET E**

Convert each decimal fraction to a fraction.

1. 0.6

0.6 $\longrightarrow$ tenths position

Reading: six tenths $\longrightarrow \dfrac{6}{10}$

Reduce: $0.6 = \dfrac{6}{10} = \dfrac{3}{5}$

2. 21.903

21.90<u>3</u> ⟶ thousandths position

Reading: twenty-one and nine hundred three thousandths ⟶ $21\dfrac{903}{1000}$

☐ CONVERTING FRACTIONS TO DECIMAL FRACTIONS

☆ SAMPLE SET F

Convert the following fractions to decimals. If the division is nonterminating, round to 2 decimal places.

1. $\dfrac{3}{4}$

```
       .75
    4)3.00
      2 8
        20
        20
         0
```

Divide 3 by 4.

$\dfrac{3}{4} = 0.75$

2. $\dfrac{1}{5}$

```
       .2
    5)1.0
      1 0
        0
```

Divide 1 by 5.

$\dfrac{1}{5} = 0.2$

3. $\dfrac{5}{6}$

```
       .833
    6)5.000
      4 8
        20
        18
        20
```

Divide by 6.

This recurring remainder indicates that the division is nonterminating.

$\dfrac{5}{6} = 0.833\ \ldots$

We are to round to 2 decimal places.

$\dfrac{5}{6} = 0.83$ to 2 decimal places.

Continued

4. $5\dfrac{1}{8}$

Note that $5\dfrac{1}{8} = 5 + \dfrac{1}{8}$.

$$
\begin{array}{r}
.125 \\
8\overline{)\,1.000} \\
\underline{8} \\
20 \\
\underline{16} \\
40 \\
\underline{40} \\
0
\end{array}
$$

Convert $\dfrac{1}{8}$ to a decimal.

$\dfrac{1}{8} = .125$

Thus, $5\dfrac{1}{8} = 5 + \dfrac{1}{8} = 5 + .125 = 5.125$.

5. $0.16\dfrac{1}{4}$

This is a complex decimal. The "6" is in the hundredths position. The number $0.16\dfrac{1}{4}$ is read as "sixteen and one-fourth hundredths."

$$
0.16\dfrac{1}{4} = \dfrac{16\dfrac{1}{4}}{100} = \dfrac{\dfrac{16 \cdot 4 + 1}{4}}{100} = \dfrac{\dfrac{65}{4}}{\dfrac{100}{1}}
$$

$$
= \dfrac{\overset{13}{\cancel{65}}}{4} \cdot \dfrac{1}{\underset{20}{\cancel{100}}} = \dfrac{13 \times 1}{4 \times 20} = \dfrac{13}{80}
$$

Now, convert $\dfrac{13}{80}$ to a decimal.

$$
\begin{array}{r}
.1625 \\
80\overline{)\,13.0000} \\
\underline{8\,0} \\
5\,00 \\
\underline{4\,80} \\
160 \\
\underline{160} \\
400 \\
\underline{400} \\
0
\end{array}
$$

$0.16\dfrac{1}{4} = 0.1625.$

Section R.6 EXERCISES

For problems 1–20, perform each indicated operation.

1. $1.84 + 7.11$
2. $15.015 - 6.527$
3. $4.904 - 2.67$
4. $156.33 - 24.095$
5. $.0012 + 1.53 + 5.1$
6. $44.98 + 22.8 - 12.76$
7. $5.0004 - 3.00004 + 1.6837$
8. $1.11 + 12.1212 - 13.131313$
9. $4.26 \cdot 3.2$
10. $2.97 \cdot 3.15$
11. $23.05 \cdot 1.1$
12. $5.009 \cdot 2.106$
13. $0.1 \cdot 3.24$
14. $100 \cdot 12.008$
15. $1000 \cdot 12.008$
16. $10{,}000 \cdot 12.008$
17. $75.642 \div 18.01$
18. $51.811 \div 1.97$
19. $0.0000448 \div 0.014$
20. $0.129516 \div 1004$

For problems 21–25, convert each decimal fraction to a fraction.

21. 0.06
22. 0.115
23. 3.7
24. 48.1162
25. 712.00004

For problems 26–30, convert each fraction to a decimal fraction. If the decimal form is nonterminating, round to 3 decimal places.

26. $\dfrac{5}{8}$
27. $\dfrac{9}{20}$
28. $15 \div 22$
29. $\dfrac{7}{11}$
30. $\dfrac{2}{9}$

R.7 Percent

Section Overview

- ❑ **THE MEANING OF PERCENT**
- ❑ **CONVERTING A FRACTION TO A PERCENT**
- ❑ **CONVERTING A DECIMAL TO A PERCENT**
- ❑ **CONVERTING A PERCENT TO A DECIMAL**

❑ THE MEANING OF PERCENT

The word *percent* comes from the Latin word "per centum," "per" meaning "for each," and "centum" meaning "hundred."

Percent (%)

> **Percent** means "for each hundred" or "for every hundred." The symbol % is used to represent the word percent.

Thus, $1\% = \dfrac{1}{100}$ or $1\% = 0.01$.

❑ CONVERTING A FRACTION TO A PERCENT

We can see how a fraction can be converted to a percent by analyzing the method that $\dfrac{3}{5}$ is converted to a percent. In order to convert $\dfrac{3}{5}$ to a percent, we need to introduce $\dfrac{1}{100}$ (since percent means for each hundred).

$$\frac{3}{5} = \frac{3}{5} \cdot \frac{100}{100} \qquad \text{Multiply the fraction by 1.}$$

$$= \frac{3}{5} \cdot 100 \cdot \frac{1}{100} \qquad \text{Since } \frac{100}{100} = 100 \cdot \frac{1}{100}.$$

$$= \frac{300}{5} \cdot \frac{1}{100} \qquad \text{Divide 300 by 5.}$$

$$= 60 \cdot \frac{1}{100} \qquad \text{Multiply the fractions.}$$

$$= 60\% \qquad \text{Replace } \frac{1}{100} \text{ with the \% symbol.}$$

Fraction to Percent

> To convert a fraction to a percent, multiply the fraction by 1 in the form $100 \cdot \frac{1}{100}$, then replace $\frac{1}{100}$ with the % symbol.

☆ **SAMPLE SET A**

Convert each fraction to a percent.

1. $\dfrac{1}{4} = \dfrac{1}{4} \cdot 100 \cdot \dfrac{1}{100}$

$= \dfrac{100}{4} \cdot \dfrac{1}{100}$

$= 25 \cdot \dfrac{1}{100}$

$= 25\%$

2. $\dfrac{8}{5} = \dfrac{8}{5} \cdot 100 \cdot \dfrac{1}{100}$

$= \dfrac{800}{5} \cdot \dfrac{1}{100}$

$= 160\%$

3. $\dfrac{4}{9} = \dfrac{4}{9} \cdot 100 \cdot \dfrac{1}{100}$

$= \dfrac{400}{9} \cdot \dfrac{1}{100}$

$= (44.4 \ . \ . \ .) \cdot \dfrac{1}{100}$

$= (44.\overline{4}) \cdot \dfrac{1}{100}$

$= 44.\overline{4}\%$

❏ CONVERTING A DECIMAL TO A PERCENT

We can see how a decimal is converted to a percent by analyzing the method that 0.75 is converted to a percent. We need to introduce $\frac{1}{100}$.

$$0.75 = 0.75 \cdot 100 \cdot \frac{1}{100} \qquad \text{Multiply the decimal by 1.}$$

$$= 75 \cdot \frac{1}{100}$$

$$= 75\% \qquad \text{Replace } \frac{1}{100} \text{ with the \% symbol.}$$

Decimal to Percent

> To convert a fraction to a percent, multiply the decimal by 1 in the form $100 \cdot \frac{1}{100}$, then replace $\frac{1}{100}$ with the % symbol. This amounts to moving the decimal point 2 places to the right.

☆ SAMPLE SET B

Convert each decimal to a percent.

1. $0.62 = 0.62 \cdot 100 \cdot \frac{1}{100}$

$= 62 \cdot \frac{1}{100}$

$= 62\%$

Notice that the decimal point in the original number has been moved to the right 2 places.

2. $8.4 = 8.4 \cdot 100 \cdot \frac{1}{100}$

$= 840 \cdot \frac{1}{100}$

$= 840\%$

Notice that the decimal point in the original number has been moved to the right 2 places.

3. $0.47623 = 0.47623 \cdot 100 \cdot \frac{1}{100}$

$= 47.623 \cdot \frac{1}{100}$

$= 47.623\%$

Notice that the decimal point in the original number has been moved to the right 2 places.

❏ CONVERTING A PERCENT TO A DECIMAL

We can see how a percent is converted to a decimal by analyzing the method that 12% is converted to a decimal. We need to introduce $\frac{1}{100}$.

$$12\% = 12 \cdot \frac{1}{100} \qquad \text{Replace } \% \text{ with } \frac{1}{100}.$$

$$= \frac{12}{100} \qquad \text{Multiply the fractions.}$$

$$= 0.12 \qquad \text{Divide 12 by 100.}$$

Percent to Decimal

To convert a percent to a decimal, replace the % symbol with $\frac{1}{100}$, then divide the number by 100. This amounts to moving the decimal point 2 places to the left.

☆ SAMPLE SET C

Convert each percent to a decimal.

1. $48\% = 48 \cdot \frac{1}{100}$

$$= \frac{48}{100}$$

$$= 0.48$$

Notice that the decimal point in the original number has been moved to the left 2 places.

2. $659\% = 659 \cdot \frac{1}{100}$

$$= \frac{659}{100}$$

$$= 6.59$$

Notice that the decimal point in the original number has been moved to the left 2 places.

3. $0.4113\% = 0.4113 \cdot \frac{1}{100}$

$$= \frac{0.4113}{100}$$

$$= 0.004113$$

Notice that the decimal point in the original number has been moved to the left 2 places.

Section R.7 EXERCISES

For problems 1–12, convert each fraction to a percent.

1. $\dfrac{2}{5}$

2. $\dfrac{7}{8}$

3. $\dfrac{1}{8}$

4. $\dfrac{5}{16}$

5. $15 \div 22$

6. $\dfrac{2}{11}$

7. $\dfrac{2}{9}$

8. $\dfrac{16}{45}$

9. $\dfrac{27}{55}$

10. $\dfrac{7}{27}$

11. 15

12. 8

For problems 13–22, convert each decimal to a percent.

13. 0.36
14. 0.42
15. 0.446
16. 0.1298
17. 4.25
18. 5.875
19. 86.98
20. 21.26
21. 14
22. 12

For problems 23–30, convert each percent to a decimal.

23. 35%
24. 76%
25. 18.6%
26. 67.2%
27. 9.0145%
28. 3.00156%
29. 0.00005%
30. 0.00034%

1

Basic Properties of Real Numbers

After completing this chapter, you should

Section 1.1 Symbols and Notations
- understand the difference between variables and constants
- be familiar with the symbols of operation, equality, and inequality
- be familiar with grouping symbols
- be able to correctly use the order of operations

Section 1.2 The Real Number Line and the Real Numbers
- be familiar with the real number line and the real numbers
- understand the ordering of the real numbers

Section 1.3 Properties of the Real Numbers
- understand the closure, commutative, associative, and distributive properties
- understand the identity and inverse properties

Section 1.4 Exponents
- understand exponential notation
- be able to read exponential notation
- understand how to use exponential notation with the order of operations

Section 1.5 Rules of Exponents
- understand the product and quotient rules for exponents
- understand the meaning of zero as an exponent

Section 1.6 The Power Rules for Exponents
- understand the power rules for powers, products, and quotients

1.1 Symbols and Notations

Section Overview

- ❏ **VARIABLES AND CONSTANTS**
- ❏ **SYMBOLS OF OPERATION, EQUALITY, AND INEQUALITY**
- ❏ **GROUPING SYMBOLS**
- ❏ **THE ORDER OF OPERATIONS**

❏ VARIABLES AND CONSTANTS

A basic characteristic of algebra is the use of symbols (usually letters) to represent numbers.

Variable

A letter or symbol that represents any member of a collection of two or more numbers is called a **variable.**

Constant

A letter or symbol that represents a specific number, known or unknown, is called a **constant.**

In the following examples, the letter x is a variable since it can be any member of the collection of numbers {35, 25, 10}. The letter h is a constant since it can assume only the value 5890.

1. Suppose that the streets on your way from home to school have speed limits of 35 mph, 25 mph, and 10 mph. In algebra we can let the letter x represent our speed as we travel from home to school. The maximum value of x depends on what section of street we are on. The letter x can assume any one of the *various* values 35, 25, 10.
2. Suppose that in writing a term paper for a geography class we need to specify the height of Mount Kilimanjaro. If we do not happen to know the height of the mountain, we can represent it (at least temporarily) on our paper with the letter h. Later, we look up the height in a reference book and find it to be 5890 meters. The letter h can assume only the one value, 5890, and no others. The value of h is *constant.*

❏ SYMBOLS OF OPERATION, EQUALITY, AND INEQUALITY

Binary Operation

A **binary operation** on a collection of numbers is a process that assigns a number to two given numbers in the collection. The binary operations used in algebra are addition, subtraction, multiplication, and division.

Symbols of Operation

If we let x and y each represent a number, we have the following notations:

Addition	$x + y$
Subtraction	$x - y$
Multiplication	$x \cdot y$ $(x)(y)$ $x(y)$ $(x)y$ xy
Division	$\dfrac{x}{y}$ x/y $x \div y$ $y\overline{)x}$

☆ SAMPLE SET A

1. $a + b$ represents the *sum* of a and b.
2. $4 + y$ represents the *sum* of 4 and y.

3. $8 - x$ represents the *difference* of 8 and x.

4. $6x$ represents the *product* of 6 and x.

5. ab represents the *product* of a and b.

6. $h3$ represents the *product* of h and 3.

7. $(14.2)a$ represents the *product* of 14.2 and a.

8. $(8)(24)$ represents the *product* of 8 and 24.

9. $5 \cdot 6(b)$ represents the *product* of 5, 6, and b.

10. $\dfrac{6}{x}$ represents the *quotient* of 6 and x.

★ **PRACTICE SET A**

1. Represent the product of 29 and x five different ways.

If we let a and b represent two numbers, then a and b are related in exactly one of three ways:

Equality Symbol
Inequality Symbols

$a = b$	a and b are equal
$a > b$	a is strictly greater than b
$a < b$	a is strictly less than b

Some variations of these symbols include

$a \neq b$	a is not equal to b
$a \geq b$	a is greater than or equal to b
$a \leq b$	a is less than or equal to b

The last five of the above symbols are inequality symbols. We can *negate* (change to the opposite) any of the above statements by drawing a line through the relation symbol (as in $a \neq b$), as shown below:

a is not greater than b can be expressed as either

$a \ngtr b$ or $a \leq b$.

a is not less than b can be expressed as either

$a \nless b$ or $a \geq b$.

$a < b$ and $a \neq b$ both indicate that a is less than b.

❑ **GROUPING SYMBOLS**

Grouping symbols are used to indicate that a particular collection of numbers and meaningful operations are to be grouped together and considered as one number. The grouping symbols commonly used in algebra are

( )
[ ]
{ }
——

Parentheses:	( )
Brackets:	[ ]
Braces:	{ }
Bar:	——

In a computation in which more than one operation is involved, grouping symbols help tell us which operations to perform first. If possible, we perform operations inside grouping symbols first.

1. $(4 + 17) - 6 = 21 - 6 = 15$

2. $8(3 + 6) = 8(9) = 72$

3. $5[8 + (10 - 4)] = 5[8 + 6] = 5[14] = 70$

4. $2\{3[4(17 - 11)]\} = 2\{3[4(6)]\} = 2\{3[24]\} = 2\{72\} = 144$

5. $\dfrac{9(5 + 1)}{24 + 3}.$ The fraction bar separates the two groups of numbers $9(5 + 1)$ and $24 + 3$. Perform the operations in the numerator and denominator separately.

$$\frac{9(5 + 1)}{24 + 3} = \frac{9(6)}{24 + 3} = \frac{54}{24 + 3} = \frac{54}{27} = 2$$

★ **PRACTICE SET B**

Use the grouping symbols to help perform the following operations.

1. $3(1 + 8)$ **2.** $4[2(11 - 5)]$ **3.** $6\{2[2(10 - 9)]\}$ **4.** $\dfrac{1 + 19}{2 + 3}$

The following examples show how to use algebraic notation to write each expression.

1. 9 minus y becomes $9 - y$
2. 46 times x becomes $46x$
3. 7 times $(x + y)$ becomes $7(x + y)$
4. 4 divided by 3, times z becomes $\left(\dfrac{4}{3}\right)z$
5. $(a - b)$ times $(b - a)$ divided by (2 times a) becomes $\dfrac{(a - b)(b - a)}{2a}$
6. Introduce a variable (*any* letter will do but here we'll let x represent the number) and use appropriate algebraic symbols to write the statement: A number plus 4 is strictly greater than 6. The answer is $x + 4 > 6$.

❑ **THE ORDER OF OPERATIONS**

Suppose we wish to find the value of $16 + 4 \cdot 9$. We could

1. add 16 and 4, then multiply this sum by 9.

   $16 + 4 \cdot 9 = 20 \cdot 9 = 180$

2. multiply 4 and 9, then add 16 to this product.

   $16 + 4 \cdot 9 = 16 + 36 = 52$

We now have two values for one number. To determine the correct value we must use the standard **order of operations.**

Order of Operations

1. Perform all operations inside grouping symbols, beginning with the innermost set.
2. Perform all multiplications and divisions, as you come to them, moving left-to-right.
3. Perform all additions and subtractions, as you come to them, moving left-to-right.

As we proceed in our study of algebra, we will come upon another operation, exponentiation, that will need to be inserted before multiplication and division. (See Section 1.4.)

☆ SAMPLE SET C

Use the order of operations to find the value of each number.

1. $16 + 4 \cdot 9$ **Multiply first.**
$= 16 + 36$ **Now add.**
$= 52$

2. $(27 - 8) + 7(6 + 12)$ **Combine within parentheses.**
$= 19 + 7(18)$ **Multiply.**
$= 19 + 126$ **Now add.**
$= 145$

3. $8 + 2[4 + 3(6 - 1)]$ **Begin with the innermost set of grouping symbols, (   ).**
$= 8 + 2[4 + 3(5)]$ **Now work within the next set of grouping symbols, [   ].**
$= 8 + 2[4 + 15]$
$= 8 + 2[19]$
$= 8 + 38$
$= 46$

4. $\dfrac{6 + 4[2 + 3(19 - 17)]}{18 - 2[2(3) + 2]} = \dfrac{6 + 4[2 + 3(2)]}{18 - 2[6 + 2]}$

$= \dfrac{6 + 4[2 + 6]}{18 - 2[8]}$

$= \dfrac{6 + 4[8]}{18 - 16}$

$= \dfrac{6 + 32}{2}$

$= \dfrac{38}{2}$

$= 19$

★ PRACTICE SET C

Use the order of operations to find each value.

1. $25 + 8(3)$ **2.** $2 + 3(18 - 5 \cdot 2)$ **3.** $4 + 3[2 + 3(1 + 8 \div 4)]$

4. $\dfrac{19 + 2\{5 + 2[18 + 6(4 + 1)]\}}{5 \cdot 6 - 3(5) - 2}$

Answers to Practice Sets are on p. 7.

Section 1.1 EXERCISES

For problems 1–26, use the order of operations to find each value.

1. $2 + 3(6)$

2. $18 - 7(8 - 3)$

3. $8 \cdot 4 \div 16 + 5$

4. $(21 + 4) \div 5 \cdot 2$

5. $3(8 + 2) \div 6 + 3$

6. $6(4 + 1) \div (16 \div 8) - 15$

7. $6(4 - 1) + 8(3 + 7) - 20$

8. $(8)(5) + 2(14) + (1)(10)$

9. $61 - 22 + 4[3(10) + 11]$

10. $\dfrac{(1 + 16 - 3)}{7} + 5(12)$

11. $\dfrac{8(6 + 20)}{8} + \dfrac{3(6 + 16)}{22}$

12. $18 \div 2 + 55$

13. $21 \div 7 \div 3$

14. $85 \div 5 \cdot 5 - 85$

15. $(300 - 25) \div (6 - 3)$

16. $4 \cdot 3 + 8 \cdot 28 - (3 + 17) + 11(6)$

17. $2\{(7 + 7) + 6[4(8 + 2)]\}$

18. $0 + 10(0) + 15[4(3) + 1]$

19. $6.1(2.2 + 1.8)$

20. $\dfrac{5.9}{2} + 0.6$

21. $(4 + 7)(8 - 3)$

22. $(10 + 5)(10 + 5) - 4(60 - 4)$

23. $\left(\dfrac{5}{12} - \dfrac{1}{4}\right) + \left(\dfrac{1}{6} + \dfrac{2}{3}\right)$

24. $4\left(\dfrac{3}{5} - \dfrac{8}{15}\right) + 9\left(\dfrac{1}{3} + \dfrac{1}{4}\right)$

25. $\dfrac{0}{5} + \dfrac{0}{1} + 0[2 + 4(0)]$

26. $0 \cdot 9 + 4 \cdot 0 \div 7 + 0[2(2 - 2)]$

For problems 27–29, state whether the given statements are the same or different.

27. $x \geq y$ and $x > y$

28. $x < y$ and $x \neq y$

29. $x = y$ and $y = x$

30. Represent the product of 3 and x five different ways.

31. Represent the sum of a and b two different ways.

For problems 32–40, rewrite each phrase using algebraic notation.

32. Ten minus three

33. x plus sixteen

34. 51 divided by a

35. 81 times x

36. 3 times $(x + y)$

37. $(x + b)$ times $(x + 7)$

38. 3 times x times y

39. x divided by (7 times b)

40. $(a + b)$ divided by $(a + 4)$

For problems 41–44, introduce a variable (any letter will do) and use appropriate algebraic symbols to write the given statement.

41. A number minus eight equals seventeen.

42. Five times a number, minus one, equals zero.

43. A number divided by six is greater than or equal to forty-four.

44. Sixteen minus twice a number equals five.

Determine whether the statements for problems 45–50 are true or false.

45. $6 - 4(4)(1) \leq 10$

46. $5(4 + 2 \cdot 10) \geq 110$

47. $8 \cdot 6 - 48 \leq 0$

48. $\dfrac{20 + 4 \cdot 3}{16} < 5$

49. $2[6(1 + 4) - 8] > 3(11 + 6)$

50. $6[4 + 8 + 3(26 - 15)] \neq 3[7(10 - 4)]$

51. The number of different ways 5 people can be arranged in a row is $5 \cdot 4 \cdot 3 \cdot 2 \cdot 1$. How many ways is this?

52. A box contains 10 computer chips. Three chips are to be chosen at random. The number of ways this can be done is

$$\frac{10 \cdot 9 \cdot 8 \cdot 7 \cdot 6 \cdot 5 \cdot 4 \cdot 3 \cdot 2 \cdot 1}{3 \cdot 2 \cdot 1 \cdot 7 \cdot 6 \cdot 5 \cdot 4 \cdot 3 \cdot 2 \cdot 1}$$

How many ways is this?

53. The probability of obtaining four of a kind in a five-card poker hand is

$$\frac{13 \cdot 48}{(52 \cdot 51 \cdot 50 \cdot 49 \cdot 48) \div (5 \cdot 4 \cdot 3 \cdot 2 \cdot 1)}$$

What is this probability?

54. Three people are on an elevator in a five story building. If each person randomly selects a floor on which to get off, the probability that at least two people get off on the same floor is

$$1 - \frac{5 \cdot 4 \cdot 3}{5 \cdot 5 \cdot 5}$$

What is this probability?

★ **Answers to Practice Sets (1.1)**

A. **1.** $29 \cdot x$, $29x$, $(29)(x)$, $29(x)$, $(29)x$

B. **1.** 27   **2.** 48   **3.** 24   **4.** 4

C. **1.** 49   **2.** 26   **3.** 37   **4.** 17

1.2 The Real Number Line and the Real Numbers

Section
Overview

☐ **THE REAL NUMBER LINE**
☐ **THE REAL NUMBERS**
☐ **ORDERING THE REAL NUMBERS**

☐ THE REAL NUMBER LINE

Real Number Line

In our study of algebra, we will use several collections of numbers. The **real number line** allows us to *visually* display the numbers in which we are interested.

A line is composed of infinitely many points. To each point we can associate a unique number, and with each number we can associate a particular point.

Coordinate

The number associated with a point on the number line is called the **coordinate** of the point.

Graph

The point on a line that is associated with a particular number is called the **graph** of that number.

We construct the real number line as follows:

Construction of the Real Number Line

1. Draw a horizontal line.

2. Choose any point on the line and label it 0. This point is called the **origin.**

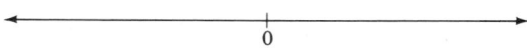

3. Choose a convenient length. This length is called "1 unit." Starting at 0, mark this length off in both directions, being careful to have the lengths look like they are about the same.

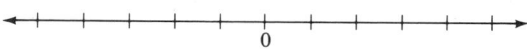

We now define a real number.

Real Number

A **real number** is any number that is the coordinate of a point on the real number line.

The collection of these infinitely many numbers is called the *collection of real numbers.* The real numbers whose graphs are to the right of 0 are called the **positive real numbers.** The real numbers whose graphs appear to the left of 0 are called the **negative real numbers.**

Positive and Negative Real Numbers

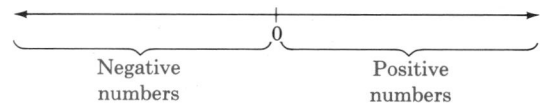

The number 0 is neither positive nor negative.

❏ THE REAL NUMBERS

The collection of real numbers has many subcollections. The subcollections that are of most interest to us are listed below along with their notations and graphs.

Natural Numbers

The **natural numbers (N):** $\{1,2,3, \ldots\}$

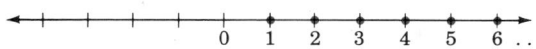

Whole Numbers

The **whole numbers (W):** $\{0,1,2,3, \ldots\}$

Notice that every natural number is a whole number.

Integers

The **integers (Z):** $\{\ldots, -3, -2, -1, 0, 1, 2, 3, \ldots\}$

Notice that every whole number is an integer.

Rational Numbers

The **rational numbers (Q):** Rational numbers are real numbers that can be written in the form a/b, where a and b are integers, and $b \neq 0$.

Fractions

Rational numbers are commonly called **fractions.**

Division by 1

Since b can equal 1, every integer is a rational number: $\dfrac{a}{1} = a$.

Division by 0

Recall that $10/2 = 5$ since $2 \cdot 5 = 10$. However, if $10/0 = x$, then $0 \cdot x = 10$. But $0 \cdot x = 0$, not 10. This suggests that no quotient exists.

Now consider $0/0 = x$. If $0/0 = x$, then $0 \cdot x = 0$. But this means that x could be any number, that is, $0/0 = 4$ since $0 \cdot 4 = 0$, or $0/0 = 28$ since $0 \cdot 28 = 0$. This suggests that the quotient is indeterminant.

$x/0$ Is Undefined or Indeterminant

Division by 0 is undefined or indeterminant.

Do not divide by 0.

Rational numbers have decimal representations that either terminate or do not terminate but contain a repeating block of digits. Some examples are:

$$\underbrace{\frac{3}{4} = 0.75}_{\text{Terminating}} \qquad \underbrace{\frac{15}{11} = 1.36363636 \ldots}_{\text{Nonterminating, but repeating}}$$

Some rational numbers are graphed below.

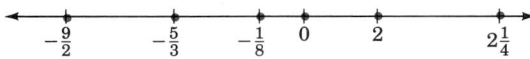

Irrational Numbers

The **irrational numbers (Ir):** Irrational numbers are numbers that cannot be written as the quotient of two integers. They are numbers whose decimal representations are nonterminating and nonrepeating. Some examples are

$$4.01001000100001 \ldots \qquad \pi = 3.1415927 \ldots$$

Notice that the collections of rational numbers and irrational numbers have no numbers in common.

When graphed on the number line, the rational and irrational numbers account for every point on the number line. Thus each point on the number line has a coordinate that is either a rational or an irrational number.

In summary, we have

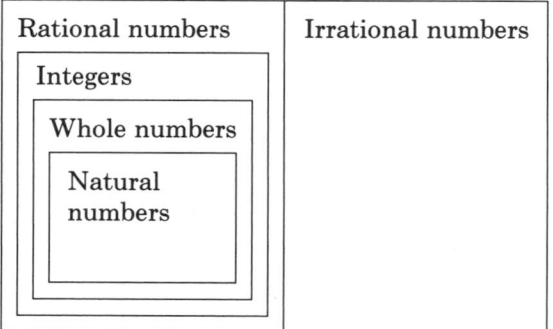

☆ SAMPLE SET A

The summary chart illustrates that

1. Every natural number is a real number.
2. Every whole number is a real number.
3. No integer is an irrational number.

★ PRACTICE SET A

1. Is every natural number a whole number?
2. Is every whole number an integer?
3. Is every integer a rational number?
4. Is every rational number a real number?
5. Is every integer a natural number?
6. Is there an integer that is a natural number?

☐ ORDERING THE REAL NUMBERS

Ordering the Real Numbers

A real number b is said to be greater than a real number a, denoted $b > a$, if the graph of b is to the right of the graph of a on the number line.

☆ SAMPLE SET B

As we would expect, $5 > 2$ since 5 is to the right of 2 on the number line. Also, $-2 > -5$ since -2 is to the right of -5 on the number line.

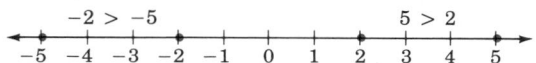

★ **PRACTICE SET B**

1. Are all positive numbers greater than 0?
2. Are all positive numbers greater than all negative numbers?
3. Is 0 greater than all negative numbers?
4. Is there a largest positive number? Is there a smallest negative number?
5. How many real numbers are there? How many real numbers are there between 0 and 1?

☆ **SAMPLE SET C**

1. What integers can replace x so that the following statement is true?

$$-4 \leq x < 2$$

This statement indicates that the number represented by x is between -4 and 2. Specifically, -4 is less than or equal to x, and at the same time, x is strictly less than 2. This statement is an example of a compound inequality.

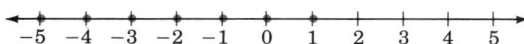

The integers are $-4, -3, -2, -1, 0, 1$.

2. Draw a number line that extends from -3 to 7. Place points at all whole numbers between and including -2 and 6.

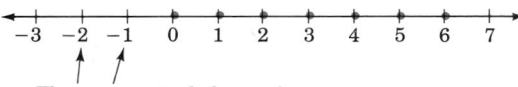

These are not whole numbers

3. Draw a number line that extends from -4 to 6 and place points at all real numbers greater than or equal to 3 but strictly less than 5.

> It is customary to use a **closed circle** to indicate that a point is included in the graph and an **open circle** to indicate that a point is not included.

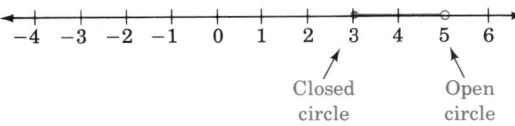

Closed   Open
circle   circle

★ **PRACTICE SET C**

1. What whole numbers can replace x so that the following statement is true?

$$-3 \leq x < 3$$

2. Draw a number line that extends from -5 to 3 and place points at all numbers greater than or equal to -4 but strictly less than 2.

Answers to Practice Sets are on p. 13.

Section 1.2 EXERCISES

For problems 1–7, next to each real number, note all collections to which it belongs by writing N for natural numbers, W for whole numbers, Z for integers, Q for rational numbers, Ir for irrational numbers, and R for real numbers. Some numbers may require more than one letter.

1. $\dfrac{1}{2}$

2. -12

3. 0

4. $-24\dfrac{7}{8}$

5. $86.3333\ldots$

6. $49.125125125\ldots$

7. -15.07

For problems 8–10, draw a number line that extends from -3 to 3. Locate each real number on the number line by placing a point (closed circle) at its approximate location.

8. $1\dfrac{1}{2}$

9. -2

10. $-\dfrac{1}{8}$

11. Is 0 a positive number, negative number, neither, or both?

12. An integer is an even integer if it can be divided by 2 without a remainder; otherwise the number is odd. Draw a number line that extends from -5 to 5 and place points at all negative even integers and at all positive odd integers.

13. Draw a number line that extends from -5 to 5. Place points at all integers strictly greater than -3 but strictly less than 4.

For problems 14–16, draw a number line that extends from -5 to 5. Place points at all real numbers between and including each pair of numbers.

14. -5 and -2

15. -3 and 4

16. -4 and 0

17. Draw a number line that extends from -5 to 5. Is it possible to locate any numbers that are strictly greater than 3 but also strictly less than -2?

For the pairs of real numbers shown in problems 18–22, write the appropriate relation symbol ($<$, $>$, $=$) in place of the *.

18. $-5 * -1$

19. $-3 * 0$

20. $-4 * 7$

21. $6 * -1$

22. $-\dfrac{1}{4} * -\dfrac{3}{4}$

23. Is there a largest real number? If so, what is it?

24. Is there a largest integer? If so, what is it?

25. Is there a largest two-digit integer? If so, what is it?

26. Is there a smallest integer? If so, what is it?

27. Is there a smallest whole number? If so, what is it?

For problems 28–32, what numbers can replace x so that the following statements are true?

28. $-1 \le x \le 5$, x an integer

29. $-7 < x < -1$, x an integer

30. $-3 \le x \le 2$, x a natural number

31. $-15 < x \le -1,$ $\quad x$ a natural number

32. $-5 \le x < 5,$ $\quad x$ a whole number

33. The temperature in the desert today was ninety-five degrees. Represent this temperature by a rational number.

34. The temperature today in Colorado Springs was eight degrees below zero. Represent this temperature with a real number.

35. Is every integer a rational number?

36. Is every rational number an integer?

37. Can two rational numbers be added together to yield an integer? If so, give an example.

For problems 38–44, on the number line, how many units (intervals) are there between?

38. 0 and 2?
39. -5 and 0?
40. 0 and 6?
41. -8 and 0?
42. -3 and 4?
43. m and n, $m > n$?
44. $-a$ and $-b$, $-b > -a$?

EXERCISES FOR REVIEW

(1.1) **45.** Find the value of $6 + 3(15 - 8) - 4$.
(1.1) **46.** Find the value of $5(8 - 6) + 3(5 + 2 \cdot 3)$.
(1.1) **47.** Are the statements $y < 4$ and $y \ge 4$ the same or different?
(1.1) **48.** Use algebraic notation to write the statement "six times a number is less than or equal to eleven."
(1.1) **49.** Is the statement $8(15 - 3 \cdot 4) - 3 \cdot 7 \ge 3$ true or false?

★ Answers to Practice Sets (1.2)

A. 1. yes 2. yes 3. yes 4. yes 5. no 6. yes

B. 1. yes 2. yes 3. yes 4. no, no 5. infinitely many, infinitely many

C. 1. 0, 1, 2 2.

1.3 Properties of the Real Numbers

Section Overview	☐ **THE CLOSURE PROPERTIES** ☐ **THE COMMUTATIVE PROPERTIES** ☐ **THE ASSOCIATIVE PROPERTIES** ☐ **THE DISTRIBUTIVE PROPERTIES** ☐ **THE IDENTITY PROPERTIES** ☐ **THE INVERSE PROPERTIES**

Property

A **property** of a collection of objects is a characteristic that describes the collection. We shall now examine some of the properties of the collection of real numbers. The properties we will examine are expressed in terms of addition and multiplication.

☐ THE CLOSURE PROPERTIES

The Closure Properties

If a and b are real numbers, then $a + b$ is a unique real number, and $a \cdot b$ is a unique real number.

For example, 3 and 11 are real numbers; $3 + 11 = 14$ and $3 \cdot 11 = 33$, and both 14 and 33 are real numbers. Although this property seems obvious, some collections are not closed under certain operations. For example,

1. The real numbers are not closed under division since, although 5 and 0 are real numbers, $5/0$ and $0/0$ are not real numbers.
2. The natural numbers are not closed under subtraction since, although 8 is a natural number, $8 - 8$ is not. ($8 - 8 = 0$ and 0 is not a natural number.)

❏ THE COMMUTATIVE PROPERTIES

Let a and b represent real numbers.

The Commutative Properties

COMMUTATIVE PROPERTY OF ADDITION	COMMUTATIVE PROPERTY OF MULTIPLICATION
$a + b = b + a$	$a \cdot b = b \cdot a$

The commutative properties tell us that two numbers can be added or multiplied in any order without affecting the result.

☆ SAMPLE SET A

The following are examples of the commutative properties.

1. $3 + 4 = 4 + 3$ Both equal 7.
2. $5 + x = x + 5$ Both represent the same sum.
3. $4 \cdot 8 = 8 \cdot 4$ Both equal 32.
4. $y7 = 7y$ Both represent the same product.
5. $5(a + 1) = (a + 1)5$ Both represent the same product.
6. $(x + 4)(y + 2) = (y + 2)(x + 4)$ Both represent the same product.

★ PRACTICE SET A

Fill in the ( ) with the proper number or letter so as to make the statement true. Use the commutative properties.

1. $6 + 5 = (\quad) + 6$ 2. $m + 12 = 12 + (\quad)$ 3. $9 \cdot 7 = (\quad) \cdot 9$
4. $6a = a(\quad)$ 5. $4(k - 5) = (\quad)4$ 6. $(9a - 1)(\quad) = (2b + 7)(9a - 1)$

❏ THE ASSOCIATIVE PROPERTIES

Let a, b, and c represent real numbers.

The Associative Properties

ASSOCIATIVE PROPERTY OF ADDITION	ASSOCIATIVE PROPERTY OF MULTIPLICATION
$(a + b) + c = a + (b + c)$	$(ab)c = a(bc)$

The associative properties tell us that we may group together the quantities as we please without affecting the result.

☆ SAMPLE SET B

The following examples show how the associative properties can be used.

1. $(2 + 6) + 1 = 2 + (6 + 1)$
   $8 + 1 = 2 + 7$
   $\qquad 9 = 9$ **Both equal 9.**

2. $(3 + x) + 17 = 3 + (x + 17)$ **Both represent the same sum.**

3. $(2 \cdot 3) \cdot 5 = 2 \cdot (3 \cdot 5)$
   $\quad 6 \cdot 5 = 2 \cdot 15$
   $\qquad 30 = 30$ **Both equal 30.**

4. $(9y)4 = 9(y4)$ **Both represent the same product.**

★ PRACTICE SET B

Fill in the (   ) to make each statement true. Use the associative properties.

1. $(9 + 2) + 5 = 9 + ($ $)$ 2. $x + (5 + y) = ($ $) + y$ 3. $(11a)6 = 11($ $)$
4. $[(7m - 2)(m + 3)](m + 4) = (7m - 2)[($ $)($ $)]$

☆ SAMPLE SET C

Simplify (rearrange into a simpler form): $5x6b8ac4$.

According to the commutative property of multiplication, we can make a series of consecutive switches and get all the numbers together and all the letters together.

$5 \cdot 6 \cdot 8 \cdot 4 \cdot x \cdot b \cdot a \cdot c$

$960xbac$ **Multiply the numbers.**

$960abcx$ **By convention, we will, when possible, write all letters in alphabetical order.**

★ PRACTICE SET C

Simplify each of the following quantities.

1. $3a7y9d$ 2. $6b8acz4 \cdot 5$ 3. $4p6qr3(a + b)$

❑ THE DISTRIBUTIVE PROPERTIES

When we were first introduced to multiplication we saw that it was developed as a description for repeated addition.

$4 + 4 + 4 = 3 \cdot 4$

Notice that there are three 4's, that is, 4 appears 3 *times*. Hence, 3 times 4.

We know that algebra is generalized arithmetic. We can now make an important generalization.

When a number a is added repeatedly n times, we have

$$\underbrace{a + a + a + \cdots + a}_{a \text{ appears } n \text{ times}}$$

> Then, using multiplication as a description for repeated addition, we can replace
>
> $$\underbrace{a + a + a + \cdots + a}_{n \text{ times}} \quad \text{with} \quad na.$$

For example:

1. $x + x + x + x$ can be written as $4x$ since x is repeatedly added 4 times.

   $$x + x + x + x = 4x$$

2. $r + r$ can be written as $2r$ since r is repeatedly added 2 times.

   $$r + r = 2r$$

The distributive property involves both multiplication and addition. Let's rewrite $4(a + b)$. We proceed by reading $4(a + b)$ as a multiplication: 4 times the quantity $(a + b)$. This directs us to write

$$4(a + b) = (a + b) + (a + b) + (a + b) + (a + b)$$
$$= a + b + a + b + a + b + a + b$$

Now we use the commutative property of addition to collect all the a's together and all the b's together.

$$4(a + b) = \underbrace{a + a + a + a}_{4 \ a\text{'s}} + \underbrace{b + b + b + b}_{4 \ b\text{'s}}$$

Now, using multiplication as a description for repeated addition, we have

$$4(a + b) = 4a + 4b$$

We have **distributed** the 4 over the sum to both a and b.

$$4(a + b) = 4a + 4b$$

The Distributive Property

> $$a(b + c) = a \cdot b + a \cdot c \qquad (b + c)a = a \cdot b + a \cdot c$$

The distributive property is useful when we cannot or do not wish to **perform** operations inside parentheses.

☆ **SAMPLE SET D**

Use the distributive property to rewrite each of the following quantities.

1. $2(5 + 7) = 2 \cdot 5 + 2 \cdot 7$ **Both equal 24.**

2. $6(x + 3) = 6 \cdot x + 6 \cdot 3$ **Both represent the same number.**
   $$= 6x + 18$$

3. $(z + 5)y = zy + 5y = yz + 5y$

★ **PRACTICE SET D**

1. What property of real numbers justifies

   $a(b + c) = (b + c)a$?

Use the distributive property to rewrite each of the following quantities.

2. $3(2 + 1)$ **3.** $(x + 6)7$ **4.** $4(a + y)$ **5.** $(9 + 2)a$ **6.** $a(x + 5)$

7. $1(x + y)$

Answers to Practice Sets are on p. 19.

❏ THE IDENTITY PROPERTIES

Additive Identity

The number 0 is called the **additive identity** since when it is added to any real number, it preserves the identity of that number. Zero is the only additive identity. For example, $6 + 0 = 6$.

Multiplicative Identity

The number 1 is called the **multiplicative identity** since when it multiplies any real number, it preserves the identity of that number. One is the only multiplicative identity. For example $6 \cdot 1 = 6$.

We summarize the identity properties as follows.

ADDITIVE IDENTITY PROPERTY	MULTIPLICATIVE IDENTITY PROPERTY
If a is a real number, then	If a is a real number, then
$a + 0 = a$ and $0 + a = a$	$a \cdot 1 = a$ and $1 \cdot a = a$

❏ THE INVERSE PROPERTIES

Additive Inverses

When two numbers are added together and the result is the additive identity, 0, the numbers are called **additive inverses** of each other. For example, when 3 is added to -3, the result is 0, that is, $3 + (-3) = 0$. The numbers 3 and -3 are additive inverses of each other.

Multiplicative Inverses

When two numbers are multiplied together and the result is the multiplicative identity, 1, the numbers are called **multiplicative inverses** of each other. For example, when 6 and $\frac{1}{6}$ are multiplied together, the result is 1, that is, $6 \cdot \frac{1}{6} = 1$. The numbers 6 and $\frac{1}{6}$ are multiplicative inverses of each other.

We summarize the inverse properties as follows.

The Inverse Properties

1. If a is any real number, then there is a unique real number $-a$, such that
$$a + (-a) = 0 \quad \text{and} \quad -a + a = 0$$
The numbers a and $-a$ are called **additive inverses** of each other.

2. If a is any nonzero real number, then there is a unique real number $\frac{1}{a}$ such that
$$a \cdot \frac{1}{a} = 1 \quad \text{and} \quad \frac{1}{a} \cdot a = 1$$
The numbers a and $\frac{1}{a}$ are called **multiplicative inverses** of each other.

Expanding Quantities

When we perform operations such as $6(a + 3) = 6a + 18$, we say we are **expanding** the quantity $6(a + 3)$.

Section 1.3 EXERCISES

Use the commutative property of addition and multiplication to write expressions for an equal number for problems 1–22. You need not perform any calculations.

1. $x + 3$

2. $5 + y$

3. $10x$

4. $18z$

5. $r6$

6. ax

7. xc

8. $7(2 + b)$

9. $6(s + 1)$

10. $(8 + a)(x + 6)$

11. $(x + 16)(a + 7)$

12. $(x + y)(x - y)$

13. $0.06m$

14. $8\star$

15. $5(6h + 1)$

16. $m(a + 2b)$

17. $k(10a - b)$

18. $(21c)(0.008)$

19. $(-16)(4)$

20. $(5)(b - 6)$

21. $\square \cdot \bigcirc$

22. $\star \cdot \diamond$

Simplify using the commutative property of multiplication for problems 23–31. You need not use the distributive property.

23. $9x2y$

24. $5a6b$

25. $2a3b4c$

26. $5x10y5z$

27. $1u3r2z5m1n$

28. $6d4e1f2(g + 2h)$

29. $\left(\dfrac{1}{2}\right) d \left(\dfrac{1}{4}\right) e \left(\dfrac{1}{2}\right) a$

30. $3(a + 6)2(a - 9)6b$

31. $1(x + 2y)(6 + z)9(3x + 5y)$

For problems 32–51, use the distributive property to expand the quantities.

32. $2(y + 9)$

33. $b(r + 5)$

34. $m(u + a)$

35. $k(j + 1)$

36. $x(2y + 5)$

37. $z(x + 9w)$

38. $(1 + d)e$

39. $(8 + 2f)g$

40. $c(2a + 10b)$

41. $15x(2y + 3z)$

42. $8y(12a + b)$

43. $z(x + y + m)$

44. $(a + 6)(x + y)$

45. $(x + 10)(a + b + c)$

46. $1(x + y)$

47. $1(a + 16)$

48. ▦ $0.48(0.34a + 0.61)$

49. ▦ $21.5(16.2a + 3.8b + 0.7c)$

50. $5\star(2\square + 3\diamond)$

51. $2z_t(L_m + 8k)$

(1.1) 52. Find the value of $4 \cdot 2 + 5(2 \cdot 4 - 6 \div 3) - 2 \cdot 5$.

(1.1) 53. Is the statement $3(5 \cdot 3 - 3 \cdot 5) + 6 \cdot 2 - 3 \cdot 4 < 0$ true or false?

(1.2) 54. Draw a number line that extends from -2 to 2 and place points at all integers between and including -2 and 3.

(1.2) 55. Replace the $*$ with the appropriate relation symbol $(<, >)$. $-7 * -3$.

(1.2) 56. What whole numbers can replace x so that the statement $-2 \le x < 2$ is true?

★ **Answers to Practice Sets (1.3)**

A. **1.** 5　**2.** m　**3.** 7　**4.** 6　**5.** $(k-5)$　**6.** $(2b+7)$

B. **1.** $2+5$　**2.** $x+5$　**3.** $a \cdot 6$　**4.** $(m+3)(m+4)$

C. **1.** $189ady$　**2.** $960abcz$　**3.** $72pqr(a+b)$

D. **1.** the commutative property of multiplication　**2.** $6+3$　**3.** $7x+42$　**4.** $4a+4y$　**5.** $9a+2a$
6. $ax+5a$　**7.** $x+y$

1.4 Exponents

Section Overview

❑ **EXPONENTIAL NOTATION**
❑ **READING EXPONENTIAL NOTATION**
❑ **THE ORDER OF OPERATIONS**

❑ EXPONENTIAL NOTATION

In Section 1.3 we were reminded that multiplication is a description for repeated addition. A natural question is "Is there a description for *repeated* multiplication?" The answer is yes. The notation that describes repeated multiplication is **exponential notation.**

Factors

In multiplication, the numbers being multiplied together are called **factors.** In repeated multiplication, all the factors are the same. In nonrepeated multiplication, none of the factors are the same. For example,

$18 \cdot 18 \cdot 18 \cdot 18$　**Repeated multiplication of 18. All four factors, 18, are the same.**

$x \cdot x \cdot x \cdot x \cdot x$　**Repeated multiplication of x. All five factors, x, are the same.**

$3 \cdot 7 \cdot a$　**Nonrepeated multiplication. None of the factors are the same.**

Exponential notation is used to show repeated multiplication of the same factor. The notation consists of using a *superscript on the factor that is repeated.* The superscript is called an **exponent.**

Exponential Notation

If x is any real number and n is a natural number, then

$$x^n = \underbrace{x \cdot x \cdot x \cdot \ldots \cdot x}_{n \text{ factors of } x}$$

An exponent records the number of identical factors in a multiplication.

Note that the definition for exponential notation only has meaning for natural number exponents. We will extend this notation to include other numbers as exponents later.

☆ SAMPLE SET A

1. $7 \cdot 7 \cdot 7 \cdot 7 \cdot 7 \cdot 7 = 7^6$.   The repeated factor is 7. The exponent 6 records the fact that 7 appears 6 times in the multiplication.

2. $x \cdot x \cdot x \cdot x = x^4$.   The repeated factor is x. The exponent 4 records the fact that x appears 4 times in the multiplication.

3. $(2y)(2y)(2y) = (2y)^3$.   The repeated factor is $2y$. The exponent 3 records the fact that the factor $2y$ appears 3 times in the multiplication.

4. $2yyy = 2y^3$.   The repeated factor is y. The exponent 3 records the fact that the factor y appears 3 times in the multiplication.

5. $(a+b)(a+b)(a-b)(a-b)(a-b) = (a+b)^2(a-b)^3$.   The repeated factors are $(a+b)$ and $(a-b)$, $(a+b)$ appearing 2 times and $(a-b)$ appearing 3 times.

★ PRACTICE SET A

Write each of the following using exponents.

1. $a \cdot a \cdot a \cdot a$ **2.** $(3b)(3b)(5c)(5c)(5c)$ **3.** $2 \cdot 2 \cdot 7 \cdot 7 \cdot 7 \cdot (a-4)(a-4)$

4. $8xxxyzzzzz$

> **CAUTION**
>
> It is extremely important to realize and remember that an exponent applies only to the factor to which it is directly connected.

☆ SAMPLE SET B

1. $8x^3$ means $8 \cdot xxx$ and *not* $8x8x8x$. The exponent 3 applies only to the factor x since it is only to the factor x that the 3 is connected.

2. $(8x)^3$ means $(8x)(8x)(8x)$ since the parentheses indicate that the exponent 3 is directly connected to the factor $8x$. Remember that the grouping symbols ( ) indicate that the quantities inside are to be considered as one single number.

3. $34(a+1)^2$ means $34 \cdot (a+1)(a+1)$ since the exponent 2 applies only to the factor $(a+1)$.

★ PRACTICE SET B

Write each of the following without exponents.

1. $4a^3$ **2.** $(4a)^3$

☆ **SAMPLE SET C**

> Select a number to show that $(2x)^2$ is not always equal to $2x^2$.
> Suppose we choose x to be 5. Consider both $(2x)^2$ and $2x^2$.
>
> | $(2x)^2$ | | $2x^2$ |
> |---|---|---|
> | $(2 \cdot 5)^2$ | | $2 \cdot 5^2$ |
> | $(10)^2$ | | $2 \cdot 25$ |
> | 100 | $\neq$ | 50 |
>
> Notice that $(2x)^2 = 2x^2$ only when $x = 0$.

★ **PRACTICE SET C**

1. Select a number to show that $(5x)^2$ is not always equal to $5x^2$.

❏ READING EXPONENTIAL NOTATION

Base
Exponent
Power

> In x^n,
>
> x is the **base**
> n is the **exponent**
> The number represented by x^n is called a **power**.
>
> The term x^n is read as "x to the nth power," or more simply as "x to the nth."

x to the nth Power

x Squared and x Cubed

The symbol x^2 is often read as "x squared," and x^3 is often read as "x cubed." A natural question is "Why are geometric terms appearing in the exponent expression?" The answer for x^3 is this: x^3 means $x \cdot x \cdot x$. In geometry, the volume of a rectangular box is found by multiplying the length by the width by the depth. A cube has the same length on each side. If we represent this length by the letter x, then the volume of the cube is $x \cdot x \cdot x$, which, of course, is described by x^3. (Can you think of why x^2 is read as x squared?)

Cube with
   length $= x$
   width $= x$
   depth $= x$

Volume $= xxx = x^3$

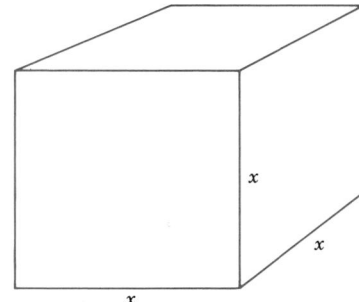

❏ THE ORDER OF OPERATIONS

In Section 1.1 we were introduced to the order of operations. It was noted that we would insert another operation before multiplication and division. We can do that now.

The Order of Operations

1. Perform all operations inside grouping symbols beginning with the innermost set.
2. Perform all exponential operations as you come to them, moving left-to-right.
3. Perform all multiplications and divisions as you come to them, moving left-to-right.
4. Perform all additions and subtractions as you come to them, moving left-to-right.

☆ **SAMPLE SET D**

Use the order of operations to simplify each of the following.

1. $2^2 + 5 = 4 + 5 = 9$

2. $5^2 + 3^2 + 10 = 25 + 9 + 10 = 44$

3. $2^2 + (5)(8) - 1 = 4 + (5)(8) - 1$
$$= 4 + 40 - 1$$
$$= 43$$

4. $7 \cdot 6 - 4^2 + 1^5 = 7 \cdot 6 - 16 + 1$
$$= 42 - 16 + 1$$
$$= 27$$

5. $(2 + 3)^3 + 7^2 - 3(4 + 1)^2 = (5)^3 + 7^2 - 3(5)^2$
$$= 125 + 49 - 3(25)$$
$$= 125 + 49 - 75$$
$$= 99$$

6. $[4(6 + 2)^3]^2 = [4(8)^3]^2$
$$= [4(512)]^2$$
$$= [2048]^2$$
$$= 4,194,304$$

7. $6(3^2 + 2^2) + 4^2 = 6(9 + 4) + 4^2$
$$= 6(13) + 4^2$$
$$= 6(13) + 16$$
$$= 78 + 16$$
$$= 94$$

8. $\dfrac{6^2 + 2^2}{4^2 + 6 \cdot 2^2} + \dfrac{1^3 + 8^2}{10^2 - (19)(5)} = \dfrac{36 + 4}{16 + 6 \cdot 4} + \dfrac{1 + 64}{100 - 95}$

$$= \dfrac{36 + 4}{16 + 24} + \dfrac{1 + 64}{100 - 95}$$

$$= \dfrac{40}{40} + \dfrac{65}{5}$$

$$= 1 + 13$$

$$= 14$$

★ **PRACTICE SET D**

Use the order of operations to simplify the following.

1. $3^2 + 4 \cdot 5$ **2.** $2^3 + 3^3 - 8 \cdot 4$ **3.** $1^4 + (2^2 + 4)^2 \div 2^3$

4. $[6(10 - 2^3)]^2 - 10^2 - 6^2$

5. $\dfrac{5^2 + 6^2 - 10}{1 + 4^2} + \dfrac{0^4 - 0^5}{7^2 - 6 \cdot 2^3}$

Answers to Practice Sets are on p. 25.

Section 1.4 EXERCISES

For problems 1–21, write each of the quantities using exponential notation.

1. b to the fourth

2. a squared

3. x to the eighth

4. (-3) cubed

5. 5 times s squared

6. 3 squared times y to the fifth

7. a cubed minus $(b + 7)$ squared

8. $(21 - x)$ cubed plus $(x + 5)$ to the seventh

9. $xxxxx$

10. $(8)(8)xxxx$

11. $2 \cdot 3 \cdot 3 \cdot 3 \cdot 3 xxyyyyyy$

12. $2 \cdot 2 \cdot 5 \cdot 6 \cdot 6 \cdot 6 xyyzzzwwww$

13. $7xx(a + 8)(a + 8)$

14. $10xyy(c + 5)(c + 5)(c + 5)$

15. $4x4x4x4x4x$

16. $(9a)(9a)(9a)(9a)$

17. $(-7)(-7)(-7)aabbba(-7)baab$

18. $(a - 10)(a - 10)(a + 10)$

19. $(z + w)(z + w)(z + w)(z - w)(z - w)$

20. $(2y)(2y)2y2y$

21. $3xyxxy - (x + 1)(x + 1)(x + 1)$

For problems 22–31, expand the quantities so that no exponents appear.

22. 4^3

23. 6^2

24. $7^3 y^2$

25. $8x^3 y^2$

26. $(18x^2 y^4)^2$

27. $(9a^3 b^2)^3$

28. $5x^2(2y^3)^3$

29. $10a^3b^2(3c)^2$

30. $(a+10)^2(a^2+10)^2$

31. $(x^2-y^2)(x^2+y^2)$

For problems 32–34, select a number (or numbers) to show that

32. $(5x)^2$ is not generally equal to $5x^2$.

33. $(7x)^2$ is not generally equal to $7x^2$.

34. $(a+b)^2$ is not generally equal to a^2+b^2.

35. For what real number is $(6a)^2$ equal to $6a^2$?

36. For what real numbers, a and b, is $(a+b)^2$ equal to a^2+b^2?

Use the order of operations to simplify the quantities for problems 37–55.

37. 3^2+7

38. 4^3-18

39. $5^2+2(40)$

40. $8^2+3+5(2+7)$

41. $2^5+3(8+1)$

42. $3^4+2^4(1+5)^3$

43. $(6^2-4^2)\div 5$

44. $2^2(10-2^3)$

45. $(3^4-4^3)\div 17$

46. $(4+3)^2+1\div(2\cdot 5)$

47. $(2^4+2^5-2^3\cdot 5)^2\div 4^2$

48. $1^6+0^8+5^2(2+8)^3$

49. $(7)(16)-9^2+4(1^1+3^2)$

50. $\dfrac{2^3-7}{5^2}$

51. $\dfrac{(1+6)^2+2}{19}$

52. $\dfrac{6^2-1}{5}+\dfrac{4^3+(2)(3)}{10}$

53. $\dfrac{5[8^2-9(6)]}{2^5-7}+\dfrac{7^2-4^2}{2^4-5}$

54. $\dfrac{(2+1)^3+2^3+1^3}{6^2}-\dfrac{15^2-[2(5)]^2}{5\cdot 5^2}$

55. $\dfrac{6^3-2\cdot 10^2}{2^2}+\dfrac{18(2^3+7^2)}{2(19)-3^3}$

EXERCISES FOR REVIEW

(1.1) 56. Use algebraic notation to write the statement "a number divided by eight, plus five, is equal to ten."

(1.2) 57. Draw a number line that extends from -5 to 5 and place points at all real numbers that are strictly greater than -3 but less than or equal to 2.

(1.2) 58. Is every integer a whole number?

(1.3) 59. Use the commutative property of multiplication to write a number equal to the number yx.

(1.3) 60. Use the distributive property to expand $3(x + 6)$.

★ **Answers to Practice Sets (1.4)**

A. 1. a^4 2. $(3b)^2(5c)^4$ 3. $2^2 \cdot 7^3(a - 4)^2$ 4. $8x^3yz^5$

B. 1. $4aaa$ 2. $(4a)(4a)(4a)$

C. 1. Select $x = 3$. Then $(5 \cdot 3)^2 = (15)^2 = 225$, but $5 \cdot 3^2 = 5 \cdot 9 = 45$. $225 \neq 45$.

D. 1. 29   2. 3   3. 9   4. 8   5. 3

1.5 Rules of Exponents

Section Overview

- ☐ **THE PRODUCT RULE FOR EXPONENTS**
- ☐ **THE QUOTIENT RULE FOR EXPONENTS**
- ☐ **ZERO AS AN EXPONENT**

We will begin our study of the rules of exponents by recalling the definition of exponents.

Definition of Exponents

If x is any real number and n is a natural number, then

$$x^n = \underbrace{x \cdot x \cdot x \cdot \ldots \cdot x}_{n \text{ factors of } x}$$

An exponent records the number of identical factors in a multiplication.

Base
Exponent
Power

In x^n,

x is the **base**
n is the **exponent**
The number represented by x^n is called a **power.**

The term x^n is read as "x to the nth."

☐ THE PRODUCT RULE FOR EXPONENTS

The first rule we wish to develop is the rule for multiplying two exponential quantities having the *same base* and natural number exponents. The following examples suggest this rule:

1. $x^2 \cdot x^4 = \underbrace{xx}_{} \cdot \underbrace{xxxx}_{} = \underbrace{xxxxxx}_{} = x^6$
   $ 2 \ + \ \ 4 \ \ = \ \ \ 6$
   $ \text{factors} \quad \text{factors}$

2. $a \cdot a^2 = \underbrace{a}_{} \cdot \underbrace{aa}_{} = \underbrace{aaa}_{} = a^3$
   $ 1 + 2 \ = \ \ 3$
   $ \text{factors} \ \ \text{factors}$

$x^n \cdot x^m = x^{n+m}$

PRODUCT RULE FOR EXPONENTS

If x is a real number and n and m are natural numbers,

$$x^n x^m = x^{n+m}$$

To multiply two exponential quantities having the same base, add the exponents. Keep in mind that the exponential quantities being multiplied *must* have the *same base* for this rule to apply.

☆ SAMPLE SET A

Find the following products. All exponents are natural numbers.

1. $x^3 \cdot x^5 = x^{3+5} = x^8$

2. $a^6 \cdot a^{14} = a^{6+14} = a^{20}$

3. $y^5 \cdot y = y^5 \cdot y^1 = y^{5+1} = y^6$

4. $(x-2y)^8(x-2y)^5 = (x-2y)^{8+5} = (x-2y)^{13}$

5. $x^3 y^4 \neq (xy)^{3+4}$ Since the bases are not the same, the product rule does not apply.

★ PRACTICE SET A

Find each product.

1. $x^2 \cdot x^5$ **2.** $x^9 \cdot x^4$ **3.** $y^6 \cdot y^4$ **4.** $c^{12} \cdot c^8$ **5.** $(x+2)^3 \cdot (x+2)^5$

☆ SAMPLE SET B

We can use the first rule of exponents (and the others that we will develop) along with the properties of real numbers.

1. $2x^3 \cdot 7x^5 = \boxed{2 \cdot 7 \cdot x^{3+5}} = 14x^8$

We used the commutative and associative properties of multiplication. In practice, we use these properties "mentally" (as signified by the drawing of the box). We don't actually write the second step.

2. $4y^3 \cdot 6y^2 = \boxed{4 \cdot 6 \cdot y^{3+2}} = 24y^5$

3. $9a^2 b^6 (8ab^4 2b^3) = \boxed{9 \cdot 8 \cdot 2a^{2+1}b^{6+4+3}} = 144a^3 b^{13}$

4. $5(a+6)^2 \cdot 3(a+6)^8 = \boxed{5 \cdot 3(a+6)^{2+8}} = 15(a+6)^{10}$

5. $4x^3 \cdot 12 \cdot y^2 = 48x^3 y^2$

6. $4a^\triangle \cdot 5a^\star = 20a^{\triangle + \star}$

The bases are the same, so we add the exponents. Although we don't know exactly what number $\triangle + \star$ is, the notation $\triangle + \star$ indicates the addition.

★ **PRACTICE SET B**

Perform each multiplication in one step.

1. $3x^5 \cdot 2x^2$ **2.** $6y^3 \cdot 3y^4$ **3.** $4a^3b^2 \cdot 9a^2b$ **4.** $x^4 \cdot 4y^2 \cdot 2x^2 \cdot 7y^6$

5. $(x-y)^3 \cdot 4(x-y)^2$ **6.** $8x^4y^2xx^3y^5$ **7.** $2aaa^3(ab^2a^3)b6ab^2$ **8.** $a^n \cdot a^m \cdot a^r$

☐ **THE QUOTIENT RULE FOR EXPONENTS**

The second rule we wish to develop is the rule for dividing two exponential quantities having the same base and natural number exponents.

The following examples suggest a rule for dividing two exponential quantities having the same base and natural number exponents.

1. $\dfrac{x^5}{x^2} = \dfrac{xxxxx}{xx} = \dfrac{(xx)\,xxx}{(xx)} = xxx = x^3.$ Notice that $5 - 2 = 3$.

2. $\dfrac{a^8}{a^3} = \dfrac{aaaaaaaa}{aaa} = \dfrac{(aaa)\,aaaaa}{(aaa)} = aaaaa = a^5.$ Notice that $8 - 3 = 5$.

QUOTIENT RULE FOR EXPONENTS

If x is a real number and n and m are natural numbers,

$$\dfrac{x^n}{x^m} = x^{n-m}, \qquad x \neq 0.$$

$\dfrac{x^n}{x^m} = x^{n-m}$

To divide two exponential quantities having the same nonzero base, subtract the exponent of the denominator from the exponent of the numerator. Keep in mind that the exponential quantities being divided *must* have the *same base* for this rule to apply.

☆ **SAMPLE SET C**

Find the following quotients. All exponents are natural numbers.

1. $\dfrac{x^5}{x^2} = \boxed{x^{5-2}} = x^3$ The part in the box is usually done mentally.

2. $\dfrac{(a+7)^{14}}{(a+7)^8} = \boxed{(a+7)^{14-8}} = (a+7)^6$

3. $\dfrac{27a^3b^6c^2}{3a^2bc} = \boxed{9a^{3-2}b^{6-1}c^{2-1}} = 9ab^5c$

4. $\dfrac{15x^\square}{3x^\triangle} = 5x^{\square-\triangle}$

The bases are the same, so we subtract the exponents. Although we don't know exactly what $\square - \triangle$ is, the notation $\square - \triangle$ indicates the subtraction.

★ PRACTICE SET C

Find each quotient

1. $\dfrac{y^9}{y^5}$ **2.** $\dfrac{a^7}{a}$ **3.** $\dfrac{(x+6)^5}{(x+6)^3}$ **4.** $\dfrac{26x^4y^6z^2}{13x^2y^2z}$

When we make the subtraction, $n - m$, in the division $\dfrac{x^n}{x^m}$, there are three possibilities for the values of the exponents:

1. The exponent of the numerator is greater than the exponent of the denominator, that is, $n > m$. Thus, the exponent, $n - m$, is a natural number.
2. The exponents are the same, that is, $n = m$. Thus, the exponent, $n - m$, is zero, a whole number.
3. The exponent of the denominator is greater than the exponent of the numerator, that is, $n < m$. Thus, the exponent, $n - m$, is an integer.

☐ ZERO AS AN EXPONENT

In Sample Set C, the exponents of the numerators were greater than the exponents of the denominators. Let's study the case when the exponents are the same.

When the exponents are the same, say n, the subtraction $n - n$ produces 0. Thus, by the second rule of exponents, $\dfrac{x^n}{x^n} = x^{n-n} = x^0$.

But what real number, if any, does x^0 represent? Let's think for a moment about our experience with division in arithmetic. We know that any nonzero number divided by itself is one.

$$\frac{8}{8} = 1, \qquad \frac{43}{43} = 1, \qquad \frac{258}{258} = 1$$

Since the letter x represents some nonzero real number, so does x^n. Thus, $\dfrac{x^n}{x^n}$ represents some nonzero real number divided by itself. Then $\dfrac{x^n}{x^n} = 1$.

But we have also established that if $x \neq 0$, $\dfrac{x^n}{x^n} = x^0$. We now have that $\dfrac{x^n}{x^n} = x^0$ and $\dfrac{x^n}{x^n} = 1$. This implies that $x^0 = 1$, $x \neq 0$.

Exponents can now be natural numbers and zero. We have enlarged our collection of numbers that can be used as exponents from the collection of natural numbers to the collection of whole numbers.

$x^0 = 1, x \neq 0$

ZERO AS AN EXPONENT

If $x \neq 0$, $x^0 = 1$

Any number, other than 0, raised to the power of 0, is 1. 0^0 has no meaning (it does not represent a number).

☆ SAMPLE SET D

Find each value. Assume the base is not zero.

1. $6^0 = 1$ **2.** $247^0 = 1$

3. $(2a + 5)^0 = 1$ **4.** $4y^0 = 4 \cdot 1 = 4$

5. $\dfrac{y^6}{y^6} = y^0 = 1$ **6.** $\dfrac{2x^2}{x^2} = 2x^0 = 2 \cdot 1 = 2$

7. $\dfrac{5(x+4)^8(x-1)^5}{5(x+4)^3(x-1)^5} = (x+4)^{8-3}(x-1)^{5-5}$

$$= (x+4)^5(x-1)^0$$
$$= (x+4)^5$$

★ PRACTICE SET D

Find each value. Assume the base is not zero.

1. $\dfrac{y^7}{y^3}$ **2.** $\dfrac{6x^4}{2x^3}$ **3.** $\dfrac{14a^7}{7a^2}$ **4.** $\dfrac{26x^2y^5}{4xy^2}$ **5.** $\dfrac{36a^4b^3c^8}{8ab^3c^6}$

6. $\dfrac{51(a-4)^3}{17(a-4)}$ **7.** $\dfrac{52a^7b^3(a+b)^8}{26a^2b(a+b)^8}$ **8.** $\dfrac{a^n}{a^3}$ **9.** $\dfrac{14x^ry^pz^q}{2x^ry^hz^5}$

Answers to Practice Sets are on p. 31.

We will study the case where the exponent of the denominator is greater than the exponent of the numerator in Section 2.6.

Section 1.5 EXERCISES

Use the product rule and quotient rule of exponents to simplify problems 1–59. Assume that all bases are nonzero and that all exponents are whole numbers.

1. $3^2 \cdot 3^3$ **2.** $5^2 \cdot 5^4$

3. $9^0 \cdot 9^2$ **4.** $7^3 \cdot 7^0$

5. $2^4 \cdot 2^5$ **6.** x^5x^4

7. x^2x^3 **8.** a^9a^7

9. y^5y^7 **10.** $m^{10}m^2$

11. k^8k^3 **12.** $y^3y^4y^6$

13. $3x^2 \cdot 2x^5$ **14.** $a^2a^3a^8$

15. $4y^4 \cdot 5y^6$ **16.** $2a^3b^2 \cdot 3ab$

17. $12xy^3z^2 \cdot 4x^2y^2z \cdot 3x$

18. $(3ab)(2a^2b)$ **19.** $(4x^2)(8xy^3)$

20. $(2xy)(3y)(4x^2y^5)$ **21.** $\left(\dfrac{1}{4}\,a^2b^4\right)\left(\dfrac{1}{2}\,b^4\right)$

22. $\left(\dfrac{3}{8}\right)\left(\dfrac{16}{21}\,x^2y^3\right)(x^3y^2)$

23. $\dfrac{8^5}{8^3}$

24. $\dfrac{6^4}{6^3}$

25. $\dfrac{2^9}{2^4}$

26. $\dfrac{4^{16}}{4^{13}}$

27. $\dfrac{x^5}{x^3}$

28. $\dfrac{y^4}{y^3}$

29. $\dfrac{y^9}{y^4}$

30. $\dfrac{k^{16}}{k^{13}}$

31. $\dfrac{x^4}{x^2}$

32. $\dfrac{y^5}{y^2}$

33. $\dfrac{m^{16}}{m^9}$

34. $\dfrac{a^9b^6}{a^5b^2}$

35. $\dfrac{y^3w^{10}}{yw^5}$

36. $\dfrac{m^{17}n^{12}}{m^{16}n^{10}}$

37. $\dfrac{x^5y^7}{x^3y^4}$

38. $\dfrac{15x^{20}y^{24}z^4}{5x^{19}yz}$

39. $\dfrac{e^{11}}{e^{11}}$

40. $\dfrac{6r^4}{6r^4}$

41. $\dfrac{x^0}{x^0}$

42. $\dfrac{a^0b^0}{c^0}$

43. $\dfrac{8a^4b^0}{4a^3}$

44. $\dfrac{24x^4y^4z^0w^8}{9xyw^7}$

45. $t^2(y^4)$

46. $x^3\left(\dfrac{x^6}{x^2}\right)$

47. $a^4b^6\left(\dfrac{a^{10}b^{16}}{a^5b^7}\right)$

48. $3a^2b^3\left(\dfrac{14a^2b^5}{2b}\right)$

49. $\dfrac{(x+3y)^{11}(2x-1)^4}{(x+3y)^3(2x-1)}$

50. $\dfrac{40x^5z^{10}(z-x^4)^{12}(x+z)^2}{10z^7(z-x^4)^5}$

51. $x^n x^r$

52. $a^x b^y c^{5z}$

53. $x^n \cdot x^{n+3}$

54. $\dfrac{x^{n+3}}{x^n}$

55. $\dfrac{x^{n+2}x^3}{x^4x^n}$

56. $a^{\star}a^{\bigcirc}$

57. $m^{\diamond}m^{\star}m^{\triangle}$

58. $y^{\triangle}y^{\triangledown}$

59. $a^{\triangle}a^{\triangledown}b^{\square}b^{\diamond}$

EXERCISES FOR REVIEW

(1.2) **60.** What natural numbers can replace x so that the statement $-5 < x \le 3$ is true?

(1.3) **61.** Use the distributive property to expand $4x(2a + 3b)$.

(1.4) **62.** Express $xxxyyyy(a + b)(a + b)$ using exponents.

(1.4) **63.** Find the value of $4^2 + 3^2 \cdot 2^3 - 10 \cdot 8$.

(1.4) **64.** Find the value of $\dfrac{4^2 + (3+2)^2 - 1}{2^3 \cdot 5} + \dfrac{2^4(3^2 - 2^3)}{4^2}$.

★ **Answers to Practice Sets (1.5)**

A. 1. $x^{2+5} = x^7$ 2. $x^{9+4} = x^{13}$ 3. $y^{6+4} = y^{10}$ 4. $c^{12+8} = c^{20}$ 5. $(x+2)^{3+5} = (x+2)^8$

B. 1. $6x^7$ 2. $18y^7$ 3. $36a^5b^3$ 4. $56x^6y^8$ 5. $4(x-y)^5$ 6. $8x^8y^7$ 7. $12a^{10}b^5$ 8. a^{n+m+r}

C. 1. y^4 2. a^6 3. $(x+6)^2$ 4. $2x^2y^4z$

D. 1. $y^{7-3} = y^4$ 2. $3x^{4-3} = 3x$ 3. $2a^{7-2} = 2a^5$ 4. $\dfrac{13}{2}xy^3$ 5. $\dfrac{9}{2}a^3c^2$ 6. $3(a-4)^2$

   7. $2a^5b^2$ 8. a^{n-3} 9. $7y^{p-h}z^{q-5}$

1.6 The Power Rules for Exponents

Section Overview

☐ **THE POWER RULE FOR POWERS**
☐ **THE POWER RULE FOR PRODUCTS**
☐ **THE POWER RULE FOR QUOTIENTS**

☐ THE POWER RULE FOR POWERS

The following examples suggest a rule for raising a power to a power:

1. $(a^2)^3 = a^2 \cdot a^2 \cdot a^2$

   Using the product rule we get

   $(a^2)^3 = a^{2+2+2}$
   $(a^2)^3 = a^{3 \cdot 2}$
   $(a^2)^3 = a^6$
2. $(x^9)^4 = x^9 \cdot x^9 \cdot x^9 \cdot x^9$
   $(x^9)^4 = x^{9+9+9+9}$
   $(x^9)^4 = x^{4 \cdot 9}$
   $(x^9)^4 = x^{36}$

POWER RULE FOR POWERS

If x is a real number and n and m are natural numbers,

$(x^n)^m = x^{n \cdot m}$

$(x^n)^m = x^{n \cdot m}$

To raise a power to a power, multiply the exponents.

☆ **SAMPLE SET A**

Simplify each expression using the power rule for powers. All exponents are natural numbers.

1. $(x^3)^4 = \boxed{x^{3 \cdot 4}}\ x^{12}$ **The box represents a step done mentally.**

2. $(y^5)^3 = \boxed{y^{5 \cdot 3}} = y^{15}$

3. $(d^{20})^6 = \boxed{d^{20 \cdot 6}} = d^{120}$

4. $(x^\square)^\triangle = x^{\square\triangle}$

   Although we don't know exactly what number $\square\triangle$ is, the notation $\square\triangle$ indicates the multiplication.

★ PRACTICE SET A

Simplify each expression using the power rule for powers.

1. $(x^5)^4$ **2.** $(y^7)^7$

❏ THE POWER RULE FOR PRODUCTS

The following examples suggest a rule for raising a product to a power:

1. $(ab)^3 = ab \cdot ab \cdot ab$ Use the commutative property of multiplication.
   $= aaabbb$
   $= a^3b^3$
2. $(xy)^5 = xy \cdot xy \cdot xy \cdot xy \cdot xy$
   $= xxxxx \cdot yyyyy$
   $= x^5y^5$
3. $(4xyz)^2 = 4xyz \cdot 4xyz$
   $= 4 \cdot 4 \cdot xx \cdot yy \cdot zz$
   $= 16x^2y^2z^2$

POWER RULE FOR PRODUCTS

If x and y are real numbers and n is a natural number,

$(xy)^n = x^ny^n$

$(xy)^n = x^ny^n$

To raise a product to a power, apply the exponent to each and every factor.

☆ SAMPLE SET B

Make use of either or both the power rule for products and power rule for powers to simplify each expression.

1. $(ab)^7 = a^7b^7$

2. $(axy)^4 = a^4x^4y^4$

3. $(3ab)^2 = 3^2a^2b^2 = 9a^2b^2$ Don't forget to apply the exponent to the 3!

4. $(2st)^5 = 2^5s^5t^5 = 32s^5t^5$

5. $(ab^3)^2 = a^2(b^3)^2 = a^2b^6$ We used two rules here. First, the power rule for products. Second, the power rule for powers.

6. $(7a^4b^2c^8)^2 = 7^2(a^4)^2(b^2)^2(c^8)^2$
   $= 49a^8b^4c^{16}$

7. If $6a^3c^7 \neq 0$, then $(6a^3c^7)^0 = 1$ Recall that $x^0 = 1$ for $x \neq 0$.

8. $[2(x + 1)^4]^6 = 2^6(x + 1)^{24}$
   $= 64(x + 1)^{24}$

★ PRACTICE SET B

Make use of either or both the power rule for products and the power rule for powers to simplify each expression.

1. $(ax)^4$ **2.** $(3bxy)^2$ **3.** $[4t(s - 5)]^3$ **4.** $(9x^3y^5)^2$

5. $(1a^5b^8c^3d)^6$ **6.** $[(a + 8)(a + 5)]^4$ **7.** $[(12c^4u^3(w - 3)^2]^5$

8. $[10t^4y^7j^3d^2v^6n^4g^8(2 - k)^{17}]^4$ **9.** $(x^3x^5y^2y^6)^9$ **10.** $(10^6 \cdot 10^{12} \cdot 10^5)^{10}$

❏ THE POWER RULE FOR QUOTIENTS

The following example suggests a rule for raising a quotient to a power.

$$\left(\frac{a}{b}\right)^3 = \frac{a}{b} \cdot \frac{a}{b} \cdot \frac{a}{b} = \frac{a \cdot a \cdot a}{b \cdot b \cdot b} = \frac{a^3}{b^3}$$

$\left(\dfrac{x}{y}\right)^n = \dfrac{x^n}{y^n}$

> **POWER RULE FOR QUOTIENTS**
>
> If x and y are real numbers and n is a natural number,
>
> $$\left(\frac{x}{y}\right)^n = \frac{x^n}{y^n}, \qquad y \neq 0$$

To raise a quotient to a power, distribute the exponent to both the numerator and denominator.

☆ SAMPLE SET C

Make use of the power rule for quotients, the power rule for products, the power rule for powers, or a combination of these rules to simplify each expression. All exponents are natural numbers.

1. $\left(\dfrac{x}{y}\right)^6 = \dfrac{x^6}{y^6}$

2. $\left(\dfrac{a}{c}\right)^2 = \dfrac{a^2}{c^2}$

3. $\left(\dfrac{2x}{b}\right)^4 = \dfrac{(2x)^4}{b^4} = \dfrac{2^4 x^4}{b^4} = \dfrac{16x^4}{b^4}$

4. $\left(\dfrac{a^3}{b^5}\right)^7 = \dfrac{(a^3)^7}{(b^5)^7} = \dfrac{a^{21}}{b^{35}}$

5. $\left(\dfrac{3c^4 r^2}{2^3 g^5}\right)^3 = \dfrac{3^3 c^{12} r^6}{2^9 g^{15}} = \dfrac{27 c^{12} r^6}{2^9 g^{15}}$ or $\dfrac{27 c^{12} r^6}{512 g^{15}}$

6. $\left[\dfrac{(a-2)}{(a+7)}\right]^4 = \dfrac{(a-2)^4}{(a+7)^4}$

7. $\left[\dfrac{6x(4-x)^4}{2a(y-4)^6}\right]^2 = \dfrac{6^2 x^2 (4-x)^8}{2^2 a^2 (y-4)^{12}} = \dfrac{36 x^2 (4-x)^8}{4a^2 (y-4)^{12}} = \dfrac{9x^2 (4-x)^8}{a^2 (y-4)^{12}}$

8. $\left(\dfrac{a^3 b^5}{a^2 b}\right)^3 = (a^{3-2} b^{5-1})^3$ We can simplify within the parentheses. We have a rule that tells us to proceed this way.

$\qquad = (ab^4)^3$
$\qquad = a^3 b^{12}$

$\left(\dfrac{a^3 b^5}{a^2 b}\right)^3 = \dfrac{a^9 b^{15}}{a^6 b^3} = a^{9-6} b^{15-3} = a^3 b^{12}$ We could have actually used the power rule for quotients first. Distribute the exponent, then simplify using the other rules.
   It is probably better, for the sake of consistency, to work inside the parentheses first.

9. $\left(\dfrac{a^r b^s}{c^t}\right)^w = \dfrac{a^{rw} b^{sw}}{c^{tw}}$

★ PRACTICE SET C

Make use of the power rule for quotients, the power rule for products, the power rule for powers, or a combination of these rules to simplify each expression.

1. $\left(\dfrac{a}{c}\right)^5$ **2.** $\left(\dfrac{2x}{3y}\right)^3$ **3.** $\left(\dfrac{x^2y^4z^7}{a^5b}\right)^9$ **4.** $\left[\dfrac{2a^4(b-1)}{3b^3(c+6)}\right]^4$ **5.** $\left(\dfrac{8a^3b^2c^6}{4a^2b}\right)^3$

6. $\left[\dfrac{(9+w)^2}{(3+w)^5}\right]^{10}$ **7.** $\left[\dfrac{5x^4(y+1)}{5x^4(y+1)}\right]^6$ **8.** $\left(\dfrac{16x^3v^4c^7}{12x^2vc^6}\right)^0$

Answers to Practice Sets are on p. 35.

Section 1.6 EXERCISES

Use the power rules for exponents to simplify problems 1–54. Assume that all bases are nonzero and that all variable exponents are natural numbers.

1. $(ac)^5$ **2.** $(nm)^7$

3. $(2a)^3$ **4.** $(2a)^5$

5. $(3xy)^4$ **6.** $(2xy)^5$

7. $(3ab)^4$ **8.** $(6mn)^2$

9. $(7y^3)^2$ **10.** $(3m^3)^4$

11. $(5x^6)^3$ **12.** $(5x^2)^3$

13. $(10a^2b)^2$ **14.** $(8x^2y^3)^2$

15. $(x^2y^3z^5)^4$ **16.** $(2a^5b^{11})^0$

17. $(x^3y^2z^4)^5$ **18.** $(m^6n^2p^5)^5$

19. $(a^4b^7c^6d^8)^8$ **20.** $(x^2y^3z^9w^7)^3$

21. $(9xy^3)^0$ **22.** $\left(\dfrac{1}{2}f^2r^6s^5\right)^4$

23. $\left(\dfrac{1}{8}c^{10}d^8e^4f^9\right)^2$ **24.** $\left(\dfrac{3}{5}a^3b^5c^{10}\right)^3$

25. $(xy)^4(x^2y^4)$ **26.** $(2a^2)^4(3a^5)^2$

27. $(a^2b^3)^3(a^3b^3)^4$ **28.** $(h^3k^5)^2(h^2k^4)^3$

29. $(x^4y^3z)^4(x^5yz^2)^2$ **30.** $(ab^3c^2)^5(a^2b^2c)^2$

31. $\dfrac{(6a^2b^8)^2}{(3ab^5)^2}$ **32.** $\dfrac{(a^3b^4)^5}{(a^4b^4)^3}$

33. $\dfrac{(x^6y^5)^3}{(x^2y^3)^5}$ **34.** $\dfrac{(a^8b^{10})^3}{(a^7b^5)^3}$

35. $\dfrac{(m^5n^6p^4)^4}{(m^4n^5p)^4}$ **36.** $\dfrac{(x^8y^3z^2)^5}{(x^6yz)^6}$

37. $\dfrac{(10x^4y^5z^{11})^3}{(xy^2)^4}$ **38.** $\dfrac{(9a^4b^5)(2b^2c)}{(3a^3b)(6bc)}$

39. $\dfrac{(2x^3y^3)^4(5x^6y^8)^2}{(4x^5y^3)^2}$ **40.** $\left(\dfrac{3x}{5y}\right)^2$ **47.** $\dfrac{(x^{n+2})^3}{x^{2n}}$ **48.** $(xy)^\triangle$

41. $\left(\dfrac{3ab}{4xy}\right)^3$ **42.** $\left(\dfrac{x^2y^2}{2z^3}\right)^5$ **49.** $(2ab)^\star$ **50.** $\dfrac{(3a^\triangle b^\triangledown)^\square}{(5xy^\Diamond)^\star}$

43. $\left(\dfrac{3a^2b^3}{c^4}\right)^3$ **44.** $\left(\dfrac{4^2a^3b^7}{b^5c^4}\right)^2$ **51.** $\dfrac{10m^\triangle}{5m^\star}$ **52.** $\dfrac{4^3a^\triangle a^\square}{4a^\triangledown}$

45. $\left[\dfrac{x^2(y-1)^3}{(x+6)}\right]^4$ **46.** $(x^nt^{2m})^4$ **53.** $\left(\dfrac{4x^\triangle}{2y^\triangledown}\right)^\square$ **54.** $\left(\dfrac{16a^3b^\star}{5a^\triangle b^\triangledown}\right)^0$

EXERCISES FOR REVIEW

(1.2) **55.** Is there a smallest integer? If so, what is it?

(1.3) **56.** Use the distributive property to expand $5a(2x+8)$.

(1.4) **57.** Find the value of $\dfrac{(5-3)^2+(5+4)^3+2}{4^2-2\cdot5-1}$.

(1.5) **58.** Assuming the bases are not zero, find the value of $(4a^2b^3)(5ab^4)$.

(1.5) **59.** Assuming the bases are not zero, find the value of $\dfrac{36x^{10}y^8z^3w^0}{9x^5y^2z}$.

★ Answers to Practice Sets (1.6)

A. **1.** x^{20} **2.** y^{49}

B. **1.** a^4x^4 **2.** $9b^2x^2y^2$ **3.** $64t^3(s-5)^3$ **4.** $81x^6y^{10}$ **5.** $a^{30}b^{48}c^{18}d^6$ **6.** $(a+8)^4(a+5)^4$

  7. $125^5c^{20}u^{15}(w-3)^{10}$ **8.** $10^4t^{16}y^{28}j^{12}d^8v^{24}n^{16}g^{32}(2-k)^{68}$ **9.** $(x^8y^8)^9 = x^{72}y^{72}$ **10.** 10^{230}

C. **1.** $\dfrac{a^5}{c^5}$ **2.** $\dfrac{8x^3}{27y^3}$ **3.** $\dfrac{x^{18}y^{36}z^{63}}{a^{45}b^9}$ **4.** $\dfrac{16a^{16}(b-1)^4}{81b^{12}(c+6)^4}$ **5.** $8a^3b^3c^{18}$ **6.** $\dfrac{(9+w)^{20}}{(3+w)^{50}}$

  7. 1, if $x^4(y+1)\neq0$ **8.** 1, if $x^2vc^6\neq0$

Chapter 1   SUMMARY OF KEY CONCEPTS

Variables and Constants (1.1)
A *variable* is a letter or symbol that represents any member of a collection of two or more numbers. A *constant* is a letter or symbol that represents a specific number.

Binary Operation (1.1)
A *binary operation* is a process that assigns two numbers to a single number. $+$, $-$, $\times$, $\div$ are binary operations.

Grouping Symbols (1.1)
Grouping symbols are used to indicate that a particular collection of numbers and meaningful operations is to be considered as a single number ($5 \div 0$ is not meaningful). Grouping symbols can also direct us in operations when more than two operations are to be performed. Common algebraic grouping symbols are

Parentheses: (   )
Brackets: [   ]
Braces: {   }
Bar: ——

Order of Operations (1.1, 1.4)
When two or more operations are to be performed on a collection of numbers, the correct value can be obtained only by using the correct order of operations.

The Real Number Line (1.2)
The *real number line* allows us to visually display some of the numbers in which we are interested.

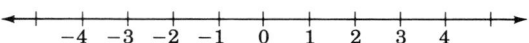

Coordinate and Graph (1.2)
The number associated with a point on the number line is called the *coordinate* of the point. The point associated with a number is called the *graph* of the number.

Real Number (1.2)
A *real number* is any number that is the coordinate of a point on the real number line.

Types of Real Numbers (1.2)
The collection of real numbers has many subcollections. The ones of most interest to us are

the *natural numbers*: $\{1, 2, 3, \ldots\}$
the *whole numbers*: $\{0, 1, 2, 3, \ldots\}$
the *integers*: $\{\ldots, -3, -2, -1, 0, 1, 2, 3, \ldots\}$
the *rational numbers*:   {all numbers that can be expressed as the quotient of two integers}
the *irrational numbers*:   {all numbers that have nonending and nonrepeating decimal representations}

Properties of Real Numbers (1.3)
Closure: If a and b are real numbers, then $a + b$ and $a \cdot b$ are unique real numbers.
Commutative: $a + b = b + a$ and $a \cdot b = b \cdot a$
Associative: $a + (b + c) = (a + b) + c$ and $a(b \cdot c) = (a \cdot b)c$
Distributive: $a(b + c) = a \cdot b + a \cdot c$
Additive identity: 0 is the additive identity. $a + 0 = a$ and $0 + a = a$.
Multiplicative identity: 1 is the multiplicative identity. $a \cdot 1 = a$ and $1 \cdot a = a$.
Additive inverse: For each real number a there is exactly one number $-a$ such that $a + (-a) = 0$ and $(-a) + a = 0$.
Multiplicative identity: For each nonzero real number a there is exactly one nonzero real number $\frac{1}{a}$ such that $a \cdot \frac{1}{a} = 1$ and $\frac{1}{a} \cdot a = 1$.

Exponents (1.4)
Exponents record the number of identical factors that appear in a multiplication.

$$\underbrace{x \cdot x \cdot x \cdot \ldots \cdot x}_{n \text{ factors of } x} = x^n$$

Rules of Exponents
(1.5, 1.6)

If x is a real number and n and m are natural numbers, then

$$x^n \cdot x^m = x^{n+m}$$

$$\frac{x^n}{x^m} = x^{n-m}, \qquad x \neq 0$$

$$x^0 = 1, \qquad x \neq 0$$

$$(x^n)^m = x^{n \cdot m}$$

$$\left(\frac{x}{y}\right)^n = \frac{x^n}{y^n}, \qquad y \neq 0$$

EXERCISE SUPPLEMENT

Section 1.1

For problems 1–13, simplify the expressions.

1. $12 + 7(4 + 3)$

2. $9(4 - 2) + 6(8 + 2) - 3(1 + 4)$

3. $6[1 + 8(7 + 2)]$

4. $26 \div 2 - 10$

5. $\dfrac{(4 + 17 + 1) + 4}{14 - 1}$

6. $51 \div 3 \div 7$

7. $(4 + 5)(4 + 6) - (4 + 7)$

8. $8(2 \cdot 12 \div 13) + 2 \cdot 5 \cdot 11 - [1 + 4(1 + 2)]$

9. $\dfrac{3}{4} + \dfrac{1}{12}\left(\dfrac{3}{4} - \dfrac{1}{2}\right)$

10. $48 - 3\left[\dfrac{1 + 17}{6}\right]$

11. $\dfrac{29 + 11}{6 - 1}$

12. $\dfrac{\dfrac{88}{11} + \dfrac{99}{9} + 1}{\dfrac{54}{9} - \dfrac{22}{11}}$

13. $\dfrac{8 \cdot 6}{2} + \dfrac{9 \cdot 9}{3} - \dfrac{10 \cdot 4}{5}$

For problems 14–17, write the appropriate relation symbol ($=, <, >$) in place of the *.

14. $22 * 6$

15. $9[4 + 3(8)] * 6[1 + 8(5)]$

16. $3(1.06 + 2.11) * 4(11.01 - 9.06)$

17. $2 * 0$

For problems 18–20, state whether the letters or symbols are the same or different.

18. $<$ and $\neq$

19. $>$ and $\not<$

20. $a = b$ and $b = a$

21. Represent the sum of c and d two different ways.

For problems 22–28, use algebraic notation.

22. 8 plus 9

23. 62 divided by f

24. 8 times $(x + 4)$

25. 6 times x, minus 2

26. $x + 1$ divided by $x - 3$

27. $y + 11$ divided by $y + 10$, minus 12

28. zero minus a times b

Section 1.2

29. Is every natural number a whole number?

30. Is every rational number a real number?

For problems 31–35, locate the numbers on a number line by placing a point at their (approximate) position.

31. 2

32. 3.6

33. $-1\dfrac{3}{8}$

34. 0

35. $-4\dfrac{1}{2}$

36. Draw a number line that extends from 10 to 20. Place a point at all odd integers.

37. Draw a number line that extends from -10 to 10. Place a point at all negative odd integers and at all even positive integers.

38. Draw a number line that extends from -5 to 10. Place a point at all integers that are greater than or equal to -2 but strictly less than 5.

39. Draw a number line that extends from -10 to 10. Place a point at all real numbers that are strictly greater than -8 but less than or equal to 7.

40. Draw a number line that extends from -10 to 10. Place a point at all real numbers between and including -6 and 4.

For problems 41–44, write the appropriate relation symbol ($=, <, >$).

41. $-3 \quad 0$

42. $-1 \quad 1$

43. $-8 \quad -5$

44. $-5 \quad -5\dfrac{1}{2}$

45. Is there a smallest two digit integer? If so, what is it?

46. Is there a smallest two digit real number? If so, what is it?

For problems 47–49, what integers can replace x so that the statements are true?

47. $4 \le x \le 7$

48. $-3 \le x < 1$

49. $-3 < x \le 2$

50. The temperature today in Los Angeles was eighty-two degrees. Represent this temperature by a real number.

51. The temperature today in Marblehead was six degrees below zero. Represent this temperature by a real number.

52. On the number line, how many units between -3 and 2?

53. On the number line, how many units between -4 and 0?

Section 1.3

54. $a + b = b + a$ is an illustration of the _____ property of addition.

55. $st = ts$ is an illustration of the _____ property of _____.

Use the commutative properties of addition and multiplication to write equivalent expressions for problems 56–63.

56. $y + 12$

57. $a + 4b$

58. $6x$

59. $2(a - 1)$

60. $(-8)(4)$

61. $(6)(-9)(-2)$

62. $(x + y)(x - y)$

63. $\triangle \cdot \diamond$

Simplify problems 64–68 using the commutative property of multiplication. You need not use the distributive property.

64. $8x3y$

65. $16ab2c$

66. $4axyc4d4e$

67. $3(x + 2)5(x - 1)0(x + 6)$

68. $8b(a - 6)9a(a - 4)$

For problems 69–76, use the distributive property to expand the expressions.

69. $3(a + 4)$

70. $a(b + 3c)$

71. $2g(4h + 2k)$

72. $(8m + 5n)6p$

73. $3y(2x + 4z + 5w)$

74. $(a + 2)(b + 2c)$

75. $(x + y)(4a + 3b)$

76. $10a_z(b_z + c)$

Section 1.4

For problems 77–85, write the expressions using exponential notation.

77. x to the fifth.

78. $(y + 2)$ cubed.

79. $(a + 2b)$ squared minus $(a + 3b)$ to the fourth.

80. x cubed plus 2 times $(y - x)$ to the seventh.

81. $aaaaaaa$

82. $2 \cdot 2 \cdot 2 \cdot 2$

83. $(-8)(-8)(-8)(-8)xxxyyyyy$

84. $(x - 9)(x - 9) + (3x + 1)(3x + 1)(3x + 1)$

85. $2zzyzyyy + 7zzyz(a - 6)^2(a - 6)$

For problems 86–91, expand the terms so that no exponents appear.

86. x^3

87. $3x^3$

88. 7^3x^2

89. $(4b)^2$

90. $(6a^2)^3(5c - 4)^2$

91. $(x^3 + 7)^2(y^2 - 3)^3(z + 10)$

92. Choose values for a and b to show that
  (a) $(a + b)^2$ is not always equal to $a^2 + b^2$.
  (b) $(a + b)^2$ may be equal to $a^2 + b^2$.

93. Choose a value for x to show that
  (a) $(4x)^2$ is not always equal to $4x^2$.
  (b) $(4x)^2$ may be equal to $4x^2$.

Sections 1.5 and 1.6

Simplify problems 94 – 125.

94. $4^2 + 8$

95. $6^3 + 5(30)$

96. $1^8 + 0^{10} + 3^2(4^2 + 2^3)$

97. $12^2 + 0.3(11)^2$

98. $\dfrac{3^4 + 1}{2^2 + 4^2 + 3^2}$

99. $\dfrac{6^2 + 3^2}{2^2 + 1} + \dfrac{(1+4)^2 - 2^3 - 1^4}{2^5 - 4^2}$

100. $a^4 a^3$

101. $2b^5 2b^3$

102. $4a^3 b^2 c^8 \cdot 3ab^2 c^0$

103. $(6x^4 y^{10})(xy^3)$

104. $(3xyz^2)(2x^2 y^3)(4x^2 y^2 z^4)$

105. $(3a)^4$

106. $(10xy)^2$

107. $(x^2 y^4)^6$

108. $(a^4 b^7 c^7 z^{12})^9$

109. $\left(\dfrac{3}{4} x^8 y^6 z^0 a^{10} b^{15}\right)^2$

110. $\dfrac{x^8}{x^5}$

111. $\dfrac{14a^4 b^6 c^7}{2ab^3 c^2}$

112. $\dfrac{11x^4}{11x^4}$

113. $x^4 \cdot \dfrac{x^{10}}{x^3}$

114. $a^3 b^7 \cdot \dfrac{a^9 b^6}{a^5 b^{10}}$

115. $\dfrac{(x^4 y^6 z^{10})^4}{(xy^5 z^7)^3}$

116. $\dfrac{(2x-1)^{13}(2x+5)^5}{(2x-1)^{10}(2x+5)}$

117. $\left(\dfrac{3x^2}{4y^3}\right)^2$

118. $\dfrac{(x+y)^9(x-y)^4}{(x+y)^3}$

119. $x^n \cdot x^m$

120. $a^{n+2} a^{n+4}$

121. $6b^{2n+7} \cdot 8b^{5n+2}$

122. $\dfrac{18x^{4n+9}}{2x^{2n+1}}$

123. $(x^{5t} y^{4r})^7$

124. $(a^{2n} b^{3m} c^{4p})^{6r}$

125. $\dfrac{u^w}{u^k}$

For problems 1–4, simplify each of the expressions.

1. **(1.1)** $8(6-3) - 5 \cdot 4 + 3(8)(2) \div 4 \cdot 3$

1. _____

2. **(1.1)** $\{2(1+7)^2\}^0$

2. _____

3. **(1.1)** $\dfrac{1^8 + 4^0 + 3^3(1+4)}{2^2(2+15)}$

3. _____

4. **(1.1)** $\dfrac{2 \cdot 3^4 - 10^2}{4-3} + \dfrac{5(2^2 + 3^2)}{11-6}$

4. _____

5. **(1.1)** Write the appropriate relation symbol ($>$, $<$) in place of the $*$.

$5(2+11) * 2(8-3) - 2$

5. _____

For problems 6 and 7, use algebraic notation.

6. **(1.1)** $(x-1)$ times ($3x$ plus 2).

6. _____

7. **(1.1)** A number divided by twelve is less than or equal to the same number plus four.

7. _____

8. **(1.2)** Locate the approximate position of -1.6 on the number line.

8. _____

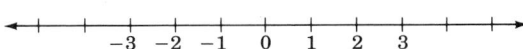

9. **(1.2)** Is 0 a positive number, a negative number, neither, or both?

9. _____

10. **(1.2)** Draw a portion of the number line and place points at all even integers strictly between 14 and 20.

10. _____

11. **(1.2)** Draw a portion of the number line and place points at all real numbers strictly greater than -1 but less than or equal to 4.

11. _____

12. **(1.2)** What whole numbers can replace x so that the following statement is true? $-4 \le x \le 5$.

12. _____

13. **(1.2)** Is there a largest real number between and including 6 and 10? If so, what is it?

13. _____

14. **(1.3)** Use the commutative property of multiplication to write $m(a+3)$ in an equivalent form.

14. _____

15._____

15. **(1.3)** Use the commutative properties to simplify $3a4b8cd$.

16._____

16. **(1.3)** Use the commutative properties to simplify $4(x-9)2y(x-9)3y$.

17. **(1.4)** Simplify 4 squared times x cubed times y to the fifth.

17._____

18. **(1.4)** Simplify $(3)(3)(3)aabbbbabba(3)a$.

18._____

For problems 19–23, use the rules of exponents to simplify each of the expressions.

19. **(1.5, 1.6)** $(3ab^2)^2(2a^3b)^3$

19._____

20. **(1.5, 1.6)** $\dfrac{x^{10}y^{12}}{x^2y^5}$

20._____

21. **(1.5, 1.6)** $\dfrac{52x^7y^{10}(y-x^4)^{12}(y+x)^5}{4y^6(y-x^4)^{10}(y+x)}$

21._____

22. **(1.5, 1.6)** $(x^ny^{3m}z^{2p})^4$

22._____

23. $\dfrac{(5x+4)^{\circ}}{(3x^2-1)^{\circ}}$

23._____

24. $\dfrac{x^{\nabla}x^{\square}y^{\triangle}}{x^{\triangle}y^{\nabla}}$

24._____

25. **(1.5, 1.6)** What word is used to describe the letter or symbol that **represents** an unspecified member of a particular collection of two or more numbers that are clearly defined?

25._____

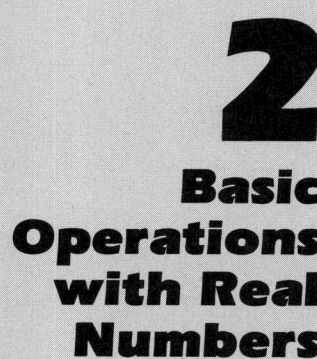

2

Basic Operations with Real Numbers

After completing this chapter, you should

Section 2.1 Signed Numbers
- be familiar with positive and negative numbers and with the concept of opposites

Section 2.2 Absolute Value
- understand the geometric and algebraic definitions of absolute value

Section 2.3 Addition of Signed Numbers
- be able to add numbers with like signs and unlike signs
- understand addition with zero

Section 2.4 Subtraction of Signed Numbers
- understand the definition of subtraction
- be able to subtract signed numbers

Section 2.5 Multiplication and Division of Signed Numbers
- be able to multiply and divide signed numbers

Section 2.6 Negative Exponents
- understand the concepts of reciprocals and negative exponents
- be able to work with negative exponents

Section 2.7 Scientific Notation
- be able to convert a number from standard form to scientific form and from scientific form to standard form
- be able to work with numbers in scientific notation

2.1 Signed Numbers

Section Overview	❑ POSITIVE AND NEGATIVE NUMBERS ❑ OPPOSITES

❑ POSITIVE AND NEGATIVE NUMBERS

When we studied the number line in Section 1.2 we noted that

> Each point on the number line corresponds to a real number, and each real number is located at a unique point on the number line.

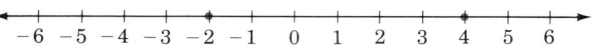

Each real number has a *sign* inherently associated with it. A real number is said to be a **positive number** if it is located to the right of 0 on the number line. It is a **negative** number if it is located to the left of 0 on the number line.

Positive and Negative Numbers

THE NOTATION OF SIGNED NUMBERS

A number is denoted as **positive** if it is directly preceded by a "+" sign *or* no sign at all.

A number is denoted as **negative** if it is directly preceded by a "−" sign.

+ and − Notation

The "+" and "−" signs now have two meanings:

+ can denote the operation of addition or a positive number.
− can denote the operation of subtraction or a negative number.

Read the "−" Sign as "Negative"

To avoid any confusion between "sign" and "operation," it is preferable to read the sign of a number as "positive" or "negative."

☆ SAMPLE SET A

1. -8 should be read as "negative eight" rather than "minus eight."

2. $4 + (-2)$ should be read as "four plus negative two" rather than "four plus minus two."

3. $-6 + (-3)$ should be read as "negative six plus negative three" rather than "minus six plus minus three."

4. $-15 - (-6)$ should be read as "negative fifteen minus negative six" rather than "minus fifteen minus minus six."

5. $-5 + 7$ should be read as "negative five plus seven" rather than "minus five plus seven."

6. $0 - 2$ should be read as "zero minus two."

★ PRACTICE SET A

Write each expression in words.

1. $4 + 10$ **2.** $7 + (-4)$ **3.** $-9 + 2$ **4.** $-16 - (+8)$

5. $-1 - (-9)$ **6.** $0 + (-7)$

☐ OPPOSITES

Opposites

On the number line, each real number has an image on the opposite side of 0. For this reason we say that each real number has an opposite. **Opposites** are the same distance from zero but have opposite signs.

The opposite of a real number is denoted by placing a negative sign directly in front of the number. Thus, if a is any real number, then $-a$ is its opposite. *Notice that the letter a is a variable.* Thus, "a" need not be positive, and "$-a$" need not be negative.

If a is a real number, $-a$ is opposite a on the number line and a is opposite $-a$ on the number line.

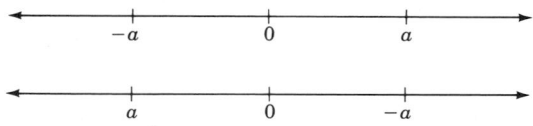

$-(-a)$ is opposite $-a$ on the number line. This implies that $-(-a) = a$.

This property of opposites suggests the double-negative property for real numbers.

THE DOUBLE-NEGATIVE PROPERTY

If a is a real number, then

$-(-a) = a$

$-(-a) = a$

☆ SAMPLE SET B

1. If $a = 3$, then $-a = -3$ and $-(-a) = -(-3) = 3$.

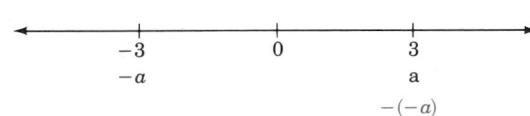

2. If $a = -4$, then $-a = -(-4) = 4$ and $-(-a) = a = -4$.

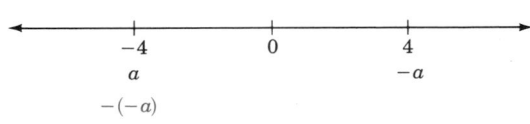

★ PRACTICE SET B

Find the opposite of each real number.

1. 8   **2.** 17   **3.** -6 **4.** -15 **5.** $-(-1)$ **6.** $-[-(-7)]$

7. Suppose that a is a positive number. What type of number is $-a$?

8. Suppose that a is a negative number. What type of number is $-a$?

9. Suppose we do not know the sign of the number m. Can we say that $-m$ is positive, negative, or that we do not know?

Answers to Practice Sets are on p. 47.

Section 2.1 EXERCISES

1. A number is denoted as positive if it is directly preceded by ⎯⎯⎯⎯⎯⎯⎯⎯⎯⎯⎯⎯⎯ .

2. A number is denoted as negative if it is directly preceded by ⎯⎯⎯⎯⎯⎯⎯⎯⎯⎯⎯⎯⎯ .

For problems 3–8, how should the real numbers be read? (Write in words.)

3. -5

4. -3

5. 12

6. 10

7. $-(-4)$

8. $-(-1)$

For problems 9–14, write the expressions in words.

9. $5 + 7$

10. $2 + 6$

11. $11 + (-2)$

12. $1 + (-5)$

13. $6 - (-8)$

14. $0 - (-15)$

Rewrite problems 15–42 in a simpler form.

15. $-(-8)$

16. $-(-5)$

17. $-(-2)$

18. $-(-9)$

19. $-(-1)$

20. $-(-4)$

21. $-[-(-3)]$

22. $-[-(-10)]$

23. $-[-(-6)]$

24. $-[-(-15)]$

25. $-\{-[-(-26)]\}$

26. $-\{-[-(-11)]\}$

27. $-\{-[-(-31)]\}$

28. $-\{-[-(-14)]\}$

29. $-[-(12)]$

30. $-[-(2)]$

31. $-[-(17)]$

32. $-[-(42)]$

33. $5 - (-2)$

34. $6 - (-14)$

35. $10 - (-6)$

36. $18 - (-12)$

37. $31 - (-1)$

38. $54 - (-18)$

39. $6 - (-3) - (-4)$

40. $2 - (-1) - (-8)$

41. $15 - (-6) - (-5)$

42. $24 - (-8) - (-13)$

EXERCISES FOR REVIEW

(1.4) **43.** There is only one real number for which $(5a)^2 = 5a^2$. What is the number?

(1.5) **44.** Simplify $(3xy)(2x^2y^3)(4x^2y^4)$.

(1.5) **45.** Simplify $x^{n+3} \cdot x^5$.

(1.6) **46.** Simplify $(a^3b^2c^4)^4$.

(1.6) **47.** Simplify $\left(\dfrac{4a^2b}{3xy^3}\right)^2$.

★ **Answers to Practice Sets (2.1)**

A. **1.** four plus ten   **2.** seven plus negative four   **3.** negative nine plus two
4. negative sixteen minus positive eight   **5.** negative one minus negative nine
6. zero plus negative seven

B. **1.** -8 **2.** -17 **3.** 6   **4.** 15   **5.** -1, since $-(-1) = 1$ **6.** 7
7. If a is positive, $-a$ is negative.   **8.** If a is negative, $-a$ is positive.   **9.** We must say that we do not know.

2.2 Absolute Value

Section Overview

☐ **GEOMETRIC DEFINITION OF ABSOLUTE VALUE**
☐ **ALGEBRAIC DEFINITION OF ABSOLUTE VALUE**

☐ GEOMETRIC DEFINITION OF ABSOLUTE VALUE

Absolute Value — Geometric Approach

> The **absolute value** of a number a, denoted $|a|$, is the distance from a to 0 on the number line.

Absolute value speaks to the question of "how far," and not "which way." The phrase how far implies length, and length is always a nonnegative (zero or positive) quantity. Thus, the absolute value of a number is a nonnegative number. This is shown in the following examples:

1. $|4| = 4$

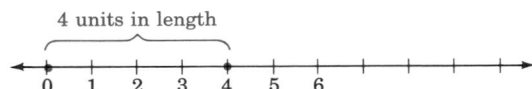

2. $|-4| = 4$

3. $|0| = 0$
4. $-|5| = -5$.   The quantity on the left side of the equal sign is read as "negative the absolute value of 5." The absolute value of 5 is 5. Hence, negative the absolute value of 5 is -5.
5. $-|-3| = -3$.   The quantity on the left side of the equal sign is read as "negative the absolute value of -3." The absolute value of -3 is 3. Hence, negative the absolute value of -3 is $-(3) = -3$.

☐ ALGEBRAIC DEFINITION OF ABSOLUTE VALUE

The problems in the first example may help to suggest the following algebraic definition of absolute value. The definition is interpreted below. Examples follow the interpretation.

Absolute Value — Algebraic Approach

> The **absolute value** of a number a is
> $$|a| = \begin{cases} a & \text{if} \quad a \geq 0 \\ -a & \text{if} \quad a < 0 \end{cases}$$

The algebraic definition takes into account the fact that the number a could be either positive or zero (≥ 0) or negative (< 0).

1. If the number a is positive or zero (≥ 0), the first part of the definition applies. The first part of the definition tells us that if the number enclosed in the absolute bars is a nonnegative number, the absolute value of the number is the number itself.
2. If the number a is negative (< 0), the second part of the definition applies. The second part of the definition tells us that if the number enclosed within the absolute value bars is a negative number, the absolute value of the number is the opposite of the number. The opposite of a negative number is a positive number.

☆ SAMPLE SET A

Use the algebraic definition of absolute value to find the following values.

1. $|8|$. The number enclosed within the absolute value bars is a nonnegative number so the first part of the definition applies. This part says that the absolute value of 8 is 8 itself.

   $|8| = 8$

2. $|-3|$. The number enclosed within absolute value bars is a negative number so the second part of the definition applies. This part says that the absolute value of -3 is the opposite of -3, which is $-(-3)$. By the double-negative property, $-(-3) = 3$.

   $|-3| = 3$

★ PRACTICE SET A

Use the algebraic definition of absolute value to find the following values.

1. $|7|$ 2. $|9|$ 3. $|-12|$ 4. $|-5|$ 5. $-|8|$

6. $-|1|$ 7. $-|-52|$ 8. $-|-31|$

Answers to the Practice Set are on p. 50.

Section 2.2 EXERCISES

For problems 1–54, determine each of the values.

1. $|5|$

2. $|3|$

3. $|6|$

4. $|14|$

5. $|-8|$

6. $|-10|$

7. $|-16|$

8. $-|8|$

9. $-|12|$

10. $-|47|$

11. $-|9|$

12. $|-9|$

13. $|-1|$

14. $|-4|$

15. $-|3|$

16. $-|7|$

17. $-|-14|$

18. $-|-19|$

19. $-|-28|$

20. $-|-31|$

21. $-|-68|$

22. $|0|$

23. $|-26|$

24. $-|-26|$

25. $-|-(-8)|$

26. $-|-(-4)|$

27. $-|-(-1)|$

28. $-|-(-7)|$

29. $-(-|4|)$

30. $-(-|2|)$

31. $-(-|-6|)$

32. $-(-|-42|)$

33. $|-|-3||$

34. $|-|-15||$

35. $|-|-12||$

36. $|-|-29||$

37. $|6-|-2||$

38. $|18-|-11||$

39. $|5-|-1||$

40. $|10-|-3||$

41. $|-(17-|-12|)|$

42. $|-(46-|-24|)|$

43. $|5|-|-2|$

44. $|-2|^3$

45. $|-(2\cdot3)|$

46. $|-2|+|-9|$

47. $(|-6|+|4|)^2$

48. $(|-1|-|1|)^3$

49. $(|4|+|-6|)^2-(|-2|)^3$

50. $-[|-10|-6]^2$

51. $-[-(-|-4|+|-3|)^3]^2$

52. A Mission Control Officer at Cape Canaveral makes the statement "lift-off, T minus 50 seconds." How long before lift-off?

53. Due to a slowdown in the industry, a Silicon Valley computer company finds itself in debt $2,400,000. Use absolute value notation to describe this company's debt.

54. A particular machine is set correctly if upon action its meter reads 0 units. One particular machine has a meter reading of -1.6 upon action. How far is this machine off its correct setting?

EXERCISES FOR REVIEW

(1.1) **55.** Write the following phrase using algebraic notation: "four times $(a + b)$."

(1.2) **56.** Is there a smallest natural number? If so, what is it?

(1.3) **57.** Name the property of real numbers that makes $5 + a = a + 5$ a true statement.

(1.5) **58.** Find the quotient of $\dfrac{x^6 y^8}{x^4 y^3}$.

(2.1) **59.** Simplify $-(-4)$.

★ Answers to Practice Set (2.2)

A. **1.** 7   **2.** 9   **3.** 12   **4.** 5   **5.** -8 **6.** -1 **7.** -52 **8.** -31

2.3 Addition of Signed Numbers

Section Overview

☐ **ADDITION OF NUMBERS WITH LIKE SIGNS**
☐ **ADDITION WITH ZERO**
☐ **ADDITION OF NUMBERS WITH UNLIKE SIGNS**

☐ ADDITION OF NUMBERS WITH LIKE SIGNS

Let us add the two positive numbers 2 and 3. We perform this addition on the number line as follows.

We begin at 0, the origin.
Since 2 is positive, we move 2 units to the right.
Since 3 is positive, we move 3 more units to the right.
We are now located at 5.
Thus, $2 + 3 = 5$.

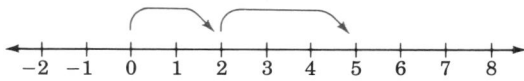

Summarizing, we have

(2 positive units) $+$ (3 positive units) $=$ (5 positive units)

Now let us add the two negative numbers -2 and -3. We perform this addition on the number line as follows.

We begin at 0, the origin.
Since -2 is negative, we move 2 units to the left.
Since -3 is negative, we move 3 more units to the left.
We are now located at -5.

Thus, $(-2) + (-3) = -5$.

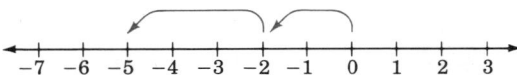

Summarizing, we have

(2 negative units) + (3 negative units) = (5 negative units)

These two examples suggest that

(positive number) + (positive number) = (positive number)

(negative number) + (negative number) = (negative number)

Adding Numbers with the Same Sign

To add two real numbers that have the same sign, add the absolute values of the numbers and associate the common sign with the sum.

☆ SAMPLE SET A

Find the sums.

1. $3 + 7$

$\left.\begin{array}{l} |3| = 3 \\ |7| = 7 \end{array}\right\}$

Add these absolute values.

$3 + 7 = 10$

The common sign is "+."

$3 + 7 = +10$ or $3 + 7 = 10$

2. $(-4) + (-9)$

$\left.\begin{array}{l} |-4| = 4 \\ |-9| = 9 \end{array}\right\}$

Add these absolute values.

$4 + 9 = 13$

The common sign is "−."

$(-4) + (-9) = -13$

★ PRACTICE SET A

Find the sums.

1. $8 + 6$ **2.** $41 + 11$ **3.** $(-4) + (-8)$ **4.** $(-36) + (-9)$

5. $-14 + (-20)$ **6.** $-\dfrac{2}{3} + \left(-\dfrac{5}{3}\right)$ **7.** $-2.8 + (-4.6)$

❏ ADDITION WITH ZERO

Notice that

Addition with 0

(0) + (a positive number) = (that same positive number)

(0) + (a negative number) = (that same negative number)

The Additive Identity Is 0

Since adding 0 to a real number leaves that number unchanged, 0 is **called the additive identity.**

❏ ADDITION OF NUMBERS WITH UNLIKE SIGNS

Now let us perform the addition $2 + (-6)$. These two numbers have unlike signs. This type of addition can also be illustrated using the number line.

We begin at 0, the origin.
Since 2 is positive, we move 2 units to the right.
Since -6 is negative, we move, from the 2, 6 units to the left.
We are now located at -4.

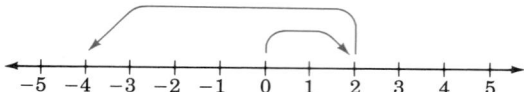

A rule for adding two numbers that have unlike signs is suggested by noting that if the signs are disregarded, 4 can be obtained from 2 and 6 by *subtracting* 2 from 6. But 2 and 6 are precisely the absolute values of 2 and -6. Also, notice that the sign of the number with the larger absolute value is negative and that the sign of the resulting sum is negative.

Adding Numbers with Unlike Signs

> To add two real numbers that have unlike signs, subtract the smaller absolute value from the larger absolute value and associate the sign of the number with the larger absolute value with this difference.

☆ SAMPLE SET B

Find the following sums.

1. $7 + (-2)$

$$\underbrace{|7| = 7}$$
Larger absolute value.
Sign is "+".

$$\underbrace{|-2| = 2}$$
Smaller absolute value.

Subtract absolute values: $7 - 2 = 5.$
Attach the proper sign: "+".

$7 + (-2) = +5$ or $7 + (-2) = 5$

2. $3 + (-11)$

$$\underbrace{|3| = 3}$$
Smaller absolute value.

$$\underbrace{|-11| = 11}$$
Larger absolute value.
Sign is "−".

Subtract absolute values: $11 - 3 = 8.$
Attach the proper sign: "−".

$3 + (-11) = -8$

3. The morning temperature on a winter's day in Lake Tahoe was -12 degrees. The afternoon temperature was 25 degrees warmer. What was the afternoon temperature?

We need to find $-12 + 25$.

$$\underbrace{|-12| = 12}_{\text{Smaller absolute value.}} \qquad \underbrace{|25| = 25}_{\substack{\text{Larger absolute value.} \\ \text{Sign is ``+''.}}}$$

Subtract absolute values: $\quad 25 - 12 = 13$.
Attach the proper sign: $\quad$ "+".

$-12 + 25 = 13$

Thus, the afternoon temperature is 13 degrees.

4. ▦ Add $-147 + 84$.

		Display Reads
Type	147	147
Press	$+/-$	-147
Press	$+$	-147
Type	84	84
Press	$=$	-63

★ **PRACTICE SET B**

Find the sums.

1. $4 + (-3)$ **2.** $-3 + 5$ **3.** $15 + (-18)$ **4.** $0 + (-6)$

5. $-26 + 12$ **6.** $35 + (-78)$ **7.** $15 + (-10)$ **8.** $1.5 + (-2)$

9. $-8 + 0$ **10.** $0 + (0.57)$ **11.** $-879 + 454$ **12.** $-1345.6 + (-6648.1)$

Answers to Practice Sets are on p. 55.

Section 2.3 EXERCISES

Find the sums for problems 1–57.

1. $4 + 12$ **5.** $(-3) + (-12)$

2. $8 + 6$ **6.** $(-6) + (-20)$

3. $6 + 2$ **7.** $(-4) + (-8)$

4. $7 + 9$ **8.** $(-11) + (-8)$

9. $(-16) + (-8)$

10. $(-2) + (-15)$

11. $14 + (-3)$

12. $21 + (-4)$

13. $14 + (-6)$

14. $18 + (-2)$

15. $10 + (-8)$

16. $40 + (-31)$

17. $(-3) + (-12)$

18. $(-6) + (-20)$

19. $10 + (-2)$

20. $8 + (-15)$

21. $-2 + (-6)$

22. $-11 + (-14)$

23. $-9 + (-6)$

24. $-1 + (-1)$

25. $-16 + (-9)$

26. $-22 + (-1)$

27. $0 + (-12)$

28. $0 + (-4)$

29. $0 + (24)$

30. $-6 + 1 + (-7)$

31. $-5 + (-12) + (-4)$

32. $-5 + 5$

33. $-7 + 7$

34. $-14 + 14$

35. $4 + (-4)$

36. $9 + (-9)$

37. $84 + (-61)$

38. $13 + (-56)$

39. $452 + (-124)$

40. $636 + (-989)$

41. $1811 + (-935)$

42. $-373 + (-14)$

43. $-1221 + (-44)$

44. $-47.03 + (-22.71)$

45. $-1.998 + (-4.086)$

46. $[(-3) + (-4)] + [(-6) + (-1)]$

47. $[(-2) + (-8)] + [(-3) + (-7)]$

48. $[(-3) + (-8)] + [(-6) + (-12)]$

49. $[(-8) + (-6)] + [(-2) + (-1)]$

50. $[4 + (-12)] + [12 + (-3)]$

51. $[5 + (-16)] + [4 + (-11)]$

52. $[2 + (-4)] + [17 + (-19)]$

53. $[10 + (-6)] + [12 + (-2)]$

54. $9 + [(-4) + 7]$

55. $14 + [(-3) + 5]$

56. $[2 + (-7)] + (-11)$

57. $[14 + (-8)] + (-2)$

58. In order for a small business to break even on a project, it must have sales of $21,000. If the amount of sales was $15,000, how much money did this company fall short?

59. Suppose a person has $56.00 in his checking account. He deposits $100.00 into his checking account by using the automatic teller machine. He then writes a check for $84.50. If an error causes the deposit not to be listed into this person's account, what is this person's checking balance?

60. A person borrows $7.00 on Monday and then $12.00 on Tuesday. How much has this person borrowed?

61. A person borrows $11.00 on Monday and then pays back $8.00 on Tuesday. How much does this person owe?

EXERCISES FOR REVIEW

(1.4)   **62.** Simplify $\dfrac{4(7^2 - 6 \cdot 2^3)}{2^2}$.

(1.5)   **63.** Simplify $\dfrac{35a^6b^2c^5}{7b^2c^4}$.

(1.6)   **64.** Simplify $\left(\dfrac{12a^8b^5}{4a^5b^2}\right)^3$.

(2.2)   **65.** Determine the value of $|-8|$.

(2.2)   **66.** Determine the value of $(|2| + |4|^2) + |-5|^2$.

★ **Answers to Practice Sets (2.3)**

A.   **1.** 14   **2.** 52   **3.** -12 **4.** -45 **5.** -34 **6.** $-\dfrac{7}{3}$ **7.** -7.4

B.   **1.** 1   **2.** 2   **3.** -3 **4.** -6 **5.** -14 **6.** -43 **7.** 5   **8.** -0.5 **9.** -8 **10.** 0.57
   11. -425 **12.** -7993.7

2.4 Subtraction of Signed Numbers

❑ **DEFINITION OF SUBTRACTION**
❑ **SUBTRACTION OF SIGNED NUMBERS**

❑ DEFINITION OF SUBTRACTION

We know from our experience with arithmetic that the subtraction $5 - 2$ produces 3, that is, $5 - 2 = 3$. Illustrating this process on the number line suggests a rule for subtracting signed numbers.

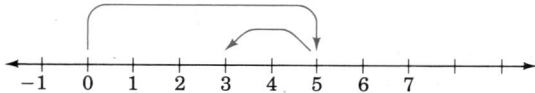

We begin at 0, the origin.
Since 5 is positive, we move 5 units to the right.
Then, we move *2 units to the left* to get to 3. (This reminds us of addition with a negative number.)

This illustration suggests that $5 - 2$ is the same as $5 + (-2)$.
   This leads us directly to the definition of subtraction.

Definition of Subtraction

> If a and b are real numbers, $a - b$ is the same as $a + (-b)$, where $-b$ is the opposite of b.

❑ SUBTRACTION OF SIGNED NUMBERS

The preceding definition suggests the rule for subtracting signed numbers.

Subtraction of Signed Numbers

> To perform the subtraction $a - b$, add the opposite of b to a, that is, change the sign of b and add.

☆ **SAMPLE SET A**

Perform the subtractions.

1. $5 - 3 = 5 + (-3) = 2$

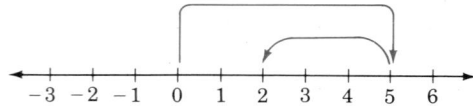

2. $4 - 9 = 4 + (-9) = -5$

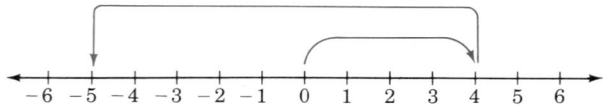

3. $-4 - 6 = -4 + (-6) = -10$

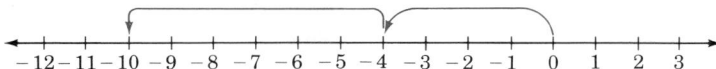

4. $-3 - (-12) = -3 + 12 = 9$

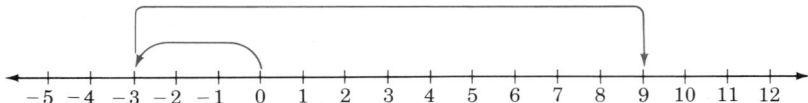

5. $0 - (-15) = 0 + 15 = 15$

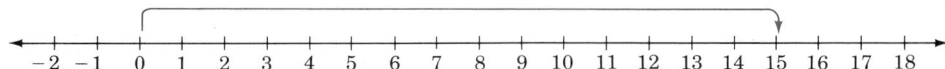

6. The high temperature today in Lake Tahoe was 26°F. The low temperature tonight is expected to be -7°F. How many degrees is the temperature expected to drop?

We need to find the difference between 26 and -7.

$$26 - (-7) = 26 + 7 = 33$$

Thus, the expected temperature drop is 33°F.

7. $-6 - (-5) - 10 = -6 + 5 + (-10)$
$$= (-6 + 5) + (-10)$$
$$= -1 + (-10)$$
$$= -11$$

★ **PRACTICE SET A**

Perform the subtractions.

1. $9 - 6$ **2.** $6 - 9$ **3.** $0 - 7$ **4.** $1 - 14$ **5.** $-8 - 12$

6. $-21 - 6$ **7.** $-6 - (-4)$ **8.** $8 - (-10)$ **9.** $1 - (-12)$

10. $86 - (-32)$ **11.** $0 - 16$ **12.** $0 - (-16)$ **13.** $0 - (8)$

14. $5 - (-5)$ **15.** $24 - (-(-24))$

Answers to the Practice Set are on p. 59.

Section 2.4 EXERCISES

For exercises 1–54, perform the indicated operations.

1. $8 - 3$

2. $12 - 7$

26. $-12 - (-11)$

3. $5 - 6$

4. $14 - 30$

27. $0 - 6$

5. $2 - 15$

6. $5 - 18$

28. $0 - 15$

7. $1 - 7$

8. $4 - 11$

29. $0 - (-7)$

9. $-6 - 5$

10. $-8 - 14$

30. $0 - (-10)$

11. $-1 - 12$

12. $-4 - 4$

31. $67 - 38$

13. $-6 - 8$

14. $-1 - 12$

32. $142 - 85$

15. $-5 - (-3)$

33. $816 - 1140$

16. $-11 - (-8)$

34. $105 - 421$

17. $-7 - (-12)$

35. $-550 - (-121)$

18. $-2 - (-10)$

36. $-15.016 - (4.001)$

19. $-4 - (-15)$

37. $-26 + 7 - 52$

20. $-11 - (-16)$

38. $-15 - 21 - (-2)$

21. $-1 - (-6)$

39. $-104 - (-216) - (-52)$

22. $-8 - (-14)$

40. $-0.012 - (-0.111) - (0.035)$

23. $-15 - (-10)$

41. $[5 + (-6)] - [2 + (-4)]$

24. $-11 - (-4)$

42. $[2 + (-8)] - [5 + (-7)]$

25. $-16 - (-8)$

43. $[4 + (-11)] - [2 + (-10)]$

44. $[9 + (-6)] - [4 + (-12)]$

45. $(11 - 8) - (1 - 6)$

46. $(5 - 12) - (4 - 10)$

47. $(1 - 10) - (2 - 15)$

48. $(0 - 8) - (4 - 12)$

49. $(-4 + 7) - (2 - 5)$

50. $(-6 + 2) - (5 - 11)$

51. $[-8 + (-5 + 3)] - [9 - (-3 - 5)]$

52. $[-4 + (-1 + 6)] - [7 - (-6 - 1)]$

53. $[2 - (-6 + 10)] - [1 - (2 - 11)]$

54. $[5 - (-2 - 5)] - [2 - (-1 - 4)]$

55. When a particular machine is operating properly, its meter will read 34. If a broken bearing in the machine causes the meter reading to drop by 45 units, what is the meter reading?

56. The low temperature today in Denver was $-4°$F and the high was $42°$F. What is the temperature difference?

EXERCISES FOR REVIEW

(1.3) **57.** Use the distributive property to expand $4x(5y + 11)$.

(1.6) **58.** Simplify $\dfrac{2(3x^2y^2)^3(2x^4y^3)^0}{27x^4y^3}$. Assume $x \neq 0$, $y \neq 0$.

(2.2) **59.** Simplify $|-(4^2 + 2^2 - 3^2)|$.

(2.3) **60.** Find the sum. $-8 + (-14)$.

(2.3) **61.** Find the sum. $3 + (-6)$.

★ **Answers to Practice Set (2.4)**

A. **1.** 3   **2.** -3 **3.** -7 **4.** -13 **5.** -20 **6.** -27 **7.** -2 **8.** 18   **9.** 13   **10.** 118   **11.** -16 **12.** 16   **13.** -8 **14.** 10   **15.** 0

2.5 Multiplication and Division of Signed Numbers

Section Overview

- ☐ **MULTIPLICATION OF SIGNED NUMBERS**
- ☐ **DIVISION OF SIGNED NUMBERS**

☐ MULTIPLICATION OF SIGNED NUMBERS

Let us consider first the product of two positive numbers.

Multiply: $3 \cdot 5$.
$3 \cdot 5$ means $5 + 5 + 5 = 15$.

This suggests that

(positive number) $\cdot$ (positive number) = positive number.

$(+)(+) = +$

More briefly, $(+)(+) = +$.

Now consider the product of a positive number and a negative number.

Multiply: $(3)(-5)$.
$(3)(-5)$ means $(-5) + (-5) + (-5) = -15$.

This suggests that

(positive number) $\cdot$ (negative number) = negative number

$(+)(-) = -$

More briefly, $(+)(-) = -$.

By the commutative property of multiplication, we get

(negative number) $\cdot$ (positive number) = negative number

$(-)(+) = -$

More briefly, $(-)(+) = -$.

The sign of the product of two negative numbers can be determined using the following illustration: Multiply -2 by, respectively, 4, 3, 2, 1, 0, $-1, -2, -3, -4$. Notice that when the multiplier decreases by 1, the product increases by 2.

$$\left. \begin{array}{l} 4(-2) = -8 \\ 3(-2) = -6 \\ 2(-2) = -4 \\ 1(-2) = -2 \end{array} \right\} \rightarrow \quad \text{As we know, } (+)(-) = -.$$

$$0(-2) = \;\;\; 0 \quad \rightarrow \quad \text{As we know, } 0 \cdot (\text{any number}) = 0.$$

$(-)(-) = +$

$$\left. \begin{array}{l} -1(-2) = \;\;\; 2 \\ -2(-2) = \;\;\; 4 \\ -3(-2) = \;\;\; 6 \\ -4(-2) = \;\;\; 8 \end{array} \right\} \rightarrow \quad \text{This pattern suggests } (-)(-) = + \quad .$$

We have the following rules for multiplying signed numbers.

Rules for Multiplying Signed Numbers

To multiply two real numbers that have

1. the **same sign,** multiply their absolute values. The product is positive.

$$(+)(+) = +$$
$$(-)(-) = +$$

2. **opposite signs,** multiply their absolute values. The product is negative.

$$(+)(-) = -$$
$$(-)(+) = -$$

☆ SAMPLE SET A

Find the following products.

1. $8 \cdot 6$

$$\left.\begin{array}{l} |8| = 8 \\ |6| = 6 \end{array}\right\}$$ **Multiply these absolute values.**

$8 \cdot 6 = 48$

Since the numbers have the same sign, the product is positive.

$8 \cdot 6 = +48$ or $8 \cdot 6 = 48$

2. $(-8)(-6)$

$$\left.\begin{array}{l} |-8| = 8 \\ |-6| = 6 \end{array}\right\}$$ **Multiply these absolute values.**

$8 \cdot 6 = 48$

Since the numbers have the same sign, the product is positive.

$(-8)(-6) = +48$ or $(-8)(-6) = 48$

3. $(-4)(7)$

$$\left.\begin{array}{l} |-4| = 4 \\ |7| = 7 \end{array}\right\}$$ **Multiply these absolute values.**

$4 \cdot 7 = 28$

Since the numbers have opposite signs, the product is negative.

$(-4)(7) = -28$

4. $6(-3)$

$$\left.\begin{array}{l} |6| = 6 \\ |-3| = 3 \end{array}\right\}$$ **Multiply these absolute values.**

$6 \cdot 3 = 18$

Since the numbers have opposite signs, the product is negative.

$6(-3) = -18$

★ PRACTICE SET A

Find the following products.

1. $3(-8)$ **2.** $4(16)$ **3.** $(-6)(-5)$ **4.** $(-7)(-2)$

5. $(-1)(4)$ **6.** $(-7)7$

❏ DIVISION OF SIGNED NUMBERS

We can determine the sign pattern for division by relating division to multiplication. Division is defined in terms of multiplication in the following way.

> If $b \cdot c = a$, then $\dfrac{a}{b} = c$, $b \neq 0$.

For example, since $3 \cdot 4 = 12$, it follows that $\dfrac{12}{3} = 4$.

Notice the pattern:

Since $\underbrace{3 \cdot 4}_{b \cdot c = a} = 12$, it follows that $\underbrace{\dfrac{12}{3}}_{\frac{a}{b} = c} = 4$

The sign pattern for division follows from the sign pattern for multiplication.

1. Since $\underbrace{(+)(+)}_{b \cdot c \;=\; a} = +$, it follows that $\underbrace{\dfrac{(+)}{(+)}}_{\frac{a}{b} \;=\; c} = +$, that is,

$\dfrac{(+)}{(+)} = +$

> $\dfrac{\text{(positive number)}}{\text{(positive number)}} = \text{positive number}$

2. Since $\underbrace{(-)(-)}_{b \cdot c \;=\; a} = +$, it follows that $\underbrace{\dfrac{(+)}{(-)}}_{\frac{a}{b} \;=\; c} = -$, that is,

$\dfrac{(+)}{(-)} = -$

> $\dfrac{\text{(positive number)}}{\text{(negative number)}} = \text{negative number}$

3. Since $\underbrace{(+)(-)}_{b \cdot c \;=\; a} = -$, it follows that $\underbrace{\dfrac{(-)}{(+)}}_{\frac{a}{b} \;=\; c} = -$, that is,

$\dfrac{(-)}{(+)} = -$

> $\dfrac{\text{(negative number)}}{\text{(positive number)}} = \text{negative number}$

4. Since $\underbrace{(-)(+)}_{b \cdot c \;=\; a} = -$, it follows that $\underbrace{\dfrac{(-)}{(-)}}_{\frac{a}{b} \;=\; c} = +$, that is

$\dfrac{(-)}{(-)} = +$

> $\dfrac{\text{(negative number)}}{\text{(negative number)}} = \text{positive number}$

We have the following rules for dividing signed numbers.

Rules for Dividing Signed
Numbers

To divide two real numbers that have

1. the **same sign,** divide their absolute values. The quotient is positive.

$$\frac{(+)}{(+)} = + \qquad \frac{(-)}{(-)} = +$$

2. **opposite signs,** divide their absolute values. The quotient is negative.

$$\frac{(-)}{(+)} = - \qquad \frac{(+)}{(-)} = -$$

☆ SAMPLE SET B

Find the following quotients.

1. $\dfrac{-10}{2}$

$\left.\begin{array}{r}|-10| = 10 \\ |2| = 2\end{array}\right\}$ Divide these absolute values.

$\dfrac{10}{2} = 5$

$\dfrac{-10}{2} = -5$ Since the numbers have opposite signs, the quotient is negative.

2. $\dfrac{-35}{-7}$

$\left.\begin{array}{r}|-35| = 35 \\ |-7| = 7\end{array}\right\}$ Divide these absolute values.

$\dfrac{35}{7} = 5$

$\dfrac{-35}{-7} = 5$ Since the numbers have the same signs, the quotient is positive.

3. $\dfrac{18}{-9}$

$\left.\begin{array}{r}|18| = 18 \\ |-9| = 9\end{array}\right\}$ Divide these absolute values.

$\dfrac{18}{9} = 2$

$\dfrac{18}{-9} = -2$ Since the numbers have opposite signs, the quotient is negative.

★ PRACTICE SET B

Find the following quotients.

1. $\dfrac{-24}{-6}$ **2.** $\dfrac{30}{-5}$ **3.** $\dfrac{-54}{27}$ **4.** $\dfrac{51}{17}$

☆ SAMPLE SET C

1. Find the value of $\dfrac{-6(4-7)-2(8-9)}{-(4+1)+1}$.

Using the order of operations and what we know about signed numbers, we get

$$\frac{-6(4-7)-2(8-9)}{-(4+1)+1} = \frac{-6(-3)-2(-1)}{-(5)+1}$$

$$= \frac{18+2}{-5+1}$$

$$= \frac{20}{-4}$$

$$= -5$$

2. Find the value of $z = \dfrac{x-u}{s}$ if $x = 57$, $u = 51$, and $s = 2$.

Substituting these values we get

$$z = \frac{57-51}{2} = \frac{6}{2} = 3$$

★ PRACTICE SET C

1. Find the value of $\dfrac{-7(4-8)+2(1-11)}{-5(1-6)-17}$.

2. Find the value of $P = \dfrac{n(n-3)}{2n}$, if $n = 5$.

Answers to Practice Sets are on p. 67.

Section 2.5 EXERCISES

Find the value of each of the following expressions.

1. $(-2)(-8)$ **2.** $(-3)(-9)$ **9.** $(3)(-12)$ **10.** $(4)(-18)$

3. $(-4)(-8)$ **4.** $(-5)(-2)$ **11.** $8(-4)$ **12.** $5(-6)$

5. $(-6)(-9)$ **6.** $(-3)(-11)$ **13.** $9(-2)$ **14.** $7(-8)$

7. $(-8)(-4)$ **8.** $(-1)(-6)$ **15.** $(-6)4$ **16.** $(-7)6$

17. $(-10)9$ **18.** $(-4)12$

19. $(10)(-6)$ **20.** $(-6)(4)$

21. $(-2)(6)$ **22.** $(-8)(7)$

23. $\dfrac{21}{7}$ **24.** $\dfrac{42}{6}$

25. $\dfrac{-39}{3}$ **26.** $\dfrac{-20}{10}$

27. $\dfrac{-45}{-5}$ **28.** $\dfrac{-16}{-8}$

29. $\dfrac{25}{-5}$ **30.** $\dfrac{36}{-4}$

31. $8 - (-3)$ **32.** $14 - (-20)$

33. $20 - (-8)$ **34.** $-4 - (-1)$

35. $0 - 4$ **36.** $0 - (-1)$

37. $-6 + 1 - 7$ **38.** $15 - 12 - 20$

39. $1 - 6 - 7 + 8$ **40.** $2 + 7 - 10 + 2$

41. $3(4 - 6)$ **42.** $8(5 - 12)$

43. $-3(1 - 6)$ **44.** $-8(4 - 12) + 2$

45. $-4(1 - 8) + 3(10 - 3)$

46. $-9(0 - 2) + 4(8 - 9) + 0(-3)$

47. $6(-2 - 9) - 6(2 + 9) + 4(-1 - 1)$

48. $\dfrac{3(4 + 1) - 2(5)}{-2}$

49. $\dfrac{4(8 + 1) - 3(-2)}{-4 - 2}$

50. $\dfrac{-1(3 + 2) + 5}{-1}$

51. $\dfrac{-3(4 - 2) + (-3)(-6)}{-4}$

52. $-1(4 + 2)$

53. $-1(6 - 1)$

54. $-(8 + 21)$

55. $-(8 - 21)$

56. $-(10 - 6)$

57. $-(5 - 2)$

58. $-(7-11)$

59. $-(8-12)$

60. $-3[(-1+6)-(2-7)]$

61. $-2[(4-8)-(5-11)]$

62. $-5[(-1+5)+(6-8)]$

63. $-[(4-9)+(-2-8)]$

64. $-3[-2(1-5)-3(-2+6)]$

65. $-2[-5(-10+11)-2(5-7)]$

66. $P=R-C.$ Find P if $R=2000$ and $C=2500$.

67. $z=\dfrac{x-u}{s}.$ Find z if $x=23$, $u=25$, and $s=1$.

68. $z=\dfrac{x-u}{s}.$ Find z if $x=410$, $u=430$, and $s=2.5$.

69. $m=\dfrac{2s+1}{T}.$ Find m if $s=-8$ and $T=5$.

70. $m=\dfrac{2s+1}{T}.$ Find m if $s=-10$ and $T=-5$.

71. $F=(p_1-p_2)r^4\cdot 9.$ Find F if $p_1=10$, $p_2=8$, $r=3$.

72. $F=(p_1-p_2)r^4\cdot 9.$ Find F if $p_1=12$, $p_2=7$, $r=2$.

73. $P=n(n-1)(n-2).$ Find P if $n=-4$.

74. $P=n(n-1)(n-2)(n-3).$ Find P if $n=-5$.

75. $P=\dfrac{n(n-2)(n-4)}{2n}.$ Find P if $n=-6$.

EXERCISES FOR REVIEW

(1.2) **76.** What natural numbers can replace x so that the statement $-4 < x \le 3$ is true?

(1.5) **77.** Simplify $\dfrac{(x + 2y)^5(3x - 1)^7}{(x + 2y)^3(3x - 1)^6}$.

(1.6) **78.** Simplify $(x^n y^{3t})^5$.

(2.3) **79.** Find the sum. $-6 + (-5)$.

(2.4) **80.** Find the difference. $-2 - (-8)$.

★ Answers to Practice Sets (2.5)

A. **1.** -24 **2.** 64 **3.** 30 **4.** 14 **5.** -4 **6.** -49

B. **1.** 4 **2.** -6 **3.** -2 **4.** 3

C. **1.** 1 **2.** 1

2.6 Negative Exponents

Section Overview

☐ **RECIPROCALS**
☐ **NEGATIVE EXPONENTS**
☐ **WORKING WITH NEGATIVE EXPONENTS**

☐ RECIPROCALS

Reciprocals

Two real numbers are said to be *reciprocals* of each other if their product is 1. Every nonzero real number has exactly one reciprocal, as shown in the examples below. Zero has no reciprocal.

1. $4 \cdot \dfrac{1}{4} = 1$.   This means that 4 and $\dfrac{1}{4}$ are reciprocals.

2. $6 \cdot \dfrac{1}{6} = 1$.   Hence, 6 and $\dfrac{1}{6}$ are reciprocals.

3. $-2 \cdot \dfrac{-1}{2} = 1$.   Hence, -2 and $-\dfrac{1}{2}$ are reciprocals.

4. $a \cdot \dfrac{1}{a} = 1$.   Hence, a and $\dfrac{1}{a}$ are reciprocals if $a \ne 0$.

5. $x \cdot \dfrac{1}{x} = 1$.   Hence, x and $\dfrac{1}{x}$ are reciprocals if $x \ne 0$.

6. $x^3 \cdot \dfrac{1}{x^3} = 1$.   Hence, x^3 and $\dfrac{1}{x^3}$ are reciprocals if $x \ne 0$.

☐ NEGATIVE EXPONENTS

We can use the idea of reciprocals to find a meaning for negative exponents.
    Consider the product of x^3 and x^{-3}. Assume $x \ne 0$.

$$x^3 \cdot x^{-3} = x^{3+(-3)} = x^0 = 1$$

Thus, since the product of x^3 and x^{-3} is 1, x^3 and x^{-3} must be reciprocals.

$x^{-3} = \dfrac{1}{x^3}$

We also know that $x^3 \cdot \dfrac{1}{x^3} = 1$. (See problem 6 above.) Thus, x^3 and $\dfrac{1}{x^3}$ are also reciprocals. Then, since x^{-3} and $\dfrac{1}{x^3}$ are both reciprocals of x^3 and a real number can have only one reciprocal, it must be that $x^{-3} = \dfrac{1}{x^3}$.

We have used -3 as the exponent, but the process works as well for all other negative integers. We make the following definition.

$x^{-n} = \dfrac{1}{x^n}$

> If n is any natural number and x is any nonzero real number, then
>
> $$x^{-n} = \dfrac{1}{x^n}$$

☆ SAMPLE SET A

Write each of the following so that only positive exponents appear.

1. $x^{-6} = \dfrac{1}{x^6}$ **2.** $a^{-1} = \dfrac{1}{a^1} = \dfrac{1}{a}$ **3.** $7^{-2} = \dfrac{1}{7^2} = \dfrac{1}{49}$

4. $(3a)^{-6} = \dfrac{1}{(3a)^6}$ **5.** $(5x - 1)^{-24} = \dfrac{1}{(5x - 1)^{24}}$ **6.** $(k + 2z)^{-(-8)} = (k + 2z)^8$

★ PRACTICE SET A

Write each of the following using only positive exponents.

1. y^{-5} **2.** m^{-2} **3.** 3^{-2} **4.** 5^{-1}

5. 2^{-4} **6.** $(xy)^{-4}$ **7.** $(a + 2b)^{-12}$ **8.** $(m - n)^{-(-4)}$

> **CAUTION**
>
> It is important to note that a^{-n} is not necessarily a negative number. For example,
>
> $3^{-2} = \dfrac{1}{3^2} = \dfrac{1}{9}$ $3^{-2} \neq -9$

☐ WORKING WITH NEGATIVE EXPONENTS

The problems of Sample Set A suggest the following rule for working with exponents:

Moving Factors Up and Down

> In a fraction, a *factor* can be moved from the numerator to the denominator or from the denominator to the numerator by changing the sign of the exponent.

☆ SAMPLE SET B

Write each of the following so that only positive exponents appear.

1. $x^{-2}y^5$.　　　·　　　The *factor* x^{-2} can be moved from the numerator to the denominator by changing the exponent -2 to $+2$.

$$x^{-2}y^5 = \frac{y^5}{x^2}$$

2. a^9b^{-3}.　　　The *factor* b^{-3} can be moved from the numerator to the denominator by changing the exponent -3 to $+3$.

$$a^9b^{-3} = \frac{a^9}{b^3}$$

3. $\dfrac{a^4b^2}{c^{-6}}$.　　　This fraction can be written without any negative exponents by moving the *factor* c^{-6} into the numerator. We must change the -6 to $+6$ to make the move legitimate.

$$\frac{a^4b^2}{c^{-6}} = a^4b^2c^6$$

4. $\dfrac{1}{x^{-3}y^{-2}z^{-1}}$.　　　This fraction can be written without negative exponents by moving all the *factors* from the denominator to the numerator. Change the sign of each exponent: -3 to $+3$, -2 to $+2$, -1 to $+1$.

$$\frac{1}{x^{-3}y^{-2}z^{-1}} = x^3y^2z^1 = x^3y^2z$$

★ PRACTICE SET B

Write each of the following so that only positive exponents appear.

1. $x^{-4}y^7$　　　**2.** $\dfrac{a^2}{b^{-4}}$　　　**3.** $\dfrac{x^3y^4}{z^{-8}}$

4. $\dfrac{6m^{-3}n^{-2}}{7k^{-1}}$　　　**5.** $\dfrac{1}{a^{-2}b^{-6}c^{-8}}$　　　**6.** $\dfrac{3a(a-5b)^{-2}}{5b(a-4b)^5}$

☆ SAMPLE SET C

1. Rewrite $\dfrac{24a^7b^9}{2^3a^4b^{-6}}$ in a simpler form.

   Notice that we are dividing powers with the same base. We'll proceed by using the rules of exponents.

   $$\dfrac{24a^7b^9}{2^3a^4b^{-6}} = \dfrac{24a^7b^9}{8a^4b^{-6}} = 3a^{7-4}b^{9-(-6)}$$

   $$= 3a^3b^{9+6}$$
   $$= 3a^3b^{15}$$

2. Write $\dfrac{9a^5b^3}{5x^3y^2}$ so that no denominator appears.

   We can eliminate the denominator by moving all factors that make up the denominator to the numerator.

   $$9a^5b^35^{-1}x^{-3}y^{-2}$$

3. Find the value of $\dfrac{1}{10^{-2}} + \dfrac{3}{4^{-3}}$.

   We can evaluate this expression by eliminating the negative exponents.

   $$\dfrac{1}{10^{-2}} + \dfrac{3}{4^{-3}} = 1 \cdot 10^2 + 3 \cdot 4^3$$

   $$= 1 \cdot 100 + 3 \cdot 64$$
   $$= 100 + 192$$
   $$= 292$$

★ PRACTICE SET C

1. Rewrite $\dfrac{36x^8b^3}{3^2x^{-2}b^{-5}}$ in a simpler form.

2. Write $\dfrac{2^4m^{-3}n^7}{4^{-1}x^5}$ in a simpler form and one in which no denominator appears.

3. Find the value of $\dfrac{2}{5^{-2}} + 6^{-2} \cdot 2^3 \cdot 3^2$.

Answers to Practice Sets are on p. 74.

Section 2.6 EXERCISES

Write the following expressions using only positive exponents. Assume all variables are nonzero.

1. x^{-2}

2. x^{-4}

3. x^{-7}

4. a^{-8}

5. a^{-10}

6. b^{-12}

7. b^{-14}

8. y^{-1}

9. y^{-5}

10. $(x+1)^{-2}$

11. $(x-5)^{-3}$

12. $(y-4)^{-6}$

13. $(a+9)^{-10}$

14. $(r+3)^{-8}$

15. $(a-1)^{-12}$

16. x^3y^{-2}

17. x^7y^{-5}

18. a^4b^{-1}

19. a^7b^{-8}

20. $a^2b^3c^{-2}$

21. $x^3y^2z^{-6}$

22. $x^3y^{-4}z^2w$

23. $a^7b^{-9}zw^3$

24. $a^3b^{-1}zw^2$

25. $x^5y^{-5}z^{-2}$

26. $x^4y^{-8}z^{-3}w^{-4}$

27. $a^{-4}b^{-6}c^{-1}d^4$

28. $x^9y^{-6}z^{-1}w^{-5}r^{-2}$

29. $4x^{-6}y^2$

30. $5x^2y^2z^{-5}$

31. $7a^{-2}b^2c^2$

32. $4x^3(x+1)^2y^{-4}z^{-1}$

33. $7a^2(a-4)^3b^{-6}c^{-7}$

34. $18b^{-6}(b^2-3)^{-5}c^{-4}d^5e^{-1}$

35. $7(w+2)^{-2}(w+1)^3$

36. $2(a-8)^{-3}(a-2)^5$

37. $(x^2+3)^3(x^2-1)^{-4}$

38. $(x^4+2x-1)^{-6}(x+5)^4$

39. $(3x^2-4x-8)^{-9}(2x+11)^{-3}$

40. $(5y^2+8y-6)^{-2}(6y-1)^{-7}$

41. $7a(a^2-4)^{-2}(b^2-1)^{-2}$

42. $(x-5)^{-4}3b^2c^4(x+6)^8$

43. $(y^3+1)^{-1}5y^3z^{-4}w^{-2}(y^3-1)^{-2}$

44. $5x^3(2x^{-7})$

45. $3y^{-3}(9x)$

46. $6a^{-4}(2a^{-6})$

47. $4a^2b^2a^{-5}b^{-2}$

48. $5^{-1}a^{-2}b^{-6}b^{-11}c^{-3}c^9$

49. $2^3x^2\,2^{-3}x^{-2}$

50. $7a^{-3}b^{-9} \cdot 5a^6bc^{-2}c^4$

51. $(x+5)^2(x+5)^{-6}$

52. $(a-4)^3(a-4)^{-10}$

53. $8(b+2)^{-8}(b+2)^{-4}(b+2)^3$

54. $3a^5b^{-7}(a^2+4)^{-3}\,6a^{-4}b(a^2+4)^{-1}(a^2+4)$

55. $-4a^3b^{-5}(2a^2b^7c^{-2})$

56. $-2x^{-2}y^{-4}z^4(-6x^3y^{-3}z)$

57. $(-5)^2(-5)^{-1}$

58. $(-9)^{-3}(9)^3$

59. $(-1)^{-1}(-1)^{-1}$

60. $(4)^2(2)^{-4}$

61. $\dfrac{1}{a^{-4}}$ **62.** $\dfrac{1}{a^{-1}}$

63. $\dfrac{4}{x^{-6}}$ **64.** $\dfrac{7}{x^{-8}}$

65. $\dfrac{23}{y^{-1}}$ **66.** $\dfrac{6}{a^2b^{-4}}$

67. $\dfrac{3c^5}{a^3b^{-3}}$

68. $\dfrac{16a^{-2}b^{-6}c}{2yz^{-5}w^{-4}}$

69. $\dfrac{24y^2z^{-8}}{6a^2b^{-1}c^{-9}d^3}$

70. $\dfrac{3^{-1}b^5(b+7)^{-4}}{9^{-1}a^{-4}(a+7)^2}$

71. $\dfrac{36a^6b^5c^8}{3^2a^3b^7c^9}$

72. $\dfrac{45a^4b^2c^6}{15a^2b^7c^8}$

73. $\dfrac{3^3x^4y^3z}{3^2xy^5z^5}$

74. $\dfrac{21x^2y^2z^5w^4}{7xyz^{12}w^{14}}$

75. $\dfrac{33a^{-4}b^{-7}}{11a^3b^{-2}}$

84. $\dfrac{3x^2y^{-2}(x-5)}{9^{-1}(x+5)^3}$

76. $\dfrac{51x^{-5}y^{-3}}{3xy}$

85. $\dfrac{14a^2b^2c^{-12}(a^2+21)^{-4}}{4^{-2}a^2b^{-1}(a+6)^3}$

77. $\dfrac{2^6x^{-5}y^{-2}a^{-7}b^5}{2^{-1}x^{-4}y^{-2}b^6}$

For problems 86–103, evaluate each numerical expression.

86. 4^{-1} **87.** 7^{-1}

88. 6^{-2} **89.** 2^{-5}

78. $\dfrac{(x+3)^3(y-6)^4}{(x+3)^5(y-6)^{-8}}$

90. 3^{-4} **91.** $6 \cdot 3^{-3}$

92. $4 \cdot 9^{-2}$ **93.** $28 \cdot 14^{-1}$

79. $\dfrac{4x^3}{y^7}$

94. $2^{-3}(3^{-2})$ **95.** $2^{-1} \cdot 3^{-1} \cdot 4^{-1}$

80. $\dfrac{5x^4y^3}{a^3}$

96. $10^{-2}+3(10^{-2})$ **97.** $(-3)^{-2}$

81. $\dfrac{23a^4b^5c^{-2}}{x^{-6}y^5}$

98. $(-10)^{-1}$ **99.** $\dfrac{3}{2^{-3}}$

82. $\dfrac{2^3b^5c^2d^{-9}}{4b^4cx}$

100. $\dfrac{4^{-1}}{5^{-2}}$ **101.** $\dfrac{2^4-7}{4^{-1}}$

83. $\dfrac{10x^3y^{-7}}{3x^5z^2}$

102. $\dfrac{2^{-1}+4^{-1}}{2^{-2}+4^{-2}}$ **103.** $\dfrac{21^0-2^6}{2\cdot6-13}$

For problems 104–126, write each expression so that only positive exponents appear.

104. $(a^6)^{-2}$ **105.** $(a^5)^{-3}$

106. $(x^7)^{-4}$ **107.** $(x^4)^{-8}$

108. $(b^{-2})^7$ **109.** $(b^{-4})^{-1}$

110. $(y^{-3})^{-4}$

111. $(y^{-9})^{-3}$

112. $(a^{-1})^{-1}$

113. $(b^{-1})^{-1}$

114. $(a^0)^{-1}, \quad a \neq 0$

115. $(m^0)^{-1}, \quad m \neq 0$

116. $(x^{-3}y^7)^{-4}$

117. $(x^6y^6z^{-1})^2$

118. $(a^{-5}b^{-1}c^0)^6$ **119.** $\left(\dfrac{y^3}{x^{-4}}\right)^5$

120. $\left(\dfrac{a^{-8}}{b^{-6}}\right)^3$ **121.** $\left(\dfrac{2a}{b^3}\right)^4$

122. $\left(\dfrac{3b}{a^2}\right)^{-5}$ **123.** $\left(\dfrac{5^{-1}a^3b^{-6}}{x^{-2}y^9}\right)^2$

124. $\left(\dfrac{4m^{-3}n^6}{2m^{-5}n}\right)^3$ **125.** $\left(\dfrac{r^5s^{-4}}{m^{-8}n^7}\right)^{-4}$

126. $\left(\dfrac{h^{-2}j^{-6}}{k^{-4}p}\right)^{-5}$

EXERCISES FOR REVIEW

(1.6)   **127.** Simplify $(4x^5y^3z^0)^3$

(2.3)   **128.** Find the sum. $-15 + 3$.

(2.4)   **129.** Find the difference. $8 - (-12)$.

(2.5)   **130.** Simplify $(-3)(-8) + 4(-5)$.

(2.5)   **131.** Find the value of m if $m = \dfrac{-3k - 5t}{kt + 6}$ when $k = 4$ and $t = -2$.

★ Answers to Practice Sets (2.6)

A. 1. $\dfrac{1}{y^5}$ 2. $\dfrac{1}{m^2}$ 3. $\dfrac{1}{9}$ 4. $\dfrac{1}{5}$ 5. $\dfrac{1}{16}$ 6. $\dfrac{1}{(xy)^4}$ 7. $\dfrac{1}{(a + 2b)^{12}}$ 8. $(m - n)^4$

B. 1. $\dfrac{y^7}{x^4}$ 2. a^2b^4 3. $x^3y^4z^8$ 4. $\dfrac{6k}{7m^3n^2}$ 5. $a^2b^6c^8$ 6. $\dfrac{3a}{5b(a - 5b)^2(a - 4b)^5}$

C. 1. $4x^{10}b^8$ 2. $64m^{-3}n^7x^{-5}$ 3. 52

2.7 Scientific Notation

❏ **STANDARD FORM TO SCIENTIFIC FORM**
❏ **SCIENTIFIC FORM TO STANDARD FORM**
❏ **WORKING WITH NUMBERS IN SCIENTIFIC NOTATION**

❏ STANDARD FORM TO SCIENTIFIC FORM

Very large numbers such as 43,000,000,000,000,000,000 (the number of different possible configurations of Rubik's cube) and very small numbers such as 0.00000000000000000000340 (the mass of the amino acid tryptophan) are extremely inconvenient to write and read. Such numbers can be expressed more conveniently by writing them as part of a power of 10.

To see how this is done, let us start with a somewhat smaller number such as 2480. Notice that

$$\underbrace{2480}_{\text{Standard form}} = 248.0 \times 10^1 = 24.80 \times 10^2 = \underbrace{2.480 \times 10^3}_{\text{Scientific form}}$$

Scientific Form

The last form is called the **scientific form** of the number. There is *one* nonzero digit to the left of the decimal point and the absolute value of the exponent on 10 records the number of places the original decimal point was moved to the *left*.

$$0.00059 = \frac{0.0059}{10} = \frac{0.0059}{10^1} = 0.0059 \times 10^{-1}$$
$$= \frac{0.059}{100} = \frac{0.059}{10^2} = 0.059 \times 10^{-2}$$
$$= \frac{0.59}{1000} = \frac{0.59}{10^3} = 0.59 \times 10^{-3}$$
$$= \frac{5.9}{10,000} = \frac{5.9}{10^4} = 5.9 \times 10^{-4}$$

There is *one* nonzero digit to the left of the decimal point and the absolute value of the exponent of 10 records the number of places the original decimal point was moved to the *right*.

Scientific Notation

Numbers written in scientific form are also said to be written using scientific notation. In **scientific notation,** a number is written as the product of a number between and including 1 and 10 (1 is included, 10 is not) and some power of 10.

Writing a Number in Scientific Notation

To write a number in scientific notation:

1. Move the decimal point so that there is one nonzero digit to its left.
2. Multiply the result by a power of 10 using an exponent whose absolute value is the number of places the decimal point was moved. Make the exponent positive if the decimal point was moved to the left and negative if the decimal point was moved to the right.

☆ SAMPLE SET A

Write the numbers in scientific notation.

1. 981   The number 981 is actually 981., and it is followed by a decimal point. In integers, the decimal point at the end is usually omitted.

$981 = 981. = 9.81 \times 10^2$ The decimal point is now two places to the left of its original position, and the power of 10 is 2.

2. $54.066 = 5.4066 \times 10^1 = 5.4066 \times 10$ The decimal point is one place to the left of its original position, and the power of 10 is 1.

3. $0.000000000004632 = 4.632 \times 10^{-12}$ The decimal point is twelve places to the right of its original position, and the power of 10 is -12.

4. $0.027 = 2.7 \times 10^{-2}$ The decimal point is two places to the right of its original position, and the power of 10 is -2.

★ PRACTICE SET A

Write the following numbers in scientific notation.

1. 346   **2.** 72.33   **3.** 5387.7965   **4.** 87,000,000

5. 179,000,000,000,000,000,000   **6.** 100,000   **7.** 1,000,000   **8.** 0.0086

9. 0.000098001   **10.** 0.000000000000000054   **11.** 0.0000001   **12.** 0.00000001

❏ SCIENTIFIC FORM TO STANDARD FORM

A number written in scientific notation can be converted to standard form by reversing the process shown in Sample Set A.

Converting from Scientific Notation
Positive Exponent
Negative Exponent

> To convert a number written in scientific notation to a number in standard form, move the decimal point the number of places prescribed by the exponent on the 10. Move the decimal point to the right when you have a positive exponent, and move the decimal point to the left when you have a negative exponent.

☆ SAMPLE SET B

1. 4.673×10^4.   The exponent of 10 is 4 so we must move the decimal point to the right 4 places (adding 0's if necessary).

$4.6730 \times 10^4 = 46730$

2. 2.9×10^7.   The exponent of 10 is 7 so we must move the decimal point to the right 7 places (adding 0's if necessary).

$2.9 \times 10^7 = 29000000$

3. 1×10^{27}. The exponent of 10 is 27 so we must move the decimal point to the right 27 places (adding 0's without a doubt).

$1 \times 10^{27} = 1,000,000,000,000,000,000,000,000,000$

4. 4.21×10^{-5}. The exponent of 10 is -5 so we must move the decimal point to the left 5 places (adding 0's if necessary).

$4.21 \times 10^{-5} = 0.0000421$

5. 1.006×10^{-18}. The exponent of 10 is -18 so we must move the decimal point to the left 18 places (adding 0's if necessary).

$1.006 \times 10^{-18} = 0.000000000000000001006$

★ PRACTICE SET B

Convert the following numbers to standard form.

1. 9.25×10^2 **2.** 4.01×10^5 **3.** 1.2×10^{-1} **4.** 8.88×10^{-5}

☐ WORKING WITH NUMBERS IN SCIENTIFIC NOTATION

Multiplying Numbers Using Scientific Notation

There are many occasions (particularly in the sciences) when it is necessary to find the product of two numbers written in scientific notation. This is accomplished by using two of the basic rules of algebra.

Suppose we wish to find $(a \times 10^n)(b \times 10^m)$. Since the only operation is multiplication, we can use the commutative property of multiplication to rearrange the numbers.

$(a \times 10^n)(b \times 10^m) = (a \times b)(10^n \times 10^m)$

Then, by the rules of exponents, $10^n \times 10^m = 10^{n+m}$. Thus,

$(a \times 10^n)(b \times 10^m) = (a \times b) \times 10^{n+m}$

The product of $(a \times b)$ may not be between 1 and 10, so $(a \times b) \times 10^{n+m}$ may not be in scientific form. The decimal point in $(a \times b)$ may have to be moved. An example of this situation is in Sample Set C, problem 2.

☆ SAMPLE SET C

1. $(2 \times 10^3)(4 \times 10^8) = (2 \times 4)(10^3 \times 10^8)$
$= 8 \times 10^{3+8}$
$= 8 \times 10^{11}$

2. $(5 \times 10^{17})(8.1 \times 10^{-22}) = (5 \times 8.1)(10^{17} \times 10^{-22})$
$= 40.5 \times 10^{17-22}$
$= 40.5 \times 10^{-5}$

We need to move the decimal point one place to the *left* to put this number in scientific notation.

Continued

Thus, we must also change the exponent of 10.

40.5×10^{-5}
$4.05 \times 10^1 \times 10^{-5}$
$4.05 \times (10^1 \times 10^{-5})$
$4.05 \times (10^{1-5})$
4.05×10^{-4}

Thus,

$$(5 \times 10^{17})(8.1 \times 10^{-22}) = 4.05 \times 10^{-4}$$

★ PRACTICE SET C

Perform each multiplication.

1. $(3 \times 10^5)(2 \times 10^{12})$ **2.** $(1 \times 10^{-4})(6 \times 10^{24})$ **3.** $(5 \times 10^{18})(3 \times 10^6)$

4. $(2.1 \times 10^{-9})(3 \times 10^{-11})$

Answers to Practice Sets are on p. 81.

Section 2.7 EXERCISES

Convert the numbers used in problems 1–30 to scientific notation.

1. Mount Kilimanjaro is the highest mountain in Africa. It is 5890 meters high.

2. The planet Mars is about 222,900,000,000 meters from the sun.

3. There is an irregularly shaped galaxy, named NGC 4449, that is about 250,000,000,000,000,000,000,000 meters from earth.

4. The farthest object astronomers have been able to see (as of 1981) is a quasar named 3C427. There seems to be a haze beyond this quasar that appears to mark the visual boundary of the universe. Quasar 3C427 is at a distance of 110,000,000,000,000,000,000,000,000 meters from the earth.

5. The smallest known insects are about the size of a typical grain of sand. They are about 0.0002 meters in length (2 ten-thousandths of a meter).

6. Atoms such as hydrogen, carbon, nitrogen, and oxygen are about 0.0000000001 meter across.

7. The island of Manhattan, in New York, is about 57,000 square meters in area.

8. The second largest moon of Saturn is Rhea. Rhea has a surface area of about 735,000 square meters, roughly the same surface area as Australia.

9. A star, named Epsilon Aurigae B, has a diameter (distance across) of 2,800,000,000,000 meters. This diameter produces a surface area of about 24,630,000,000,000,000,000,000,000 square meters. This star is what astronomers call a red giant and it is the largest red giant known. If Epsilon Aurigae were placed at the sun's position, its surface would extend out to the planet Uranus.

10. The volume of the planet Venus is 927,590,000,000,000,000,000 cubic meters.

11. The average mass of a newborn American female is about 3360 grams.

12. The largest brain ever measured was that of a sperm whale. It had a mass of 9200 grams.

13. The mass of the Eiffel tower in Paris, France, is 8,000,000 grams.

14. In 1981, a Japanese company built the largest oil tanker to date. The ship has a mass of about 510,000,000,000 grams. This oil tanker is more than 6 times as massive as the U.S. aircraft carrier, U.S.S. *Nimitz*.

15. In the constellation of Virgo, there is a cluster of about 2500 galaxies. The combined mass of these galaxies is 150,000,000,000,000,000,000,-000,000,000,000,000,000,000,000,000,-000,000,000,000 grams.

16. The mass of an amoeba is about 0.000004 gram.

17. Cells in the human liver have masses of about 0.000000008 gram.

18. The human sperm cell has a mass of about 0.000000000017 gram.

19. The principal protein of muscle is myosin. Myosin has a mass of 0.00000000000000000103 gram.

20. Amino acids are molecules that combine to make up protein molecules. The amino acid tryptophan has a mass of 0.0000000000000000000000340 gram.

21. An atom of the chemical element bromine has 35 electrons. The mass of a bromine atom is 0.00000000000000000000000031 gram.

22. Physicists are performing experiments that they hope will determine the mass of a small particle called a neutrino. It is suspected that neutrinos have masses of about 0.00000000000000000000000000000001 gram.

23. The approximate time it takes for a human being to die of asphyxiation is 316 seconds.

24. On the average, the male housefly lives 1,468,800 seconds (17 days).

25. Aluminum-26 has a half-life of 740,000 years.

26. Manganese-53 has a half-life of 59,918,000,000,000 seconds (1,900,000 years).

27. In its orbit around the sun, the earth moves a distance one and one half feet in about 0.0000316 second.

28. A pi-meson is a subatomic particle that has a half-life of about 0.0000000261 second.

29. A subatomic particle called a neutral pion has a half-life of about 0.0000000000000001 second.

30. Near the surface of the earth, the speed of sound is 1195 feet per second.

For problems 31–41, convert the numbers from scientific notation to standard decimal form.

31. The sun is about 1×10^8 meters from earth.

32. The mass of the earth is about 5.98×10^{27} grams.

33. Light travels about 5.866×10^{12} miles in one year.

34. One year is about 3×10^7 seconds.

35. Rubik's cube has about 4.3×10^{19} different configurations.

36. A photon is a particle of light. A 100-watt light bulb emits 1×10^{20} photons every second.

37. There are about 6×10^7 cells in the retina of the human eye.

38. A car traveling at an average speed will travel a distance about equal to the length of the smallest fingernail in 3.16×10^{-4} seconds.

39. A ribosome of *E. coli* has a mass of about 4.7×10^{-19} grams.

40. A mitochondrion is the energy-producing element of a cell. A mitochondrion is about 1.5×10^{-6} meters in diameter.

41. There is a species of frogs in Cuba that attain a length of at most 1.25×10^{-2} meters.

Perform the following operations.

42. $(2 \times 10^4)(3 \times 10^5)$

43. $(4 \times 10^2)(8 \times 10^6)$

44. $(6 \times 10^{14})(6 \times 10^{-10})$

45. $(3 \times 10^{-5})(8 \times 10^7)$

46. $(2 \times 10^{-1})(3 \times 10^{-5})$

47. $(9 \times 10^{-5})(1 \times 10^{-11})$

48. $(3.1 \times 10^4)(3.1 \times 10^{-6})$

49. $(4.2 \times 10^{-12})(3.6 \times 10^{-20})$

50. $(1.1 \times 10^6)^2$

51. $(5.9 \times 10^{14})^2$

52. $(1.02 \times 10^{-17})^2$

53. $(8.8 \times 10^{-50})^2$

54. If Mount Kilimanjaro was 1,000,000 times as high as it really is, how high would it be? (See problem 1.)

55. If the planet Mars was 300,000 times as far from the sun as it really is, how far from the sun would it be? (See problem 2.)

56. If 800,000,000 of the smallest insects known were lined up head to tail, how far would they stretch? (See problem 5.)

57. If Rhea, the moon of Saturn, had a surface area 0.00000000002 of its real surface area, what would that surface area be? (See problem 8.)

58. If the star Epsilon Aurigae B had a surface area 0.005 of its real surface area, what would that surface area be? (See problem 9.)

59. If the mass of all the galaxies in the constellation Virgo was only 0.0000000000000000000000003 of its real mass, what would that mass be? (See problem 15.)

60. What is the mass of 15,000,000,000,000 bromine atoms? (See problem 21.)

EXERCISES FOR REVIEW

(1.2) **61.** What integers can replace x so that the statement $-6 < x < -2$ is true?

(1.5) **62.** Simplify $(5x^2y^4)(2xy^5)$

(2.2) **63.** Determine the value of $-[-(-|-5|)]$.

(2.6) **64.** Write $\dfrac{x^3y^{-5}}{z^{-4}}$ so that only positive exponents appear.

(2.6) **65.** Write $(2z + 1)^3(2z + 1)^{-5}$ so that only positive exponents appear.

★ **Answers to Practice Sets (2.7)**

A. **1.** 3.46×10^2 **2.** 7.233×10 **3.** 5.3877965×10^3 **4.** 8.7×10^7 **5.** 1.79×10^{20}
6. 1.0×10^5 **7.** 1.0×10^6 **8.** 8.6×10^{-3} **9.** 9.8001×10^{-5} **10.** 5.4×10^{-17}
11. 1.0×10^{-7} **12.** 1.0×10^{-8}

B. **1.** 925 **2.** 401000 **3.** 0.12 **4.** 0.0000888

C. **1.** 6×10^{17} **2.** 6×10^{20} **3.** 1.5×10^{25} **4.** 6.3×10^{-20}

Chapter 2   SUMMARY OF KEY CONCEPTS

Positive and Negative Numbers (2.1)

A number is denoted as *positive* if it is directly preceded by a "+" sign or no sign at all. A number is denoted as negative if it is directly preceded by a "−" sign.

Opposites (2.1)

Opposites are numbers that are the same distance from zero on the number line but have opposite signs.

Double-Negative Property (2.1)

$-(-a) = a$

Absolute Value (Geometric) (2.2)

The absolute value of a number a, denoted $|a|$, is the distance from a to 0 on the number line.

Absolute Value (Algebraic) (2.2)

$$|a| = \begin{cases} a & \text{if } a \geq 0 \\ -a & \text{if } a < 0 \end{cases}$$

Addition of Signed Numbers (2.3)

To add two numbers with

like signs, add the absolute values of the numbers and associate the common sign with the sum.

unlike signs, subtract the smaller absolute value from the larger absolute value and associate the sign of the larger absolute value with the difference.

Addition with 0 (2.3)

$0 +$ any number = that particular number, that is, $0 + a = a$ for any real number a.

Additive Identity (2.3)

Since adding 0 to a real number leaves that number unchanged, 0 is called the additive identity.

Definition of Subtraction (2.4)

$a - b = a + (-b)$

Subtraction of Signed Numbers (2.4)

To perform the subtraction $a - b$, add the opposite of b to a, that is, change the sign of b and add.

Multiplication and Division of Signed Numbers (2.5)

$(+)(+) = +$ $\dfrac{(+)}{(+)} = +$ $\dfrac{(+)}{(-)} = -$

$(-)(-) = +$

$(+)(-) = -$

$(-)(+) = -$ $\dfrac{(-)}{(-)} = +$ $\dfrac{(-)}{(+)} = -$

Reciprocals (2.6)

Two numbers are reciprocals of each other if their product is 1. The numbers 4 and $\dfrac{1}{4}$ are reciprocals since $(4)\left(\dfrac{1}{4}\right) = 1$.

Negative Exponents (2.6)

If n is any natural number and x is any nonzero real number, then $x^{-n} = \dfrac{1}{x^n}$.

Writing a Number in Scientific Notation (2.7)

To write a number in scientific notation:

1. Move the decimal point so that there is one nonzero digit to its left.
2. Multiply the result by a power of 10 using an exponent whose absolute value is the number of places the decimal point was moved. Make the exponent positive if the decimal point was moved to the left and negative if the decimal point was moved to the right.

Converting from Scientific Notation:

positive exponent (2.7)

To convert a number written in scientific notation to a number in standard form when there is a *positive* exponent as the power of 10, move the decimal point to the *right* the number of places prescribed by the exponent on the 10.

negative exponent (2.7)

To convert a number written in scientific notation to a number in standard form when there is a *negative* exponent as the power of 10, move the decimal point to the *left* the number of places prescribed by the exponent on the 10.

EXERCISE SUPPLEMENT

Section 2.1

For problems 1–5, find $-a$ if a is

1. 27   **2.** -15 **3.** $-\dfrac{8}{9}$

4. $-(-3)$ **5.** k

Section 2.2

Simplify problems 6–10.

6. $|8|$ **7.** $|-3|$ **8.** $-|16|$

9. $-(-|12|)$ **10.** $-|0|$

Sections 2.3–2.5

Simplify problems 11–25.

11. $4 + (-6)$ **12.** $-16 + (-8)$

13. $3 - (-14)$ **14.** $(-5)(2)$

15. $(-6)(-3)$ **16.** $(-1)(-4)$

17. $(4)(-3)$ **18.** $\dfrac{-25}{5}$

19. $\dfrac{-100}{-10}$ **20.** $16 - 18 + 5$

21. $\dfrac{(-2)(-4) + 10}{-5}$ **22.** $\dfrac{-3(-8 + 4) - 12}{4(3 + 6) - 2(-8)}$

23. $\dfrac{-1(-3 - 2) - 4(-4)}{-13 + 10}$

24. $-(2 - 10)$ **25.** $0 - 6(-4)(-2)$

Section 2.5

Find the value of each expression for problems 26–28.

26. $P = R - C$. Find P if $R = 3000$ and $C = 3800$.

27. $z = \dfrac{x - u}{s}$. Find z if $x = 22$, $u = 30$, and $s = 8$.

28. $P = n(n - 1)(n - 2)$. Find P if $n = -3$.

Section 2.6

Write the expressions for problems 29–55 using only positive exponents.

29. a^{-1}

30. c^{-6}

31. $a^3 b^{-2} c^{-5}$

32. $(x + 5)^{-2}$

33. $x^3 y^2 (x - 3)^{-7}$

34. $4^{-2} a^{-3} b^{-4} c^5$

35. $2^{-1} x^{-1}$

36. $(2x + 9)^{-3} 7 x^4 y^{-5} z^{-2} (3x - 1)^2 (2x + 5)^{-1}$

37. $(-2)^{-1}$ **38.** $\dfrac{1}{x^{-4}}$

39. $\dfrac{7x}{y^{-3} z^{-2}}$ **40.** $\dfrac{4c^{-2}}{b^{-6}}$

41. $\dfrac{3^{-2} a^{-5} b^{-9} c^2}{x^2 y^{-4} z^{-1}}$ **42.** $\dfrac{(z - 6)^{-2}}{(z + 6)^{-4}}$

43. $\dfrac{16 a^5 b^{-2}}{-2 a^3 b^{-5}}$ **44.** $\dfrac{-44 x^3 y^{-6} z^{-8}}{-11 x^{-2} y^{-7} z^{-8}}$

45. 8^{-2} **46.** 9^{-1}

47. 2^{-5} **48.** $(x^3)^{-2}$

49. $(a^2 b)^{-3}$ **50.** $(x^{-2})^{-4}$

51. $(c^{-1})^{-4}$ **52.** $(y^{-1})^{-1}$

53. $(x^3 y^{-4} z^{-2})^{-6}$ **54.** $\left(\dfrac{x^{-6}}{y^{-2}}\right)^{-5}$

55. $\left(\dfrac{2 b^{-7} c^{-8} d^4}{x^{-2} y^3 z}\right)^{-4}$

Section 2.7

Write problems 56–66 using scientific notation.

56. 8739   **57.** 73567

58. 21,000   **59.** 746,000

60. 8866846   **61.** 0.0387

62. 0.0097   **63.** 0.376

64. 0.0000024

65. 0.000000000000537

66. 46,000,000,000,000,000

Convert problems 67–79 from scientific form to standard form.

67. 3.87×10^5 **68.** 4.145×10^4

69. 6.009×10^7 **70.** 1.80067×10^6

71. 3.88×10^{-5} **72.** 4.116×10^{-2}

73. 8.002×10^{-12} **74.** 7.36490×10^{-14}

75. 2.101×10^{15} **76.** 6.7202×10^{26}

77. 1×10^6 **78.** 1×10^7

79. 1×10^9

Find the products for problems 80–90. Write the result in scientific notation.

80. $(1 \times 10^5)(2 \times 10^3)$

81. $(3 \times 10^6)(7 \times 10^7)$

82. $(2 \times 10^{14})(8 \times 10^{19})$

83. $(9 \times 10^2)(3 \times 10^{75})$

84. $(1 \times 10^4)(1 \times 10^5)$

85. $(8 \times 10^{-3})(3 \times 10^{-6})$

86. $(9 \times 10^{-5})(2 \times 10^{-1})$

87. $(3 \times 10^{-2})(7 \times 10^2)$

88. $(7.3 \times 10^4)(2.1 \times 10^{-8})$

89. $(1.06 \times 10^{-16})(2.815 \times 10^{-12})$

90. $(9.3806 \times 10^{52})(1.009 \times 10^{-31})$

Simplify the expressions for problems 1–9.

1.

2.

3.

4.

5.

6.

7.

8.

9.

10.

11.

12.

13.

14.

1. (2.1) $-\{-[(-6)]\}$

2. (2.2) $-|-15|$

3. (2.4) $-[|-12|-10]^2$

4. (2.4) $-5(-6) + 4(-8) - |-5|$

5. (2.5) $\dfrac{3(-8) - (-2)(-4 - 5)}{(-2)(-3)}$

6. (2.5) $-|7| - (2)^2 + (-2)^2$

7. (2.5) $\dfrac{-6(2)(-2)}{-(-5 - 3)}$

8. (2.5) $\dfrac{-3\{[(-2 - 3)][-2]\}}{-3(4 - 2)}$

9. (2.5) If $z = \dfrac{x - u}{s}$, find z if $x = 14$, $u = 20$, and $s = 2$.

When simplifying the terms for problems 10–20, write each so that only positive exponents appear.

10. (2.6) $\dfrac{1}{-(-5)^{-3}}$

11. (2.6) $\dfrac{5x^3y^{-2}}{z^{-4}}$

12. (2.6) $2^{-2}m^6(n - 4)^{-3}$

13. (2.6) $4a^{-6}(2a^{-5})$

14. (2.6) $\dfrac{6^{-1}x^3y^{-5}x^{-3}}{y^{-5}}$

15. _____

15. (2.6) $\dfrac{(k-6)^2(k-6)^{-4}}{(k-6)^3}$

16. _____

16. (2.6) $\dfrac{(y+1)^3(y-3)^4}{(y+1)^5(y-3)^{-8}}$

17. _____

17. (2.6) $\dfrac{(3^{-6})(3^2)(3^{-10})}{(3^{-5})(3^{-9})}$

18. _____

18. (2.6) $(a^4)^{-3}$

19. _____

19. (2.6) $\left[\dfrac{r^6 s^{-2}}{m^{-5}n^4}\right]^{-4}$

20. _____

20. (2.6) $(c^0)^{-2}, \qquad c \neq 0$

21. _____

21. (2.7) Write 0.000271 using scientific notation.

22. (2.7) Write 8.90×10^5 in standard form.

23. _____

23. (2.7) Find the value of $(3 \times 10^5)(2 \times 10^{-2})$.

24. _____

24. (2.7) Find the value of $(4 \times 10^{-16})^2$.

25. _____

25. (2.7) If k is a negative integer, is $-k$ a positive or negative integer?

86

3

Algebraic Expressions and Equations

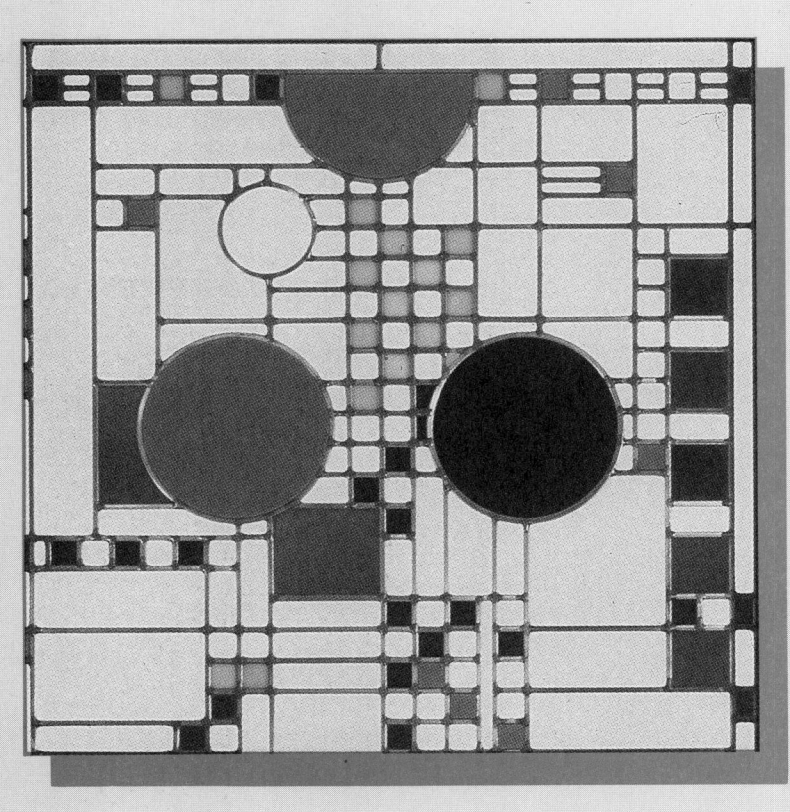

After completing this chapter, you should

Section 3.1 Algebraic Expressions
- be familiar with algebraic expressions
- understand the difference between a term and a factor
- be familiar with the concept of common factors
- know the function of a coefficient

Section 3.2 Equations
- understand the meaning of an equation
- be able to perform numerical evaluations

Section 3.3 Classification of Expressions and Equations
- be familiar with polynomials
- be able to classify polynomials and polynomial equations

Section 3.4 Combining Polynomials Using Addition and Subtraction
- understand the concept of like terms
- be able to combine like terms
- be able to simplify expressions containing parentheses

Section 3.5 Combining Polynomials Using Multiplication
- be able to multiply a polynomial by a monomial
- be able to simplify $+(a + b)$ and $-(a + b)$
- be able to multiply a polynomial by a polynomial

Section 3.6 Special Binomial Products
- be able to expand $(a + b)^2$, $(a - b)^2$, and $(a + b)(a - b)$

Section 3.7 Terminology Associated with Equations
- be able to identify the independent and dependent variables of an equation
- be able to specify the domain of an equation

3.1 Algebraic Expressions

**Section
Overview**

☐ **ALGEBRAIC EXPRESSIONS**
☐ **TERMS AND FACTORS**
☐ **COMMON FACTORS**
☐ **COEFFICIENTS**

☐ ALGEBRAIC EXPRESSIONS

Algebraic Expression

> An **algebraic expression** is a number, a letter, or a collection of numbers and letters along with meaningful signs of operation.

Expressions

Algebraic expressions are often referred to simply as *expressions,* as in the following examples:

1. $x + 4$ is an expression.
2. $7y$ is an expression.
3. $\dfrac{x - 3x^2 y}{7 + 9x}$ is an expression.
4. The number 8 is an expression. 8 can be written with explicit signs of operation by writing it as $8 + 0$ or $8 \cdot 1$.

> $3x^2 + 6 = 4x - 1$ is *not* an expression, it is an *equation.* We will study equations in the next section.

☐ TERMS AND FACTORS

Terms

> In an algebraic expression, the quantities joined by "+" signs are called **terms.**

In some expressions it will appear that terms are joined by "−" signs. We must keep in mind that subtraction is addition of the negative, that is, $a - b = a + (-b)$.

An important concept that all students of algebra must be aware of is the difference between *terms* and *factors.*

Factors

> Any numbers or symbols that are multiplied together are **factors** of their product.

Terms are parts of *sums* and are therefore joined by addition (or subtraction) signs. Factors are parts of *products* and are therefore joined by multiplication signs.

☆ SAMPLE SET A

Identify the terms in the following expressions.

1. $3x^4 + 6x^2 + 5x + 8$.   This expression has four terms: $3x^4$, $6x^2$, $5x$, and 8.

2. $15y^8$.   In this expression there is only one term. The term is $15y^8$.

3. $14x^5 y + (a + 3)^2$.   In this expression there are two terms: the terms are $14x^5 y$ and $(a + 3)^2$. Notice that the term $(a + 3)^2$ is itself composed of two like factors, each of which is composed of the two terms, a and 3.

4. $m^3 - 3$.   Using our definition of subtraction, this expression can be written in the form $m^3 + (-3)$. Now we can see that the terms are m^3 and -3.

Rather than rewriting the expression when a subtraction occurs, we can identify terms more quickly by associating the $+$ or $-$ sign with the individual quantity.

5. $p^4 - 7p^3 - 2p - 11$.   Associating the sign with the individual quantities we see that the terms of this expression are p^4, $-7p^3$, $-2p$, and -11.

★ PRACTICE SET A

1. Let's say it again. The difference between terms and factors is that terms are joined by _____ signs and factors are joined by _____ signs.

List the terms in the following expressions.

2. $4x^2 - 8x + 7$ 3. $2xy + 6x^2 + (x - y)^4$ 4. $5x^2 + 3x - 3xy^7 + (x - y)(x^3 - 6)$

☆ SAMPLE SET B

Identify the factors in each term.

1. $9a^2 - 6a - 12$ contains three terms. Some of the factors in each term are

first term: 9 and a^2, or, 9 and a and a
second term: -6 and a
third term: -12 and 1, or, 12 and -1

2. $14x^5y + (a + 3)^2$ contains two terms. Some of the factors of these terms are

first term: 14, x^5, y
second term: $(a + 3)$ and $(a + 3)$

★ PRACTICE SET B

1. In the expression $8x^2 - 5x + 6$, list the factors of the

first term:
second term:
third term:

2. In the expression $10 + 2(b + 6)(b - 18)^2$, list the factors of the

first term:
second term:

❑ COMMON FACTORS

Common Factors

Sometimes, when we observe an expression carefully, we will notice that some particular factor appears in every term. When we observe this, we say we are observing *common factors*. We use the phrase *common factors* since the particular factor we observe is common to all the terms in the expression. The factor appears in each and every term in the expression.

☆ SAMPLE SET C

Name the common factors in each expression.

1. $5x^3 - 7x^3 + 14x^3$.   The factor x^3 appears in each and every term. The expression x^3 is a common factor.

2. $4x^2 + 7x$.   The factor x appears in each term. The term $4x^2$ is actually $4xx$. Thus, x is a common factor.

3. $12xy^2 - 9xy + 15$.   The only factor common to all three terms is the number 3. (Notice that $12 = 3 \cdot 4, 9 = 3 \cdot 3, 15 = 3 \cdot 5$.)

4. $3(x + 5) - 8(x + 5)$.   The factor $(x + 5)$ appears in each term. So, $(x + 5)$ is a common factor.

5. $45x^3(x - 7)^2 + 15x^2(x - 7) - 20x^2(x - 7)^5$.   The number 5, the x^2, and the $(x - 7)$ appear in each term. Also, $5x^2(x - 7)$ is a factor (since each of the individual quantities is joined by a multiplication sign). Thus, $5x^2(x - 7)$ is a common factor.

6. $10x^2 + 9x - 4$.   There is no factor that appears in each and every term. Hence, there are no common factors in this expression.

★ PRACTICE SET C

List, if any appear, the common factors in the following expressions.

1. $x^2 + 5x^2 - 9x^2$ 2. $4x^2 - 8x^3 + 16x^4 - 24x^5$ 3. $4(a + 1)^3 + 10(a + 1)$

4. $9ab(a - 8) - 15a(a - 8)^2$ 5. $14a^2b^2c(c - 7)(2c + 5) + 28c(2c + 5)$ 6. $6(x^2 - y^2) + 19x(x^2 + y^2)$

☐ COEFFICIENTS

Coefficient

In algebra, as we now know, a letter is often used to represent some quantity. Suppose we represent some quantity by the letter x. The notation $5x$ means $x + x + x + x + x$. We can now see that we have five of these quantities. In the expression $5x$, the number 5 is called the *numerical coefficient* of the quantity x. Often, the numerical coefficient is just called the coefficient. The **coefficient** of a quantity records how many of that quantity there are.

☆ SAMPLE SET D

1. $12x$ means there are 12 x's.

2. $4ab$ means there are four ab's.

3. $10(x - 3)$ means there are ten $(x - 3)$'s.

4. $1y$ means there is one y. We usually write just y rather than $1y$ since it is clear just by looking that there is only one y.

5. $7a^3$ means there are seven a^3's.

6. $5ax$ means there are five ax's. It could also mean there are $5a$ x's. This example shows us that it is important for us to be very clear as to which quantity we are working with. When we see the expression $5ax$ we must ask ourselves "Are we working with the quantity ax or the quantity x?".

7. $6x^2y^9$ means there are six x^2y^9's. It could also mean there are $6x^2$ y^9's. It could even mean there are $6y^9$ x^2's.

8. $5x^3(y - 7)$ means there are five $x^3(y - 7)$'s. It could also mean there are $5x^3$ $(x - 7)$'s. It could also mean there are $5(x - 7)$ x^3's.

★ PRACTICE SET D

What does the *coefficient* of a quantity tell us?

It is important to keep in mind the difference between *coefficients* and *exponents*.

The Difference Between
Coefficients and Exponents

Coefficients record the number of like *terms* in an algebraic expression.

$$\underbrace{x + x + x + x}_{\text{4 terms}} = \underset{\text{coefficient is 4}}{4x}$$

Exponents record the number of like *factors* in a term.

$$\underbrace{x \cdot x \cdot x \cdot x}_{\text{4 factors}} = \underset{\text{exponent is 4}}{x^4}$$

In a term, the *coefficient* of a particular group of factors is the remaining group of factors.

☆ SAMPLE SET E

1. $3x$.   The coefficient of x is 3.

2. $6a^3$.   The coefficient of a^3 is 6.

3. $9(4 - a)$.   The coefficient of $(4 - a)$ is 9.

4. $\dfrac{3}{8} xy^4$.   The coefficient of xy^4 is $\dfrac{3}{8}$.

5. $3x^2y$.   The coefficient of x^2y is 3; the coefficient of y is $3x^2$; and the coefficient of 3 is x^2y.

6. $4(x + y)^2$.   The coefficient of $(x + y)^2$ is 4; the coefficient of 4 is $(x + y)^2$; and the coefficient of $(x + y)$ is $4(x + y)$ since $4(x + y)^2$ can be written as $4(x + y)(x + y)$.

★ PRACTICE SET E

Determine the coefficients.

1. In the term $6x^3$, the coefficient of
  (a) x^3 is _____.
  (b) 6 is _____.
2. In the term $3x(y - 1)$, the coefficient of
  (a) $x(y - 1)$ is _____.
  (b) $(y - 1)$ is _____.
  (c) $3(y - 1)$ is _____.
  (d) x is _____.

  (e) 3 is _____.
  (f) The numerical coefficient is _____.
3. In the term $10ab^4$, the coefficient of
  (a) ab^4 is _____.
  (b) b^4 is _____.
  (c) a is _____.
  (d) 10 is _____.
  (e) $10ab^3$ is _____.

Answers to Practice Sets are on p. 93.

Section 3.1 EXERCISES

1. What is an algebraic expression?

2. Why is the number 14 considered to be an expression?

3. Why is the number x considered to be an expression?

For the expressions in problems 4–20, write the number of terms that appear and then list the terms.

4. $2x + 1$

5. $6x - 10$

6. $2x^3 + x - 15$

7. $5x^2 + 6x - 2$

8. $3x$

9. $5cz$

10. 2

11. 61

12. x

13. $4y^3$

14. $17ab^2$

15. $a + 1$

16. $(a + 1)$

17. $2x + x + 7$

18. $2x + (x + 7)$

19. $(a + 1) + (a - 1)$

20. $a + 1 + (a - 1)$

For problems 21–35, list, if any should appear, the common factors in the expressions.

21. $x^2 + 5x^2 - 2x^2$

22. $11y^3 - 33y^3$

23. $45ab^2 + 9b^2$

24. $6x^2y^3 + 18x^2$

25. $2(a + b) - 3(a + b)$

26. $8a^2(b + 1) - 10a^2(b + 1)$

27. $14ab^2c^2(c + 8) + 12ab^2c^2$

28. $4x^2y + 5a^2b$

29. $9a(a - 3)^2 + 10b(a - 3)$

30. $15x^2 - 30xy^2$

31. $12a^3b^2c - 7(b + 1)(c - a)$

32. $0.06ab^2 + 0.03a$

33. $5.2(a + 7)^2 + 17.1(a + 7)$

34. $\dfrac{3}{4} x^2y^2z^2 + \dfrac{3}{8} x^2z^2$

35. $\dfrac{9}{16} (a^2 - b^2) + \dfrac{9}{32} (b^2 - a^2)$

For problems 36–50, note how many:

36. a's in $4a$?

37. z's in $12z$?

38. x^2's in $5x^2$? **39.** y^3's in $6y^3$?

40. xy's in $9xy$? **41.** a^2b's in $10a^2b$?

42. $(a + 1)$'s in $4(a + 1)$?

43. $(9 + y)$'s in $8(9 + y)$?

44. y^2's in $3x^3y^2$?

45. $12x$'s in $12x^2y^5$?

46. $(a + 5)$'s in $2(a + 5)$?

47. $(x - y)$'s in $5x(x - y)$?

48. $(x + 1)$'s in $8(x + 1)$?

49. 2's in $2x^2(x - 7)$?

50. $3(a + 8)$'s in $6x^2(a + 8)^3(a - 8)$?

For problems 51–60, a term will be given followed by a group of its factors. List the coefficient of the given group of factors.

51. $7y$; y

52. $10x$; x

53. $5a$; 5

54. $12a^2b^3c^2r^7$; $a^2c^2r^7$

55. $6x^2b^2(c - 1)$; $c - 1$

56. $10x(x + 7)^2$; $10(x + 7)$

57. $9a^2b^5$; $3ab^3$

58. $15x^4y^4(z + 9a)^3$; $(z + 9a)$

59. $(-4)a^6b^2$; ab

60. $(-11a)(a + 8)^3(a - 1)$; $(a + 8)^2$

EXERCISES FOR REVIEW

(1.6) **61.** Simplify $\left[\dfrac{2x^8(x - 1)^5}{x^4(x - 1)^2} \right]^4$.

(2.2) **62.** Supply the missing phrase. Absolute value speaks to the question of _____ and not "which way."

(2.5) **63.** Find the value of $-[-6(-4 - 2) + 7(-3 + 5)]$.

(2.6) **64.** Find the value of $\dfrac{2^5 - 4^2}{3^{-2}}$.

(2.7) **65.** Express 0.0000152 using scientific notation.

★ **Answers to Practice Sets (3.1)**

A. **1.** addition, multiplication **2.** $4x^2$, $-8x$, 7 **3.** $2xy$, $6x^2$, $(x - y)^4$ **4.** $5x^2$, $3x$, $-3xy^7$, $(x - y)(x^3 - 6)$

B. **1.** 8, x, x; -5, x; 6 and 1 or 3 and 2 **2.** 10 and 1 or 5 and 2; 2, $b + 6$, $b - 18$, $b - 18$

C. **1.** x^2 **2.** $4x^2$ **3.** $2(a + 1)$ **4.** $3a(a - 8)$ **5.** $14c(2c + 5)$ **6.** no common factor

D. how many of that quantity there are

E. **1.** (a) 6 (b) x^3 **2.** (a) 3 (b) $3x$ (c) x (d) $3(y - 1)$ (e) $x(y - 1)$ (f) 3
  3. (a) 10 (b) $10a$ (c) $10b^4$ (d) ab^4 (e) b

3.2 Equations

☐ EQUATIONS

Equation

> An **equation** is a statement that two algebraic expressions are equal.

An equation is composed of three parts.

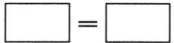

Each of the boxes represents an algebraic expression. An equation consists of two expressions separated by an equal sign. The equal sign makes the statement that the two expressions are equivalent, that is, they represent the same value. For example:

1. $f = 32a$.   The equation expresses the relationship between the variables f and a. It states that the value of f is always 32 times that of a.
2. $y = 6x + 8$.   The equation expresses the relationship between the variables x and y. It states that the value of y is always 8 more than 6 times the value of x.

☐ NUMERICAL EVALUATION

Numerical Evaluation

> **Numerical evaluation** is the process of determining a value by substituting numbers for letters.

In various areas (business, statistics, physics, chemistry, astronomy, sociology, psychology, etc.), particular equations occur quite frequently. Such equations are called **formulas.** Numerical evaluation is used frequently with formulas.

Formulas

☆ **SAMPLE SET A**

---

1. $f = 32a$.   Determine the value of f if $a = 2$.

$f = 32(2)$ **Replace a by 2.**
$\quad = 64$

2. $p = \dfrac{10,000}{v}$.   This chemistry equation expresses the relationship between the pressure p of a gas and the volume v of the gas. Determine the value of p if $v = 500$.

$p = \dfrac{10,000}{500}$ **Replace v by 500.**

$\quad = 20$

On the Calculator

Type	10000
Press	$\boxed{\div}$
Type	500
Press	$\boxed{=}$
Display reads:	20

3. $z = \dfrac{x - u}{s}$.   This statistics equation expresses the relationship between the variables z, x, u, and s. Determine the value of z if $x = 41$, $u = 45$, and $s = 1.3$. Round to two decimal places.

$z = \dfrac{41 - 45}{1.3}$

$\quad = \dfrac{-4}{1.3}$

$\quad = -3.08$

On the Calculator

Type	41
Press	$-$
Type	45
Press	$=$
Press	$\div$
Type	1.3
Press	$=$
Display reads:	-3.076923

We'll round to -3.08

4. $p = 5w^3 + w^2 - w - 1$.   This equation expresses the relationship between p and w. Determine the value of p if $w = 5$.

$p = 5(5)^3 + (5)^2 - (5) - 1$
$\quad = 5(125) + 25 - (5) - 1$
$\quad = 625 + 25 - 5 - 1$
$\quad = 644$

On the Calculator

Type	5
Press	y^x
Type	3
Press	$=$
Press	$\times$
Type	5
Press	$=$
Press	$+$
Type	5
Press	x^2
Press	$-$
Type	5
Press	$-$
Type	1
Press	$=$
Display reads:	644

★ PRACTICE SET A

1. $f = 32a.$ Determine the value of f if $a = 6$.

2. $p = \dfrac{10{,}000}{v}.$ Determine the value of p if $v = 250$.

3. $F = \dfrac{9}{5}C + 32.$ Determine the value of F if $C = 10$.

4. $y = -9x - 14.$ Determine the value of y if $x = -3$.

5. $m = 5p^3 - 2p + 7.$ Determine the value of m if $p = -2$.

Answers to the Practice Set are on p. 100.

Section 3.2 EXERCISES

For problems 1–12, observe the equations and state the relationship being expressed.

1. $x = 6y$

2. $y = x + 4$

3. $e = g - 9$

4. $y = x - 7$

5. $3t = 6s$

6. $u = v^5$

7. $r = \dfrac{2}{9}s$

8. $b = \dfrac{3}{4}a$

9. $f = 0.97k + 55$

10. $w = 4z^3 - 21$

11. $q^2 = 9x^8 + 2y$

12. $I = m^2qb^5 + 3.115p$

Use numerical evaluation on the equations for problems 13–45.

13. Geometry (circumference of a circle)
$C = 2\pi r.$ Find C if π is approximated by 3.14 and $r = 5$.

14. Geometry (area of a rectangle)
$A = lw.$ Find A if $l = 15$ and $w = 9$.

15. Electricity (current in a circuit)
$I = \dfrac{E}{R}.$ Find I if $E = 21$ and $R = 7$.

16. Electricity (current in a circuit)
$I = \dfrac{E}{R}.$ Find I if $E = 106$ and $R = 8$.

17. Business (simple interest)
$I = prt.$ Find I if $p = 3000, r = .12,$ and $t = 1$.

18. Business (simple interest)
$I = prt.$ Find I if $p = 250, r = 0.07,$ and $t = 6$.

19. Geometry (area of a parallelogram)
$A = \dfrac{1}{2}bh.$ Find A if $b = 16$ and $h = 6$.

20. Geometry (area of a triangle)
$A = \dfrac{1}{2}bh.$ Find A if $b = 25$ and $h = 10$.

21. Geometry (perimeter of a rectangle)
$P = 2l + 2w.$ Find P if $l = 3$ and $w = 1$.

22. Geometry (perimeter of a rectangle)
$P = 2l + 2w.$ Find P if $l = 74$ and $w = 16$.

23. Geometry (perimeter of a rectangle)
$P = 2l + 2w.$ Find P if $l = 8\dfrac{1}{4}$ and $w = 12\dfrac{8}{9}$.

24. Physics (force)
$F = 32m.$ Find F if $m = 6$.

25. Physics (force)
$F = 32m.$ Find F if $m = 14$.

26. Physics (force)
$F = 32m.$ Find F if $m = \dfrac{1}{16}$.

27. Physics (force)
$F = 32m.$ Find F if $m = 6.42$.

28. Physics (momentum)
$p = mv.$ Find p if $m = 18$ and $v = 5$.

29. Physics (momentum)
$p = mv.$ Find p if $m = 44$ and $v = 9$.

30. Physics (momentum)
$p = mv.$ Find p if $m = 9.18$ and $v = 16.5$.

31. Physics (energy)

$E = \dfrac{1}{2} mv^2$.   Find E if $m = 12$ and $v = 5$.

32. Physics (energy)

$E = \dfrac{1}{2} mv^2$.   Find E if $m = 8$ and $v = 15$.

33. Physics (energy)

$E = \dfrac{1}{2} mv^2$.   Find E if $m = 24.02$ and $v = 7$.

34. Astronomy (Kepler's law of planetary motion)

$P^2 = ka^3$.   Find P^2 if $k = 1$ and $a = 4$.

35. ⊞ Astronomy (Kepler's law of planetary motion) $P^2 = ka^3$.   Find P^2 if $k = 8$ and $a = 31$.

36. ⊞ Astronomy (Kepler's law of planetary motion) $P^2 = ka^3$.   Find P^2 if $k = 4$ and $a = 5.1$.

(*Hint:* On the calculator, Type 5.1, Press $\boxed{y^x}$, Type 3, Press $\boxed{=}$, Press $\boxed{\times}$, Type 4, Press $\boxed{=}$.)

37. ⊞ Astronomy (Kepler's law of planetary motion) $P^2 = ka^3$.   Find P^2 if $k = 53.7$ and $a = 0.7$.

38. Business (profit, revenue, and cost)

$P = R - C$.   Find P if $R = 3100$ and $C = 2500$.

39. Business (profit, revenue, and cost)

$P = R - C$.   Find P if $R = 4240$ and $C = 3590$.

40. Geometry (area of a circle)

$A = \pi r^2$.   Find A if π is approximately 3.14 and $r = 3$.

41. Geometry (area of a circle)

$A = \pi r^2$.   Find A if π is approximately 3.14 and $r = 11$.

42. $t = 21x + 6$.   Find t if $x = 3$.

43. $t = 21x + 6$.   Find t if $x = 97$.

44. ⊞ $E = mc^2$.   Find E if

$m = 2$ and $c = 186,000$.

(*Hint:* The number 10 that occurs on the display a few spaces away from the other number on the display is the exponent of 10 in the scientific notation form of the number.)

45. ⊞ $E = mc^2$.   Find E if

$m = 5$ and $c = 186,000$.

46. An object travels on a horizontal line. The distance it travels is represented by d and is measured in meters. The equation relating time of travel, t, and distance of travel, d, is

$$d = t^2 - 4t + 20$$

Determine the distance traveled by the object if it has been in motion for 6 seconds.

47. In medicine, there are several rules of thumb used by physicians to determine a child's dose, D_c, of a particular drug. One such rule, Young's Rule, relates a child's dose of a drug to an adult's dose of that drug, D_a. Young's Rule is

$$D_c = \frac{t}{t + 12} \cdot D_a$$

where t is the child's age in years. What dose should be given to a child 8 years old if the corresponding adult dosage is 15 units?

48. A hemispherical water tank of radius 6 feet has water dripping into it. The equation relating the volume, V, of water in the tank at any time is $V = 6\pi h^2 - \frac{\pi}{3} h^3$, where h represents the depth of the water. Using 3.14 to approximate the irrational number π, determine the volume of water in the tank when the depth of the water is 3 feet.

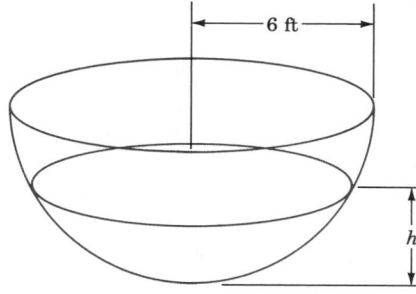

49. The equation $W = 3.51L - 192$ has been established by the International Whaling Commission to relate the weight, W (in long tons), of a mature blue whale to its length, L (in feet). The equation is only used when $L \geq 70$. When

$$0 < L < 70$$

blue whales are considered immature. At birth, a blue whale is approximately 24 feet long. Determine the weight of a blue whale that measures 83 feet in length.

50. A relationship exists between the length of a cantilever beam and the amount it is deflected when a weight is attached to its end. If a cantilever beam 20 feet long has a 600 pound weight attached to its end, the equation relating beam length and amount of deflection is

$$d = \frac{60x^2 - x^3}{16,000}$$

where d is the amount of deflection measured in inches and x is the length from the supported part of the beam to some point on the beam at which the amount of deflection is measured. Find the amount of deflection of the beam 17 feet from the supported end.

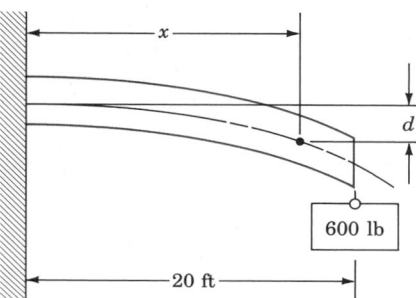

51. There is a relationship between the length of a suspension bridge cable that is secured between two vertical supports and the amount of sag of the cable. If we represent the length of the cable by c, the horizontal distance between the vertical supports by d, and the amount of sag by s, the equation is $c = d + \frac{8s^2}{3d} - \frac{32s^4}{5d^3}$. If the horizontal distance between the two vertical supports is 190 feet and the amount of sag in a cable that is suspended between the two supports is 20 feet, what is the length of the cable?

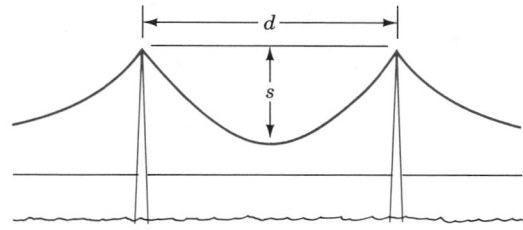

EXERCISES FOR REVIEW

(1.5)   **52.** Simplify $(4x^3y^8)(3x^2y)$.

(2.2)   **53.** Simplify $-|-8|$.

(2.6)   **54.** Find the value of $4^{-2} \cdot 8^2 - 3^2$.

(3.1)   **55.** For the expression $5(a + b) + 2x^2$, write the number of terms that appear and then write the terms themselves.

(3.1)   **56.** How many xy^3's are there in $5x^2y^5$?

★ **Answers to Practice Set (3.2)**

A.   **1.** 192   **2.** 40   **3.** 50   **4.** 13   **5.** -29

3.3 Classification of Expressions and Equations

Section Overview

- ❏ **POLYNOMIALS**
- ❏ **CLASSIFICATION OF POLYNOMIALS**
- ❏ **CLASSIFICATION OF POLYNOMIAL EQUATIONS**

❏ POLYNOMIALS

Polynomials

Let us consider the collection of all algebraic expressions that do not contain variables in the denominators of fractions and where all exponents on the variable quantities are whole numbers. Expressions in this collection are called **polynomials.**

Some expressions that *are* polynomials are

1. $3x^4$

2. $\dfrac{2}{5}\,x^2y^6$.   A fraction occurs, but no variable appears in the denominator.

3. $5x^3 + 3x^2 - 2x + 1$

Some expressions that *are not* polynomials are

1. $\dfrac{3}{x} - 16$.   A variable appears in the denominator.

2. $4x^2 - 5x + x^{-3}$.   A negative exponent appears on a variable.

❏ CLASSIFICATION OF POLYNOMIALS

Polynomials can be classified using two criteria: the number of terms and degree of the polynomial.

	Number of Terms	Name	Example	Comment
Monomial	One	Monomial	$4x^2$	*mono* means "one" in Greek.
Binomial	Two	Binomial	$4x^2 - 7x$	*bi* means "two" in Latin.
Trinomial	Three	Trinomial	$4x^2 - 7x + 3$	*tri* means "three" in Greek.
Polynomial	Four or more	Polynomial	$4x^3 - 7x^2 + 3x - 1$	*poly* means "many" in Greek.

Degree of a Term Containing One Variable

The **degree of a term** containing only *one* variable is the value of the exponent of the variable. Exponents appearing on numbers do not affect the degree of the term. We consider only the exponent of the variable. For example:

1. $5x^3$ is a monomial of degree 3.
2. $60a^5$ is a monomial of degree 5.
3. $21b^2$ is a monomial of degree 2.
4. 8 is a monomial of degree 0. We say that a nonzero number is a term of 0 degree since it could be written as $8x^0$. Since $x^0 = 1$ $(x \neq 0)$, $8x^0 = 8$. The exponent on the variable is 0 so it must be of degree 0. (By convention, the number 0 has no degree.)
5. $4x$ is a monomial of the first degree. $4x$ could be written as $4x^1$. The exponent on the variable is 1 so it must be of the first degree.

Degree of a Term Containing Several Variables

The degree of a term containing *more* than one variable is the *sum* of the exponents of the variables, as shown below.

1. $4x^2y^5$ is a monomial of degree $2 + 5 = 7$. This is a 7th degree monomial.
2. $37ab^2c^6d^3$ is a monomial of degree $1 + 2 + 6 + 3 = 12$. This is a 12th degree monomial.
3. $5xy$ is a monomial of degree $1 + 1 = 2$. This is a 2nd degree monomial.

Degree of a Polynomial

The **degree of a polynomial** is the degree of the *term* of highest degree; for example:

1. $2x^3 + 6x - 1$ is a trinomial of degree 3. The first term, $2x^3$, is the term of the highest degree. Therefore, its degree is the degree of the polynomial.
2. $7y - 10y^4$ is a binomial of degree 4.
3. $a - 4 + 5a^2$ is a trinomial of degree 2.
4. $2x^6 + 9x^4 - x^7 - 8x^3 + x - 9$ is a polynomial of degree 7.
5. $4x^3y^5 - 2xy^3$ is a binomial of degree 8. The degree of the first term is 8.
6. $3x + 10$ is a binomial of degree 1.

Linear
Quadratic
Cubic

Polynomials of the first degree are called **linear** polynomials.
Polynomials of the second degree are called **quadratic** polynomials.
Polynomials of the third degree are called **cubic** polynomials.
Polynomials of the fourth degree are called **fourth degree** polynomials.
Polynomials of the nth degree are called **nth degree** polynomials.
Nonzero constants are polynomials of the **0th** degree.

Some examples of these polynomials follow:

1. $4x - 9$ is a linear polynomial.
2. $3x^2 + 5x - 7$ is a quadratic polynomial.
3. $8y - 2x^3$ is a cubic polynomial.
4. $16a^2 - 32a^5 - 64$ is a 5th degree polynomial.
5. $x^{12} - y^{12}$ is a 12th degree polynomial.
6. $7x^5y^7z^3 - 2x^4y^7z + x^3y^7$ is a 15th degree polynomial. The first term is of degree $5 + 7 + 3 = 15$.
7. 43 is a 0th degree polynomial.

❏ CLASSIFICATION OF POLYNOMIAL EQUATIONS

As we know, an equation is composed of two algebraic expressions separated by an equal sign. If the two expressions happen to be polynomial expressions, then we can classify the equation according to its degree. Classification of equations by degree is useful since equations of the same degree have the same type of graph. (We will study graphs of equations in Chapter 6.)

The degree of an equation is the degree of the highest degree expression.

☆ SAMPLE SET A

1. $x + 7 = 15$.   This is a linear equation since it is of degree 1, the degree of the expression on the left of the "=" sign.

2. $5x^2 + 2x - 7 = 4$ is a quadratic equation since it is of degree 2.

3. $9x^3 - 8 = 5x^2 + 1$ is a cubic equation since it is of degree 3. The expression on the left of the "=" sign is of degree 3.

4. $y^4 - x^4 = 0$ is a 4th degree equation.

5. $a^5 - 3a^4 = -a^3 + 6a^4 - 7$ is a 5th degree equation.

6. $y = \dfrac{2}{3} x + 3$ is a linear equation.

7. $y = 3x^2 - 1$ is a quadratic equation.

8. $x^2y^2 - 4 = 0$ is a 4th degree equation. The degree of $x^2y^2 - 4$ is $2 + 2 = 4$.

★ PRACTICE SET A

Classify the following equations in terms of their degree.

1. $3x + 6 = 0$

2. $9x^2 + 5x - 6 = 3$

3. $25y^3 + y = 9y^2 - 17y + 4$

4. $x = 9$

5. $y = 2x + 1$

6. $3y = 9x^2$

7. $x^2 - 9 = 0$

8. $y = x$

9. $5x^7 = 3x^5 - 2x^8 + 11x - 9$

Answers to the Practice Set are on p. 104.

Section 3.3 EXERCISES

For problems 1–25, classify each polynomial as a monomial, binomial, or trinomial. State the degree of each polynomial and write the numerical coefficient of each term.

1. $5x + 7$

2. $16x + 21$

3. $4x^2 + 9$

4. $7y^3 + 8$

5. $a^4 + 1$

6. $2b^5 - 8$

7. $5x$

8. $7a$

9. $5x^3 + 2x + 3$

10. $17y^4 + y^5 - 9$

11. $41a^3 + 22a^2 + a$

12. $6y^2 + 9$

13. $2c^6 + 0$

14. $8x^2 - 0$

15. $9g$

16. $5xy + 3x$

17. $3yz - 6y + 11$

18. $7ab^2c^2 + 2a^2b^3c^5 + a^{14}$

19. $x^4y^3z^2 + 9z$

20. $5a^3b$

21. $6 + 3x^2y^5b$

22. $-9 + 3x^2 + 2xy6z^2$

23. 5

24. $3x^2y^0z^4 + 12z^3, \quad y \neq 0$

25. $4xy^3z^5w^0, \quad w \neq 0$

Classify each of the equations for problems 26–45 by degree. If the term linear, quadratic, or cubic applies, state it.

26. $4x + 7 = 0$ **27.** $3y - 15 = 9$

28. $y = 5s + 6$ **29.** $y = x^2 + 2$

30. $4y = 8x + 24$ **31.** $9z = 12x - 18$

32. $y^2 + 3 = 2y - 6$

33. $y - 5 + y^3 = 3y^2 + 2$

34. $x^2 + x - 4 = 7x^2 - 2x + 9$

35. $2y + 5x - 3 + 4xy = 5xy + 2y$

36. $3x - 7y = 9$

37. $8a + 2b = 4b - 8$

38. $2x^5 - 8x^2 + 9x + 4 = 12x^4 + 3x^3 + 4x^2 + 1$

39. $x - y = 0$

40. $x^2 - 25 = 0$

41. $x^3 - 64 = 0$

42. $x^{12} - y^{12} = 0$

43. $x + 3x^5 = x + 2x^5$

44. $3x^2y^4 + 2x - 8y = 14$

45. $10a^2b^3c^6d^0e^4 + 27a^3b^2b^4b^3b^2c^5 = 1, d \neq 0$

46. The expression $\dfrac{4x^3}{9x-7}$ is not a polynomial because

47. The expression $\dfrac{a^4}{7-a}$ is not a polynomial because

48. Is every algebraic expression a polynomial ex-

pression? If not, give an example of an algebraic expression that is not a polynomial expression.

49. Is every polynomial expression an algebraic expression? If not, give an example of a polynomial expression that is not an algebraic expression.

50. How do we find the degree of a term that contains more than one variable?

EXERCISES FOR REVIEW

(1.1) **51.** Use algebraic notation to write "eleven minus three times a number is five."

(1.6) **52.** Simplify $(x^4y^2z^3)^5$.

(2.5) **53.** Find the value of z if $z = \dfrac{x-u}{s}$ and $x = 55$, $u = 49$, and $s = 3$.

(3.1) **54.** List, if any should appear, the common factors in the expression $3x^4 + 6x^3 - 18x^2$.

(3.2) **55.** State (by writing it) the relationship being expressed by the equation $y = 3x + 5$.

★ **Answers to Practice Set (3.3)**

A. **1.** first, or linear   **2.** quadratic   **3.** cubic   **4.** linear   **5.** linear   **6.** quadratic
  7. quadratic   **8.** linear   **9.** eighth degree

3.4 Combining Polynomials Using Addition and Subtraction

Section Overview
- ☐ **LIKE TERMS**
- ☐ **COMBINING LIKE TERMS**
- ☐ **SIMPLIFYING EXPRESSIONS CONTAINING PARENTHESES**

☐ LIKE TERMS

Like Terms

Terms whose variable parts, including the exponents, are identical are called **like terms**. Like terms is an appropriate name since terms with identical variable parts and different numerical coefficients represent different amounts of the same quantity. As long as we are dealing with quantities of the same type we can combine them using addition and subtraction.

Simplifying an Algebraic Expression

An algebraic expression can be *simplified* by combining like terms.

Combine the like terms.

1. 6 houses + 4 houses = 10 houses. 6 and 4 of the same type give 10 of that type.

2. 6 houses + 4 houses + 2 motels = 10 houses + 2 motels. 6 and 4 of the same type give 10 of that type. Thus, we have 10 of one type and 2 of another type.

3. Suppose we let the letter x represent "house." Then, $6x + 4x = 10x$. 6 and 4 of the same type give 10 of that type.

4. Suppose we let x represent "house" and y represent "motel."

$6x + 4x + 2y = 10x + 2y$

★ **PRACTICE SET A**

Like terms with the same numerical coefficient represent equal amounts of the same quantity.

Like terms with different numerical coefficients represent _____.

❑ **COMBINING LIKE TERMS**

Since like terms represent amounts of the same quantity, they may be combined, that is, like terms may be added together.

Simplify each of the following polynomials by combining like terms.

1. $2x + 5x + 3x$.   There are 2 x's, then 5 more, then 3 more. This makes a total of 10 x's.

$2x + 5x + 3x = 10x$

2. $7x + 8y - 3x$.   From 7 x's, we lose 3 x's. This makes 4 x's. The 8 y's represent a quantity different from the x's and therefore will not combine with them.

$7x + 8y - 3x = 4x + 8y$

3. $4a^3 - 2a^2 + 8a^3 + a^2 - 2a^3$. $4a^3$, $8a^3$, and $-2a^3$ represent quantities of the same type.

$4a^3 + 8a^3 - 2a^3 = 10a^3$

$-2a^2$ and a^2 represent quantities of the same type.

$-2a^2 + a^2 = -a^2$

Thus,

$4a^3 - 2a^2 + 8a^3 + a^2 - 2a^3 = 10a^3 - a^2$

★ **PRACTICE SET B**

Simplify each of the following expressions.

1. $4y + 7y$ 2. $3x + 6x + 11x$ 3. $5a + 2b + 4a - b - 7b$

4. $10x^3 - 4x^3 + 3x^2 - 12x^3 + 5x^2 + 2x + x^3 + 8x$ 5. $2a^5 - a^5 + 1 - 4ab - 9 + 9ab - 2 - 3 - a^5$

☐ SIMPLIFYING EXPRESSIONS CONTAINING PARENTHESES

Simplifying Expressions
Containing Parentheses

When parentheses occur in expressions, they must be removed before the expression can be simplified. Parentheses can be removed using the distributive property.

Distributive Property

$a(b + c) = ab + ac$

☆ SAMPLE SET C

Simplify each of the following expressions by using the distributive property and combining like terms.

1. $6a + 5(a + 3)$ Multiply.

$6a + 5a + 15$ Combine the like terms $6a$ and $5a$.
$11a + 15$

2. $4x + 9(x^2 - 6x - 2) + 5$ Remove parentheses.

$4x + 9x^2 - 54x - 18 + 5$ Combine like terms.
$-50x + 9x^2 - 13$

By convention, the terms in an expression are placed in descending order with the highest degree term appearing first. Numerical terms are placed at the right end of the expression. The commutative property of addition allows us to change the order of the terms.

$9x^2 - 50x - 13$

3. $2 + 2[5 + 4(1 + a)]$

Eliminate the innermost set of parentheses first.

$2 + 2[5 + 4 + 4a]$

By the order of operations, simplify inside the parentheses before multiplying (by the 2).

$2 + 2[9 + 4a]$ Remove this set of parentheses.
$2 + 18 + 8a$ Combine like terms.
$20 + 8a$ Write in descending order.
$8a + 20$

4. $x(x - 3) + 6x(2x + 3)$

Use the rule for multiplying powers with the same base.

$x^2 - 3x + 12x^2 + 18x$ Combine like terms.
$13x^2 + 15x$

★ PRACTICE SET C

Simplify each of the following expressions by using the distributive property and combining like terms.

1. $4(x + 6) + 3(2 + x + 3x^2) - 2x^2$ **2.** $7(x + x^3) - 4x^3 - x + 1 + 4(x^2 - 2x^3 + 7)$

3. $5(a + 2) + 6a - 7 + (8 + 4)(a + 3a + 2)$ **4.** $x(x + 3) + 4x^2 + 2x$

5. $a^3(a^2 + a + 5) + a(a^4 + 3a^2 + 4) + 1$ **6.** $2[8 - 3(x - 3)]$

7. $x^2 + 3x + 7[x + 4x^2 + 3(x + x^2)]$

Answers to Practice Sets are on p. 110.

Section 3.4 EXERCISES

For problems $1-73$, simplify each of the algebraic expressions.

1. $x + 3x$

2. $4x + 7x$

3. $9a + 12a$

4. $5m - 3m$

5. $10x - 7x$

6. $7y - 9y$

7. $6k - 11k$

8. $3a + 5a + 2a$

9. $9y + 10y + 2y$

10. $5m - 7m - 2m$

11. $h - 3h - 5h$

12. $a + 8a + 3a$

13. $7ab + 4ab$

14. $8ax + 2ax + 6ax$

15. $3a^2 + 6a^2 + 2a^2$

16. $14a^2b + 4a^2b + 19a^2b$

17. $10y - 15y$

18. $7ab - 9ab + 4ab$

19. $210ab^4 + 412ab^4 + 100a^4b$ (Look closely at the exponents.)

20. $5x^2y^0 + 3x^2y + 2x^2y + 1,$ $y \neq 0$ (Look closely at the exponents.)

21. $8w^2 - 12w^2 - 3w^2$

22. $6xy - 3xy + 7xy - 18xy$

23. $7x^3 - 2x^2 - 10x + 1 - 5x^2 - 3x^3 - 12 + x$

24. $21y - 15x + 40xy - 6 - 11y + 7 - 12x - xy$

25. $1x + 1y - 1x - 1y + x - y$

26. $5x^2 - 3x - 7 + 2x^2 - x$

27. $-2z^3 + 15z + 4z^3 + z^2 - 6z^2 + z$

28. $18x^2y - 14x^2y - 20x^2y$

29. $-9w^5 - 9w^4 - 9w^5 + 10w^4$

30. $2x^4 + 4x^3 - 8x^2 + 12x - 1 - 7x^3 - 1x^4 - 6x + 2$

31. $17d^3r + 3d^3r - 5d^3r + 6d^2r + d^3r - 30d^2r + 3 - 7 + 2$

32. $a^0 + 2a^0 - 4a^0, \qquad a \neq 0$

33. $4x^0 + 3x^0 - 5x^0 + 7x^0 - x^0, \qquad x \neq 0$

34. $2a^3b^2c + 3a^2b^2c^0 + 4a^2b^2 - a^3b^2c, \qquad c \neq 0$

35. $3z - 6z + 8z$

36. $3z^2 - z + 3z^3$

37. $6x^3 + 12x + 5$

38. $3(x + 5) + 2x$

39. $7(a + 2) + 4$

40. $y + 5(y + 6)$

41. $2b + 6(3 - 5b)$

42. $5a - 7c + 3(a - c)$

43. $8x - 3x + 4(2x + 5) + 3(6x - 4)$

44. $2z + 4ab + 5z - ab + 12(1 - ab - z)$

45. $(a + 5)4 + 6a - 20$

46. $(4a + 5b - 2)3 + 3(4a + 5b - 2)$

47. $(10x + 3y^2)4 + 4(10x + 3y^2)$

48. $2(x - 6) + 5$

49. $1(3x + 15) + 2x - 12$

50. $1(2 + 9a + 4a^2) + a^2 - 11a$

51. $1(2x - 6b + 6a^2b + 8b^2) + 1(5x + 2b - 3a^2b)$

52. After observing problems 49–51, can you make a conjecture about $1(a + b)$?

$1(a + b) =$

53. Using the result of problem 52, is it correct to write

$(a + b) = a + b$?

54. $3(2a + 2a^2) + 8(3a + 3a^2)$

55. $x(x + 2) + 2(x^2 + 3x - 4)$

56. $A(A + 7) + 4(A^2 + 3a + 1)$

57. $b(2b^3 + 5b^2 + b + 6) - 6b^2 - 4b + 2$

58. $4a - a(a + 5)$

59. $x - 3x(x^2 - 7x - 1)$

60. $ab(a - 5) - 4a^2b + 2ab - 2$

61. $xy(3xy + 2x - 5y) - 2x^2y^2 - 5x^2y + 4xy^2$

62. $3h[2h + 5(h + 2)]$

63. $2k[5k + 3(1 + 7k)]$

64. $8a[2a - 4ab + 9(a - 5 - ab)]$

65. $6\{m + 5n[n + 3(n - 1)] + 2n^2\} - 4n^2 - 9m$

66. $5[4(r - 2s) - 3r - 5s] + 12s$

67. $8\{9[b - 2a + 6c(c + 4) - 4c^2] + 4a + b\} - 3b$

68. $5[4(6x - 3) + x] - 2x - 25x + 4$

69. $3xy^2(4xy + 5y) + 2xy^3 + 6x^2y^3 + 4y^3 - 12xy^3$

70. $9a^3b^7(a^3b^5 - 2a^2b^2 + 6) - 2a(a^2b^7 - 5a^5b^{12} + 3a^4b^9) - a^3b^7$

71. $-8(3a + 2)$

73. $-4xy^2[7xy - 6(5 - xy^2) + 3(-xy + 1) + 1]$

72. $-4(2x - 3y)$

EXERCISES FOR REVIEW

(1.5) **74.** Simplify $\left(\dfrac{x^{10}y^8z^2}{x^2y^6}\right)^3$.

(2.5) **75.** Find the value of $\dfrac{-3(4 - 9) - 6(-3) - 1}{2^3}$.

(2.6) **76.** Write the expression $\dfrac{42x^2y^5z^3}{21x^4y^7}$ so that no denominator appears.

(3.1) **77.** How many $(2a + 5)$'s are there in $3x(2a + 5)$?

(3.3) **78.** Simplify $3(5n + 6m^2) - 2(3n + 4m^2)$.

★ Answers to Practice Sets (3.4)

A. different amounts of the same quantity

B. **1.** $11y$ **2.** $20x$ **3.** $9a - 6b$ **4.** $-5x^3 + 8x^2 + 10x$ **5.** $5ab - 13$

C. **1.** $7x^2 + 7x + 30$ **2.** $-5x^3 + 4x^2 + 6x + 29$ **3.** $59a + 27$ **4.** $5x^2 + 5x$
    5. $2a^5 + a^4 + 8a^3 + 4a + 1$ **6.** $-6x + 34$ **7.** $50x^2 + 31x$

3.5 Combining Polynomials Using Multiplication

Section Overview	☐ MULTIPLYING A POLYNOMIAL BY A MONOMIAL ☐ SIMPLIFYING $+(a + b)$ AND $-(a + b)$ ☐ MULTIPLYING A POLYNOMIAL BY A POLYNOMIAL

☐ MULTIPLYING A POLYNOMIAL BY A MONOMIAL

Multiplying a polynomial by a monomial is a direct application of the distributive property.

Distributive Property

$a(b + c) = ab + ac$

The distributive property suggests the following rule.

Multiplying a Polynomial by a Monomial

To multiply a polynomial by a monomial, multiply *every* term of the polynomial by the monomial and then add the resulting products together.

☆ **SAMPLE SET A**

1. $3(x + 9) = 3 \cdot x + 3 \cdot 9$
$= 3x + 27$

2. $6(x^3 - 2x) = 6(x^3 + (-2x)) = 6 \cdot x^3 + 6(-2x)$
$= 6x^3 - 12x$

3. $(x - 7)x = x \cdot x + x(-7)$
$= x^2 - 7x$

4. $8a^2(3a^4 - 5a^3 + a) = 8a^2 \cdot 3a^4 + 8a^2(-5a^3) + 8a^2 \cdot a$
$= 24a^6 - 40a^5 + 8a^3$

5. $4x^2y^7z(x^5y + 8y^2z^2) = 4x^2y^7z \cdot x^5y + 4x^2y^7z \cdot 8y^2z^2$
$= 4x^7y^8z + 32x^2y^9z^3$

6. $10ab^2c(125a^2) = 1250a^3b^2c$

7. $(9x^2z + 4w)(5zw^3) = 9x^2z \cdot 5zw^3 + 4w \cdot 5zw^3$
$= 45x^2z^2w^3 + 20zw^4$
$= 45x^2w^3z^2 + 20w^4z$

★ **PRACTICE SET A**

Determine the following products.

1. $3(x + 8)$ **2.** $(2 + a)4$ **3.** $(a^2 - 2b + 6)2a$ **4.** $8a^2b^3(2a + 7b + 3)$

5. $4x(2x^5 + 6x^4 - 8x^3 - x^2 + 9x - 11)$ **6.** $(3a^2b)(2ab^2 + 4b^3)$

7. $5mn(m^2n^2 + m + n^0), \quad n \neq 0$ **8.** 🖩 $6.03(2.11a^3 + 8.00a^2b)$

❏ **SIMPLIFYING $+(a + b)$ AND $-(a + b)$**

$+(a + b)$
and
$-(a + b)$

Oftentimes, we will encounter multiplications of the form

$+1(a + b)$ or $-1(a + b)$

These terms will actually appear as

$+(a + b)$ and $-(a + b)$

Using the distributive property, we can remove the parentheses.

$+(a + b) = +1(a + b) = (+1)(a) + (+1)(b)$
$= a + b$

The parentheses have been removed and the sign of each term has remained the same.

$-(a + b) = -1(a + b) = (-1)(a) + (-1)(b)$
$= -a - b$

The parentheses have been removed and the sign of each term has been changed to its opposite.

1. To remove a set of parentheses preceded by a "+" sign, simply remove the parentheses and leave the sign of each term the same.
2. To remove a set of parentheses preceded by a "−" sign, remove the parentheses and change the sign of each term to its opposite sign.

☆ **SAMPLE SET B**

Simplify the expressions.

1. $(6x − 1)$.   This set of parentheses is preceded by a "+" sign (implied). We simply drop the parentheses.

$$(6x − 1) = 6x − 1$$

2. $(14a^2b^3 − 6a^3b^2 + ab^4) = 14a^2b^3 − 6a^3b^2 + ab^4$

3. $−(21a^2 + 7a − 18)$.   This set of parentheses is preceded by a "−" sign. We can drop the parentheses as long as we change the sign of *every* term inside the parentheses to its opposite sign.

$$−(21a^2 + 7a − 18) = −21a^2 − 7a + 18$$

4. $−(7y^3 − 2y^2 + 9y + 1) = −7y^3 + 2y^2 − 9y − 1$

★ **PRACTICE SET B**

Simplify by removing the parentheses.

1. $(2a + 3b)$

2. $(a^2 − 6a + 10)$

3. $−(x + 2y)$

4. $−(5m − 2n)$

5. $−(−3s^2 − 7s + 9)$

❏ **MULTIPLYING A POLYNOMIAL BY A POLYNOMIAL**

Since we can consider an expression enclosed within parentheses as a single quantity, we have, by the distributive property,

$$(a + b)(c + d) = (a + b)c + (a + b)d$$
$$= ac + bc + ad + bd$$

For convenience we will use the commutative property of addition to write this expression so that the first two terms contain a and the second two contain b.

$$(a + b)(c + d) = ac + ad + bc + bd$$

This method is commonly called the **FOIL method.**

F   First terms
O   Outer terms
I   Inner terms
L   Last terms

$$(a + b)(c + d) = ac + ad + bc + bd$$

F   O   I   L

For example, $(a + b)(2 + 3)$ means $(a + b)$ repeated as a term, first 2 times and then 3 times.

$$(a + b)(2 + 3) = \underbrace{(a + b) + (a + b)}_{2\ terms} + \underbrace{(a + b) + (a + b) + (a + b)}_{3\ terms}$$

Rearranging,

$$= a + a + b + b + a + a + a + b + b + b$$
$$= 2a + 2b + 3a + 3b$$

Combining like terms,

$$= 5a + 5b$$

This use of the distributive property suggests the following rule.

Multiplying a Polynomial by a Polynomial

> To multiply two polynomials together, multiply *every* term of one polynomial by *every* term of the other polynomial.

☆ SAMPLE SET C

Perform the following multiplications and simplify.

1. $(a + 6)(a + 3) = a \cdot a + a \cdot 3 + 6 \cdot a + 6 \cdot 3$
$\qquad\qquad\quad = a^2 + 3a + 6a + 18$
$\qquad\qquad\quad = a^2 + 9a + 18$

F: $\quad a \cdot a$
O: $\quad 3 \cdot a$
I: $\quad 6 \cdot a$
L: $\quad 6 \cdot 3$

With some practice, the second and third terms can be combined mentally.

2. $(x + y)(2x + 4y) = x \cdot 2x + x \cdot 4y + y \cdot 2x + y \cdot 4y$
$\qquad\qquad\qquad = 2x^2 + 4xy + 2xy + 4y^2$
$\qquad\qquad\qquad = 2x^2 + 6xy + 4y^2$

F: $\quad x \cdot 2x$
O: $\quad x \cdot 4y$
I: $\quad y \cdot 2x$
L: $\quad y \cdot 4y$

3. $(x^2 + 4)(x^2 + 7x + 2) = x^2 \cdot x^2 + x^2 \cdot 7x + x^2 \cdot 2 + 4 \cdot x^2 + 4 \cdot 7x + 4 \cdot 2$
$\qquad\qquad\qquad\qquad = x^4 + 7x^3 + 2x^2 + 4x^2 + 28x + 8$
$\qquad\qquad\qquad\qquad = x^4 + 7x^3 + 6x^2 + 28x + 8$

4. $(a - 4)(a - 3) = a \cdot a + a(-3) - 4 \cdot a - 4(-3)$
$\qquad\qquad\quad = a^2 - 3a - 4a + 12$
$\qquad\qquad\quad = a^2 - 7a + 12$

F: $\quad a \cdot a$
O: $\quad a \cdot (-3)$
I: $\quad (-4) \cdot a$
L: $\quad (-4)(-3)$

5. $(m - 3)^2 = (m - 3)(m - 3)$
$\qquad\qquad = m \cdot m + m(-3) - 3 \cdot m - 3(-3)$
$\qquad\qquad = m^2 - 3m - 3m + 9$
$\qquad\qquad = m^2 - 6m + 9$

6. $(x + 5)^3 = (x + 5)(x + 5)(x + 5)$ **Associate the first two factors.**
$\qquad\qquad = [(x + 5)(x + 5)](x + 5)$
$\qquad\qquad = [x^2 + 5x + 5x + 25](x + 5)$
$\qquad\qquad = [x^2 + 10x + 25](x + 5)$
$\qquad\qquad = x^2 \cdot x + x^2 \cdot 5 + 10x \cdot x + 10x \cdot 5 + 25 \cdot x + 25 \cdot 5$
$\qquad\qquad = x^3 + 5x^2 + 10x^2 + 50x + 25x + 125$
$\qquad\qquad = x^3 + 15x^2 + 75x + 125$

★ **PRACTICE SET C**

Find the following products and simplify.

1. $(a + 1)(a + 4)$ **2.** $(m - 9)(m - 2)$ **3.** $(2x + 4)(x + 5)$ **4.** $(x + y)(2x - 3y)$

5. $(3a^2 - 1)(5a^2 + a)$ **6.** $(2x^2y^3 + xy^2)(5x^3y^2 + x^2y)$ **7.** $(a + 3)(a^2 + 3a + 6)$

8. $(a + 4)(a + 4)$ **9.** $(r - 7)(r - 7)$ **10.** $(x + 6)^2$ **11.** $(y - 8)^2$

☆ **SAMPLE SET D**

Perform the following additions and subtractions.

1. $3x + 7 + (x - 3)$. We must first remove the parentheses. They are preceded by a "+" sign, so we remove them and leave the sign of each term the same.

$3x + 7 + x - 3$ Combine like terms.
$4x + 4$

2. $5y^3 + 11 - (12y^3 - 2)$. We first remove the parentheses. They are preceded by a "−" sign, so we remove them and change the sign of each term inside them.

$5y^3 + 11 - 12y^3 + 2$ Combine like terms.
$-7y^3 + 13$

3. Add $4x^2 + 2x - 8$ to $3x^2 - 7x - 10$.

$(4x^2 + 2x - 8) + (3x^2 - 7x - 10)$
$4x^2 + 2x - 8 + 3x^2 - 7x - 10$
$7x^2 - 5x - 18$

4. Subtract $8x^2 - 5x + 2$ from $3x^2 + x - 12$.

$(3x^2 + x - 12) - (8x^2 - 5x + 2)$
$3x^2 + x - 12 - 8x^2 + 5x - 2$
$-5x^2 + 6x - 14$

Be very careful *not* to write this problem as

$3x^2 + x - 12 - 8x^2 - 5x + 2$

This form has us subtracting only the very first term, $8x^2$, rather than the entire expression. Use parentheses.

Another incorrect form is

$8x^2 - 5x + 2 - (3x^2 + x - 12)$

This form has us performing the subtraction in the wrong order.

★ **PRACTICE SET D**

Perform the following additions and subtractions.

1. $6y^2 + 2y - 1 + (5y^2 - 18)$

2. $(9m - n) - (10m + 12n)$

3. Add $2r^2 + 4r - 1$ to $3r^2 - r - 7$.

4. Subtract $4s - 3$ from $7s + 8$.

Answers to Practice Sets are on p. 119.

Section 3.5 EXERCISES

For problems 1–88, perform the multiplications and combine any like terms.

1. $7(x + 6)$

2. $4(y + 3)$

23. $m(m - 4)$

3. $6(y + 4)$

4. $8(m + 7)$

24. $k(k - 11)$

5. $5(a - 6)$

6. $2(x - 10)$

25. $3x(x + 2)$

7. $3(4x + 2)$

8. $6(3x + 4)$

26. $4y(y + 7)$

9. $9(4y - 3)$

10. $5(8m - 6)$

27. $6a(a - 5)$

11. $-9(a + 7)$

12. $-3(b + 8)$

28. $9x(x - 3)$

13. $-4(x + 2)$

14. $-6(y + 7)$

29. $3x(5x + 4)$

15. $-3(a - 6)$

16. $-9(k - 7)$

30. $4m(2m + 7)$

17. $-5(2a + 1)$

18. $-7(4x + 2)$

31. $2b(b - 1)$

19. $-3(10y - 6)$

20. $-8(4y - 11)$

32. $7a(a - 4)$

21. $x(x + 6)$

22. $y(y + 7)$

33. $3x^2(5x^2 + 4)$

34. $9y^3(3y^2 + 2)$

35. $4a^4(5a^3 + 3a^2 + 2a)$

36. $2x^4(6x^3 - 5x^2 - 2x + 3)$

37. $-5x^2(x + 2)$

38. $-6y^3(y + 5)$

39. $2x^2y(3x^2y^2 - 6x)$

40. $8a^3b^2c(2ab^3 + 3b)$

41. $b^5x^2(2bx - 11)$

42. $4x(3x^2 - 6x + 10)$

43. $9y^3(2y^4 - 3y^3 + 8y^2 + y - 6)$

44. $-a^2b^3(6ab^4 + 5ab^3 - 8b^2 + 7b - 2)$

45. $(a + 4)(a + 2)$

46. $(x + 1)(x + 7)$

47. $(y + 6)(y - 3)$

48. $(t + 8)(t - 2)$

49. $(i - 3)(i + 5)$

50. $(x - y)(2x + y)$

51. $(3a - 1)(2a - 6)$

52. $(5a - 2)(6a - 8)$

53. $(6y + 11)(3y + 10)$

54. $(2t + 6)(3t + 4)$

55. $(4 + x)(3 - x)$

56. $(6 + a)(4 + a)$

57. $(x^2 + 2)(x + 1)$

58. $(x^2 + 5)(x + 4)$

59. $(3x^2 - 5)(2x^2 + 1)$

60. $(4a^2b^3 - 2a)(5a^2b - 3b)$

61. $(6x^3y^4 + 6x)(2x^2y^3 + 5y)$

62. $5(x - 7)(x - 3)$

63. $4(a + 1)(a - 8)$

64. $a(a - 3)(a + 5)$

65. $x(x + 1)(x + 4)$

66. $x^2(x + 5)(x + 7)$

67. $y^3(y - 3)(y - 2)$

68. $2a^2(a + 4)(a + 3)$

69. $5y^6(y + 7)(y + 1)$

70. $ab^2(a^2 - 2b)(a + b^4)$

71. $x^3y^2(5x^2y^2 - 3)(2xy - 1)$

72. $6(a^2 + 5a + 3)$

73. $8(c^3 + 5c + 11)$

74. $3a^2(2a^3 - 10a^2 - 4a + 9)$

75. $6a^3b^3(4a^2b^6 + 7ab^8 + 2b^{10} + 14)$

76. $(a - 4)(a^2 + a - 5)$

77. $(x - 7)(x^2 + x - 3)$

78. $(2x + 1)(5x^3 + 6x^2 + 8)$

79. $(7a^2 + 2)(3a^5 - 4a^3 - a - 1)$

80. $(x + y)(2x^2 + 3xy + 5y^2)$

81. $(2a + b)(5a^2 + 4a^2b - b - 4)$

82. $(x + 3)^2$ **83.** $(x + 1)^2$

84. $(x - 5)^2$ **85.** $(a + 2)^2$

86. $(a - 9)^2$ **87.** $-(3x - 5)^2$

88. $-(8t + 7)^2$

For problems 89–107, perform the indicated operations and combine like terms.

89. $3x^2 + 5x - 2 + (4x^2 - 10x - 5)$

90. $-2x^3 + 4x^2 + 5x - 8 + (x^3 - 3x^2 - 11x + 1)$

91. $-5x - 12xy + 4y^2 + (-7x + 7xy - 2y^2)$

92. $(6a^2 - 3a + 7) - 4a^2 + 2a - 8$

93. $(5x^2 - 24x - 15) + x^2 - 9x + 14$

94. $(3x^3 - 7x^2 + 2) + (x^3 + 6)$

95. $(9a^2b - 3ab + 12ab^2) + ab^2 + 2ab$

96. $6x^2 - 12x + (4x^2 - 3x - 1) + 4x^2 - 10x - 4$

97. $5a^3 - 2a - 26 + (4a^3 - 11a^2 + 2a) - 7a + 8a^3 + 20$

98. $2xy - 15 - (5xy + 4)$

99. Add $4x + 6$ to $8x - 15$.

100. Add $5y^2 - 5y + 1$ to $-9y^2 + 4y - 2$.

101. Add $3(x + 6)$ to $4(x - 7)$.

102. Add $-2(x^2 - 4)$ to $5(x^2 + 3x - 1)$.

103. Add four times $5x + 2$ to three times $2x - 1$.

104. Add five times $-3x + 2$ to seven times $4x + 3$.

106. Subtract $6x^2 - 10x + 4$ from $3x^2 - 2x + 5$.

105. Add -4 times $9x + 6$ to -2 times $-8x - 3$.

107. Subtract $a^2 - 16$ from $a^2 - 16$.

EXERCISES FOR REVIEW

(1.6)   **108.** Simplify $\left(\dfrac{15x^2y^6}{5xy^2}\right)^4$.

(2.7)   **109.** Express the number 198,000 using scientific notation.

(3.1)   **110.** How many $4a^2x^3$'s are there in $-16a^4x^5$?

(3.3)   **111.** State the degree of the polynomial $4xy^3 + 3x^5y - 5x^3y^3$, and write the numerical coefficient of each term.

(3.4)   **112.** Simplify $3(4x - 5) + 2(5x - 2) - (x - 3)$.

★ Answers to Practice Sets (3.5)

A. **1.** $3x + 24$ **2.** $4a + 8$ **3.** $2a^3 - 4ab + 12a$ **4.** $16a^3b^3 + 56a^2b^4 + 24a^2b^3$
 5. $8x^6 + 24x^5 - 32x^4 - 4x^3 + 36x^2 - 44x$ **6.** $6a^3b^3 + 12a^2b^4$ **7.** $5m^3n^3 + 5m^2n + 5mn$
 8. $12.7233a^3 + 48.24a^2b$

B. **1.** $2a + 3b$ **2.** $a^2 - 6a + 10$ **3.** $-x - 2y$ **4.** $-5m + 2n$ **5.** $3s^2 + 7s - 9$

C. **1.** $a^2 + 5a + 4$ **2.** $m^2 - 11m + 18$ **3.** $2x^2 + 14x + 20$ **4.** $2x^2 - xy - 3y^2$
 5. $15a^4 + 3a^3 - 5a^2 - a$ **6.** $10x^5y^5 + 7x^4y^4 + x^3y^3$ **7.** $a^3 + 6a^2 + 15a + 18$ **8.** $a^2 + 8a + 16$
 9. $r^2 - 14r + 49$ **10.** $x^2 + 12x + 36$ **11.** $y^2 - 16y + 64$

D. **1.** $11y^2 + 2y - 19$ **2.** $-m - 13n$ **3.** $5r^2 + 3r - 8$ **4.** $3s + 11$

3.6 Special Binomial Products

Section Overview

☐ **EXPANDING $(a + b)^2$ AND $(a - b)^2$**
☐ **EXPANDING $(a + b)(a - b)$**

Three binomial products occur so frequently in algebra that we designate them as *special binomial products*. We have seen them before (Sections 2.7 and 3.5), but we will study them again because of their importance as time saving devices and in solving equations (which we will study in a later chapter).

These special products can be shown as the *squares of a binomial*

$(a + b)^2$ and $(a - b)^2$

and as the *sum and difference of two terms.*

$(a + b)(a - b)$

There are two simple rules that allow us to easily expand (multiply out) these binomials. They are well worth memorizing, as they will save a lot of time in the future.

❏ EXPANDING $(a + b)^2$ AND $(a - b)^2$

Squaring a Binomial

To square a binomial:*

1. Square the first term.
2. Take the product of the two terms and double it.
3. Square the last term.
4. Add the three results together.

$(a + b)^2 = a^2 + 2ab + b^2$
$(a - b)^2 = a^2 - 2ab + b^2$

❏ EXPANDING $(a + b)(a - b)$

Sum and Difference of Two Terms

To expand the sum and difference of two terms:†

1. Square the first term and square the second term.
2. Subtract the square of the second term from the square of the first term.

$(a + b)(a - b) = a^2 - b^2$

---

* See problems 56 and 57 at the end of this section.

† See problem 58.

☆ SAMPLE SET A

1. $(x + 4)^2$

Square the first term: x^2.
The product of both terms is $4x$. Double it: $8x$.
Square the last term: 16.
Add them together: $x^2 + 8x + 16$.

$(x + 4)^2 = x^2 + 8x + 16$

Note that $(x + 4)^2 \neq x^2 + 4^2$. The $8x$ term is missing!

2. $(a - 8)^2$

Square the first term: a^2.
The product of both terms is $-8a$. Double it: $-16a$.
Square the last term: 64.
Add them together: $a^2 + (-16a) + 64$.

$(a - 8)^2 = a^2 - 16a + 64$

Notice that the sign of the last term in this expression is "+." This will always happen since the last term results from a number being *squared*. Any nonzero number times itself is always positive.

$(+)(+) = +$ and $(-)(-) = +$

The sign of the second term in the trinomial will always be the sign that occurs *inside* the parentheses.

3. $(y - 1)^2$

Square the first term: y^2.
The product of both terms is $-y$. Double it: $-2y$.
Square the last term: $+1$.
Add them together: $y^2 + (-2y) + 1$.

$(y - 1)^2 = y^2 - 2y + 1$
minus   plus

4. $(5x + 3)^2$

Square the first term: $25x^2$.
The product of both terms is $15x$. Double it: $30x$.
Square the last term: 9.
Add them together: $25x^2 + 30x + 9$.

$$(5x + 3)^2 = 25x^2 + 30x + 9$$
$$\underset{\text{plus}}{\curvearrowright} \quad \underset{\text{plus}}{\uparrow}$$

5. $(7b - 2)^2$

Square the first term: $49b^2$.
The product of both terms is $-14b$. Double it: $-28b$.
Square the last term: 4.
Add them together: $49b^2 + (-28b) + 4$.

$$(7b - 2)^2 = 49b^2 - 28b + 4$$
$$\underset{\text{minus}}{\curvearrowright} \quad \underset{\text{plus}}{\uparrow}$$

6. $(x + 6)(x - 6)$

Square the first term: x^2.
Subtract the square of the second term (36) from the square of the first term: $x^2 - 36$.

$$(x + 6)(x - 6) = x^2 - 36$$

7. $(4a - 12)(4a + 12)$

Square the first term: $16a^2$.
Subtract the square of the second term (144) from the square of the first term: $16a^2 - 144$.

$$(4a - 12)(4a + 12) = 16a^2 - 144$$

8. $(6x + 8y)(6x - 8y)$

Square the first term: $36x^2$.
Subtract the square of the second term ($64y^2$) from the square of the first term: $36x^2 - 64y^2$.

$$(6x + 8y)(6x - 8y) = 36x^2 - 64y^2$$

★ **PRACTICE SET A**

Find the following products.

1. $(x + 5)^2$ **2.** $(x + 7)^2$ **3.** $(y - 6)^2$ **4.** $(3a + b)^2$

5. $(9m - n)^2$ **6.** $(10x - 2y)^2$ **7.** $(12a - 7b)^2$ **8.** $(5h - 15k)^2$

Answers to the Practice Set are on p. 123.

Section 3.6 EXERCISES

For problems 1–55, find the products.

1. $(x + 3)^2$ **2.** $(x + 5)^2$ **3.** $(x + 8)^2$ **4.** $(x + 6)^2$

5. $(y + 9)^2$

6. $(y + 1)^2$

7. $(a - 4)^2$

8. $(a - 6)^2$

9. $(a - 7)^2$

10. $(b + 10)^2$

11. $(b + 15)^2$

12. $(a - 10)^2$

13. $(x - 12)^2$

14. $(x + 20)^2$

15. $(y - 20)^2$

16. $(3x + 5)^2$

17. $(4x + 2)^2$

18. $(6x - 2)^2$

19. $(7x - 2)^2$

20. $(5a - 6)^2$

21. $(3a - 9)^2$

22. $(3w - 2z)^2$

23. $(5a - 3b)^2$

24. $(6t - 7s)^2$

25. $(2h - 8k)^2$

26. $\left(a + \dfrac{1}{2}\right)^2$

27. $\left(a + \dfrac{1}{3}\right)^2$

28. $\left(x + \dfrac{3}{4}\right)^2$

29. $\left(x + \dfrac{2}{5}\right)^2$

30. $\left(x - \dfrac{2}{3}\right)^2$

31. $\left(y - \dfrac{5}{6}\right)^2$

32. $\left(y + \dfrac{2}{3}\right)^2$

33. $(x + 1.3)^2$

34. $(x + 5.2)^2$

35. $(a + 0.5)^2$

36. $(a + 0.08)^2$

37. $(x - 3.1)^2$

38. $(y - 7.2)^2$

39. $(b - 0.04)^2$

40. $(f - 1.006)^2$

41. $(x + 5)(x - 5)$

42. $(x + 6)(x - 6)$

43. $(x + 1)(x - 1)$

44. $(t - 1)(t + 1)$

45. $(f + 9)(f - 9)$

46. $(y - 7)(y + 7)$

47. $(2y + 3)(2y - 3)$

48. $(5x + 6)(5x - 6)$

49. $(2a - 7b)(2a + 7b)$

50. $(7x + 3t)(7x - 3t)$

51. $(5h - 2k)(5h + 2k)$

52. $\left(x + \dfrac{1}{3}\right)\left(x - \dfrac{1}{3}\right)$

53. $\left(a + \dfrac{2}{9}\right)\left(a - \dfrac{2}{9}\right)$

54. $\left(x + \dfrac{7}{3}\right)\left(x - \dfrac{7}{3}\right)$

55. $\left(2b + \dfrac{6}{7}\right)\left(2b - \dfrac{6}{7}\right)$

56. Expand $(a + b)^2$ to prove it is equal to $a^2 + 2ab + b^2$.

57. Expand $(a - b)^2$ to prove it is equal to $a^2 - 2ab + b^2$.

58. Expand $(a + b)(a - b)$ to prove it is equal to $a^2 - b^2$.

59. Fill in the missing label in the equation below.

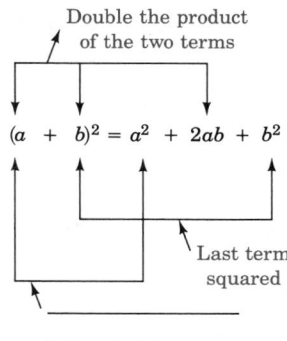

60. Label the parts of the equation below.

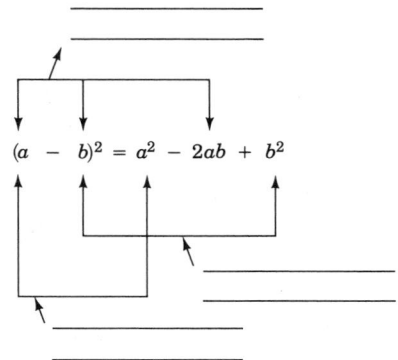

61. Label the parts of the equation below.

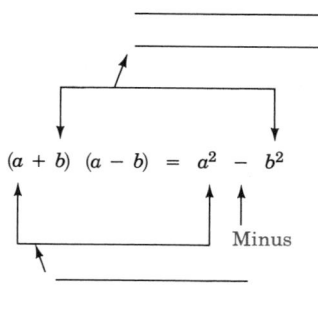

EXERCISES FOR REVIEW

(1.5) **62.** Simplify $(x^3 y^0 z^4)^5$.

(2.6) **63.** Find the value of $10^{-1} \cdot 2^{-3}$.

(3.5) **64.** Find the product. $(x + 6)(x - 7)$.

(3.5) **65.** Find the product. $(5m - 3)(2m + 3)$.

(3.5) **66.** Find the product. $(a + 4)(a^2 - 2a + 3)$.

★ **Answers to Practice Set (3.6)**

A. 1. $x^2 + 10x + 25$ **2.** $x^2 + 14x + 49$ **3.** $y^2 - 12y + 36$ **4.** $9a^2 + 6ab + b^2$
   5. $81m^2 - 18mn + n^2$ **6.** $100x^2 - 40xy + 4y^2$ **7.** $144a^2 - 168ab + 49b^2$ **8.** $25h^2 - 150hk + 225k^2$

3.7 Terminology Associated with Equations

❏ **INDEPENDENT AND DEPENDENT VARIABLES**
❏ **THE DOMAIN OF AN EQUATION**

❏ INDEPENDENT AND DEPENDENT VARIABLES

Independent and Dependent Variables

In an equation, any variable whose value can be freely assigned is said to be an **independent variable.** Any variable whose value is determined once the other values have been assigned is said to be a **dependent variable.** Two examples will help illustrate these concepts.

1. Consider the equation $y = 2x - 7$. If we are free to choose values for x, then x would be considered the independent variable. Since the value of y depends on the value of x, y would be the dependent variable.
2. Consider the equation $m = -4gk^2$. If we are free to choose values for both g and k, then g and k would be considered independent variables. Since the value of m depends on the values chosen for g and k, m would be the dependent variable.

❏ THE DOMAIN OF AN EQUATION

Domain

The process of replacing letters with numbers is called numerical evaluation. The collection of numbers that can replace the independent variable in an equation and yield a meaningful result is called the **domain** of the equation. The domain of an equation may be the entire collection of real numbers or may be restricted to some subcollection of the real numbers. The restrictions may be due to particular applications of the equation or to problems of computability.

☆ SAMPLE SET A

Find the domain of each of the following equations.

1. $y = \dfrac{2}{x}$, where x is the independent variable.

Any number except 0 can be substituted for x and yield a meaningful result. Hence, the domain is the collection of all real numbers except 0.

2. $d = 55t$, where t is the independent variable and the equation relates time, t, and distance, d.

It makes little sense to replace t by a negative number, so the domain is the collection of all real numbers greater than or equal to 0.

3. $k = \dfrac{2w}{w - 4}$, where the independent variable is w.

The letter w can be replaced by any real number except 4 since that will produce a division by 0. Hence, the domain is the collection of all real numbers except 4.

4. $a = 5b^2 + 2b - 6$, where the independent variable is b.

We can replace b by any real number and the expression $5b^2 + 2b - 6$ is computable. Hence, the domain is the collection of all real numbers.

★ **PRACTICE SET A**

Find the domain of each of the following equations. Assume that the independent variable is the variable that appears in the expression on the right side of the "=" sign.

1. $y = 5x + 10$ **2.** $y = \dfrac{5}{x}$ **3.** $y = \dfrac{3 + x}{x}$ **4.** $y = \dfrac{9}{x - 6}$ **5.** $m = \dfrac{1}{n + 2}$

6. $s = \dfrac{4}{9} t^2$, where this equation relates the distance an object falls, s, to the time, t, it has had to fall.

7. $g = \dfrac{4h - 7}{21}$

Answers to the Practice Set are below.

Section 3.7 EXERCISES

For problems 1–12, find the domain of the equations. Assume that the independent variable is the variable that appears in the expression to the right of the equal sign.

1. $y = 4x + 7$ **2.** $y = 3x - 5$

3. $y = x^2 + 2x - 9$ **4.** $y = 8x^3 - 6$

5. $y = 11x$ **6.** $s = 7t$

7. $y = \dfrac{3}{x}$ **8.** $y = \dfrac{2}{x}$

9. $m = \dfrac{-16}{h}$ **10.** $k = \dfrac{4t^2}{t - 1}$

11. $t = \dfrac{5}{s - 6}$ **12.** $y = \dfrac{12}{x + 7}$

EXERCISES FOR REVIEW

(1.3) **13.** Name the property of real numbers that makes $4yx^2 = 4x^2y$ a true statement.

(1.5) **14.** Simplify $\dfrac{x^{5n+6}}{x^4}$.

(2.2) **15.** Supply the missing phrase. Absolute value speaks to the question of _____ and not "which way."

(3.6) **16.** Find the product. $(x - 8)^2$.

(3.6) **17.** Find the product. $(4x + 3)(4x - 3)$.

★ **Answers to Practice Set (3.7)**

A. 1. all real numbers   **2.** all real numbers except 0   **3.** all real numbers except 0   **4.** all real numbers except 6   **5.** all real numbers except -2 **6.** all real numbers greater than or equal to 0   **7.** all real numbers

Algebraic Expressions **(3.1)**	An *algebraic expression* (often called simply an expression) is a number, a letter, or a collection of numbers and letters along with meaningful signs of operation. ($5 \div 0$ is not meaningful.)
Terms **(3.1)**	In an algebraic expression, the quantities joined by "+" signs are *terms*.
Distinction Between Terms and Factors **(3.1)**	*Terms* are parts of sums and are therefore separated by addition signs. *Factors* are parts of products and are therefore separated by multiplication signs.
Common Factors **(3.1)**	In an algebraic expression, a factor that appears in *every* term, that is, a factor that is common to each term, is called a *common factor*.
Coefficients **(3.1)**	The *coefficient* of a quantity records how many of that quantity there are. The coefficient of a group of factors is the remaining group of factors.
Distinction Between Coefficients and Exponents **(3.1)**	*Coefficients* record the number of like terms in an expression.

$$\underbrace{x + x + x}_{\text{3 terms}} = \underset{\text{coefficient is 3}}{3x}$$

Exponents record the number of like factors in an expression

$$\underbrace{x \cdot x \cdot x}_{\text{3 factors}} = \underset{\text{exponent is 3}}{x^3}$$

Equation **(3.2)**	An *equation* is a statement that two expressions are equal.
Numerical Evaluation **(3.2)**	*Numerical evaluation* is the process of determining a value by substituting numbers for letters.
Polynomials **(3.3)**	A polynomial is an algebraic expression that does not contain variables in the denominators of fractions and in which all exponents on variable quantities are whole numbers.
	A *monomial* is a polynomial consisting of only one term. A *binomial* is a polynomial consisting of two terms. A *trinomial* is a polynomial consisting of three terms.
Degree of a Polynomial **(3.3)**	The degree of a term containing one variable is the value of the exponent on the variable. The degree of a term containing more than one variable is the sum of the exponents on the variables. The degree of a polynomial is the degree of the term of the highest degree.
Linear Quadratic Cubic Polynomials **(3.3)**	Polynomials of the first degree are *linear* polynomials. Polynomials of the second degree are *quadratic* polynomials. Polynomials of the third degree are *cubic* polynomials.
Like Terms **(3.4)**	*Like terms* are terms in which the variable parts, including the exponents, are identical.
Descending Order **(3.4)**	By convention, and when possible, the terms of an expression are placed in descending order with the highest degree term appearing first. $5x^3 - 2x^2 + 10x - 15$ is in descending order.
Multiplying a Polynomial by a Monomial **(3.5)**	To multiply a polynomial by a monomial, multiply every term of the polynomial by the monomial and then add the resulting products together.

$$7(x - 3) = 7x - 7 \cdot 3 = 7x - 21$$

Simplifying $+(a + b)$ and $-(a + b)$ **(3.5)**	$+(a + b) = a + b$ $-(a + b) = -a - b$

Multiplying a Polynomial by a Polynomial **(3.5)**	To multiply polynomials together, multiply every term of one polynomial by every term of the other polynomial.

$$(x + 3)(x - 4) = x^2 - 4x + 3x - 12$$
$$= x^2 - x - 12$$

Special Products **(3.6)**	$(a + b)^2 = a^2 + 2ab + b^2$ *Note:* $(a + b)^2 \neq a^2 + b^2$ $(a - b)^2 = a^2 - 2ab + b^2$ $(a - b)^2 \neq a^2 - b^2$ $(a + b)(a - b) = a^2 - b^2$
Independent and Dependent Variables **(3.7)**	In an equation, any variable whose value can be freely assigned is said to be an *independent variable.* Any variable whose value is determined once the other values have been assigned is said to be a *dependent variable.*
Domain **(3.7)**	The collection of numbers that can be used as replacements for the independent variable in an expression or equation and yield a meaningful result is called the *domain* of the expression or equation.

EXERCISE SUPPLEMENT

Section 3.1

For problems 1–4, write the number of terms that appear, then write the terms.

1. $4x^2 + 7x + 12$

2. $14y^6$

3. $c + 8$

4. 8

List, if any should appear, the common factors for problems 5–11.

5. $a^2 + 4a^2 + 6a^2$

6. $9y^4 - 18y^4$

7. $12x^2y^3 + 36y^3$

8. $6(a + 4) + 12(a + 4)$

9. $4(a + 2b) + 6(a + 2b)$

10. $17x^2y(z + 4) + 51y(z + 4)$

11. $6a^2b^3c + 5x^2y$

For problems 12–16, answer the question of how many.

12. x's in $9x$?

13. $(a + b)$'s in $12(a + b)$?

14. a^4's in $6a^4$?

15. c^3's in $2a^2bc^3$?

16. $(2x + 3y)^2$'s in $5(x + 2y)(2x + 3y)^3$?

For problems 17–20, a term will be given followed by a group of its factors. List the coefficient of the given group of factors.

17. $8z, z$

18. $16a^3b^2c^4, c^4$

19. $7y(y + 3), 7y$

20. $(-5)a^5b^5c^5, bc$

Section 3.2

For problems 21–25, observe the equations and write the relationship being expressed.

21. $a = 3b$

22. $r = 4t + 11$

23. $f = \dfrac{1}{2}m^2 + 6g$

24. $x = 5y^3 + 2y + 6$

25. $P^2 = ka^3$

Use numerical evaluation to evaluate the equations for problems 26–31.

26. $C = 2\pi r$.   Find C if π is approximated by 3.14 and $r = 6$.

27. $I = \dfrac{E}{R}$.   Find I if $E = 20$ and $R = 2$.

28. $I = prt$.   Find I if $p = 1000$, $r = 0.06$, and $t = 3$.

29. $E = mc^2$.   Find E if $m = 120$ and $c = 186,000$.

30. $z = \dfrac{x - u}{s}$.   Find z if $x = 42$, $u = 30$, and $s = 12$.

31. $R = \dfrac{24C}{P(n + 1)}$.   Find R if $C = 35$, $P = 300$, and $n = 19$.

Section 3.3

For problems 32–38, classify each of the polynomials as a monomial, binomial, or trinomial. State the degree of each polynomial and write the numerical coefficient of each term.

32. $2a + 9$

33. $4y^3 + 3y + 1$

34. $10a^4$

35. 147

36. $4xy + 2yz^2 + 6x$

37. $9ab^2c^2 + 10a^3b^2c^5$

38. $(2xy^3)^0$, $xy^3 \neq 0$

39. Why is the expression $\dfrac{4x}{3x - 7}$ not a polynomial?

40. Why is the expression $5a^{3/4}$ not a polynomial?

For problems 41–45, classify each of the equations by degree. If the term linear, quadratic, or cubic applies, use it.

41. $3y + 2x = 1$

42. $4a^2 - 5a + 8 = 0$

43. $y - x - z + 4w = 21$

44. $5x^2 + 2x^2 - 3x + 1 = 19$

45. $(6x^3)^0 + 5x^2 = 7$

Sections 3.4 – 3.6

Simplify the algebraic expressions for problems 46 – 120.

46. $4a^2b + 8a^2b - a^2b$

47. $21x^2y^3 + 3xy + x^2y^3 + 6$

48. $7(x + 1) + 2x - 6$

49. $2(3y^2 + 4y + 4) + 5y^2 + 3(10y + 2)$

50. $5[3x + 7(2x^2 + 3x + 2) + 5] - 10x^2 + 4(3x^2 + x)$

51. $8\{3[4y^3 + y + 2] + 6(y^3 + 2y^2)\} - 24y^3 - 10y^2 - 3$

52. $4a^2bc^3 + 5abc^3 + 9abc^3 + 7a^2bc^2$

53. $x(2x + 5) + 3x^2 - 3x + 3$

54. $4k(3k^2 + 2k + 6) + k(5k^2 + k) + 16$

55. $2\{5[6(b + 2a + c^2)]\}$

56. $9x^2y(3xy + 4x) - 7x^3y^2 - 30x^3y + 5y(x^3y + 2x)$

57. $3m[5 + 2m(m + 6m^2)] + m(m^2 + 4m + 1)$

58. $2r[4(r + 5) - 2r - 10] + 6r(r + 2)$

59. $abc(3abc + c + b) + 6a(2bc + bc^2)$

60. $s^{10}(2s^5 + 3s^4 + 4s^3 + 5s^2 + 2s + 2) - s^{15} + 2s^{14} + 3s(s^{12} + 4s^{11}) - s^{10}$

61. $6a^4(a^2 + 5)$

62. $2x^2y^4(3x^2y + 4xy + 3y)$

63. $5m^6(2m^7 + 3m^4 + m^2 + m + 1)$

64. $a^3b^3c^4(4a + 2b + 3c + ab + ac + bc^2)$

65. $(x + 2)(x + 3)$

66. $(y + 4)(y + 5)$

67. $(a + 1)(a + 3)$

68. $(3x + 4)(2x + 6)$

69. $4xy - 10xy$

70. $5ab^2 - 3(2ab^2 + 4)$

71. $7x^4 - 15x^4$

72. $5x^2 + 2x - 3 - 7x^2 - 3x - 4 - 2x^2 - 11$

73. $4(x - 8)$

74. $7x(x^2 - x + 3)$

75. $-3a(5a - 6)$

76. $4x^2y^2(2x - 3y - 5) - 16x^3y^2 - 3x^2y^3$

77. $-5y(y^2 - 3y - 6) - 2y(3y^2 + 7) + (-2)(-5)$

78. $-[-(-4)]$

79. $-[-(-\{-[-(5)]\})]$

80. $x^2 + 3x - 4 - 4x^2 - 5x - 9 + 2x^2 - 6$

81. $4a^2b - 3b^2 - 5b^2 - 8q^2b - 10a^2b - b^2$

82. $2x^2 - x - (3x^2 - 4x - 5)$

83. $3(a - 1) - 4(a + 6)$

84. $-6(a + 2) - 7(a - 4) + 6(a - 1)$

85. Add $-3x + 4$ to $5x - 8$.

86. Add $4(x^2 - 2x - 3)$ to $-6(x^2 - 5)$.

87. Subtract 3 times $(2x - 1)$ from 8 times $(x - 4)$.

88. $(x + 4)(x - 6)$

89. $(x - 3)(x - 8)$

90. $(2a - 5)(5a - 1)$

91. $(8b + 2c)(2b - c)$

92. $(a - 3)^2$

93. $(3 - a)^2$

94. $(x - y)^2$

95. $(6x - 4)^2$

96. $(3a - 5b)^2$

97. $(-x - y)^2$

98. $(k + 6)(k - 6)$

99. $(m + 1)(m - 1)$

100. $(a - 2)(a + 2)$

101. $(3c + 10)(3c - 10)$

102. $(4a + 3b)(4a - 3b)$

103. $(5 + 2b)(5 - 2b)$

104. $(2y + 5)(4y + 5)$

105. $(y + 3a)(2y + a)$

106. $(6 + a)(6 - 3a)$

107. $(x^2 + 2)(x^2 - 3)$

108. $6(a - 3)(a + 8)$

109. $8(2y - 4)(3y + 8)$

110. $x(x - 7)(x + 4)$

111. $m^2n(m + n)(m + 2n)$

112. $(b + 2)(b^2 - 2b + 3)$

113. $3p(p^2 + 5p + 4)(p^2 + 2p + 7)$

114. $(a + 6)^2$

115. $(x-2)^2$

116. $(2x-3)^2$

117. $(x^2+y)^2$

118. $(2m-5n)^2$

119. $(3x^2y^3-4x^4y)^2$

120. $(a-2)^4$

Section 3.7

Find the domain of the equations for problems 121–125.

121. $y=8x+7$

122. $y=5x^2-2x+6$

123. $y=\dfrac{4}{x-2}$

124. $m=\dfrac{-2x}{h}$

125. $z=\dfrac{4x+5}{y+10}$

1. _____

2. _____

3. _____

4. _____

5. _____

6. _____

7. _____

8. _____

9. _____

10. _____

11. _____

12. _____

13. _____

14. _____

15. _____

1. (3.1) In the expression below, specify the number of terms that are present, then list them.

$$3a(a + 1) - (a + 2)(a - 3)$$

2. (3.1) List, if there are any, the common factors of

$$20x^3y^2 + 15x^3y^2z^2 + 10x^3z^2$$

3. (3.1) How many $y^2(b + 2)$'s in $8xy^2(b + 2)(b - 6)$?

4. (3.1) Write the coefficient of x^3 in $8x^3y^3z$.

5. (3.2) Find the value of P^2 if $k = 4$ and $a = 3$.

$$P^2 = ka^3$$

6. (3.3) Classify the polynomial given below as a monomial, binomial, trinomial, or none of these. Specify the degree of the polynomial and write the numerical coefficient of each term.

$$3x^3y + 4xy^4 + 8x^2y^2z^0w, \quad z \neq 0$$

Simplify the algebraic expressions for problems 7–20.

7. (3.4) $4x^2 + 3x + 2x + 11x^2 - 3$

8. (3.4) $3a[2(a + 1) + 4] - 18a$

9. (3.5) $(x + 2)(x + 4)$

10. (3.5) $(3a - 7)(2a + 10)$

11. (3.6) $(y + 3)^2$

12. (3.6) $(6a + 7y)^2$

13. (3.6) $(4x - 9y)^2$

14. (3.4 – 3.6) $3x^2(2x + 5)(3x + 1)$

15. (3.4 – 3.6) $(3a - b)(4a - 3b)$

131

16. _____

16. (3.4 – 3.6) $-6y^2(2y + 3y^2 - 4)$

17. _____

17. (3.4 – 3.6) $-4b^3(b^2 - 1)^2$

18. _____

18. (3.4 – 3.6) $(2a^3 + 3b^2)^2$

19. _____

19. (3.4 – 3.6) $6a(a - 2) - (2a^2 + a - 11)$

20. _____

20. (3.4 – 3.6) $(5h + 2k)(5h - 2k)$

21. _____

21. (3.4 – 3.6) Subtract $4a^2 - 10$ from $2a^2 + 6a + 1$.

22. _____

22. (3.4 – 3.6) Add three times $6x - 1$ to two times $-4x + 5$.

23. _____

23. (3.4 – 3.6) Evaluate $6k^2 + 2k - 7$ if $k = -1$.

24. _____

24. (3.4 – 3.6) Evaluate $-2m(m - 3)^2$ if $m = -4$.

25. _____

25. (3.7) What is the domain of $y = \dfrac{3x - 7}{x + 3}$?

4

Solving Linear Equations and Inequalities

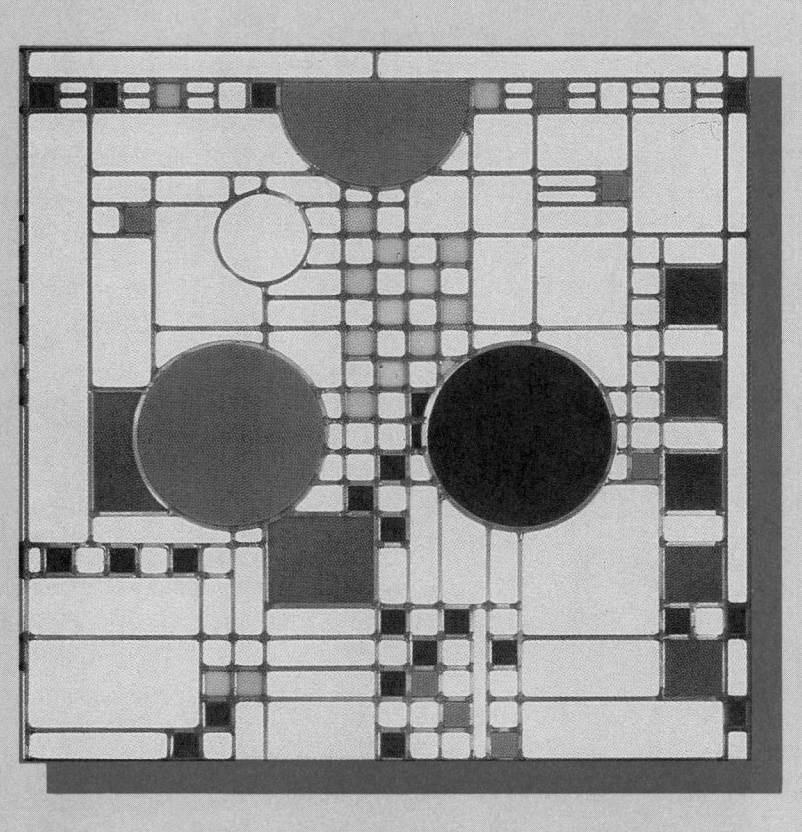

After completing this chapter, you should

Section 4.1 Solving Equations
- be able to identify various types of equations
- understand the meaning of solutions and equivalent equations
- be able to solve equations of the form $x + a = b$ and $x - a = b$
- be familiar with and able to solve literal equations

Section 4.2 Solving Equations of the Form $ax = b$ and $\dfrac{x}{a} = b$
- understand the equality property of addition and multiplication
- be able to solve equations of the form $ax = b$ and $x/a = b$

Section 4.3 Further Techniques in Equation Solving
- be comfortable with combining techniques in equation solving
- be able to recognize identities and contradictions

Section 4.4 Applications I – Translating from Verbal to Mathematical Expressions
- be able to translate from verbal to mathematical expressions

Section 4.5 Applications II – Solving Problems
- be able to solve various applied problems

Section 4.6 Linear Inequalities in One Variable
- understand the meaning of inequalities
- be able to recognize linear inequalities
- know, and be able to work with, the algebra of linear inequalities and with compound inequalities

Section 4.7 Linear Equations in Two Variables
- be able to identify the solution of a linear equation in two variables
- know that solutions to linear equations in two variables can be written as ordered pairs

4.1 Solving Equations

❑ **TYPES OF EQUATIONS**
❑ **SOLUTIONS AND EQUIVALENT EQUATIONS**
❑ **LITERAL EQUATIONS**
❑ **SOLVING EQUATIONS OF THE FORM** $x + a = b$ **AND** $x - a = b$

❑ TYPES OF EQUATIONS

Identity

Some equations are always true. These equations are called identities. **Identities** are equations that are true for all acceptable values of the variable, that is, for all values in the domain of the equation.

$5x = 5x$ is true for all acceptable values of x.
$y + 1 = y + 1$ is true for all acceptable values of y.
$2 + 5 = 7$ is true, and no substitutions are necessary.

Contradiction

Some equations are never true. These equations are called contradictions. **Contradictions** are equations that are never true regardless of the value substituted for the variable.

$x = x + 1$ is never true for any acceptable value of x.
$0 \cdot k = 14$ is never true for any acceptable value of k.
$2 = 1$ is never true.

Conditional Equation

The truth of some equations is conditional upon the value chosen for the variable. Such equations are called conditional equations. **Conditional equations** are equations that are true for at least one replacement of the variable and false for at least one replacement of the variable.

$x + 6 = 11$ is true only on the condition that $x = 5$.
$y - 7 = -1$ is true only on the condition that $y = 6$.

❑ SOLUTIONS AND EQUIVALENT EQUATIONS

Solutions and Solving an Equation

The collection of values that make an equation true are called **solutions** of the equation. An equation is **solved** when all its solutions have been found.

Equivalent Equations

Some equations have precisely the same collection of solutions. Such equations are called **equivalent equations.** The equations

$$2x + 1 = 7, \qquad 2x = 6, \qquad \text{and} \qquad x = 3$$

are equivalent equations because the only value that makes each one true is 3.

☆ **SAMPLE SET A**

Tell why each equation is an identity, a contradiction, or conditional.

1. The equation $x - 4 = 6$ is a conditional equation since it will be true only on the condition that $x = 10$.

2. The equation $x - 2 = x - 2$ is an identity since it is true for all values of x. For example,

if $x = 5$, $5 - 2 = 5 - 2$ is true
if $x = -7$, $-7 - 2 = -7 - 2$ is true

3. The equation $a + 5 = a + 1$ is a contradiction since every value of a produces a false statement. For example,

if $a = 8$, $\quad 8 + 5 = 8 + 1$ is false

if $a = -2$, $\quad -2 + 5 = -2 + 1$ is false

★ PRACTICE SET A

For each of the following equations, write "identity," "contradiction," or "conditional." If you can, find the solution by making an educated guess based on your knowledge of arithmetic.

1. $x + 1 = 10$ **2.** $y - 4 = 7$ **3.** $5a = 25$ **4.** $\dfrac{x}{4} = 9$

5. $\dfrac{18}{b} = 6$ **6.** $y - 2 = y - 2$ **7.** $x + 4 = x - 3$ **8.** $x + x + x = 3x$

9. $8x = 0$ **10.** $m - 7 = -5$

☐ LITERAL EQUATIONS

Literal Equations

Some equations involve more than one variable. Such equations are called **literal equations.**

> An equation is solved for a particular variable if that variable alone equals an expression that does not contain that particular variable.

The following equations are examples of literal equations.

1. $y = 2x + 7$.   It is solved for y.
2. $d = rt$.   It is solved for d.
3. $I = prt$.   It is solved for I.
4. $z = \dfrac{x - u}{s}$.   It is solved for z.
5. $y + 1 = x + 4$.   This equation is not solved for any particular variable since no variable is isolated.

☐ SOLVING EQUATIONS OF THE FORM $x + a = b$ AND $x - a = b$

Recall that the equal sign of an equation indicates that the number represented by the expression on the left side is the same as the number represented by the expression on the right side.

This number ↓	is the same as ↓	this number ↓
x	$=$	6
$x + 2$	$=$	8
$x - 1$	$=$	5

This suggests the following procedures:

> 1. We can obtain an equivalent equation (an equation having the same solutions as the original equation) by *adding* the *same number* to *both sides* of the equation.
> 2. We can obtain an equivalent equation by *subtracting* the *same number* from *both sides* of the equation.

We can use these results to isolate x, thus solving for x.

Solving $x + a = b$ for x

$x + a = b$	The a is associated with x by addition. Undo the association
$x + a - a = b - a$	by subtracting a from *both* sides.
$x + 0 = b - a$	$a - a = 0$ and 0 is the additive identity. $x + 0 = x$.
$x = b - a$	This equation is equivalent to the first equation, and it is solved for x.

Solving $x - a = b$ for x

$x - a = b$	The a is associated with x by subtraction. Undo the association
$x - a + a = b + a$	by adding a to *both* sides.
$x + 0 = b + a$	$-a + a = 0$ and 0 is the additive identity. $x + 0 = x$.
$x = b + a$	This equation is equivalent to the first equation, and it is solved for x.

Method for Solving $x + a = b$ and $x - a = b$ for x

> To solve the equation $x + a = b$ for x, *subtract a* from *both* sides of the equation.
>
> To solve the equation $x - a = b$ for x, *add a* to *both* sides of the equation.

☆ **SAMPLE SET B**

1. Solve $x + 7 = 10$ for x.

$x + 7 = 10$	7 is associated with x by addition. Undo the association
$x + 7 - 7 = 10 - 7$	by subtracting 7 from *both* sides.
$x + 0 = 3$	$7 - 7 = 0$ and 0 is the additive identity. $x + 0 = x$.
$x = 3$	x is isolated, and the equation $x = 3$ is equivalent to the original equation $x + 7 = 10$. Therefore, these two equations have the same solution. The solution to $x = 3$ is clearly 3. Thus, the solution to $x + 7 = 10$ is also **3**.

Check: Substitute 3 for x in the original equation.

$$x + 7 = 10$$
$$3 + 7 \overset{?}{=} 10$$
$$10 \overset{\checkmark}{=} 10$$

2. Solve $m - 2 = -9$ for m.

$m - 2 = -9$

$m - 2 + 2 = -9 + 2$

$m + 0 = -7$

$m = -7$

2 is associated with m by subtraction. Undo this association by adding 2 to *both* sides.

$-2 + 2 = 0$ and 0 is the additive identity. $m + 0 = m$.

Check: Substitute -7 for m in the original equation.

$m - 2 = -9$

$-7 - 2 \overset{?}{=} -9$

$-9 \overset{\checkmark}{=} -9$

3. [calc] Solve $y - 2.181 = -16.915$ for y.

$y - 2.181 = -16.915$

$y - 2.181 + 2.181 = -16.915 + 2.181$

$y = -14.734$

On the Calculator

Type	16.915
Press	$\boxed{+/-}$
Press	$\boxed{+}$
Type	2.181
Press	$\boxed{=}$
Display reads:	-14.734

4. Solve $y + m = s$ for y.

$y + m = s$

$y + m - m = s - m$

$y + 0 = s - m$

$y = s - m$

m is associated with y by addition. Undo the association by subtracting m from *both* sides.

$m - m = 0$ and 0 is the additive identity. $y + 0 = y$.

Check: Substitute $s - m$ for y in the original equation.

$y + m = s$

$s - m + m \overset{?}{=} s$

$s \overset{\checkmark}{=} s$ True

5. Solve $k - 3h = -8h + 5$ for k.

$k - 3h = -8h + 5$

$k - 3h + 3h = -8h + 5 + 3h$

$k + 0 = -5h + 5$

$k = -5h + 5$

$3h$ is associated with k by subtraction. Undo the association by adding $3h$ to *both* sides.

$-3h + 3h = 0$ and 0 is the additive identity. $k + 0 = k$.

★ **PRACTICE SET B**

1. Solve $y - 3 = 8$ for y.

2. Solve $x + 9 = -4$ for x.

3. Solve $m + 6 = 0$ for m.

4. Solve $g - 7.2 = 1.3$ for g.

5. Solve $f + 2d = 5d$ for f.

6. Solve $x + 8y = 2y - 1$ for x.

7. Solve $y + 4x - 1 = 5x + 8$ for y.

Answers to Practice Sets are on p. 139.

Section 4.1 EXERCISES

For problems 1–6, classify each of the equations as an identity, contradiction, or conditional equation.

  1. $m + 6 = 15$
  2. $y - 8 = -12$
  3. $x + 1 = x + 1$
  4. $k - 2 = k - 3$
  5. $g + g + g + g = 4g$
  6. $x + 1 = 0$

For problems 7–13, determine which of the literal equations have been solved for a variable. Write "solved" or "not solved."

  7. $y = 3x + 7$
  8. $m = 2k + n - 1$
  9. $4a = y - 6$
  10. $hk = 2k + h$
  11. $2a = a + 1$
  12. $5m = 2m - 7$
  13. $m = m$

For problems 14–31, solve each of the conditional equations.

14. $h - 8 = 14$ **15.** $k + 10 = 1$

16. $m - 2 = 5$ **17.** $y + 6 = -11$

18. $y - 8 = -1$ **19.** $x + 14 = 0$

20. $m - 12 = 0$ **21.** $g + 164 = -123$

22. $h - 265 = -547$ **23.** $x + 17 = -426$

24. $h - 4.82 = -3.56$

25. $y + 17.003 = -1.056$

26. $k + 1.0135 = -6.0032$

27. Solve $n + m = 4$ for n.

28. Solve $P + 3Q - 8 = 0$ for P.

29. Solve $a + b - 3c = d - 2f$ for b.

30. Solve $x - 3y + 5z + 1 = 2y - 7z + 8$ for x.

31. Solve $4a - 2b + c + 11 = 6a - 5b$ for c.

EXERCISES FOR REVIEW

(1.6)   **32.** Simplify $(4x^5y^2)^3$.

(2.6)   **33.** Write $\dfrac{20x^3y^7}{5x^5y^3}$ so that only positive exponents appear.

(3.1)   **34.** Write the number of terms that appear in the expression $5x^2 + 2x - 6 + (a + b)$, and then list them.

(3.6)   **35.** Find the product. $(3x - 1)^2$.

(3.7)   **36.** Specify the domain of the equation $y = \dfrac{5}{x - 2}$.

★ Answers to Practice Sets (4.1)

A.   **1.** conditional, $x = 9$ **2.** conditional, $y = 11$ **3.** conditional, $a = 5$ **4.** conditional, $x = 36$
   5. conditional, $b = 3$ **6.** identity   **7.** contradiction   **8.** identity   **9.** conditional, $x = 0$
   10. conditional, $m = 2$

B.   **1.** $y = 11$ **2.** $x = -13$ **3.** $m = -6$ **4.** $g = 8.5$ **5.** $f = 3d$ **6.** $x = -6y - 1$ **7.** $y = x + 9$

4.2 Solving Equations of the Form $ax = b$ and $\dfrac{x}{a} = b$

Section Overview

- ☐ **EQUALITY PROPERTY OF DIVISION AND MULTIPLICATION**
- ☐ **SOLVING $ax = b$ AND $\dfrac{x}{a} = b$ FOR x**

☐ EQUALITY PROPERTY OF DIVISION AND MULTIPLICATION

Recalling that the equal sign of an equation indicates that the number represented by the expression on the left side is the same as the number represented by the expression on the right side suggests the equality property of division and multiplication, which states:

> 1. We can obtain an equivalent equation by *dividing both sides* of the equation by the same nonzero number, that is, if $c \neq 0$, then $a = b$ is equivalent to $\dfrac{a}{c} = \dfrac{b}{c}$.
> 2. We can obtain an equivalent equation by *multiplying both sides* of the equation by the same nonzero number, that is, if $c \neq 0$, then $a = b$ is equivalent to $ac = bc$.

We can use these results to isolate x, thus solving the equation for x.

Solving $ax = b$ for x

$ax = b$ a is associated with x by multiplication. Undo the association by dividing *both* sides by a.

$\dfrac{ax}{a} = \dfrac{b}{a}$

$\dfrac{\cancel{a}x}{\cancel{a}} = \dfrac{b}{a}$

$1 \cdot x = \dfrac{b}{a}$ $\dfrac{a}{a} = 1$ and 1 is the multiplicative identity. $1 \cdot x = x$.

Solving

$\dfrac{x}{a} = b$

for x

$x = \dfrac{b}{a}$ This equation is equivalent to the first and is solved for x.

$\dfrac{x}{a} = b$ a is associated with x by division. Undo the association by multiplying *both* sides by a.

$a \cdot \dfrac{x}{a} = a \cdot b$

$\cancel{a} \cdot \dfrac{x}{\cancel{a}} = ab$

$1 \cdot x = ab$ $\dfrac{a}{a} = 1$ and 1 is the multiplicative identity. $1 \cdot x = x$.

$x = ab$ This equation is equivalent to the first and is solved for x.

❑ SOLVING $ax = b$ AND $\dfrac{x}{a} = b$ FOR x

Method for Solving

$ax = b$

and

$\dfrac{x}{a} = b$

> To solve $ax = b$ for x, *divide both sides* of the equation by a.
>
> To solve $\dfrac{x}{a} = b$ for x, *multiply both sides* of the equation by a.

☆ SAMPLE SET A

1. Solve $5x = 35$ for x.

$5x = 35$ 5 is associated with x by multiplication. Undo the association by dividing *both* sides by 5.

$\dfrac{5x}{5} = \dfrac{35}{5}$

$\dfrac{\cancel{5}x}{\cancel{5}} = 7$

$1 \cdot x = 7$ $\dfrac{5}{5} = 1$ and 1 is the multiplicative identity. $1 \cdot x = x$.

$x = 7$

Check: $5(7) \stackrel{?}{=} 35$

        $35 \stackrel{\checkmark}{=} 35$

2. Solve $\dfrac{x}{4} = 5$ for x.

$\dfrac{x}{4} = 5$ 4 is associated with x by division. Undo the association by multiplying *both* sides by 4.

$4 \cdot \dfrac{x}{4} = 4 \cdot 5$

$\cancel{4} \cdot \dfrac{x}{\cancel{4}} = 4 \cdot 5$

$1 \cdot x = 20$

$x = 20$

Check: $\dfrac{20}{4} \overset{?}{=} 5$

$5 \overset{\checkmark}{=} 5$

$\dfrac{4}{4} = 1$ **and 1 is the multiplicative identity.** $1 \cdot x = x.$

3. Solve $\dfrac{2y}{9} = 3$ for y.

Method (1)   (Use of cancelling):

$\dfrac{2y}{9} = 3$

$(\cancel{9}) \left(\dfrac{2y}{\cancel{9}} \right) = (9)(3)$

$2y = 27$

$\dfrac{\cancel{2}y}{\cancel{2}} = \dfrac{27}{2}$

$y = \dfrac{27}{2}$

9 is associated with y **by division. Undo the association by multiplying** *both* **sides by 9.**

2 is associated with y **by multiplication. Undo the association by dividing** *both* **sides by 2.**

Check: $\dfrac{\cancel{2} \left(\dfrac{27}{\cancel{2}} \right)}{9} \overset{?}{=} 3$

$\dfrac{27}{9} \overset{?}{=} 3$

$3 \overset{\checkmark}{=} 3$

Method (2)   (Use of reciprocals):

$\dfrac{2y}{9} = 3$

$\left(\dfrac{9}{2} \right) \left(\dfrac{2y}{9} \right) = \left(\dfrac{9}{2} \right)(3)$

$\left(\dfrac{9}{2} \cdot \dfrac{2}{9} \right) y = \dfrac{27}{2}$

$1 \cdot y = \dfrac{27}{2}$

$y = \dfrac{27}{2}$

Since $\dfrac{2y}{9} = \dfrac{2}{9} \, y,$ $\dfrac{2}{9}$ **is associated with** y **by multiplication.**

Then, since $\dfrac{9}{2} \cdot \dfrac{2}{9} = 1,$ **the multiplicative identity, we can**

undo the association by multiplying *both* **sides by** $\dfrac{9}{2}.$

Continued

4. Solve the literal equation $\dfrac{4ax}{m} = 3b$ for x.

$\dfrac{4ax}{m} = 3b$

m is associated with x by division. Undo the association by multiplying *both* sides by m.

$m\left(\dfrac{4ax}{m}\right) = m \cdot 3b$

$4ax = 3bm$

$4a$ is associated with x by multiplication. Undo the association by multiplying *both* sides by $4a$.

$\dfrac{4ax}{4a} = \dfrac{3bm}{4a}$

$x = \dfrac{3bm}{4a}$

Check: $\dfrac{4a\left(\dfrac{3bm}{4a}\right)}{m} \stackrel{?}{=} 3b$

$\dfrac{4a\left(\dfrac{3bm}{4a}\right)}{m} \stackrel{?}{=} 3b$

$\dfrac{3bm}{m} \stackrel{?}{=} 3b$

$3b \stackrel{\checkmark}{=} 3b$

★ PRACTICE SET A

1. Solve $6a = 42$ for a.

2. Solve $-12m = 16$ for m.

3. Solve $\dfrac{y}{8} = -2$ for y.

4. Solve $6.42x = 1.09$ for x.
   Round the result to two decimal places.

5. Solve $\dfrac{5k}{12} = 2$ for k.

6. Solve $\dfrac{-ab}{2c} = 4d$ for b.

7. Solve $\dfrac{3xy}{4} = 9xh$ for y.

8. Solve $\dfrac{2k^2mn}{5pq} = -6n$ for m.

Answers to the Practice Set are on p. 145.

Section 4.2 EXERCISES

In problems 1 – 43, solve each of the conditional equations.

1. $3x = 42$

2. $5y = 75$

3. $6x = 48$

4. $8x = 56$

5. $4x = 56$

6. $3x = 93$

7. $5a = -80$

8. $9m = -108$

9. $6p = -108$

10. $12q = -180$

11. $-4a = 16$

12. $-20x = 100$

13. $-6x = -42$

14. $-8m = -40$

15. $-3k = 126$

16. $-9y = 126$

17. $\dfrac{x}{6} = 1$

18. $\dfrac{a}{5} = 6$

19. $\dfrac{k}{7} = 6$

20. $\dfrac{x}{3} = 72$

21. $\dfrac{x}{8} = 96$

22. $\dfrac{y}{-3} = -4$

23. $\dfrac{m}{7} = -8$

24. $\dfrac{k}{18} = 47$

25. $\dfrac{f}{-62} = 103$

26. $3.06m = 12.546$

27. $5.012k = 0.30072$

28. $\dfrac{x}{2.19} = 5$

29. $\dfrac{y}{4.11} = 2.3$

30. $\dfrac{4y}{7} = 2$

31. $\dfrac{3m}{10} = -1$

32. $\dfrac{5k}{6} = 8$

33. $\dfrac{8h}{-7} = -3$

34. $\dfrac{-16z}{21} = -4$

35. Solve $pq = 7r$ for p.

36. Solve $m^2n = 2s$ for n.

37. Solve $2.8ab = 5.6d$ for b.

38. Solve $\dfrac{mnp}{2k} = 4k$ for p.

39. Solve $\dfrac{-8a^2b}{3c} = -5a^2$ for b.

40. Solve $\dfrac{3pcb}{2m} = 2b$ for pc.

41. Solve $\dfrac{8rst}{3p} = -2prs$ for t.

42. Solve $\dfrac{\square \cdot \star}{\triangle} = \diamond$ for $\square$.

43. Solve $\dfrac{3\square\triangle\triangledown}{2\triangledown} = \triangle\triangledown$ for $\square$.

EXERCISES FOR REVIEW

(1.6)   **44.** Simplify $\left(\dfrac{2x^0 y^0 z^3}{z^2}\right)^5$.

(3.3)   **45.** Classify $10x^3 - 7x$ as a monomial, binomial, or trinomial. State its degree and write the numerical coefficient of each item.

(3.4)   **46.** Simplify $3a^2 - 2a + 4a(a + 2)$.

(3.7)   **47.** Specify the domain of the equation $y = \dfrac{3}{7 + x}$.

(4.1)   **48.** Solve the conditional equation $x + 6 = -2$.

★ **Answers to Practice Set (4.2)**

A. **1.** $a = 7$ **2.** $m = -\dfrac{4}{3}$ **3.** $y = -16$ **4.** $x = 0.17$ (rounded to two decimal places)   **5.** $k = \dfrac{24}{5}$

6. $b = \dfrac{-8cd}{a}$ **7.** $y = 12h$ **8.** $m = \dfrac{-15pq}{k^2}$

4.3 Further Techniques in Equation Solving

Section Overview

- ☐ **COMBINING TECHNIQUES IN EQUATION SOLVING**
- ☐ **RECOGNIZING IDENTITIES AND CONTRADICTIONS**

☐ COMBINING TECHNIQUES IN EQUATION SOLVING

In Sections 4.1 and 4.2 we worked with techniques that involved the use of addition, subtraction, multiplication, and division to solve equations. We can combine these techniques to solve more complicated equations. To do so, it is helpful to recall that an equation is solved for a particular variable when all other numbers and/or letters have been disassociated from it and it is alone on one side of the equal sign. We will also note that

> To associate numbers and letters we use the order of operations.
>
> 1. Multiply/divide
> 2. Add/subtract
>
> To undo an association between numbers and letters we use the order of operations in reverse.
>
> 1. Add/subtract
> 2. Multiply/divide

☆ **SAMPLE SET A**

1. Solve $4x - 7 = 9$ for x.

$4x - 7 = 9$

First, undo the association between x and 7. The 7 is associated with x by subtraction. Undo the association by adding 7 to *both* sides.

$4x - 7 + 7 = 9 + 7$
$4x = 16$

Now, undo the association between x and 4. The 4 is associated with x by multiplication. Undo the association by dividing *both* sides by 4.

$\dfrac{\cancel{4}x}{\cancel{4}} = \dfrac{16}{4}$
$16 - 7 \stackrel{?}{=} 9$
$x = 4$

Check: $4(4) - 7 \stackrel{?}{=} 9$
$9 \stackrel{\checkmark}{=} 9$

2. Solve $\dfrac{3y}{4} - 5 = -11$.

$\dfrac{3y}{4} - 5 = -11$

-5 is associated with y by subtraction. Undo the association by adding 5 to *both* sides.

$\dfrac{3y}{4} - 5 + 5 = -11 + 5$

$\dfrac{3y}{4} = -6$

4 is associated with y by division. Undo the association by multiplying *both* sides by 4.

$4 \cdot \dfrac{3y}{4} = 4(-6)$

$\cancel{4} \cdot \dfrac{3y}{\cancel{4}} = 4(-6)$

$3y = -24$

3 is associated with y by multiplication. Undo the association by dividing *both* sides by 3.

$\dfrac{3y}{3} = \dfrac{-24}{3}$

$\dfrac{\cancel{3}y}{\cancel{3}} = -8$

$y = -8$

Check: $\dfrac{3(-8)}{4} - 5 \stackrel{?}{=} -11$

$\dfrac{-24}{4} - 5 \stackrel{?}{=} -11$

$-6 - 5 \stackrel{?}{=} -11$

$-11 \stackrel{\checkmark}{=} -11$

3. Solve $\dfrac{8a}{3b} + 2m = 6m - 5$ for a.

$\dfrac{8a}{3b} + 2m = 6m - 5$ $2m$ is associated with a by addition. Undo the association by subtracting $2m$ from *both* sides.

$\dfrac{8a}{3b} + 2m - 2m = 6m - 5 - 2m$

$\dfrac{8a}{3b} = 4m - 5$ $3b$ is associated with a by division. Undo the association by multiplying *both* sides by $3b$.

$(3b)\left(\dfrac{8a}{3b}\right) = 3b(4m - 5)$

$8a = 12bm - 15b$ 8 is associated with a by multiplication. Undo the multiplication by dividing *both* sides by 8.

$\dfrac{\cancel{8}a}{\cancel{8}} = \dfrac{12bm - 15b}{8}$

$a = \dfrac{12bm - 15b}{8}$

★ **PRACTICE SET A**

1. Solve $3y - 1 = 11$ for y.

2. Solve $\dfrac{5m}{2} + 6 = 1$ for m.

3. Solve $2n + 3m = 4$ for n.

4. Solve $\dfrac{9k}{2h} + 5 = p - 2$ for k.

Sometimes when solving an equation it is necessary to simplify the expressions composing it.

☆ **SAMPLE SET B**

1. Solve $4x + 1 - 3x = (-2)(4)$ for x.

$4x + 1 - 3x = (-2)(4)$
$x + 1 = -8$
$x = -9$

Check: $4(-9) + 1 - 3(-9) \overset{?}{=} -8$
$-36 + 1 + 27 \overset{?}{=} -8$
$-8 \overset{\checkmark}{=} -8$

Continued

2. Solve $3(m - 6) - 2m = -4 + 1$ for m.

$3(m - 6) - 2m = -4 + 1$
$3m - 18 - 2m = -3$
$m - 18 = -3$
$m = 15$

Check: $3(15 - 6) - 2(15) \overset{?}{=} -4 + 1$
$\quad\quad\quad 3(9) - 30 \overset{?}{=} -3$
$\quad\quad\quad 27 - 30 \overset{?}{=} -3$
$\quad\quad\quad -3 \overset{\checkmark}{=} -3$

★ **PRACTICE SET B**

Solve and check each equation.

1. $16x - 3 - 15x = 8$ for x.

2. $4(y - 5) - 3y = -1$ for y.

3. $-2(a^2 + 3a - 1) + 2a^2 + 7a = 0$ for a.

4. $5m(m - 2a - 1) - 5m^2 + 2a(5m + 3) = 10$ for a.

Often the variable we wish to solve for will appear on both sides of the equal sign. We can isolate the variable on either the left or right side of the equation by using the techniques of Sections 4.1 and 4.2.

1. Solve $6x - 4 = 2x + 8$ for x.

$6x - 4 = 2x + 8$ **To isolate x on the left side, subtract $2x$ from both sides.**

$6x - 4 - 2x = 2x + 8 - 2x$
$4x - 4 = 8$ **Add 4 to both sides.**
$4x - 4 + 4 = 8 + 4$
$4x = 12$ **Divide both sides by 4.**
$\dfrac{4x}{4} = \dfrac{12}{4}$

$x = 3$

Check: $6(3) - 4 \overset{?}{=} 2(3) + 8$
$18 - 4 \overset{?}{=} 6 + 8$
$14 \overset{\checkmark}{=} 14$

2. Solve $6(1 - 3x) + 1 = 2x - [3(x - 7) - 20]$ for x.

$6(1 - 3x) + 1 = 2x - [3(x - 7) - 20]$

$6 - 18x + 1 = 2x - [3x - 21 - 20]$
$-18x + 7 = 2x - [3x - 41]$
$-18x + 7 = 2x - 3x + 41$
$-18x + 7 = -x + 41$ **To isolate x on the right side, add $18x$ to both sides.**

$-18x + 7 + 18x = -x + 41 + 18x$
$7 = 17x + 41$ **Subtract 41 from both sides.**
$7 - 41 = 17x + 41 - 41$
$-34 = 17x$ **Divide both sides by 17.**
$\dfrac{-34}{17} = \dfrac{17x}{17}$

$-2 = x$ **Since the equation $-2 = x$ is equivalent to the equation $x = -2$, we can write the answer as $x = -2$.**

$x = -2$

Check: $6(1 - 3(-2)) + 1 \overset{?}{=} 2(-2) - [3(-2 - 7) - 20]$
$6(1 + 6) + 1 \overset{?}{=} -4 - [3(-9) - 20]$
$6(7) + 1 \overset{?}{=} -4 - [-27 - 20]$
$42 + 1 \overset{?}{=} -4 - [-47]$
$43 \overset{?}{=} -4 + 47$
$43 \overset{\checkmark}{=} 43$

★ **PRACTICE SET C**

1. Solve $8a + 5 = 3a - 5$ for a.   **2.** Solve $9y + 3(y + 6) = 15y + 21$ for y.

3. Solve $3k + 2[4(k - 1) + 3] = 63 - 2k$ for k.

❏ RECOGNIZING IDENTITIES AND CONTRADICTIONS

As we noted in Section 4.1, some equations are identities and some are contradictions. As the problems of Sample Set D will suggest,

Recognizing an Identity

Recognizing a Contradiction

1. If, when solving an equation, all the variables are eliminated and a true statement results, the equation is an **identity**.
2. If, when solving an equation, all the variables are eliminated and a false statement results, the equation is a **contradiction**.

☆ SAMPLE SET D

1. Solve $9x + 3(4 - 3x) = 12$ for x.

$9x + 3(4 - 3x) = 12$

$9x + 12 - 9x = 12$
$12 = 12$

The variable has been eliminated and the result is a true statement. The original equation is an *identity*.

2. Solve $-2(10 - 2y) - 4y + 1 = -18$ for y.

$-2(10 - 2y) - 4y + 1 = -18$

$-20 + 4y - 4y + 1 = -18$
$-19 \neq -18$

The variable has been eliminated and the result is a false statement. The original equation is a *contradiction*.

★ PRACTICE SET D

Classify each equation as an identity or a contradiction.

1. $6x + 3(1 - 2x) = 3$

2. $-8m + 4(2m - 7) = 28$

3. $3(2x - 4) - 2(3x + 1) + 14 = 0$

4. $-5(x + 6) + 8 = 3[4 - (x + 2)] - 2x$

Answers to Practice Sets are on p. 155.

Section 4.3 EXERCISES

For problems 1–37, solve each conditional equation. If the equation is not conditional, identify it as an identity or a contradiction.

1. $3x + 1 = 16$

2. $6y - 4 = 20$

3. $4a - 1 = 27$

4. $3x + 4 = 40$

5. $2y + 7 = -3$

6. $8k - 7 = -23$

7. $5x + 6 = -9$

8. $7a + 2 = -26$

9. $10y - 3 = -23$

10. $14x + 1 = -55$

11. $\dfrac{x}{9} + 2 = 6$

12. $\dfrac{m}{7} - 8 = -11$

13. $\dfrac{y}{4} + 6 = 12$

14. $\dfrac{x}{8} - 2 = 5$

15. $\dfrac{m}{11} - 15 = -19$

16. $\dfrac{k}{15} + 20 = 10$

23. $16(y - 1) + 11 = -85$

17. $6 + \dfrac{k}{5} = 5$

24. $6x + 14 = 5x - 12$

18. $1 - \dfrac{n}{2} = 6$

25. $23y - 19 = 22y + 1$

19. $\dfrac{7x}{4} + 6 = -8$

26. $-3m + 1 = 3m - 5$

20. $\dfrac{-6m}{5} + 11 = -13$

27. $8k + 7 = 2k + 1$

21. $\dfrac{3k}{14} + 25 = 22$

22. $3(x - 6) + 5 = -25$

28. $12n + 5 = 5n - 16$

29. $2(x - 7) = 2x + 5$

30. $-4(5y + 3) + 5(1 + 4y) = 0$

31. $3x + 7 = -3 - (x + 2)$

32. $4(4y + 2) = 3y + 2[1 - 3(1 - 2y)]$

33. $5(3x - 8) + 11 = 2 - 2x + 3(x - 4)$

34. $12 - (m - 2) = 2m + 3m - 2m + 3(5 - 3m)$

35. $-4 \cdot k - (-4 - 3k) = -3k - 2k - (3 - 6k) + 1$

36. $3[4 - 2(y + 2)] = 2y - 4[1 + 2(1 + y)]$

37. $-5[2m - (3m - 1)] = 4m - 3m + 2(5 - 2m) + 1$

For problems 38–50, solve the literal equations for the indicated variable. When directed, find the value of that variable for the given values of the other variables.

38. Solve $I = \dfrac{E}{R}$ for R. Find the value of R when $I = 0.005$ and $E = 0.0035$.

39. Solve $P = R - C$ for R. Find the value of R when $P = 27$ and $C = 85$.

40. Solve $z = \dfrac{x - \bar{x}}{s}$ for x. Find the value of x when $z = 1.96$, $s = 2.5$, and $\bar{x} = 15$.

41. Solve $F = \dfrac{S_x{}^2}{S_y{}^2}$ for $S_x{}^2$. $S_x{}^2$ represents a single quantity. Find the value of $S_x{}^2$ when $F = 2.21$ and $S_y{}^2 = 3.24$.

42. Solve $p = \dfrac{nRT}{V}$ for R.

43. Solve $x = 4y + 7$ for y.

44. Solve $y = 10x + 16$ for x.

45. Solve $2x + 5y = 12$ for y.

46. Solve $-9x + 3y + 15 = 0$ for y.

47. Solve $m = \dfrac{2n - h}{5}$ for n.

48. Solve $t = \dfrac{Q + 6P}{8}$ for P.

49. Solve $\star = \dfrac{\square + 9j}{\triangle}$ for j.

50. Solve $\diamond = \dfrac{\triangle + \star\square}{2\triangledown}$ for $\star$.

EXERCISES FOR REVIEW

(1.5) **51.** Simplify $(x + 3)^2(x - 2)^3(x - 2)^4(x + 3)$.
(3.6) **52.** Find the product. $(x - 7)(x + 7)$.
(3.6) **53.** Find the product. $(2x - 1)^2$.
(4.1) **54.** Solve the equation $y - 2 = -2$.

(4.2) **55.** Solve the equation $\dfrac{4x}{5} = -3$.

★ **Answers to Practice Sets (4.3)**

A. **1.** $y = 4$ **2.** $m = -2$ **3.** $n = \dfrac{4 - 3m}{2}$ **4.** $k = \dfrac{2hp - 14h}{9}$

B. **1.** $x = 11$ **2.** $y = 19$ **3.** $a = -2$ **4.** $a = \dfrac{10 + 5m}{6}$

C. **1.** $a = -2$ **2.** $y = -1$ **3.** $k = 5$

D. **1.** identity, $3 = 3$ **2.** contradiction, $-28 = 28$ **3.** identity, $0 = 0$ **4.** contradiction, $-22 = 6$

4.4 Applications I – Translating from Verbal to Mathematical Expressions

Section Overview

❑ **TRANSLATING FROM VERBAL TO MATHEMATICAL EXPRESSIONS**

❑ **TRANSLATING FROM VERBAL TO MATHEMATICAL EXPRESSIONS**

To solve a problem using algebra, we must first express the problem algebraically. To express a problem algebraically, we must scrutinize the wording of the problem to determine the variables and constants that are present and the relationships among them. Then we must translate the verbal phrases and statements to algebraic expressions and equations.

To help us translate verbal expressions to mathematics, we can use the following table as a mathematics dictionary.

MATHEMATICS DICTIONARY

Word or Phrase	Mathematical Operation
Sum, sum of, added to, increased by, more than, plus, and	$+$
Difference, minus, subtracted from, decreased by, less, less than	$-$
Product, the product of, of, multiplied by, times	$\cdot$
Quotient, divided by, ratio	$\div$
Equals, is equal to, is, the result is, becomes	$=$
A number, an unknown quantity, an unknown, a quantity	x (or any symbol)

☆ SAMPLE SET A

Translate the following phrases or sentences into mathematical expressions or equations.

1. Six more than a number.

$$\underbrace{6} \quad \underbrace{+} \quad \underbrace{x}$$
$$6 + x$$

2. Fifteen minus a number.

$$\underbrace{15} \quad \underbrace{-} \quad \underbrace{x}$$
$$15 - x$$

3. A quantity less eight.

$$\underbrace{y} \quad \underbrace{-} \quad \underbrace{8}$$
$$y - 8$$

4. Twice a number is ten.

$$\underbrace{2 \cdot} \quad \underbrace{x} \quad \underbrace{=} \quad \underbrace{10}$$
$$2x = 10$$

5. One half of a number is twenty.

$$\underbrace{\frac{1}{2}} \quad \underbrace{\cdot} \quad \underbrace{z} \quad \underbrace{=} \quad \underbrace{20}$$
$$\frac{1}{2}z = 20$$

6. Three times a number is five more than twice the same number.

$$\underbrace{3} \quad \underbrace{\cdot} \quad \underbrace{y} \quad \underbrace{=} \quad \underbrace{5} \quad \underbrace{+} \quad \underbrace{2 \cdot} \quad \underbrace{y}$$
$$3y = 5 + 2y$$

★ PRACTICE SET A

Translate the following phrases or sentences into mathematical expressions or equations.

1. Eleven more than a number.

2. Nine minus a number.

3. A quantity less twenty.

4. Four times a number is thirty two.

5. One third of a number is six.

6. Ten times a number is eight more than five times the same number.

Sometimes the structure of the sentence indicates the use of grouping symbols.

☆ **SAMPLE SET B**

Translate the following phrases or sentences into mathematical expressions or equations.

1. A number divided by five, minus ten, is fifteen.

$$(x \quad \div \quad 5) \quad - \quad 10 \quad = \quad 15$$

$$\frac{x}{5} - 10 = 15$$

Commas set off terms.

2. Eight divided by five more than a number is ten.

$$8 \quad \div \quad (5 \quad + \quad x) \quad = 10$$

The wording indicates this is to be considered as a single quantity.

$$\frac{8}{5 + x} = 10$$

3. A number multiplied by ten more than itself is twenty.

$$x \quad \cdot \quad (10 \quad + \quad x) \quad = \quad 20$$

$$x(10 + x) = 20$$

4. A number plus one is divided by three times the number minus twelve and the result is four.

$$(x + 1) \div (3 \cdot x - 12) = 4$$

$$\frac{x + 1}{3x - 12} = 4$$

Notice that since the phrase "three times the number minus twelve" does not contain a comma, we get the expression $3x - 12$. If the phrase had appeared as "three times the number, minus twelve," the result would have been

$$\frac{x + 1}{3x} - 12 = 4$$

5. Some phrases and sentences do not translate directly. We must be careful to read them properly. The word *from* often appears in such phrases and sentences. The word *from* means "a point of departure for motion." The following translation will illustrate this use.

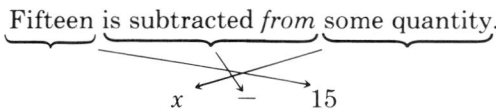

Fifteen is subtracted *from* some quantity.

$$x \quad - \quad 15$$

The word *from* indicates the motion (subtraction) is to begin at the point of "some quantity."

6. Eight less than some quantity. Notice that *less than* could be replaced with *from*.

$$x - 8$$

★ **PRACTICE SET B**

Translate the following phrases and sentences into mathematical expressions or equations.

1. A number divided by sixteen, plus one, is five.

2. Seven times two more than a number is twenty-one.

3. A number divided by two more than itself is zero.

4. A number minus five is divided by twice the number plus three and the result is seventeen.

5. Fifty-two is subtracted from some quantity.

6. An unknown quantity is subtracted from eleven and the result is five less than the unknown quantity.

Answers to Practice Sets are on p. 160.

Section 4.4 EXERCISES

For problems 1–50, translate the following phrases or sentences into mathematical expressions or equations.

1. A quantity less four.

2. Eight more than a number.

3. A number plus seven.

4. A number minus three.

5. Negative five plus an unknown quantity.

6. Negative sixteen minus some quantity.

7. Fourteen added to twice a number.

8. Ten added to three times some number.

9. One third minus an unknown quantity.

10. Twice a number is eleven.

11. Four ninths of a number is twenty-one.

12. One third of a number is two fifths.

13. Three times a number is nine more than twice the number.

14. Five times a number is that number minus two.

15. Twice a number added to six results in thirty.

16. Ten times a number less four results in sixty-six.

17. A number less twenty-five is equal to 3.019.

18. Seven more than some number is five more than twice the number.

19. When a number is divided by four, the result is sixty-eight.

20. Eleven fifteenths of two more than a number is eight.

21. One tenth of a number is that number less one.

22. Two more than twice a number is one half the number less three.

23. A number is equal to itself plus four times itself.

24. Three fifths of a quantity added to the quantity itself is thirty-nine.

25. A number plus seven is divided by two and the result is twenty-two.

26. Ten times a number minus one is divided by fourteen and the result is one.

27. A number is added to itself then divided by three. This result is then divided by three. The entire result is fifteen.

28. Ten divided by two more than a number is twenty-one.

29. Five divided by a number plus six is fourteen.

30. Twelve divided by twice a number is fifty-five.

31. Twenty divided by eight times a number added to one is nine.

32. A number divided by itself, plus one, results in seven.

33. A number divided by ten, plus four, results in twenty-four.

34. A number plus six, divided by two, is seventy-one.

35. A number plus six, divided by two, plus five, is forty-three.

36. A number multiplied by itself added to five is thirty-one.

37. A quantity multiplied by seven plus twice itself is ninety.

38. A number is increased by one and then multiplied by five times itself. The result is eighty-four.

39. A number is added to six and that result is multiplied by thirteen. This result is then divided by six times the number. The entire result is equal to fifty-nine.

40. A number is subtracted from ten and that result is multiplied by four. This result is then divided by three more than the number. The entire result is equal to six.

41. An unknown quantity is decreased by eleven. This result is then divided by fifteen. Now, one is subtracted from this result and five is obtained.

42. Ten less than some number.

43. Five less than some unknown number.

44. Twelve less than a number.

45. One less than an unknown quantity.

46. Sixteen less than some number is forty-two.

47. Eight less than some unknown number is three.

48. Seven is added to ten less than some number. The result is one.

49. Twenty-three is divided by two less than twice some number and the result is thirty-four.

50. One less than some number is multiplied by three less than five times the number and the entire result is divided by six less than the number. The result is twenty-seven less than eleven times the number.

EXERCISES FOR REVIEW

(1.2) **51.** Supply the missing word. The point on a line that is associated with a particular number is called the _____ of that number.

(1.4) **52.** Supply the missing word. An exponent records the number of identical _____ in a multiplication.

(2.2) **53.** Write the algebraic definition of the absolute value of the number a.

(4.3) **54.** Solve the equation $4y + 5 = -3$.

(4.3) **55.** Solve the equation $2(3x + 1) - 5x = 4(x - 6) + 17$.

★ **Answers to Practice Sets (4.4)**

A. **1.** $11 + x$ **2.** $9 - x$ **3.** $x - 20$ **4.** $4x = 32$ **5.** $\frac{x}{3} = 6$ **6.** $10x = 8 + 5x$

B. **1.** $\frac{x}{16} + 1 = 5$ **2.** $7(2 + x) = 21$ **3.** $\frac{x}{2 + x} = 0$ **4.** $\frac{x - 5}{2x + 3} = 17$ **5.** $x - 52$ **6.** $11 - x = x - 5$

4.5 Applications II – Solving Problems

Section Overview

☐ **SOLVING APPLIED PROBLEMS**

☐ SOLVING APPLIED PROBLEMS

Let's study some interesting problems that involve linear equations in one variable. In order to solve such problems, we apply the following five-step method:

Five-Step Method for Solving Word Problems

1. Let x (or some other letter) represent the unknown quantity.
2. Translate the words to mathematical symbols and form an equation.
3. Solve this equation.
4. Ask yourself, "Does this result seem reasonable?" Check the solution by substituting the result into the original statement of the problem.

   If the answer doesn't check, you have either solved the equation incorrectly, or you have developed the wrong equation. Check your method of solution first. If the result does not check, reconsider your equation.
5. Write the conclusion.

If it has been your experience that word problems are difficult, then follow the five-step method carefully. Most people have difficulty because they neglect step 1.

Always start by INTRODUCING A VARIABLE!

Keep in mind what the variable is representing throughout the problem.

☆ SAMPLE SET A

This year an item costs $44, an increase of $3 over last year's price. What was last year's price?

Step 1: Let x = last year's price.

Step 2: $x + 3 = 44.$ **$x + 3$ represents the $3 increase in price.**

Step 3: $x + 3 = 44$
$x + 3 - 3 = 44 - 3$
$x = 41$

Step 4: $41 + 3 \overset{\checkmark}{=} 44.$

Step 5: Last year's price was $41.

★ PRACTICE SET A

This year an item costs $23, an increase of $4 over last year's price. What was last year's price?

Step 1: Let x =

Step 2:

Step 3:

Step 4:

Step 5: Last year's price was _____.

☆ SAMPLE SET B

The perimeter (length around) of a square is 60 cm (centimeters). Find the length of a side.

Step 1: Let x = length of a side.

Step 2: We can draw a picture.

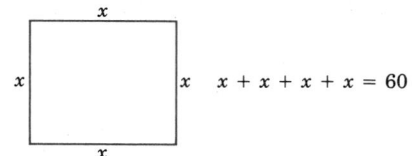

$x + x + x + x = 60$

Continued

Step 3: $x + x + x + x = 60$
            $4x = 60$ **Divide both sides by 4.**
            $x = 15.$

Step 4: $4(15) \stackrel{?}{=} 60.$
Step 5:   The length of a side is 15 cm.

★ PRACTICE SET B

The perimeter of a triangle is 54 inches. If each side has the same length, find the length of a side.

Step 1:   Let $x =$
Step 2:

Step 3:

Step 4:

Step 5:   The length of a side is _____ inches.

☆ SAMPLE SET C

Six percent of a number is 54. What is the number?

Step 1:   Let $x =$ the number
Step 2:   We must convert 6% to a decimal.

            $6\% = .06$
            $.06x = 54$ **.06x occurs because we want 6% of x.**

Step 3: $.06x = 54.$ **Divide both sides by .06.**

            $x = \dfrac{54}{.06}$

            $x = 900$

Step 4: $.06(900) \stackrel{?}{=} 54.$
Step 5:   The number is 900.

★ PRACTICE SET C

Eight percent of a number is 36. What is the number?

Step 1:   Let $x =$
Step 2:

Step 3:

Step 4:

Step 5:   The number is _____ .

☆ **SAMPLE SET D**

> An astronomer notices that one star gives off about 3.6 times as much energy as another star. Together the stars give off 55.844 units of energy. How many units of energy does each star emit?
>
> Step 1:   In this problem we have two unknowns and, therefore, we might think, two variables. However, notice that the energy given off by one star is given in terms of the other star. So, rather than introducing two variables, we introduce only one. The other unknown(s) is expressed in terms of this one. (We might call this quantity the base quantity.)
>
> Let x = number of units of energy given off by the less energetic star. Then, $3.6x$ = number of units of energy given off by the more energetic star.
>
> Step 2: $x + 3.6x = 55.844$.
>
> Step 3: $x + 3.6x = 55.844$
> $4.6x = 55.844$ **Divide both sides by 4.6. A calculator would be useful at this point.**
>
> $x = \dfrac{55.844}{4.6}$
>
> $x = 12.14$ **The wording of the problem implies *two* numbers are needed for a complete solution. We need the number of units of energy for the other star.**
>
> $3.6x = 3.6(12.14)$
> $\quad\;\; = 43.704$
>
> Step 4: $12.14 + 43.704 \not\le 55.844$.
> Step 5:   One star gives off 12.14 units of energy and the other star gives off 43.704 units of energy.

★ **PRACTICE SET D**

Garden A produces 5.8 times as many vegetables as garden B. Together the gardens produce 102 pounds of vegetables. How many pounds of vegetables does garden A produce?

Step 1:   Let x =
Step 2:

Step 3:

Step 4:

Step 5:

☆ SAMPLE SET E

Two consecutive even numbers sum to 432. What are the two numbers?

Step 1:   Let x = the smaller even number. Then $x + 2$ = the next (consecutive) even number since consecutive even numbers differ by 2 (as do consecutive odd numbers).

Step 2: $x + x + 2 = 432$.

Step 3: $x + x + 2 = 432$
$2x + 2 = 432$
$2x = 430$
$x = 215$.   Also, since $x = 215$, $x + 2 = 217$.

Step 4: $215 + 217 = 432$, but 215 and 217 are odd numbers and we are looking for even numbers. Upon checking our method of solution and reexamining our equation, we find no mistakes.

Step 5:   We must conclude that this problem has no solution. There are no two consecutive *even* numbers that sum to 432.

★ PRACTICE SET E

The sum of two consecutive even numbers is 498. What are the two numbers?

Step 1:

Step 2:

Step 3:

Step 4:

Step 5:

Answers to Practice Sets are on p. 170.

Section 4.5 EXERCISES

Solve problems 1–33. Note that some of the problems may seem to have no practical applications and may not seem very interesting. They, along with the other problems, will, however, help to develop your logic and problem-solving ability.

1. If eighteen is subtracted from some number the result is fifty-two. What is the number?

Step 1:   Let x =

Step 2:   The equation is

Step 3:   (Solve the equation.)

Step 4:   (Check)

Step 5:   The number is _____.

2. If nine more than twice a number is forty-six, what is the number?

Step 1:   Let $x =$
Step 2:   The equation is

Step 3:   (Solve the equation.)

Step 4:   (Check)

Step 5:   The number is _____ .

3. If nine less than three eighths of a number is two and one fourth, what is the number?

Step 1:   Let $x =$
Step 2:

Step 3:

Step 4:

Step 5:   The number is _____ .

4. Twenty percent of a number is 68. What is the number?

Step 1:   Let $x =$
Step 2:

Step 3:

Step 4:

Step 5:   The number is _____ .

5. Eight more than a quantity is 37. What is the original quantity?

Step 1:   Let $x =$
Step 2:

Step 3:

Step 4:

Step 5:   The original quantity is _____ .

6. If a quantity plus 85% more of the quantity is 62.9, what is the original quantity?

Step 1:   Let $x =$ original quantity.
Step 2: $\underbrace{x}_{\substack{\text{original}\\\text{quantity}}} + \underbrace{.85x}_{85\% \text{ more}} = 62.9$

Step 3:

Step 4:

Step 5:   The original quantity is _____ .

7. A company must increase production by 12% over last year's production. The new output will be 56 items. What was last year's output?

Step 1:   Let $P =$
Step 2:

Step 3:

Step 4:

Step 5:   Last year's output was _____ items.

8. A company has determined that it must increase production of a certain line of goods by $1\frac{1}{2}$ times last year's production. The new output will be 2885 items. What was last year's output?

   Step 1:
   Step 2:

   Step 3:

   Step 4:

   Step 5:   Last year's output was _____ items.

9. A proton is about 1837 times as heavy as an electron. If an electron weighs 2.68 units, how many units does a proton weigh?

   Step 1:
   Step 2:

   Step 3:

   Step 4:

   Step 5:   A proton weighs _____ units.

10. Neptune is about 30 times as far from the sun as is the Earth. If it takes light 8 minutes to travel from the sun to the Earth, how many minutes does it take to travel to Neptune?

    Step 1:
    Step 2:

    Step 3:

Step 4:

Step 5:   Light takes _____ minutes to reach Neptune.

11. The radius of the sun is about 695,202 km (kilometers). That is about 109 times as big as the radius of the Earth. What is the radius of the earth?

    Step 1:
    Step 2:

    Step 3:

    Step 4:

    Step 5:   The radius of the earth is _____ km.

12. The perimeter of a triangle is 105 cm. If each of the two legs is exactly twice the length of the base, how long is each leg?

    Step 1:   Let $x =$
               Draw a picture.

    Step 2:

    Step 3:

    Step 4:

    Step 5:   Each leg is _____ cm long. The base is _____ .

13. A lumber company has contracted to cut boards into two pieces so that one piece is three times the length of the other piece. If a board is 12 feet long, what is the length of each piece after cutting?

    Step 1:
    Step 2:

    Step 3:

    Step 4:

    Step 5:   The length of the shorter piece is _____ feet, and the length of the longer piece is _____ feet.

14. A student doing a chemistry experiment has a beaker that contains 84 ml (milliliters) of an alcohol and water solution. Her lab directions tell her that there is 4.6 times as much water as alcohol in the solution. How many milliliters of alcohol are in the solution? How many milliliters of water?

    Step 1:
    Step 2:

    Step 3:

    Step 4:

    Step 5:   There are _____ ml of alcohol in the solution. There are _____ ml of water in the solution.

15. A statistician is collecting data to help him estimate the average income of accountants in California. He needs to collect 390 pieces of data and he is $\frac{2}{3}$ done. How many pieces of data has the statistician collected?

    Step 1:
    Step 2:

    Step 3:

    Step 4:

    Step 5:   The statistician has collected _____ pieces of data.

    Suppose the statistician is 4 pieces of data short of being $\frac{2}{3}$ done. How many pieces of data has he collected?

16. A television commercial advertises that a certain type of battery will last, on the average, 20 hours longer than twice the life of another type of battery. If consumer tests show that the advertised battery lasts 725 hours, how many hours must the other type of battery last for the advertiser's claim to be valid?

    Step 1:
    Step 2:

    Step 3:

    Step 4:

    Step 5:   The other type of battery must last _____ hours for the advertiser's claim to be valid.

17. A 1000-ml flask containing a chloride solution will fill 3 beakers of the same size with 210 ml of the solution left over. How many milliliters of the chloride solution will each beaker hold?

Step 1:

Step 2:

Step 3:

Step 4:

Step 5: Each beaker will hold _____ ml of the chloride solution.

18. A star burns $\frac{2}{9}$ of its original mass then blows off $\frac{3}{7}$ of the remaining mass as a planetary nebula. If the final mass is 3 units of mass, what was the original mass?

Step 1:

Step 2:

Step 3:

Step 4:

Step 5: The original mass was _____ units of mass.

Continue using the five-step procedure for problems 19–33.

19. The sum of a number and sixteen is forty-two. What is the number?

20. When eleven is subtracted from a number, the result is 85. What is the number?

21. Three times a number is divided by 6 and the result is 10.5. What is the number?

22. When a number is multiplied by itself, the result is 144. What is the number?

23. A number is tripled, then increased by seven. The result is 48. What is the number?

24. Eight times a number is decreased by three times the number, giving a difference of 22. What is the number?

25. One number is fifteen more than another number. The sum of the two numbers is 27. What are they?

26. The length of a rectangle is 6 meters more than three times the width. The perimeter of the rectangle is 44 meters. What are the dimensions of the rectangle?

27. Seven is added to the product of 41 and some number. The result, when divided by four, is 63. What is the number?

28. The second side of a triangle is five times the length of the smallest side. The third is twice the length of the second side. The perimeter of the triangle is 48 inches. Find the length of each side.

29. Person A is four times as old as person B, who is six times as old as person C, who is twice as old as person D. How old is each person if their combined ages are 189 months?

30. Two consecutive odd integers sum to 151. What are they?

31. Three consecutive integers sum to 36. What are they?

32. Three consecutive even integers add up to 131. What are they?

(a) If two days of earth time pass, how many days actually pass on the spacecraft?

(b) If 30 years of earth time pass, how many years have actually passed on the spacecraft?

Step 1:

Step 2:

Step 3:

33. As a consequence of Einstein's theory of relativity, the rate of time passage is different for a person in a stationary position and a person in motion. (Hard to believe, but true!) To the moving observer, the rate of time passage is slower than that of the stationary observer, that is, the moving person ages slower than the stationary observer. (This fact has been proven many times by experiments with radioactive materials.) The effect is called "time dilation" and is really only noticeable when an object is traveling at near the speed of light (186,000 miles per second). Considering these ideas, try to solve the following problems:

Step 4:

Step 5: _____ years have passed on the spacecraft.

(c) If 30 years have passed on the spacecraft, how many years have passed on the earth?

Two people have identical clocks. One is standing on the earth and the other is moving in a spacecraft at 95% the speed of light, 176,700 miles per second. The moving person's rate of time passage at this speed is about 0.31 times as fast as the person standing on earth.

(d) A space traveler makes a round-trip voyage to the star Capella. The trip takes her 120 years (traveling at 176,000 miles per second). If it is the year 2000 on earth when she leaves, what earth year will it be when she returns?

EXERCISES FOR REVIEW

(3.7) **34.** Specify the domain of the equation $y = \dfrac{x-1}{x+4}$.

(4.1) **35.** Classify the equation $x + 4 = 1$ as an identity, a contradiction, or a conditional equation.

(4.1) **36.** Classify the equation $2x + 3 = 2x + 3$ as an identity, a contradiction, or a conditional equation.

(4.3) **37.** Solve the equation $4(x - 1) + 12 = -3(2x + 4)$.

(4.4) **38.** Translate the following sentence to a mathematical equation. Three less than an unknown number is multiplied by negative four. The result is two more than the original unknown number.

★ **Answers to Practice Sets (4.5)**

A. Last year's price was $19

B. The length of a side is 18 inches.

C. The number is 450.

D. Garden A produces 87 pounds of vegetables.

E. The two numbers are 248 and 250.

4.6 Linear Inequalities in One Variable

Section Overview

- ☐ **INEQUALITIES**
- ☐ **LINEAR INEQUALITIES**
- ☐ **THE ALGEBRA OF LINEAR INEQUALITIES**
- ☐ **COMPOUND INEQUALITIES**

☐ INEQUALITIES

Relationships of Inequality

We have discovered that an equation is a mathematical way of expressing the relationship of equality between quantities. Not all relationships need be relationships of equality, however. Certainly the number of human beings on earth is greater than 20. Also, the average American consumes less than 10 grams of vitamin C every day. These types of relationships are not relationships of equality, but rather, relationships of **inequality.**

☐ LINEAR INEQUALITIES

Linear Inequality

A **linear inequality** is a mathematical statement that one linear expression is greater than or less than another linear expression.

Inequality Notation

The following notation is used to express relationships of inequality:

$>$	Strictly greater than
$<$	Strictly less than
$\geq$	Greater than or equal to
$\leq$	Less than or equal to

Note that the expression $x > 12$ has infinitely many solutions. Any number strictly greater than 12 will satisfy the statement. Some solutions are 13, 15, 90, 12.1, 16.3, and 102.51.

☆ SAMPLE SET A

The following *are* linear inequalities in one variable.

1. $x \leq 12$

2. $x + 7 > 4$

3. $y + 3 \geq 2y - 7$

4. $P + 26 < 10(4P - 6)$

5. $\dfrac{2r - 9}{5} > 15$

The following *are not* linear inequalities in one variable.

6. $x^2 < 4.$ The term x^2 is quadratic, not linear.

7. $x \leq 5y + 3.$ There are two variables. This is a linear inequality in two variables.

8. $y + 1 \neq 5.$ Although the symbol $\neq$ certainly expresses an inequality, it is customary to use only the symbols $<, >, \leq, \geq$.

★ **PRACTICE SET A**

A linear equation, we know, may have exactly one solution, infinitely many solutions, or no solution. Speculate on the number of solutions of a linear inequality. (*Hint:* Consider the inequalities $x < x - 6$ and $x \geq 9$.)

A linear inequality may have _____ solutions, or no solutions.

☐ THE ALGEBRA OF LINEAR INEQUALITIES

Inequalities can be solved by basically the same methods as linear equations. There is one important exception that we will discuss in item 3 of the algebra of linear inequalities.

THE ALGEBRA OF LINEAR INEQUALITIES

Let a, b, and c represent real numbers and assume that

$a < b$ (or $a > b$)

Then, if $a < b$,

1. $a + c < b + c$ and $a - c < b - c$.

   If any real number is added to or subtracted from both sides of an inequality, the sense of the inequality remains unchanged.
2. If c is a *positive* real number, then if $a < b$,

   $ac < bc$ and $\dfrac{a}{c} < \dfrac{b}{c}$.

   If both sides of an inequality are multiplied or divided by the same positive number the sense of the inequality remains unchanged.
3. If c is a *negative* real number, then if $a < b$,

   $ac > bc$ and $\dfrac{a}{c} > \dfrac{b}{c}$.

   If both sides of an inequality are multiplied or divided by the same *negative* number, **the inequality sign must be reversed** (change direction) in order for the resulting inequality to be equivalent to the original inequality. (See problem 4 in the next set of examples.)

For example, consider the inequality $3 < 7$.

1. For $3 < 7$, if 8 is added to both sides, we get

   $3 + 8 < 7 + 8$.
   $11 < 15$ True

2. For $3 < 7$, if 8 is subtracted from both sides, we get

   $3 - 8 < 7 - 8$
   $-5 < -1$ True

3. For $3 < 7$, if both sides are multiplied by 8 (a positive number), we get

   $8(3) < 8(7)$
   $24 < 56$ True

4. For $3 < 7$, if both sides are multiplied by -8 (a negative number), we get

   $(-8)3 > (-8)7$

Notice the change in direction of the inequality sign.

$-24 > -56$ True

If we had forgotten to reverse the direction of the inequality sign we would have obtained the incorrect statement $-24 < -56$.

5. For $3 < 7$, if both sides are divided by 8 (a positive number), we get

$\dfrac{3}{8} < \dfrac{7}{8}$ True

6. For $3 < 7$, if both sides are divided by -8 (a negative number), we get

$\dfrac{3}{-8} > \dfrac{7}{-8}$ True (since $-.375 > -.875$)

☆ SAMPLE SET B

Solve the following linear inequalities. Draw a number line and place a point at each solution.

1. $3x > 15$ **Divide both sides by 3. The 3 is a positive number, so we need not reverse the sense of the inequality.**

$x > 5$

Thus, all numbers strictly greater than 5 are solutions to the inequality $3x > 15$.

Recall that an open circle indicates that the point is not included.

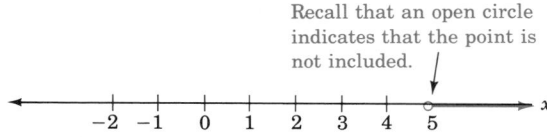

2. $2y - 1 \le 16$ **Add 1 to both sides.**
$2y \le 17$ **Divide both sides by 2.**

$y \le \dfrac{17}{2}$

Recall that a closed circle indicates that the point is included.

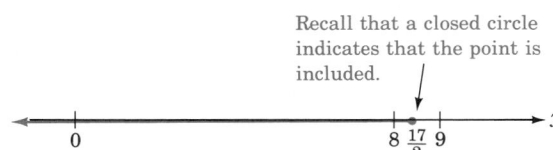

3. $-8x + 5 < 14$ **Subtract 5 from both sides.**
$-8x < 9$ **Divide both sides by -8. We must reverse the sense of the inequality since we are dividing by a negative number.**

$x > -\dfrac{9}{8}$

Continued

4. $5 - 3(y + 2) < 6y - 10$
$5 - 3y - 6 < 6y - 10$
$-3y - 1 < 6y - 10$
$-9y < -9$
$y > 1$

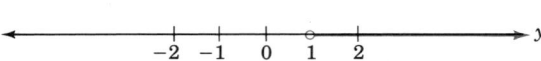

5. $\dfrac{2z + 7}{-4} \geq -6$ **Multiply by -4**

$2z + 7 \leq 24$ **Notice the change in the sense of the inequality.**
$2z \leq 17$

$z \leq \dfrac{17}{2}$

★ **PRACTICE SET B**

Solve the following linear inequalities.

1. $y - 6 \leq 5$

2. $x + 4 > 9$

3. $4x - 1 \geq 15$

4. $-5y + 16 \leq 7$

5. $7(4s - 3) < 2s + 8$

6. $5(1 - 4h) + 4 < (1 - h)2 + 6$

7. $18 \geq 4(2x - 3) - 9x$

8. $-\dfrac{3b}{16} \leq 4$

9. $\dfrac{-7z + 10}{-12} < -1$

10. $-x - \dfrac{2}{3} \leq \dfrac{5}{6}$

☐ COMPOUND INEQUALITIES

Compound Inequality

Another type of inequality is the *compound inequality*. A compound inequality is of the form:

$a < x < b$

There are actually two statements here. The first statement is $a < x$. The next statement is $x < b$. When we read this statement we say "a is less than x," then continue saying "and x is less than b."

Just by looking at the inequality we can see that the number x is between the numbers a and b. The compound inequality $a < x < b$ indicates "betweenness." Without changing the meaning, the statement $a < x$ can be read $x > a$. (Surely, if the number a is less than the number x, the number x must be greater than the number a.) Thus, we can read $a < x < b$ as "x is greater than a and at the same time is less than b." For example:

1. $4 < x < 9$.   The letter x is some number strictly between 4 and 9. Hence, x is greater than 4 and, at the same time, less than 9. The numbers 4 and 9 are not included so we use open circles at these points.

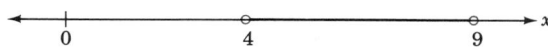

2. $-2 < z < 0$.   The z stands for some number between -2 and 0. Hence, z is greater than -2 but also less than 0.

3. $1 < x + 6 < 8$.   The expression $x + 6$ represents some number strictly between 1 and 8. Hence, $x + 6$ represents some number strictly greater than 1, but less than 8.

4. $\dfrac{1}{4} \le \dfrac{5x - 2}{6} \le \dfrac{7}{9}$.   The term $\dfrac{5x - 2}{6}$ represents some number between and including $\dfrac{1}{4}$ and $\dfrac{7}{9}$. Hence, $\dfrac{5x - 2}{6}$ represents some number greater than or equal to $\dfrac{1}{4}$ but less than or equal to $\dfrac{7}{9}$.

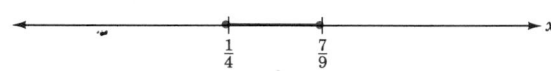

Consider problem 3 above, $1 < x + 6 < 8$. The statement says that the quantity $x + 6$ is between 1 and 8. This statement will be true for only certain values of x. For example, if $x = 1$, the statement is true since $1 < 1 + 6 < 8$. However, if $x = 4.9$, the statement is false since $1 < 4.9 + 6 < 8$ is clearly not true. The first of the inequalities is satisfied since 1 is less than 10.9, but the second inequality is not satisfied since 10.9 is not less than 8.

We would like to know for exactly which values of x the statement $1 < x + 6 < 8$ is true. We proceed by using the properties discussed earlier in this section, but now we must apply the rules to all *three* parts rather than just the two parts in a regular inequality.

☆ SAMPLE SET C

1. Solve $1 < x + 6 < 8$.

$1 - 6 < x + 6 - 6 < 8 - 6$ **Subtract 6 from all three parts.**
$-5 < x < 2$

Thus, if x is any number strictly between -5 and 2, the statement $1 < x + 6 < 8$ will be true.

Continued

2. Solve $-3 < \dfrac{-2x - 7}{5} < 8$.

$-3(5) < \dfrac{-2x - 7}{5}(5) < 8(5)$ Multiply each part by 5.

$-15 < -2x - 7 < 40$ Add 7 to all three parts.
$-8 < -2x < 47$ Divide all three parts by -2.

$4 > x > -\dfrac{47}{2}$ Remember to reverse the direction of the inequality signs.

$-\dfrac{47}{2} < x < 4$ It is customary (but not necessary) to write the inequality so that inequality arrows point to the left.

Thus, if x is any number between $-\dfrac{47}{2}$ and 4, the original inequality will be satisfied.

★ **PRACTICE SET C**

Find the values of x that satisfy the given continued inequality.

1. $4 < x - 5 < 12$ **2.** $-3 < 7y + 1 < 18$ **3.** $0 \le 1 - 6x \le 7$

4. $-5 \le \dfrac{2x + 1}{3} \le 10$ **5.** $9 < \dfrac{-4x + 5}{-2} < 14$ **6.** Does $4 < x < -1$ have a solution?

Answers to Practice Sets are on p. 179.

Section 4.6 EXERCISES

For problems 1–50, solve the inequalities.

1. $x + 7 < 12$ **7.** $2z + 8 < 7$

2. $y - 5 \le 8$ **8.** $4x - 14 > 21$

3. $y + 19 \ge 2$ **9.** $-5x \le 20$

4. $x - 5 > 16$ **10.** $-8x < 40$

5. $3x - 7 \le 8$ **11.** $-7z < 77$

6. $9y - 12 \le 6$ **12.** $-3y > 39$

13. $\frac{x}{4} \geq 12$

22. $\frac{14y}{-3} \geq -18$

14. $\frac{y}{7} > 3$

23. $\frac{21y}{-8} < -2$

15. $\frac{2x}{9} \geq 4$

24. $-3x + 7 \leq -5$

16. $\frac{5y}{2} \geq 15$

25. $-7y + 10 \leq -4$

17. $\frac{10x}{3} \leq 4$

26. $6x - 11 < 31$

18. $\frac{-5y}{4} < 8$

27. $3x - 15 \leq 30$

19. $\frac{-12b}{5} < 24$

28. $-2y + \frac{4}{3} \leq -\frac{2}{3}$

20. $\frac{-6a}{7} \leq -24$

29. $5(2x - 5) \geq 15$

21. $\frac{8x}{-5} > 6$

30. $4(x + 1) > -12$

31. $6(3x - 7) \geq 48$

39. $3x - 12 \geq 7x + 4$

32. $3(-x + 3) > -27$

40. $-2x - 7 > 5x$

33. $-4(y + 3) > 0$

41. $-x - 4 > -3x + 12$

34. $-7(x - 77) \leq 0$

42. $3 - x \geq 4$

35. $2x - 1 < x + 5$

43. $5 - y \leq 14$

36. $6y + 12 \leq 5y - 1$

44. $2 - 4x \leq -3 + x$

37. $3x + 2 \leq 2x - 5$

45. $3[4 + 5(x + 1)] < -3$

38. $4x + 5 > 5x - 11$

46. $2[6 + 2(3x - 7)] \geq 4$

47. $7[-3 - 4(x - 1)] \leq 91$

48. $-2(4x - 1) < 3(5x + 8)$

49. $-5(3x - 2) > -3(-x - 15) + 1$

50. 🖩 $-.0091x \geq 2.885x - 12.014$

51. What numbers satisfy the condition: twice a number plus one is greater than negative three?

52. What numbers satisfy the condition: eight more than three times a number is less than or equal to fourteen?

53. One number is five times larger than another number. The difference between these two numbers is less than twenty-four. What are the largest possible values for the two numbers? Is there a smallest possible value for either number?

54. The area of a rectangle is found by multiplying the length of the rectangle by the width of the rectangle. If the length of a rectangle is 8 feet, what is the largest possible measure for the width if it must be an integer (positive whole number) and the area must be less than 48 square feet?

EXERCISES FOR REVIEW

(1.6) **55.** Simplify $(x^2 y^3 z^2)^5$.

(2.2) **56.** Simplify $-[-(-|-8|)]$.

(3.5) **57.** Find the product. $(2x - 7)(x + 4)$.

(4.5) **58.** Twenty-five percent of a number is 12.32. What is the number?

(4.5) **59.** The perimeter of a triangle is 40 inches. If the length of each of the two legs is exactly twice the length of the base, how long is each leg?

★ **Answers to Practice Sets (4.6)**

A. infinitely many

B. **1.** $y \leq 11$　**2.** $x > 5$　**3.** $x \geq 4$　**4.** $y \geq \dfrac{9}{5}$　**5.** $s < \dfrac{29}{2}$　**6.** $h > \dfrac{1}{18}$　**7.** $x \geq -30$

　　8. $b \geq \dfrac{-64}{3}$　**9.** $z < -\dfrac{2}{7}$　**10.** $x \geq \dfrac{-3}{2}$

C. **1.** $9 < x < 17$　**2.** $-\dfrac{4}{7} < y < \dfrac{17}{7}$　**3.** $-1 \leq x \leq \dfrac{1}{6}$　**4.** $-8 \leq x \leq \dfrac{29}{2}$　**5.** $\dfrac{23}{4} < x < \dfrac{33}{4}$　**6.** no

4.7 Linear Equations in Two Variables

☐ **SOLUTIONS TO LINEAR EQUATIONS IN TWO VARIABLES**
☐ **ORDERED PAIRS AS SOLUTIONS**

☐ SOLUTIONS TO LINEAR EQUATIONS IN TWO VARIABLES

Solution to an Equation in Two Variables

We have discovered that an equation is a mathematical way of expressing the relationship of equality between quantities. If the relationship is between two quantities, the equation will contain two variables. We say that an equation in two variables has a solution if an ordered *pair* of values can be found such that when these two values are substituted into the equation a true statement results. This is illustrated when we observe some solutions to the equation $y = 2x + 5$.

1. $x = 4, y = 13$;   since $13 = 2(4) + 5$ is true.
2. $x = 1, y = 7$;   since $7 = 2(1) + 5$ is true.
3. $x = 0, y = 5$;   since $5 = 2(0) + 5$ is true.
4. $x = -6, y = -7$;   since $-7 = 2(-6) + 5$ is true.

☐ ORDERED PAIRS AS SOLUTIONS

It is important to keep in mind that a solution to a linear equation in two variables is an ordered pair of values, one value for each variable. A solution is not completely known until the values of *both* variables are specified.

Independent and Dependent Variables

Recall that, in an equation, any variable whose value can be freely assigned is said to be an **independent variable.** Any variable whose value is determined once the other value or values have been assigned is said to be a **dependent variable.** If, in a linear equation, the independent variable is x and the dependent variable is y, and a solution to the equation is $x = a$ and $y = b$, the solution is written as the

ORDERED PAIR (a, b)

Ordered Pair

In an **ordered pair,** (a, b), the first component, a, gives the value of the independent variable, and the second component, b, gives the value of the dependent variable.

$$(a, b)$$

Value of the independent variable / First component   Value of the dependent variable / Second component

We can use ordered pairs to show some solutions to the equation $y = 6x - 7$.

1. $(0, -7)$.   If $x = 0$ and $y = -7$, we get a true statement upon substitution and computation.

$$y = 6x - 7$$
$$-7 \stackrel{?}{=} 6(0) - 7$$
$$-7 = -7 \qquad \text{True}$$

2. (8, 41).   If $x = 8$ and $y = 41$, we get a true statement upon substitution and computation.

$y = 6x - 7$
$41 \stackrel{?}{=} 6(8) - 7$
$41 \stackrel{?}{=} 48 - 7$
$41 \stackrel{\checkmark}{=} 41$ True

3. $(-4, -31)$.   If $x = -4$ and $y = -31$, we get a true statement upon substitution and computation.

$y = 6x - 7$
$-31 \stackrel{?}{=} 6(-4) - 7$
$-31 \stackrel{?}{=} -24 - 7$
$-31 \stackrel{\checkmark}{=} -31$ True

These are only three of the infinitely many solutions to this equation.

☆ SAMPLE SET A

Find a solution to each of the following linear equations in two variables and write the solution as an ordered pair.

1. $y = 3x - 6$, if $x = 1$

Substitute 1 for x, compute, and solve for y.

$y = 3(1) - 6$
  $= 3 - 6$
  $= -3$

Hence, one solution is $(1, -3)$.

2. $y = 15 - 4x$, if $x = -10$

Substitute -10 for x, compute, and solve for y.

$y = 15 - 4(-10)$
  $= 15 + 40$
  $= 55$

Hence, one solution is $(-10, 55)$.

3. $b = -9a + 21$, if $a = 2$

Substitute 2 for a, compute, and solve for b.

$b = -9(2) + 21$
  $= -18 + 21$
  $= 3$

Hence, one solution is $(2, 3)$.

4. $5x - 2y = 1$, if $x = 0$

Substitute 0 for x, compute, and solve for y.

$5(0) - 2y = 1$
$0 - 2y = 1$
$-2y = 1$
$y = -\dfrac{1}{2}$

Hence, one solution is $\left(0, -\dfrac{1}{2}\right)$.

★ **PRACTICE SET A**

Find a solution to each of the following linear equations in two variables and write the solution as an ordered pair.

1. $y = 7x - 20$, if $x = 3$ **2.** $m = -6n + 1$, if $n = 2$ **3.** $b = 3a - 7$, if $a = 0$

4. $10x - 5y - 20 = 0$, if $x = -8$ **5.** $3a + 2b + 6 = 0$, if $a = -1$

Answers to the Practice Set are on page 186.

Section 4.7 EXERCISES

For problems 1–38, solve the linear equations in two variables.

1. $y = 8x + 14$, if $x = 1$

2. $y = -2x + 1$, if $x = 0$

3. $y = 5x + 6$, if $x = 4$

4. $x + y = 7$, if $x = 8$

5. $3x + 4y = 0$, if $x = -3$

6. $-2x + y = 1$, if $x = \frac{1}{2}$

7. $5x - 3y + 1 = 0$, if $x = -6$

8. $-4x - 4y = 4$, if $y = 7$

9. $2x + 6y = 1$, if $y = 0$

10. $-x - y = 0$, if $y = \frac{14}{3}$

11. $y = x$, if $x = 1$

21. $y = 6(x - 7)$, if $x = 2$

12. $x + y = 0$, if $x = 0$

22. $y = 2(4x + 5)$, if $x = -1$

13. $y + \dfrac{3}{4} = x$, if $x = \dfrac{9}{4}$

23. $5y = 9(x - 3)$, if $x = 2$

14. $y + 17 = x$, if $x = -12$

15. $-20y + 14x = 1$, if $x = 8$

24. $3y = 4(4x + 1)$, if $x = -3$

16. $\dfrac{3}{5}y + \dfrac{1}{4}x = \dfrac{1}{2}$, if $x = -3$

25. $-2y = 3(2x - 5)$, if $x = 6$

17. $\dfrac{1}{5}x + y = -9$, if $y = -1$

26. $-8y = 7(8x + 2)$, if $x = 0$

18. $y + 7 - x = 0$, if $x = \star$

27. $b = 4a - 12$, if $a = -7$

19. $2x + 31y - 3 = 0$, if $x = a$

20. $436x + 189y = 881$, if $x = -4231$

28. $b = -5a + 21$, if $a = -9$

29. $4b - 6 = 2a + 1$, if $a = 0$

30. $-5m + 11 = n + 1$, if $n = 4$

31. $3(t + 2) = 4(s - 9)$, if $s = 1$

32. $7(t - 6) = 10(2 - s)$, if $s = 5$

33. $y = 0x + 5$, if $x = 1$

34. $2y = 0x - 11$, if $x = -7$

35. $-y = 0x + 10$, if $x = 3$

36. $-5y = 0x - 1$, if $x = 0$

37. $y = 0(x - 1) + 6$, if $x = 1$

38. $y = 0(3x + 9) - 1$, if $x = 12$

Calculator Problems

39. An examination of the winning speeds in the Indianapolis 500 automobile race from 1961 to 1970 produces the equation $y = 1.93x + 137.60$, where x is the number of years from 1960 and y is the winning speed. Statistical methods were used to obtain the equation, and, for a given year, the equation gives only the approximate winning speed. Use the equation $y = 1.93x + 137.60$ to find the approximate winning speed in

(a) 1965   (b) 1970
(c) 1986   (d) 1990

40. In electricity theory, Ohm's law relates electrical current to voltage by the equation $y = 0.00082x$, where x is the voltage in volts and y is the current in amperes. This equation was found by statistical methods and for a given voltage yields only an approximate value for the current. Use the equation $y = 0.00082x$ to find the approximate current for a voltage of

(a) 6 volts   (b) 10 volts

41. Statistical methods have been used to obtain a relationship between the actual and reported number of German submarines sunk each month by the U.S. Navy in World War II. The equation expressing the approximate number of actual sinkings, y, for a given number of reported sinkings, x, is $y = 1.04x + 0.76$. Find the approximate number of actual sinkings of German submarines if the reported number of sinkings is

(a) 4   (b) 9   (c) 10

duces appears to be related to the speed at which the machine is running. Statistical methods found that the equation $y = 0.73x - 0.86$ is able to give the approximate number of defective items, y, for a given machine speed, x. Use this equation to find the approximate number of defective items for a machine speed of

(a) 9   (b) 12

42. Statistical methods have been used to obtain a relationship between the heart weight (in milligrams) and the body weight (in milligrams) of 10-month-old diabetic offspring of crossbred male mice. The equation expressing the approximate body weight for a given heart weight is $y = 0.213x - 4.44$. Find the approximate body weight for a heart weight of

(a) 210 mg   (b) 245 mg

45. A computer company has found, using statistical techniques, that there is a relationship between the aptitude test scores of assembly line workers and their productivity. Using data accumulated over a period of time, the equation $y = 0.89x - 41.78$ was derived. The x represents an aptitude test score and y the approximate corresponding number of items assembled per hour. Estimate the number of items produced by a worker with an aptitude score of

(a) 80   (b) 95

43. Statistical methods have been used to produce the equation $y = 0.176x - 0.64$. This equation gives the approximate red blood cell count (in millions) of a dog's blood, y, for a given packed cell volume (in millimeters), x. Find the approximate red blood cell count for a packed cell volume of

(a) 40 mm   (b) 42 mm

46. Chemists, making use of statistical techniques, have been able to express the approximate weight of potassium bromide, W, that will dissolve in 100 grams of water at T degrees centigrade. The equation expressing this relationship is $W = 0.52T + 54.2$. Use this equation to predict the potassium bromide weight that will dissolve in 100 grams of water that is heated to a temperature of

(a) 70 degrees centigrade
(b) 95 degrees centigrade

44. An industrial machine can run at different speeds. The machine also produces defective items, and the number of defective items it pro-

47. The marketing department at a large company has been able to express the relationship between the demand for a product and its price by using statistical techniques. The department found, by analyzing studies done in six different market areas, that the equation giving the approximate demand for a product (in thousands of units) for a particular price (in cents) is $y = -14.15x + 257.11$. Find the approximate number of units demanded when the price is

(a) \$0.12   (b) \$0.15

48. The management of a speed-reading program claims that the approximate speed gain (in words per minute), G, is related to the number of weeks spent in its program, W, is given by the equation $G = 26.68W - 7.44$. Predict the approximate speed gain for a student who has spent

(a) 3 weeks in the program
(b) 10 weeks in the program

EXERCISES
FOR REVIEW

(3.5)   **49.** Find the product. $(4x - 1)(3x + 5)$.
(3.6)   **50.** Find the product. $(5x + 2)(5x - 2)$.
(4.4)   **51.** Solve the equation $6[2(x - 4) + 1] = 3[2(x - 7)]$.
(4.6)   **52.** Solve the inequality $-3a - (a - 5) \geq a + 10$.
(4.6)   **53.** Solve the compound inequality $-1 < 4y + 11 < 27$.

★ **Answers to Practice Set (4.7)**

A. 1. $(3, 1)$ **2.** $(2, -11)$ **3.** $(0, -7)$ **4.** $(-8, -20)$ **5.** $\left(-1, \dfrac{-3}{2}\right)$

Identity (4.1)

An equation that is true for all acceptable values of the variable is called an *identity*. $x + 3 = x + 3$ is an identity.

Contradiction (4.1)

Contradictions are equations that are never true regardless of the value substituted for the variable. $x + 1 = x$ is a contradiction.

Conditional Equation (4.1)

An equation whose truth is conditional upon the value selected for the variable is called a *conditional equation.*

Solutions and Solving an Equation (4.1)

The collection of values that make an equation true are called the *solutions* of the equation. An equation is said to be *solved* when all its solutions have been found.

Equivalent Equations (4.1, 4.2)

Equations that have precisely the same collection of solutions are called *equivalent equations*.

An equivalent equation can be obtained from a particular equation by applying the *same* binary operation to *both* sides of the equation, that is,

1. adding or subtracting the *same* number to or from *both* sides of that particular equation.
2. multiplying or dividing *both* sides of that particular equation by the *same non-zero* number.

Literal Equation (4.1)

A *literal equation* is an equation that is composed of more than one variable.

Recognizing an Identity (4.3)

If, when solving an equation, all the variables are eliminated and a true statement results, the equation is an *identity.*

Recognizing a Contradiction (4.3)

If, when solving an equation, all the variables are eliminated and a false statement results, the equation is a *contradiction.*

Translating from Verbal to Mathematical Expressions (4.4)

When solving word problems it is absolutely necessary to know how certain words translate into mathematical symbols.

Five-Step Method for Solving Word Problems (4.5)

1. Let x (or some other letter) represent the unknown quantity.
2. Translate the words to mathematics and form an equation. A diagram may be helpful.
3. Solve the equation.
4. Check the solution by substituting the result into the original statement of the problem.
5. Write a conclusion.

Linear Inequality (4.6)

A *linear inequality* is a mathematical statement that one linear expression is greater than or less than another linear expression.

Inequality Notation (4.6)

$>$ Strictly greater than
$<$ Strictly less than
$\geq$ Greater than or equal to
$\leq$ Less than or equal to

Compound Inequality (4.6)

An inequality of the form

$$a < x < b$$

is called a *compound inequality.*

Solution to an Equation in Two Variables and Ordered Pairs (4.7)

A pair of values that when substituted into an equation in two variables produces a true statement is called a solution to the equation in two variables. These values are commonly written as an *ordered pair*. The expression (a, b) is an ordered pair. In an ordered pair, the independent variable is written first and the dependent variable is written second.

EXERCISE SUPPLEMENT

Sections 4.1 – 4.3

Solve the equations for problems 1–45.

1. $y + 3 = 11$

2. $a - 7 = 4$

3. $r - 1 = 16$

4. $a + 2 = 0$

5. $x + 6 = -4$

6. $x - 5 = -6$

7. $x + 8 = 8$

8. $y - 4 = 4$

9. $2x = 32$

10. $4x = 24$

11. $3r = -12$

12. $6m = -30$

13. $-5x = -30$

14. $-8y = -72$

15. $-x = 6$

16. $-y = -10$

17. $3x + 7 = 19$

18. $6x - 1 = 29$

19. $4x + 2 = -2$

20. $6x - 5 = -29$

21. $8x + 6 = -10$

22. $9a + 5 = -22$

23. $\dfrac{m}{6} + 4 = 8$

24. $\dfrac{b}{5} - 2 = 5$

25. $\dfrac{y}{9} = 54$

26. $\dfrac{a}{-3} = -17$

27. $\dfrac{c}{6} = 15$

28. $\dfrac{3a}{4} = 9$

29. $\dfrac{4y}{5} = -12$

30. $\dfrac{r}{4} = 7$

31. $\dfrac{6a}{-5} = 11$

32. $\dfrac{9x}{7} = 6$

33. $\dfrac{c}{2} - 8 = 0$

34. $\dfrac{m}{-5} + 4 = -1$

35. $\dfrac{x}{7} - 15 = -11$

36. $\dfrac{3x}{4} + 2 = 14$

37. $\dfrac{3r + 2}{5} = -1$

38. $\dfrac{6x - 1}{7} = -3$

39. $\dfrac{4x - 3}{6} + 2 = -6$

40. $\dfrac{y - 21}{8} = -3$

41. $4(x + 2) = 20$

42. $-2(a - 3) = 16$

43. $-7(2a - 1) = 63$

44. $3x + 7 = 5x - 21$

45. $-(8r + 1) = 33$

46. Solve $I = prt$ for t. Find the value of t when $I = 3500$, $p = 3000$, and $r = 0.05$.

47. Solve $A = LW$ for W. Find the value of W when $A = 26$ and $L = 2$.

48. Solve $\rho = mv$ for m. Find the value of m when $\rho = 4240$ and $v = 260$.

49. Solve $P = R - C$ for R. Find the value of R when $P = 480$ and $C = 210$.

50. Solve $P = \dfrac{nRT}{V}$ for n.

51. Solve $y = 5x + 8$ for x.

52. Solve $3y - 6x = 12$ for y.

53. Solve $4y + 2x + 8 = 0$ for y.

54. Solve $k = \dfrac{4m + 6}{7}$ for m.

55. Solve $t = \dfrac{10a - 3b}{2c}$ for b.

Section 4.4

For problems 56–70, translate the phrases or sentences to mathematical expressions or equations.

56. A quantity less eight.

57. A number, times four plus seven.

58. Negative ten minus some number.

59. Two fifths of a number minus five.

60. One seventh of a number plus two ninths of the number.

61. Three times a number is forty.

62. Twice a quantity plus nine is equal to the quantity plus sixty.

63. Four times a number minus five is divided by seven. The result is ten more than the number.

64. A number is added to itself five times, and that result is multiplied by eight. The entire result is twelve.

65. A number multiplied by eleven more than itself is six.

66. A quantity less three is divided by two more than the quantity itself. The result is one less than the original quantity.

67. A number is divided by twice the number, and eight times the number is added to that result. The result is negative one.

68. An unknown quantity is decreased by six. This result is then divided by twenty. Ten is subtracted from this result and negative two is obtained.

69. One less than some number is divided by five times the number. The result is the cube of the number.

70. Nine less than some number is multiplied by the number less nine. The result is the square of six times the number.

Section 4.5

For problems 71–80, find the solution.

71. This year an item costs $106, an increase of $10 over last year's price. What was last year's price?

72. The perimeter of a square is 44 inches. Find the length of a side.

73. Nine percent of a number is 77.4. What is the number?

74. Two consecutive integers sum to 63. What are they?

75. Four consecutive odd integers add to 56. What are they?

76. If twenty-one is subtracted from some number and that result is multiplied by two, the result is thirty-eight. What is the number?

77. If 37% more of a quantity is 159.1, what is the quantity?

78. A statistician is collecting data to help her estimate the number of pickpockets in a certain city. She needs 108 pieces of data and is $\dfrac{3}{4}$ done. How many pieces of data has she collected?

79. The statistician in problem 78 is eight pieces of data short of being $\dfrac{5}{6}$ done. How many pieces of data has she collected?

80. A television commercial advertises that a certain type of light bulb will last, on the average, 200 hours longer than three times the life of another type of bulb. If consumer tests show that the advertised bulb lasts 4700 hours, how many hours must the other type of bulb last for the advertiser's claim to be valid?

Section 4.6

Solve the inequalities for problems 81–100.

81. $y + 3 < 15$

82. $x - 6 \geq 12$

83. $4x + 3 > 23$

84. $5x - 14 < 1$

85. $6a - 6 \leq -27$

86. $-2y \geq 14$

87. $-8a \leq -88$

88. $\dfrac{x}{7} > -2$

89. $\dfrac{b}{-3} \leq 4$

90. $\dfrac{2a}{7} < 6$

91. $\dfrac{16c}{3} \ge -48$

92. $-4c + 3 \le 5$

93. $-11y + 4 > 15$

94. $3(4x - 5) > -6$

95. $-7(8x + 10) + 2 < -32$

96. $5x + 4 \ge 7x + 16$

97. $-x - 5 < 3x - 11$

98. $4(6x + 1) + 2 \ge -3(x - 1) + 4$

99. $-(5x + 6) + 2x - 1 < 3(1 - 4x) + 11$

100. What numbers satisfy the condition: nine less than negative four times a number is strictly greater than negative one?

Section 4.7

Solve the equations for problems 101–110.

101. $y = -5x + 4$, if $x = -3$

102. $y = -10x + 11$, if $x = -1$

103. $3a + 2b = 14$, if $b = 4$

104. $4m + 2k = 30$, if $m = 8$

105. $-4r + 5s = -16$, if $s = 0$

106. $y = -2(7x - 4)$, if $x = -1$

107. $-4a + 19 = 2(b + 6) - 5$, if $b = -1$

108. $6(t + 8) = -(a - 5)$, if $a = 10$

109. $-(a + b) = 5$, if $a = -5$

110. $-a(a + 1) = 2b + 1$, if $a = -2$

Solve the equations and inequalities for problems 1–16.

1. _____

1. (4.1) $x + 8 = 14$

2. _____

2. (4.1) $6a + 3 = -10$

3. _____

3. (4.2) $\dfrac{-3a}{8} = 6$

4. _____

4. (4.3) $\dfrac{x}{-2} + 16 = 11$

5. _____

5. (4.2) $\dfrac{y - 9}{4} + 6 = 3$

6. _____

6. (4.3) $5b - 8 = 7b + 12$

7. _____

7. (4.3) $3(2a + 4) = 2(a + 3)$

8. _____

8. (4.3) $5(y + 3) - (2y - 1) = -5$

9. _____

9. (4.2) $\dfrac{-(4x + 3 - 5x)}{3} = 2$

10. _____

10. (4.3) Solve $2p - 6q + 1 = -2$ for p.

11. _____

11. (4.2) Solve $p = \dfrac{nRT}{V}$ for T.

12. _____

12. (4.3) Solve $\dfrac{\triangle + \square}{\bigstar} = \nabla$ for $\triangle$.

13. _____

13. (4.6) $a - 8 \geq 4$

14. _____

14. (4.6) $-3a + 1 < -5$

15. _____

15. (4.6) $-2(a + 6) \leq -a + 11$

16. _____

16. (4.6) $\dfrac{-4x - 3}{3} > -9$

17. _____

18. _____

Translate the phrases or sentences into mathematical expressions or equations for problems 17–21.

17. (4.4) Three added to twice a number.

18. (4.4) Eight less than two thirds of a number.

19. _____

19. (4.4) Two more than four times a number.

20. _____

20. (4.4) A number is added to itself and this result is multiplied by the original number cubed. The result is twelve.

21. _____

21. (4.4) A number is decreased by five and that result is divided by ten more than the original number. The result is six times the original number.

22. _____

Solve problems 22–25.

22. (4.5) Eight percent of a number is 1.2. What is the number?

23. _____

23. (4.5) Three consecutive odd integers sum to 38. What are they?

24. _____

24. (4.5) Five more than three times a number is strictly less than seventeen. What is the number?

25. _____

25. (4.7) Solve $y = 8x - 11$ for y if $x = 3$, and write the solution as an ordered pair.

5

Factoring Poly-nomials

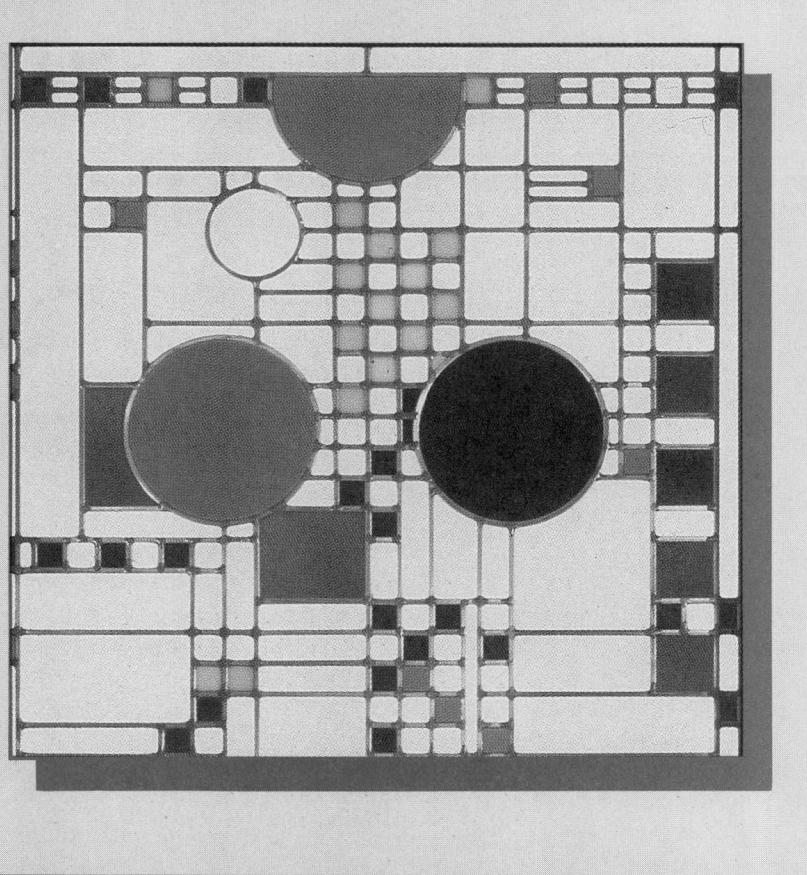

After completing this chapter, you should

Section 5.1 Finding the Factors of a Monomial
- be reminded of products of polynomials
- be able to determine a second factor of a polynomial given a first factor

Section 5.2 Factoring a Monomial from a Polynomial
- be able to factor a monomial from a polynomial

Section 5.3 The Greatest Common Factor
- understand more clearly the factorization process
- be able to determine the greatest common factor of two or more terms

Section 5.4 Factoring by Grouping
- know how to factor a polynomial using the grouping method and when to try the grouping method

Section 5.5 Factoring Two Special Products
- know the fundamental rules of factoring
- be able to factor the difference of two squares and perfect square trinomials

Section 5.6 Factoring Trinomials with Leading Coefficient 1
- be able to factor trinomials with leading coefficient 1
- become familiar with some factoring hints

Section 5.7 Factoring Trinomials with Leading Coefficient Other Than 1
- be able to factor trinomials with leading coefficient other than 1

5.1 Finding the Factors of a Monomial

□ **PRODUCTS OF POLYNOMIALS**
□ **FACTORING**

□ PRODUCTS OF POLYNOMIALS

Previously, we studied multiplication of polynomials (Section 3.5). We were given *factors* and asked to find their *product,* as shown below.

1. Given the factors 4 and 8, find the product. $4 \cdot 8 = 32$. The product is 32.
2. Given the factors $6x^2$ and $2x - 7$, find the product.

$$6x^2(2x - 7) = 12x^3 - 42x^2$$

The product is $12x^3 - 42x^2$.
3. Given the factors $x - 2y$ and $3x + y$, find the product.

$$(x - 2y)(3x + y) = 3x^2 + xy - 6xy - 2y^2$$
$$= 3x^2 - 5xy - 2y^2$$

The product is $3x^2 - 5xy - 2y^2$.
4. Given the factors $a + 8$ and $a + 8$, find the product.

$$(a + 8)^2 = a^2 + 16a + 64$$

The product is $a^2 + 16a + 64$.

□ FACTORING

Now, let's reverse the situation. We will be given the product, and we will try to find the factors. This process, which is the reverse of multiplication, is called *factoring*.

Factoring

> **Factoring** is the process of determining the factors of a given product.

☆ SAMPLE SET A

1. The number 24 is the product, and one factor is 6. What is the other factor?

We're looking for a number (   ) such that $6 \cdot ($ $) = 24$. We know from experience that (   ) = 4. As problems become progressively more complex, our experience may not give us the solution directly. We need a method for finding factors. To develop this method we can use the relatively simple problem $6 \cdot ($ $) = 24$ as a guide.

To find the number (   ), we would *divide* 24 by 6.

$$\frac{24}{6} = 4$$

The other factor is 4.

2. The product is $18x^3y^4z^2$ and one factor is $9xy^2z$. What is the other factor?

We know that since $9xy^2z$ is a factor of $18x^3y^4z^2$, there must be some quantity (   ) such that $9xy^2z \cdot ($ $) = 18x^3y^4z^2$. Dividing $18x^3y^4z^2$ by $9xy^2z$, we get

$$\frac{18x^3y^4z^2}{9xy^2z} = 2x^2y^2z$$

Thus, the other factor is $2x^2y^2z$.

Checking will convince us that $2x^2y^2z$ is indeed the proper factor.

$$(2x^2y^2z)(9xy^2z) = 18x^{2+1}y^{2+2}z^{1+1}$$
$$= 18x^3y^4z^2$$

We should try to find the quotient mentally and avoid actually writing the division problem.

3. The product is $-21a^5b^n$ and $3ab^4$ is a factor. Find the other factor.

Mentally dividing $-21a^5b^n$ by $3ab^4$, we get

$$\frac{-21a^5b^n}{3ab^4} = -7a^{5-1}b^{n-4} = -7a^4b^{n-4}$$

Thus, the other factor is $-7a^4b^{n-4}$.

★ PRACTICE SET A

1. The product is 84 and one factor is 6. What is the other factor?

2. The product is $14x^3y^2z^5$ and one factor is $7xyz$. What is the other factor?

Answers to the Practice Set are on p. 197.

Section 5.1 EXERCISES

In problems 1–34, the first quantity represents the product and the second quantity represents a factor of that product. Find the other factor.

1. 30, 6

2. 45, 9

3. $10a$, 5

4. $16a$, 8

5. $21b$, $7b$

6. $15a$, $5a$

7. $20x^3$, 4

8. $30y^4$, 6

9. $8x^4$, $4x$

10. $16y^5$, $2y$

11. $6x^2y$, $3x$

12. $9a^4b^5$, $9a^4$

13. $15x^2b^4c^7$, $5x^2bc^6$

14. $25a^3b^2c$, $5ac$

15. $18x^2b^5$, $-2xb^4$

16. $22b^8c^6d^3$, $-11b^8c^4$

17. $-60x^5b^3f^9$, $-15x^2b^2f^2$

18. $39x^4y^5z^{11}$, $3xy^3z^{10}$

19. $147a^{20}b^6c^{18}d^2$, $21a^3bd$

20. $-121a^6b^8c^{10}$, $11b^2c^5$

21. $\frac{1}{8}x^4y^3$, $\frac{1}{2}xy^3$

22. $7x^2y^3z^2$, $7x^2y^3z$

23. $5a^4b^7c^3d^2$, $5a^4b^7c^3d$

24. $14x^4y^3z^7$, $14x^4y^3z^7$

25. $12a^3b^2c^8$, $12a^3b^2c^8$

26. $6(a+1)^2(a+5)$, $3(a+1)^2$

27. $8(x+y)^3(x-2y)$, $2(x-2y)$

28. $14(a-3)^6(a+4)^2$, $2(a-3)^2(a+4)$

29. $26(x-5y)^{10}(x-3y)^{12}$, $-2(x-5y)^7(x-3y)^7$

30. $34(1-a)^4(1+a)^8$, $-17(1-a)^4(1+a)^2$

31. $(x+y)(x-y)$, $x-y$

32. $(a+3)(a-3)$, $a-3$

33. $48x^{n+3}y^{2n-1}$, $8x^3y^{n+5}$

34. $0.0024x^{4n}y^{3n+5}z^2$, $0.03x^{3n}y^5$

EXERCISES FOR REVIEW

(1.5) **35.** Simplify $(x^4y^0z^2)^3$.

(2.2) **36.** Simplify $-\{-[-(-|6|)]\}$.

(3.6) **37.** Find the product. $(2x - 4)^2$.

★ **Answers to Practice Set (5.1)**

A. **1.** 14   **2.** $2x^2yz^4$

5.2 Factoring a Monomial from a Polynomial

Section Overview

❑ **THE FACTORIZATION PROCESS**

❑ **THE FACTORIZATION PROCESS**

We introduce the process of factoring a monomial from a polynomial by examining a problem: Suppose that $12x^2 + 20x$ is the product and one of the factors is $4x$. To find the other factor we could set up the problem this way:

$$4x \cdot (\qquad) = 12x^2 + 20x$$

Since the product $12x^2 + 20x$ consists of two terms, the expression multiplying $4x$ must consist of two terms, since, by the distributive property

$$4x \cdot (\qquad) = 12x^2 + 20x$$

Now we see that this problem is simply an extension of finding the factors of a monomial.

1st term: $4x(\quad) = 12x^2$ *2nd term:* $4x(\quad) = 20x$

$(\quad) = \dfrac{12x^2}{4x}$ $(\quad) = \dfrac{20x}{4x}$

$(\quad) = 3x$ $(\quad) = 5$

Thus, $4x \cdot (3x + 5) = 12x^2 + 20x$.

Usually, these divisions can be done mentally and the terms of the factor filled in directly.

☆ **SAMPLE SET A**

1. The product is $3x^7 - 2x^6 + 4x^5 - 3x^4$ and one factor is x^4. Find the other factor.

We have the problem: x^4 times "what expression" yields $3x^7 - 2x^6 + 4x^5 - 3x^4$? Mathematically,

$$x^4 \cdot (\qquad) = 3x^7 - 2x^6 + 4x^5 - 3x^4$$

Since there are four terms in the product, there must be four terms inside the parentheses. To find each of the four terms, we'll divide (mentally) each term of the product by x^4. The resulting quotient will be the necessary term of the factor. *Continued*

1st term: $\dfrac{3x^7}{x^4} = 3x^{7-4} = 3x^3$ Place $3x^3$ into the 1st position in the (   ).

2nd term: $\dfrac{-2x^6}{x^4} = -2x^2$ Place $-2x^2$ into the 2nd position in the (   ).

3rd term: $\dfrac{4x^5}{x^4} = 4x$ Place $4x$ into the 3rd position in the (   ).

4th term: $\dfrac{-3x^4}{x^4} = -3$ Place -3 into the 4th position in the (   ).

Therefore, the other factor is $3x^3 - 2x^2 + 4x - 3$.

This result can be checked by applying the distributive property.

$$x^4 \cdot (3x^3 - 2x^2 + 4x - 3) \stackrel{?}{=} 3x^7 - 2x^6 + 4x^5 - 3x^4$$

$$x^4 \cdot (3x^3 - 2x^2 + 4x - 3) \stackrel{?}{=} 3x^7 - 2x^6 + 4x^5 - 3x^4$$

$$3x^{4+3} - 2x^{4+2} + 4x^{4+1} - 3x^4 \stackrel{?}{=} 3x^7 - 2x^6 + 4x^5 - 3x^4$$
$$3x^7 - 2x^6 + 4x^5 - 3x^4 \stackrel{\checkmark}{=} 3x^7 - 2x^6 + 4x^5 - 3x^4.$$

Thus,

$$x^4 \cdot (3x^3 - 2x^2 + 4x - 3) = 3x^7 - 2x^6 + 4x^5 - 3x^4$$

Again, if the divisions can be performed mentally, the process can proceed very quickly.

2. The product is $10x^3y^6 + 15x^3y^4 - 5x^2y^4$ and a factor is $5x^2y^4$. Find the other factor.

$$5x^2y^4 \cdot (\qquad) = 10x^3y^6 + 15x^3y^4 - 5x^2y^4$$

Since there are three terms in the product, there must be three terms inside the parentheses. To find each of these three terms, we'll divide each term of the product by $5x^2y^4$.

1st term: $\dfrac{10x^3y^6}{5x^2y^4} = 2xy^2$ Place the $2xy^2$ into the 1st position in the (   ).

2nd term: $\dfrac{15x^3y^4}{5x^2y^4} = 3x$ Place the $3x$ into the 2nd position in the (   ).

3rd term: $\dfrac{-5x^2y^4}{5x^2y^4} = -1$ Place the -1 into the 3rd position in the (   ).

The other factor is $2xy^2 + 3x - 1$, and

$$5x^2y^4 \cdot (2xy^2 + 3x - 1) = 10x^3y^6 + 15x^3y^4 - 5x^2y^4$$

3. The product is $-4a^2 - b^3 + 2c$ and a factor is -1. Find the other factor.

$$-1(\qquad) = -4a^2 - b^3 + 2c$$

Since there are three terms in the product, there must be three terms inside the parentheses. We will divide (mentally) each term of the product by -1.

1st term: $\dfrac{-4a^2}{-1} = 4a^2$ Place $4a^2$ into the 1st position inside the (   ).

2nd term: $\dfrac{-b^3}{-1} = b^3$ Place b^3 into the 2nd position inside the (   ).

3rd term: $\dfrac{2c}{-1} = -2c$ Place $-2c$ into the 3rd position inside the (   ).

The other factor is $4a^2 + b^3 - 2c$, and

$$-1(4a^2 + b^3 - 2c) = -4a^2 - b^3 + 2c$$

Without writing the -1, we get

$$-(4a^2 + b^3 - 2c) = -4a^2 - b^3 + 2c$$

4. The product is $-3a^2b^5 - 15a^3b^2 + 9a^2b^2$ and a factor is $-3a^2b^2$. Find the other factor.

$$-3a^2b^2(\qquad) = -3a^2b^5 - 15a^3b^2 + 9a^2b^2$$

Mentally dividing each term of the original trinomial by $-3a^2b^2$, we get $b^3 + 5a - 3$ as the other factor, and

$$-3a^2b^2(b^3 + 5a - 3) = -3a^2b^5 - 15a^3b^2 + 9a^2b^2$$

★ PRACTICE SET A

1. The product is $3x^2 - 6x$ and a factor is $3x$. Find the other factor.

2. The product is $5y^4 + 10y^3 - 15y^2$ and a factor is $5y^2$. Find the other factor.

3. The product is $4x^5y^3 - 8x^4y^4 + 16x^3y^5 + 24xy^7$ and a factor is $4xy^3$. Find the other factor.

4. The product is $-25a^4 - 35a^2 + 5$ and a factor is -5. Find the other factor.

5. The product is $-a^2 + b^2$ and a factor is -1. Find the other factor.

Answers to the Practice Set are on p. 202.

Section 5.2 EXERCISES

For problems 1–50, the first quantity represents the product and the second quantity a factor. Find the other factor.

1. $4x + 10$, 2

2. $6y + 18$, 3

3. $5x + 25$, 5

4. $16a + 64$, 8

5. $3a^2 + 9a$, $3a$

6. $14b^2 + 16b$, $2b$

7. $21x^2 + 28x$, $7x$

8. $45y^2 + 50y$, $5y$

9. $18a^2 - 4a$, $2a$

10. $20a^2 - 12a$, $\quad 4a$

11. $7x^2 - 14x$, $\quad 7x$

12. $6y^2 - 24y$, $\quad 6y$

13. $8a^2 + 4a$, $\quad 4a$

14. $26b^2 + 13b$, $\quad 13b$

15. $9x^2 + 6x + 18$, $\quad 6$

16. $12b^2 + 16b + 20$, $\quad 4$

17. $21x^2 + 7x - 14$, $\quad 7$

18. $35x^2 + 40x - 5$, $\quad 5$

19. $14y^2 - 28y + 14$, $\quad 14$

20. $36a^2 - 16a + 12$, $\quad 4$

21. $4y^2 - 10y - 12$, $\quad 2$

22. $6b^2 - 6b - 3$, $\quad 3$

23. $18x^3 + 20x$, $\quad 2x$

24. $40y^3 + 24y$, $\quad 4y$

25. $16x^3 - 12x^2$, $\quad 4x^2$

26. $11x^3 - 11x + 11$, $\quad 11$

27. $10a^3 + 12a^2 + 16a + 8$, $\quad 2$

28. $14b^3 + 16b^2 + 26b + 30$, $\quad 2$

29. $8a^3 - 4a^2 - 12a + 16$, $\quad 4$

30. $25x^3 - 30x^2 + 15x - 10$, $\quad 5$

31. $4x^6 + 16x^4 - 16x$, $\quad 4x$

32. $9a^5 + 6a^5 - 18a^4 + 24a^2$, $\quad 3a^2$

33. $10x^3 - 35x^2, \quad 5x^2$

34. $12x^3y^5 + 20x^3y^2, \quad 4x^3y^2$

35. $10a^4b^3 + 4a^3b^4, \quad 2a^3b^3$

36. $8a^3b^6c^8 + 12a^2b^5c^6 - 16a^2b^7c^5, \quad 4a^2b^5c^5$

37. $4x^5y^4 + x^2 + x, \quad x$

38. $14a^5b^2 - 3a^4b^4 + 7a^3, \quad a^3$

39. $64a^5b^3c^{11} + 56a^4b^4c^{10} - 48a^3b^5c^9 - 8a^3b^2c^5, \quad 8a^3b^2c^5$

40. $3h^3b^2 - 2h^6b^3 - 9h^2b + hb, \quad hb$

41. $5a + 10, \quad -5$

42. $6b + 8, \quad -2$

43. $8x^2 + 12x, \quad -4x$

44. $20a^2b^2 - 10a^2, \quad -10a^2$

45. $a + b, \quad -1$

46. $x + y, \quad -1$

47. $a - b + c, \quad -1$

48. $2x + 4y - z, \quad -1$

49. $-a - b - c, \quad -1$

50. $x^2 - x + 1, \quad -1$

EXERCISES FOR REVIEW

(3.1) **51.** How many $4y^2$'s are there in $24x^2y^3$?

(3.6) **52.** Find the product. $(2y - 3)^2$.

(4.3) **53.** Solve $2(2a - 1) - a = 7$.

(5.1) **54.** Given that $3m^2n$ is a factor of $12m^3n^4$, find the other factor.

★ **Answers to Practice Set (5.2)**

A. **1.** $x - 2$ **2.** $y^2 + 2y - 3$ **3.** $x^4 - 2x^3y + 4x^2y^2 + 6y^4$ **4.** $5a^4 + 7a^2 - 1$ **5.** $a^2 - b^2$

5.3 The Greatest Common Factor

Section Overview

☐ **FACTORING METHOD**
☐ **GREATEST COMMON FACTOR**

☐ FACTORING METHOD

In the last two types of problems (Sections 5.1 and 5.2), we knew one of the factors and were able to determine the other factor through division. Suppose, now, we're given the product without any factors. Our problem is to find the factors, if possible. This procedure and the previous two procedures are based on the distributive property.

$$a(b + c) = ab + ac$$

$$\underbrace{}_{\text{factors}} \quad \underbrace{}_{\text{product}}$$

We will use the distributive property in reverse.

$$\underbrace{ab + ac}_{\text{product}} = \underbrace{a(b + c)}_{\text{factors}}$$

We notice that in the product, a is common to both terms. (In fact, a is a common factor of both terms.) Since a is common to both terms, we will *factor it out* and write

$$a(\quad)$$

Now we need to determine what to place inside the parentheses. This is the procedure of the previous section. Divide each term of the product by the known factor a.

$$\frac{ab}{a} = b \quad \text{and} \quad \frac{ac}{a} = c$$

Thus, b and c are the required terms of the other factor. Hence,

$$ab + ac = a(b + c)$$

When factoring a monomial from a polynomial, we seek out factors that are not only common to each term of the polynomial, but factors that have these properties:

1. The numerical coefficients are the largest common numerical coefficients.
2. The variables possess the largest exponents common to all the variables.

☐ GREATEST COMMON FACTOR

Greatest Common Factor

A monomial factor that meets the above two requirements is called the **greatest common factor** of the polynomial.

☆ SAMPLE SET A

1. Factor $3x - 18$.   The greatest common factor is 3.

$3x - 18 = 3 \cdot x - 3 \cdot 6$ **Factor out 3.**

$3x - 18 = 3(\qquad)$ **Divide each term of the product by 3.**

$$\frac{3x}{3} = x \quad \text{and} \quad \frac{-18}{3} = -6$$

(Try to perform this division mentally.)

$3x - 18 = 3(x - 6)$

2. Factor $9x^3 + 18x^2 + 27x$.   Notice that $9x$ is the greatest common factor.

$9x^3 + 18x^2 + 27x = 9x \cdot x^2 + 9x \cdot 2x + 9x \cdot 3$.   **Factor out $9x$.**

$9x^3 + 18x^2 + 27x = 9x(\qquad)$ **Mentally divide $9x$ into each term of the product.**

$9x^3 + 18x^2 + 27x = 9x(x^2 + 2x + 3)$

3. Factor $10x^2y^3 - 20xy^4 - 35y^5$.   Notice that $5y^3$ is the greatest common factor. Factor out $5y^3$.

$10x^2y^3 - 20xy^4 - 35y^5 = 5y^3(\qquad)$

Mentally divide $5y^3$ into each term of the product and place the resulting quotients inside the ( ).

$10x^2y^3 - 20xy^4 - 35y^5 = 5y^3(2x^2 - 4xy - 7y^2)$

4. Factor $-12x^5 + 8x^3 - 4x^2$.   We see that the greatest common factor is $-4x^2$.

$-12x^5 + 8x^3 - 4x^2 = -4x^2(\qquad)$

Mentally dividing $-4x^2$ into each term of the product, we get

$-12x^5 + 8x^3 - 4x^2 = -4x^2(3x^3 - 2x + 1)$

★ PRACTICE SET A

1. Factor $4x - 48$.

2. Factor $6y^3 + 24y^2 + 36y$.

3. Factor $10a^5b^4 - 14a^4b^5 - 8b^6$.

4. Factor $-14m^4 + 28m^2 - 7m$.

Consider this problem: factor $Ax + Ay$. Surely, $Ax + Ay = A(x + y)$. We know from the very beginning of our study of algebra that letters represent single quantities. We also know that a quantity occurring within a set of parentheses is to be considered as a single quantity. Suppose that the letter A is representing the quantity $(a + b)$. Then we have

$$Ax + \qquad Ay = \qquad A(x + y)$$
$$(a + b)x + (a + b)y = (a + b)(x + y)$$

When we observe the expression

$$(a + b)x + (a + b)y$$

we notice that $(a + b)$ is common to both terms. Since it is common, we factor it out.

$$(a + b)(\qquad)$$

As usual, we determine what to place inside the parentheses by dividing each term of the product by $(a + b)$.

$$\frac{(a + b)x}{(a + b)} = x \qquad \text{and} \qquad \frac{(a + b)y}{(a + b)} = y$$

Thus, we get

$$(a + b)x + (a + b)y = (a + b)(x + y)$$

This is a forerunner of the factoring that will be done in Section 5.4.

☆ SAMPLE SET B

1. Factor $(x - 7)a + (x - 7)b$.   Notice that $(x - 7)$ is the greatest common factor. Factor out $(x - 7)$.

$$(x - 7)a + (x - 7)b = (x - 7)(\qquad)$$

Then, $\dfrac{(x - 7)a}{(x - 7)} = a$ and $\dfrac{(x - 7)b}{(x - 7)} = b$.

$$(x - 7)a + (x - 7)b = (x - 7)(a + b)$$

2. Factor $3x^2(x + 1) - 5x(x + 1)$.   Notice that x and $(x + 1)$ are common to both terms. Factor them out. We'll perform this factorization by letting $A = x(x + 1)$. Then we have

$$3xA - 5A = A(3x - 5)$$

But $A = x(x + 1)$, so

$$3x^2(x + 1) - 5x(x + 1) = x(x + 1)(3x - 5)$$

★ PRACTICE SET B

1. Factor $(y + 4)a + (y + 4)b$.   **2.** Factor $8m^3(n - 4) - 6m^2(n - 4)$.

Answers to Practice Sets are on p. 207.

Section 5.3 EXERCISES

For problems 1–43, factor the polynomials.

1. $9a + 18$

2. $6a + 24$

3. $8b + 12$

4. $16x + 12$

5. $4x - 6$

6. $8x - 14$

7. $21y - 28$

8. $16f - 36$

9. $12x^2 + 18x$

10. $10y^2 + 15y$

11. $8y^2 + 18$

12. $7x^2 - 21$

13. $3y^2 - 6$

14. $2x^2 - 2$

15. $6y^2 - 6y$

16. $ax^2 - a$

17. $by^2 + b$

18. $7by^2 + 14b$

19. $5a^2x^2 + 10x$

20. $24ax^2 + 28a$

21. $10x^2 + 5x - 15$

22. $12x^2 - 8x - 16$

23. $15y^3 - 24y + 9$

24. $ax^2 + ax + a$

25. $by^3 + by^2 + by + b$

26. $2y^2 + 6y + 4xy$

27. $9x^2 + 6xy + 4x$

28. $30a^2b^2 + 40a^2b^2 + 50a^2b^2$

29. $13x^2y^5c - 26x^2y^5c - 39x^2y^5$

30. $-4x^2 - 12x - 8$

31. $-6y^3 - 8y^2 - 14y + 10$

32. $Ab + Ac$

33. $Nx + Ny$

34. $Qx + Qy$

35. $Ax - Ay$

36. $(x + 4)b + (x + 4)c$

37. $(x - 9)a + (x - 9)b$

38. $(2x + 7)a + (2x + 7)b$

39. $(9a - b)w - (9a - b)x$

40. $(5 - v)X + (5 - v)Y$

41. $3x^5y^4 - 12x^3y^4 + 27x^5y^3 - 6x^2y^6$

42. $8a^3b^{15} + 24a^2b^{14} + 48a^3b^6 - 20a^3b^7 + 80a^4b^6 - 4a^3b^7 + 4a^2b$

43. $-8x^3y^2 - 3x^3y^2 + 16x^4y^3 + 2x^2y$

EXERCISES FOR REVIEW

(4.5)   **44.** A quantity plus 21% more of that quantity is 26.25. What is the original quantity?

(4.7)   **45.** Solve the equation $6(t-1) = 4(5-s)$ if $s = 2$.

(5.2)   **46.** Given that $4a^3$ is a factor of $8a^3 - 12a^2$, find the other factor.

★ **Answers to Practice Sets (5.3)**

A. **1.** $4(x-12)$ **2.** $6y(y^2 + 4y + 6)$ **3.** $2b^4(5a^5 - 7a^4b - 4b^2)$ **4.** $-7m(2m^3 - 4m + 1)$

B. **1.** $(y+4)(a+b)$ **2.** $2m^2(n-4)(4m-3)$

5.4 Factoring by Grouping

Section Overview
- ☐ **USING GROUPING TO FACTOR A POLYNOMIAL**
- ☐ **KNOWING WHEN TO TRY THE GROUPING METHOD**

☐ USING GROUPING TO FACTOR A POLYNOMIAL

Sometimes a polynomial will not have a particular factor common to every term. However, we may still be able to produce a factored form for the polynomial.

The polynomial $x^3 + 3x^2 - 6x - 18$ has no single factor that is common to every term. However, we notice that if we *group* together the first two terms and the second two terms, we see that each resulting binomial has a particular factor common to both terms.

$$\underbrace{x^3 + 3x^2}_{x^2 \text{ is common}} - \underbrace{6x - 18}_{-6 \text{ is common}}$$

Factor x^2 out of the first two terms, and factor -6 out of the second two terms.

$x^2(x+3) - 6(x+3)$

Now look closely at this binomial. Each of the two terms contains the factor $(x+3)$.

Factor out $(x+3)$.
$(x+3)(x^2-6)$ is the final factorization.

$x^3 + 3x^2 - 6x - 18 = (x+3)(x^2-6)$

☐ KNOWING WHEN TO TRY THE GROUPING METHOD

We are alerted to the idea of grouping when the polynomial we are considering has *either* of these qualities:

1. no factor common to *all* terms
2. an *even* number of terms

When factoring by grouping, the sign (+ or −) of the factor we are taking out will *usually* (but not always) be the same as the sign of the first term in that group.

☆ **SAMPLE SET A**

Factor $8a^2b^4 - 4b^4 + 14a^2 - 7$.

1. We notice there is no factor common to all terms.
2. We see there are four terms, an even number.
3. We see that terms 1 and 2 have $+4b^4$ in common (since the 1st term in the group is $+8a^2b^4$).
4. We notice that the 3rd and 4th terms have $+7$ in common (since the 1st term in the group is $+14a^2$).

$$8a^2b^4 - 4b^4 + 14a^2 - 7 = 4b^4(2a^2 - 1) + 7(2a^2 - 1)$$

$(2a^2 - 1)$ is common

$$8a^2b^4 - 4b^4 + 14a^2 - 7 = (2a^2 - 1)(4b^4 + 7)$$

★ **PRACTICE SET A**

Use the grouping method to factor the following polynomials.

1. $ax + ay + bx + by$ **2.** $2am + 8m + 5an + 20n$ **3.** $a^2x^3 + 4a^2y^3 + 3bx^3 + 12by^3$

4. $15mx + 10nx - 6my - 4ny$ **5.** $40abx - 24abxy - 35c^2x + 21c^2xy$

6. When factoring the polynomial $8a^2b^4 - 4b^4 + 14a^2 - 7$ in Sample Set A, we grouped together terms 1 and 2 and 3 and 4. Could we have grouped together terms 1 and 3 and 2 and 4? Try this.

$$8a^2b^4 - 4b^4 + 14a^2 - 7 =$$

Do we get the same result? If the results do not look precisely the same, recall the commutative property of multiplication.

Answers to the Practice Set are on p. 210.

Section 5.4 EXERCISES

For problems 1–23, use the grouping method to factor the polynomials. Some polynomials may not be factorable using the grouping method.

1. $2ab + 3a + 18b + 27$

2. $xy - 7x + 4y - 28$

3. $xy + x + 3y + 3$

4. $mp + 3mq + np + 3nq$

5. $ar + 4as + 5br + 20bs$

6. $14ax - 6bx + 21ay - 9by$

7. $12mx - 6bx + 21ay - 9by$

8. $36ak - 8ah - 27bk + 6bh$

9. $a^2b^2 + 2a^2 + 3b^2 + 6$

10. $3n^2 + 6n + 9m^3 + 12m$

11. $8y^4 - 5y^3 + 12z^2 - 10z$

12. $x^2 + 4x - 3y^2 + y$

13. $x^2 - 3x + xy - 3y$

14. $2n^2 + 12n - 5mn - 30m$

15. $4pq - 7p + 3q^2 - 21$

16. $8x^2 + 16xy - 5x - 10y$

17. $12s^2 - 27s - 8st + 18t$

18. $15x^2 - 12x - 10xy + 8y$

19. $a^4b^4 + 3a^5b^5 + 2a^2b^2 + 6a^3b^3$

20. $4a^3bc - 14a^2bc^3 + 10abc^2 - 35bc^4$

21. $5x^2y^3z + 3x^3yw - 10y^3z^2 - 6wxyz$

22. $a^3b^2cd + abc^2dx - a^2bxy - cx^2y$

23. $5m^{10}n^{17}p^3 - m^6n^7p^4 - 40m^4n^{10}qt^2 + 8pqt^2$

EXERCISES FOR REVIEW

(1.5)   **24.** Simplify $(x^5 y^3)(x^2 y)$.

(2.7)   **25.** Use scientific notation to find the product of $(3 \times 10^{-5})(2 \times 10^2)$.

(3.7)   **26.** Find the domain of the equation $y = \dfrac{6}{x + 5}$.

(4.7)   **27.** Construct the graph of the inequality $y \geq -2$.

$\longleftarrow\!\!\!\longrightarrow$

(5.3)   **28.** Factor $8a^4 b^4 + 12a^3 b^5 - 8a^2 b^3$.

★ **Answers to Practice Set (5.4)**

A. 1. $(a + b)(x + y)$ **2.** $(2m + 5n)(a + 4)$ **3.** $(a^2 + 3b)(x^3 + 4y^3)$ **4.** $(5x - 2y)(3m + 2n)$
5. $x(8ab - 7c^2)(5 - 3y)$ **6.** yes

5.5 Factoring Two Special Products

Section Overview

- ☐ **THE DIFFERENCE OF TWO SQUARES**
- ☐ **FUNDAMENTAL RULES OF FACTORING**
- ☐ **PERFECT SQUARE TRINOMIALS**

☐ THE DIFFERENCE OF TWO SQUARES

Recall that when we multiplied together the two binomials $(a + b)$ and $(a - b)$, we obtained the product $a^2 - b^2$.

$$(a + b)(a - b) = a^2 - b^2$$

Perfect Square

Notice that the terms a^2 and b^2 in the product can be produced by squaring a and b, respectively. A term that is the square of another term is called a **perfect square.** Thus, both a^2 and b^2 are perfect squares. The minus sign between a^2 and b^2 means that we are taking the *difference* of the two squares.

Since we know that $(a + b)(a - b) = a^2 - b^2$, we need only turn the equation around to find the factorization form.

$$a^2 - b^2 = (a + b)(a - b)$$

The factorization form says that we can factor $a^2 - b^2$, the difference of two squares, by finding the terms that produce the perfect squares and substituting these quantities into the factorization form.

When using real numbers (as we are), there is no factored form for the sum of two squares. That is, using real numbers,

$$a^2 + b^2 \; cannot \; \text{be factored}$$

☆ **SAMPLE SET A**

1. Factor $x^2 - 16$.   Both x^2 and 16 are perfect squares. The terms that, when squared, produce x^2 and 16 are x and 4, respectively. Thus,

   $$x^2 - 16 = (x + 4)(x - 4)$$

   We can check our factorization simply by multiplying.

   $$(x + 4)(x - 4) = x^2 - 4x + 4x - 16$$
   $$= x^2 - 16.$$

2. $49a^2b^4 - 121$.   Both $49a^2b^4$ and 121 are perfect squares. The terms that, when squared, produce $49a^2b^4$ and 121 are $7ab^2$ and 11, respectively. Substituting these terms into the factorization form we get

   $$49a^2b^4 - 121 = (7ab^2 + 11)(7ab^2 - 11)$$

   We can check our factorization by multiplying.

   $$(7ab^2 + 11)(7ab^2 - 11) = 49a^2b^4 - 11ab^2 + 11ab^2 - 121$$
   $$= 49a^2b^4 - 121$$

3. $3x^2 - 27$.   This doesn't look like the difference of two squares since we don't readily know the terms that produce $3x^2$ and 27. However, notice that 3 is common to both the terms. Factor out 3.

   $$3(x^2 - 9)$$

   Now we see that $x^2 - 9$ is the difference of two squares. Factoring the $x^2 - 9$ we get

   $$3x^2 - 27 = 3(x^2 - 9)$$
   $$= 3(x + 3)(x - 3)$$

   Be careful not to drop the factor 3.

★ **PRACTICE SET A**

If possible, factor the following binomials completely.

1. $m^2 - 25$

2. $36p^2 - 81q^2$

3. $49a^4 - b^2c^2$

4. $x^8y^4 - 100w^{12}$

5. $3x^2 - 75$

6. $a^3b^4m - am^3n^2$

❑ **FUNDAMENTAL RULES OF FACTORING**

There are two fundamental rules that we follow when factoring:

Fundamental Rules of
Factoring

1. Factor out all common monomials first.
2. Factor completely.

Factor each binomial completely.

1. $4a^8b - 36b^5$.   Factor out the common factor $4b$.

$4b(a^8 - 9b^4)$

Now we can see a difference of two squares, whereas in the original polynomial we could not. We'll complete our factorization by factoring the difference of two squares.

$$4a^8b - 36b^5 = 4b(a^8 - 9b^4)$$
$$= 4b(a^4 + 3b^2)(a^4 - 3b^2)$$

2. $x^{16} - y^8$.   Factor this difference of two squares.

$$x^{16} - y^8 = \underbrace{(x^8 + y^4)}\underbrace{(x^8 - y^4)}$$

Sum of two squares Difference of two squares
Does not factor$$Factor it!
$$= (x^8 + y^4)(x^4 + y^2)\ \underbrace{(x^4 - y^2)}$$
Factor again!
$$= (x^8 + y^4)(x^4 + y^2)(x^2 + y)(x^2 - y)$$

Finally, the factorization is complete.

These types of products appear from time to time, so be aware that you may have to factor more than once.

Factor each binomial completely.

1. $m^4 - n^4$ 2. $16y^8 - 1$

❑ PERFECT SQUARE TRINOMIALS

Recall the process of squaring a binomial.

$$(a + b)^2 = a^2 + 2ab + b^2 \qquad (a - b)^2 = a^2 - 2ab + b^2$$

Our Method Is	We Notice
Square the first term.	The first term of the product should be a perfect square.
Take the product of the two terms and double it.	The middle term of the product should be divisible by 2 (since it's multiplied by 2).
Square the last term.	The last term of the product should be a perfect square.

Perfect square trinomials *always* factor as the square of a binomial.

To recognize a perfect square trinomial, look for the following features:

1. The first and last terms are perfect squares.
2. The middle term is divisible by 2, and if we divide the middle term in half (the opposite of doubling it), we will get the product of the terms that when squared produce the first and last terms.

In other words, factoring a perfect square trinomial amounts to finding the terms that, when squared, produce the first and last terms of the trinomial, and substituting into one of the formula

$$a^2 + 2ab + b^2 = (a+b)^2 \qquad a^2 - 2ab + b^2 = (a-b)^2$$

☆ **SAMPLE SET C**

Factor each perfect square trinomial.

1. $x^2 + 6x + 9$.   This expression is a perfect square trinomial. The x^2 and 9 are perfect squares. The terms that when squared produce x^2 and 9 are x and 3, respectively.

   The middle term is divisible by 2, and $\dfrac{6x}{2} = 3x$. The $3x$ is the product of x and 3, which are the terms that produce the perfect squares.

   $x^2 + 6x + 9 = (x+3)^2$

2. $x^4 - 10x^2y^3 + 25y^6$.   This expression is a perfect square trinomial. The x^4 and $25y^6$ are both perfect squares. The terms that when squared produce x^4 and $25y^6$ are x^2 and $5y^3$, respectively.

   The middle term $-10x^2y^3$ is divisible by 2. In fact, $\dfrac{-10x^2y^3}{2} = -5x^2y^3$. Thus,

   $x^4 - 10x^2y^3 + 25y^6 = (x^2 - 5y^3)^2$

3. $x^2 + 10x + 16$.   This expression is *not* a perfect square trinomial. Although the middle term is divisible by 2, $\dfrac{10x}{2} = 5x$, the 5 and x are not the terms that when squared produce the first and last terms. (This expression would be a perfect square trinomial if the middle term were $8x$.)

4. $4a^4 + 32a^2b - 64b^2$.   This expression is *not* a perfect square trinomial since the last term $-64b^2$ is not a perfect square (since any quantity squared is always positive or zero and never negative).

   Thus, $4a^4 + 32a^2b - 64b^2$ cannot be factored using this method.

★ **PRACTICE SET C**

Factor, if possible, the following trinomials.

1. $m^2 - 8m + 16$ **2.** $k^2 + 10k + 25$ **3.** $4a^2 + 12a + 9$

4. $9x^2 - 24xy + 16y^2$ **5.** $2w^3z + 16w^2z^2 + 32wz^3$ **6.** $x^2 + 12x + 49$

Answers to Practice Sets are on p. 217.

Section 5.5 EXERCISES

For problems 1–45, factor the binomials.

1. $a^2 - 9$

2. $a^2 - 25$

3. $x^2 - 16$

4. $y^2 - 49$

5. $a^2 - 100$

6. $b^2 - 36$

7. $4a^2 - 64$

8. $2b^2 - 32$

9. $3x^2 - 27$

10. $5x^2 - 125$

11. $4a^2 - 25$

12. $9x^2 - 100$

13. $36y^2 - 25$

14. $121a^2 - 9$

15. $12a^2 - 75$

16. $10y^2 - 320$

17. $8y^2 - 50$

18. $a^2b^2 - 9$

19. $x^2y^2 - 25$

20. $x^4y^4 - 36$

21. $x^4y^4 - 9a^2$

22. $a^2b^4 - 16y^4$

23. $4a^2b^2 - 9b^2$

24. $16x^2 - 25y^2$

25. $a^2 - b^2$

26. $a^4 - b^4$

27. $x^4 - y^4$

28. $x^8 - y^2$

29. $a^8 - y^2$

30. $b^6 - y^2$

31. $b^6 - x^4$

32. $9 - x^2$

33. $25 - a^2$

34. $49 - 16a^2$

35. $100 - 36b^4$

36. $128 - 32x^2$

37. $x^4 - 16$

38. $2ab^3 - a^3b$

39. $a^4 - b^4$

40. $a^{16} - b^4$

41. $x^{12} - y^{12}$

42. $a^2c - 9c$

43. $a^3c^2 - 25ac^2$

44. $a^4b^4c^2d^2 - 36x^2y^2$

45. $49x^2y^4z^6 - 64a^4b^2c^8d^{10}$

For problems 46–78, factor, if possible, the trinomials.

46. $x^2 + 8x + 16$

47. $x^2 + 10x + 25$

48. $a^2 + 4a + 4$

49. $a^2 + 12a + 36$

50. $b^2 + 18b + 81$

51. $y^2 + 20y + 100$

52. $c^2 + 6c + 9$

53. $a^2 - 4a + 4$

54. $b^2 - 6b + 9$

55. $x^2 - 10x + 25$

56. $b^2 - 22b + 121$

57. $a^2 - 24a + 144$

58. $a^2 + 2a + 1$

59. $x^2 + 2x + 1$

60. $x^2 - 2x + 1$

61. $b^2 - 2b + 1$

62. $4a^2 + 12a + 9$

63. $9x^2 + 6x + 1$

64. $4x^2 + 28x + 49$

65. $16a^2 - 24a + 9$

66. $25a^2 - 20a + 4$

67. $9x^2 + 6xy + y^2$

68. $16x^2 + 24xy + 9y^2$

69. $36a^2 + 60ab + 25b^2$

70. $4x^2 - 12xy + 9y^2$

71. $12a^2 - 60a + 75$

72. $16x^2 + 8x + 1$

73. $32x^2 + 16x + 2$

74. $x^2 + x + 1$

75. $4a^2 + a + 9$

77. $x^5 + 8x^4 + 16x^3$

76. $9a^2 - 21a + 49$

78. $12a^3b - 48a^2b^2 + 48ab^3$

EXERCISES FOR REVIEW

(5.3)   **79.** Factor $(m - 3)x - (m - 3)y$.
(5.4)   **80.** Factor $8xm + 16xn + 3ym + 6yn$ by grouping.

★ Answers to Practice Sets (5.5)

A. **1.** $(m + 5)(m - 5)$ **2.** $9(2p - 3q)(2p + 3q)$ **3.** $(7a^2 + bc)(7a^2 - bc)$
   4. $(x^4y^2 + 10w^6)(x^4y^2 - 10w^6)$ **5.** $3(x + 5)(x - 5)$ **6.** $am(ab^2 + mn)(ab^2 - mn)$

B. **1.** $(m^2 + n^2)(m - n)(m + n)$ **2.** $(4y^4 + 1)(2y^2 + 1)(2y^2 - 1)$

C. **1.** $(m - 4)^2$ **2.** $(k + 5)^2$ **3.** $(2a + 3)^2$ **4.** $(3x - 4y)^2$ **5.** $2wz(w + 4z)^2$ **6.** not possible

5.6 Factoring Trinomials with Leading Coefficient 1

Section Overview

☐ **METHOD**
☐ **FACTORING HINTS**

☐ **METHOD**

Let's consider the product of the two binomials $(x + 4)$ and $(x + 7)$.

$$(x + 4)(x + 7) = x^2 + 7x + 4x + 28$$
$$= x^2 + 11x + 28$$

Notice that the *first term* in the resulting trinomial comes from the product of the first terms in the binomials: $x \cdot x = x^2$. The *last term* in the trinomial comes from the product of the last terms in the binomials: $4 \cdot 7 = 28$. The *middle term* comes from the addition of the outer and inner products: $7x + 4x = 11x$. Also, notice that the coefficient of the middle term is exactly the *sum* of the last terms in the binomials: $4 + 7 = 11$.

The problem we're interested in is that given a trinomial, how can we find the factors? When the leading coefficient (the coefficient of the quadratic term) is 1, the observations we made above lead us to the following method of factoring.

Method of Factoring

1. Write two sets of parentheses: (   )(   ).
2. Place a binomial into each set of parentheses. The first term of each binomial is a factor of the first term of the trinomial.
3. Determine the second terms of the binomials by determining the factors of the third term that when added together yield the coefficient of the middle term.

☆ **SAMPLE SET A**

Factor the following trinomials.

1. $x^2 + 5x + 6$

1. Write two sets of parentheses: ( )( ).
2. Place the factors of x^2 into the first position of each set of parentheses:

   $(x \quad)(x \quad)$

3. The third term of the trinomial is 6. We seek two numbers whose

   (a) product is 6 and
   (b) sum is 5.

   The required numbers are 3 and 2. Place $+3$ and $+2$ into the parentheses.

   $x^2 + 5x + 6 = (x + 3)(x + 2)$

   The factorization is complete. We'll check to be sure.

   $$(x + 3)(x + 2) = x^2 + 2x + 3x + 6$$
   $$= x^2 + 5x + 6$$

2. $y^2 - 2y - 24$

1. Write two sets of parentheses: ( )( ).
2. Place the factors of y^2 into the first position of each set of parentheses:

   $(y \quad)(y \quad)$

3. The third term of the trinomial is -24. We seek two numbers whose

   (a) product is -24 and
   (b) sum is -2.

   The required numbers are -6 and 4. Place -6 and $+4$ into the parentheses.

   $y^2 - 2y - 24 = (y - 6)(y + 4)$

   The factorization is complete. We'll check to be sure.

   $$(y - 6)(y + 4) = y^2 + 4y - 6y - 24$$
   $$= y^2 - 2y - 24$$

Notice that the other combinations of the factors of -24 (some of which are -2, 12; 3, -8; and -4, 6) do not work. For example,

$(y - 2)(y + 12) = y^2 + \boxed{10y} - 24$

$(y + 3)(y - 8) = y^2 - \boxed{5y} - 24$

$(y - 4)(y + 6) = y^2 + \boxed{2y} - 24$

In all of these equations, the middle terms are incorrect.

3. $a^2 - 11a + 30$

1. Write two sets of parentheses: ( )( ).
2. Place the factors of a^2 into the first position of each set of parentheses:

   $(a \quad)(a \quad)$

3. The third term of the trinomial is $+30$. We seek two numbers whose

(a) product is 30 and
(b) sum is -11.

The required numbers are -5 and -6. Place -5 and -6 into the parentheses.

$a^2 - 11a + 30 = (a - 5)(a - 6)$

The factorization is complete. We'll check to be sure.

$$(a - 5)(a - 6) = a^2 - 6a - 5a + 30$$
$$= a^2 - 11a + 30$$

4. $3x^2 - 15x - 42$

Before we begin, let's recall the most basic rule of factoring: *factor out common monomial factors first.* Notice that 3 is the greatest common monomial factor of *every* term. Factor out 3.

$3x^2 - 15x - 42 = 3(x^2 - 5x - 14)$

Now we can continue.

1. Write two sets of parentheses: $3(\quad)(\quad)$.
2. Place the factors of x^2 into the first position of each set of parentheses:

   $3(x\quad)(x\quad)$

3. The third term of the trinomial is -14. We seek two numbers whose

(a) product is -14 and
(b) sum is -5.

The required numbers are -7 and 2. Place -7 and $+2$ into the parentheses.

$3x^2 - 15x - 42 = 3(x - 7)(x + 2)$

The factorization is complete. We'll check to be sure.

$$3(x - 7)(x + 2) = 3(x^2 + 2x - 7x - 14)$$
$$= 3(x^2 - 5x - 14)$$
$$= 3x^2 - 15x - 42$$

★ PRACTICE SET A

Factor, if possible, the following trinomials.

1. $k^2 + 8k + 15$ **2.** $y^2 + 7y - 30$ **3.** $m^2 + 10m + 24$ **4.** $m^2 - 10m + 16$

❑ FACTORING HINTS

Factoring trinomials may take some practice, but with time and experience, you will be able to factor much more quickly.

There are some clues that are helpful in determining the factors of the third term that when added yield the coefficient of the middle term.

Factoring Hints

Look at the *sign* of the last term:

(a) If the sign is positive, we know that the two factors must have the *same* sign, since $(+)(+) = (+)$ and $(-)(-) = (+)$. The two factors will have the same sign as the sign of the middle term.

(b) If the sign is negitive, we know that two factors must have *opposite* signs, since $(+)(-) = (-)$ and $(-)(+) = (-)$.

☆ **SAMPLE SET B**

Factor $x^2 - 7x + 12$.

1. Write two sets of parentheses: ( )( ).
2. The third term of the trinomial is $+12$. The sign is positive, so the two factors of 12 we are looking for must have the same sign. They will have the sign of the middle term. The sign of the middle term is negative, so both factors of 12 are negative. They are -12 and -1, -6 and -2, or -4 and -3. Only the factors -4 and -3 add to -7, so -4 and -3 are the proper factors of 12 to be used.

$$x^2 - 7x + 12 = (x - 4)(x - 3)$$

★ **PRACTICE SET B**

Factor, if possible, the following trinomials.

1. $4k^2 + 32k + 28$

2. $3y^4 + 24y^3 + 36y^2$

3. $x^2 - xy - 6y^2$

4. $-5a^5b - 10a^4b^2 + 15a^3b^3$

Answers to Practice Sets are on p. 222.

Section 5.6 EXERCISES

For problems 1–40, factor the trinomials when possible.

1. $x^2 + 4x + 3$

4. $x^2 + 6x + 5$

2. $x^2 + 6x + 8$

5. $y^2 + 8y + 12$

3. $x^2 + 7x + 12$

6. $y^2 - 5y + 6$

7. $y^2 - 5y + 4$

8. $a^2 + a - 6$

9. $a^2 + 3a - 4$

10. $x^2 + 4x - 21$

11. $x^2 - 4x - 21$

12. $x^2 + 7x + 12$

13. $y^2 + 10y + 16$

14. $x^2 + 6x - 16$

15. $y^2 - 8y + 7$

16. $y^2 - 5y - 24$

17. $a^2 + a - 30$

18. $a^2 - 3a + 2$

19. $a^2 - 12a + 20$

20. $y^2 - 4y - 32$

21. $x^2 + 13x + 42$

22. $x^2 + 2x - 35$

23. $x^2 + 13x + 40$

24. $y^2 + 6y - 27$

25. $b^2 + 15b + 56$

26. $3a^2 + 24a + 36$
(*Hint:* Always search for a common factor.)

27. $4x^2 + 12x + 8$

28. $2a^2 - 18a + 40$

29. $5y^2 - 70y + 440$

30. $6x^2 - 54x + 48$

36. $3xm^2 + 33xm + 54x$

31. $x^3 + 6x^2 + 8x$

37. $2y^2n^2 - 10y^2n - 48y^2$

32. $x^3 - 8x^2 + 15x$

38. $4x^4 - 42x^3 + 144x^2$

33. $x^4 + 9x^3 + 14x^2$

39. $y^5 + 13y^4 + 42y^3$

34. $2a^3 + 12a^2 + 10a$

35. $4a^3 - 40a^2 + 84a$

40. $4x^2a^6 - 48x^2a^5 + 252x^2a^4$

EXERCISES FOR REVIEW

(5.4) **41.** Factor $6xy + 2ax - 3ay - a^2$.
(5.5) **42.** Factor $8a^2 - 50$.
(5.5) **43.** Factor $4x^2 + 17x - 15$.

★ Answers to Practice Sets (5.6)

A. **1.** $(k + 3)(k + 5)$ **2.** $(y + 10)(y - 3)$ **3.** $(m + 6)(m + 4)$ **4.** $(m - 8)(m - 2)$

B. **1.** $4(k + 7)(k + 1)$ **2.** $3y^2(y + 2)(y + 6)$ **3.** $(x + 2y)(x - 3y)$ **4.** $-5a^3b(a + 3b)(a - b)$

5.7 Factoring Trinomials with Leading Coefficient Other Than 1

Section Overview

❑ **THE METHOD OF FACTORIZATION**

❑ THE METHOD OF FACTORIZATION

In the last section we saw that we could easily factor trinomials of the form $x^2 + bx + c$ by finding the factors of the constant c that add to the coefficient of the linear term b, as shown in the following example:

Factor $x^2 - 4x - 21$.
The third term of the trinomial is -21. We seek two numbers whose

(a) product is -21 and
(b) sum is -4.

The required numbers are -7 and $+3$.

$$x^2 - 4x - 21 = (x - 7)(x + 3)$$

The problem of factoring the polynomial $ax^2 + bx + c$, $a \neq 1$, becomes more involved. We will study two methods of factoring such polynomials. Each method produces the same result, and you should select the method you are most comfortable with. The first method is called the *trial and error method* and requires some educated guesses. We will examine two examples (Sample Sets A and B). Then, we will study a second method of factoring. The second method is called the *collect and discard method,* and it requires less guessing than the trial and error method. Sample Set C illustrates the use of the collect and discard method.

THE TRIAL AND ERROR METHOD OF FACTORING $ax^2 + bx + c$

Trial and Error Method   Consider the product

$$\begin{aligned}
(4x + 3)(5x + 2) &= 4x \cdot\cdot 5x + 4x \cdot 2 + 3 \cdot 5x + 3 \cdot 2 \\
&= 20x^2 + \underline{8x + 15x} + 6 \\
&= 20x^2 + \overset{\frown}{23x} + 6
\end{aligned}$$

Examining the trinomial $20x^2 + 23x + 6$, we can immediately see some factors of the first and last terms.

20x^2	**6**
20x, x	6, 1
10x, 2x	3, 2
5x, 4x	

Our goal is to choose the proper combination of factors of the first and last terms that yield the middle term $23x$.

Notice that the middle term comes from the *sum* of the *outer* and *inner* products in the multiplication of the two binomials.

outer product
8x

$(4x + 3)(5x + 2)$

15x
inner product

This fact provides us a way to find the proper combination.

> Look for the combination that when *multiplied* and then *added* yields the middle term.

The proper combination we're looking for is

$20x^2$		6
$20x$, x		6, 1
$10x$, x		3, 2
$5x$, $4x$		

$$\begin{array}{c} 15x \\ +8x \\ \hline 23x \end{array}$$

☆ SAMPLE SET A

1. Factor $6x^2 + x - 12$.

Factor the first and last terms.

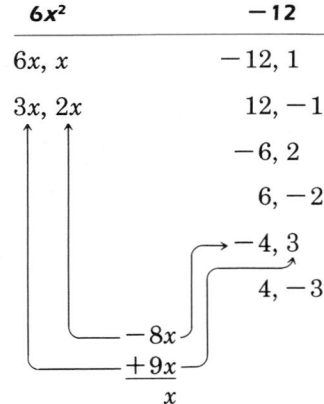

$6x^2$	-12
$6x$, x	-12, 1
$3x$, $2x$	12, -1
	-6, 2
	6, -2
	-4, 3
	4, -3

$$\begin{array}{c} -8x \\ +9x \\ \hline x \end{array}$$

Thus, $3x$ and 3 are to be multiplied, and $2x$ and -4 are to be multiplied.

$6x^2 + x - 12 = ($ $)($ $)$ **Put the factors of the leading term in immediately.**

$\qquad = (3x$ $)(2x$ $)$ **Since $3x$ and 3 are to be multiplied, they must be located in different binomials.**

$\qquad = (3x$ $)(2x + 3)$ **Place the -4 in the remaining set of parentheses.**

$\qquad = (3x - 4)(2x + 3)$

$6x^2 + x - 12 = (3x - 4)(2x + 3)$

Check: $(3x - 4)(2x + 3) = 6x^2 + 9x - 8x - 12$
$\qquad\qquad\qquad\qquad\quad = 6x^2 + x - 12$

2. Factor $8x^2 - 30x - 27$.

Find the factors of the first and last terms.

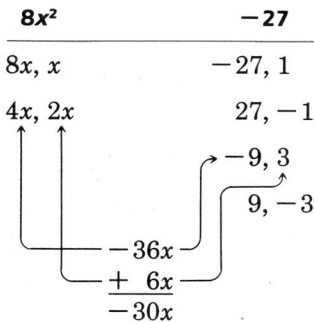

Thus, the $4x$ and -9 are to be multiplied, and $2x$ and 3 are to be multiplied.

$$8x^2 - 30x - 27 = (4x \qquad)(2x \qquad)$$
$$= (4x \qquad)(2x - 9)$$
$$= (4x + 3)(2x - 9)$$

Check: $\quad (4x + 3)(2x - 9) = 8x^2 - 36x + 6x - 27$
$$= 8x^2 - 30x - 27$$

3. Factor $15x^2 + 44x + 32$.

Before we start finding the factors of the first and last terms, notice that the constant term is $+32$. Since the product is *positive,* the two factors we are looking for *must* have the same sign. They must both be positive or both be negative. Now the middle term, $+44x$, is preceded by a positive sign. We know that the middle term comes from the *sum* of the outer and inner products. If these two numbers are to sum to a positive number, they must both be positive themselves. If they were negative, their sum would be negative. Thus, we can conclude that the two factors of $+32$ that we are looking for are both positive numbers. This eliminates several factors of 32 and lessens our amount of work.

Factor the first and last terms.

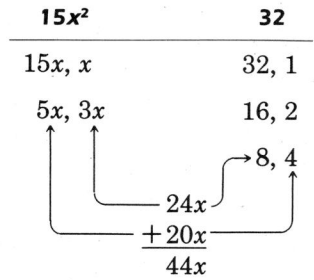

After a few trials we see that $5x$ and 4 are to be multiplied, and $3x$ and 8 are to be multiplied.

$$15x^2 + 44x + 32 = (5x + 8)(3x + 4)$$

Continued

4. Factor $18x^2 - 56x + 6$.

We see that each term is even, so we can factor out 2.

$2(9x^2 - 28x + 3)$

Notice that the constant term is positive. Thus, we know that the factors of 3 that we are looking for must have the same sign. Since the sign of the middle term is negative, both factors must be negative.

Factor the first and last terms.

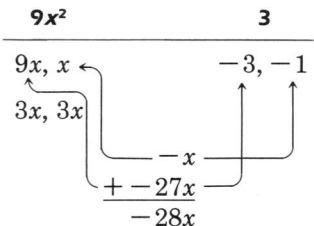

There are not many combinations to try, and we find that $9x$ and -3 are to be multiplied and x and -1 are to be multiplied.

$$18x^2 - 56x + 6 = 2(9x^2 - 28x + 3)$$
$$= 2(9x - 1)(x - 3)$$

If we had not factored the 2 out first, we would have gotten the factorization

$$18x^2 - 56x + 6 = (9x - 1)\ (2x - 6)$$
$$\underset{\text{2 is common}}{\underbrace{}}$$

The factorization is not complete since one of the factors may be factored further.

$$18x^2 - 56x + 6 = (9x - 1)(2x - 6)$$
$$= (9x - 1) \cdot 2(x - 3)$$
$$= 2(9x - 1)(x - 3) \qquad \textbf{(By the commutative property of multiplication.)}$$

The results are the same, but it is much easier to factor a polynomial after all common factors have been factored out first.

5. Factor $3x^2 + x - 14$.

There are no common factors. We see that the constant term is negative. Thus, the factors of -14 must have different signs.

Factor the first and last terms.

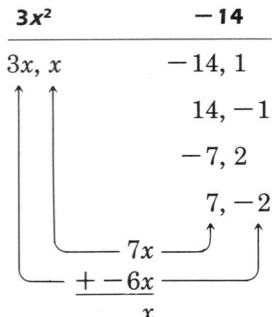

After a few trials, we see that $3x$ and -2 are to be multiplied and x and 7 are to be multiplied.

$$3x^2 + x - 14 = (3x + 7)(x - 2)$$

6. Factor $8x^2 - 26xy + 15y^2$.

We see that the constant term is positive and that the middle term is preceded by a minus sign. Hence, the factors of $15y^2$ that we are looking for must both be negative.

Factor the first and last terms.

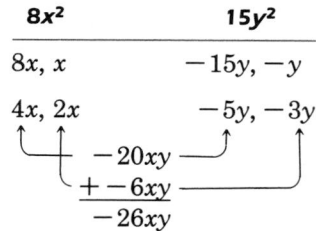

$8x^2$	$15y^2$
$8x, x$	$-15y, -y$
$4x, 2x$	$-5y, -3y$

$-20xy$
$+ -6xy$
$-26xy$

After a few trials, we see that $4x$ and $-5y$ are to be multiplied and $2x$ and $-3y$ are to be multiplied.

$$8x^2 - 26xy + 15y^2 = (4x - 3y)(2x - 5y)$$

★ PRACTICE SET A

Factor the following, if possible.

1. $2x^2 + 13x - 7$

2. $3x^2 + x - 4$

3. $4a^2 - 25a - 21$

4. $16b^2 - 22b - 3$

5. $10y^2 - 19y - 15$

6. $6m^3 + 40m^2 - 14m$

7. $14p^2 + 31pq - 10q^2$

8. $-24w^2z^2 + 14wz^3 - 2z^4$

9. $3x^2 + 6xy + 2y^2$

As you get more practice factoring these types of polynomials you become faster at picking the proper combinations. It takes a lot of practice!

There is a shortcut that may help in picking the proper combinations. This process does not always work, but it seems to hold true in many cases. After you have factored the first and last terms and are beginning to look for the proper combinations, start with the *intermediate* factors and not the extreme ones.

☆ **SAMPLE SET B**

Factor $24x^2 - 41x + 12$.

Factor the first and last terms.

$24x^2$	**12**
$24x, x$	$-12, -1$
$12x, 2x$	$-6, -2$
$8x, 3x$	$-4, -3$
$6x, 4x$	

Rather than starting with the $24x$, x and -12, -1, pick some intermediate values, $8x$ and $3x$, the $6x$ and $4x$, or the -6 and -2, or the -4 and -3.

$24x^2 - 41x + 12 = (8x - 3)(3x - 4)$

★ **PRACTICE SET B**

1. Factor $48x^2 + 22x - 15$.

2. Factor $54y^2 + 39yw - 28w^2$.

THE COLLECT AND DISCARD METHOD OF FACTORING $ax^2 + bx + c$

Collect and Discard Method

Consider the polynomial $6x^2 + x - 12$. We begin by identifying a and c. In this case, $a = 6$ and $c = -12$. We start out as we would with $a = 1$.

$6x^2 + x - 12$: $(6x\quad)(6x\quad)$

Now, compute $a \cdot c$.

$a \cdot c = (6)(-12) = -72$

Find the factors of -72 that add to 1, the coefficient of x, the linear term. The factors are 9 and -8. Include these factors in the parentheses.

$6x^2 + x - 12$: $(6x + 9)(6x - 8)$

But we have included too much. We must eliminate the surplus. Factor each parentheses.

$6x^2 + x - 12$: $3(2x + 3) \cdot 2(3x - 4)$

Discard the factors that multiply to $a = 6$. In this case, 3 and 2. We are left with the proper factorization.

$6x^2 + x - 12 = (2x + 3)(3x - 4)$

☆ **SAMPLE SET C**

1. Factor $10x^2 + 23x - 5$.

Identify $a = 10$ and $b = -5$.

$10x^2 + 23x - 5;$ $(10x$ $)(10x$ $)$

Compute

$a \cdot c = (10)(-5) = -50$

Find the factors of -50 that add to $+23$, the coefficient of x, the linear term. The factors are 25 and -2. Place these numbers into the parentheses.

$10x^2 + 23x - 5:$ $(10x + 25)(10x - 2)$

We have collected too much. Factor each set of parentheses and eliminate the surplus.

$10x^2 + 23x - 5:$ $(5)(2x + 5) \cdot (2)(5x - 1)$

Discard the factors that multiply to $a = 10$. In this case, 5 and 2.

$10x^2 + 23x - 5 = (2x + 5)(5x - 1)$

2. Factor $8x^2 - 30x - 27$.

Identify $a = 8$ and $c = -27$.

$8x^2 - 30x - 27:$ $(8x$ $)(8x$ $)$

Compute

$a \cdot c = (8)(-27) = -216$

Find the factors of -216 that add to -30, the coefficient of x, the linear term. This requires some thought. The factors are -36 and 6. Place these numbers into the parentheses.

$8x^2 - 30x - 27:$ $(8x - 36)(8x + 6)$

We have collected too much. Factor each set of parentheses and eliminate the surplus.

$8x^2 - 30x - 27:$ $(4)(2x - 9) \cdot (2)(4x + 3)$

Discard the factors that multiply to $a = 8$. In this case, 4 and 2.

$8x^2 - 30x - 27 = (2x - 9)(4x + 3)$

3. Factor $18x^2 - 5xy - 2y^2$.

Identify $a = 18$ and $c = -2$.

$18x^2 - 5xy - 2y^2:$ $(18x$ $)(18x$ $)$

Compute

$a \cdot c = (18)(-2) = -36$

Find the factors of -36 that add to -5, the coefficient of xy. In this case, -9 and 4. Place these numbers into the parentheses, affixing y to each.

$18x^2 - 5xy - 2y^2:$ $(18x - 9y)(18x + 4y)$

We have collected too much. Factor each set of parentheses and eliminate the surplus.

$18x^2 - 5xy - 2y^2:$ $(9)(2x - y) \cdot (2)(9x + 2y)$

Discard the factors that multiply to $a = 18$. In this case, 9 and 4.

$18x^2 - 5xy - 2y^2 = (2x - y)(9x + 2y)$

★ **PRACTICE SET C**

1. Factor $6x^2 + 7x - 3$.

2. Factor $14x^2 - 31x - 10$.

3. Factor $48x^2 + 22x - 15$.

4. Factor $10x^2 - 23xw + 12w^2$.

Answers to Practice Sets are on p. 232.

Section 5.7 EXERCISES

Factor problems $1-24$, if possible.

1. $x^2 + 3x + 2$

2. $x^2 + 7x + 12$

3. $2x^2 + 7x + 5$

4. $3x^2 + 4x + 1$

5. $2x^2 + 11x + 12$

6. $10x^2 + 33x + 20$

7. $3x^2 - x - 4$

8. $3x^2 + x - 4$

9. $4x^2 + 8x - 21$

10. $2a^2 - a - 3$

11. $9a^2 - 7a + 2$

12. $16a^2 + 16a + 3$

13. $16y^2 - 26y + 3$

14. $3y^2 + 14y - 5$

15. $10x^2 + 29x + 10$

16. $14y^2 + 29y - 15$

17. $81a^2 + 19a + 2$

18. $24x^2 + 34x + 5$

19. $24x^2 - 34x + 5$

20. $24x^2 - 26x - 5$

21. $24x^2 + 26x - 5$

22. $6a^2 + 13a + 6$

23. $6x^2 + 5xy + y^2$

24. $6a^2 - ay - y^2$

For problems 25 – 28, the given trinomial occurs when solving the corresponding applied problem. Factor each trinomial. You do not need to solve the problem.

25. $5r^2 - 24r - 5$.   It takes 5 hours to paddle a boat 12 miles downstream and then back. The current flows at the rate of 1 mile per hour. At what rate was the boat paddled?

26. $x^2 + 5x - 84$.   The length of a rectangle is 5 inches more than the width of the rectangle. If the area of the rectangle is 84 square inches, what are the length and width of the rectangle?

27. $x^2 + 24x - 145$.   A square measures 12 inches on each side. Another square is to be drawn around this square in such a way that the total area is 289 square inches. What is the distance from the edge of the smaller square to the edge of the larger square? (The two squares have the same center.)

28. $x^2 + 8x - 20$.   A woman wishes to construct a rectangular box that is open at the top. She wishes it to be 4 inches high and have a rectangular base whose length is three times the width. The material used for the base costs $2 per square inch, and the material used for the sides costs $1.50 per square inch. The woman will spend exactly $120 for materials. Find the dimensions of the box (length of the base, width of the base, and height).

For problems 29 – 39, factor the trinomials if possible.

29. $16x^2 - 8xy - 3y^2$

30. $6a^2 + 7ab + 2b^2$

31. $12a^2 + 7ab + 12b^2$

32. $9x^2 + 18xy + 8y^2$

33. $8a^2 + 10ab - 6b^2$

34. $12a^2 + 54a - 90$

35. $12b^4 + 30b^2a + 12a^2$

36. $30a^4b^4 - 3a^2b^2 - 6c^2$

37. $3a^6 - 3a^3b^2 - 18b^4$

38. $20a^2b^2 + 2abc^2 - 6a^2c^4$

39. $14a^2z^2 - 40a^3z^2 - 46a^4z^2$

EXERCISES FOR REVIEW

(1.6) **40.** Simplify $(a^3b^6)^4$.

(3.5) **41.** Find the product. $x^2(x - 3)(x + 4)$.

(3.6) **42.** Find the product. $(5m - 3n)^2$.

(4.2) **43.** Solve the equation $5(2x - 1) - 4(x + 7) = 0$.

(5.6) **44.** Factor $x^5 - 8x^4 + 7x^3$.

★ **Answers to Practice Sets (5.7)**

A. **1.** $(2x - 1)(x + 7)$ **2.** $(3x + 4)(x - 1)$ **3.** $(4a + 3)(a - 7)$ **4.** $(8b + 1)(2b - 3)$
   5. $(5y + 3)(2y - 5)$ **6.** $2m(3m - 1)(m + 7)$ **7.** $(7p - 2q)(2p + 5q)$ **8.** $-2z^2(4w - z)(3w - z)$
   9. not factorable

B. **1.** $(6x + 5)(8x - 3)$ **2.** $(9y - 4w)(6y + 7w)$

C. **1.** $(3x - 1)(2x + 3)$ **2.** $(7x + 2)(2x - 5)$ **3.** $(6x + 5)(8x - 3)$ **4.** $(5x - 4w)(2x - 3w)$

Factoring (5.1)

Factoring is the process of determining the factors of some product. Factoring is the reverse of multiplication.

Greatest Common Factor (5.3)

The *greatest common factor* of a polynomial is the factor that is common to every term of the polynomial and also is such that

1. The numerical coefficient is the largest number common to each term.
2. The variables possess the largest exponents that are common to all the variables.

Factoring a Monomial from a Polynomial (5.3)

If A is the greatest common factor of $Ax + Ay$, then

$$Ax + Ay = A(x + y)$$

Factoring by Grouping (5.4)

We are alerted to the idea of factoring by grouping when the polynomial we are considering

1. Has no factor common to all terms.
2. Has an even number of terms.

$$\underbrace{Ax + Ay}_{A \text{ is common}} + \underbrace{Bx + By}_{B \text{ is common}} = A(x + y) + B\underbrace{(x + y)}_{x + y \text{ is common}}$$

$$= (x + y)(A + B)$$

Special products (5.5)

$$a^2 - b^2 = (a + b)(a - b)$$
$$a^2 + 2ab + b^2 = (a + b)^2$$
$$a^2 - 2ab + b^2 = (a - b)^2$$

Fundamental Rule of Factoring (5.5)

1. Factor out all common monomials first.
2. Factor completely.

Factoring Trinomials (5.6, 5.7)

One method of factoring a trinomial is to list all the factor pairs of both of the first and last terms and then choose the combination that when multiplied and then added produces the middle term.

EXERCISE SUPPLEMENT

Section 5.1

For problems 1–20, the first quantity represents the product and the second quantity represents a factor. Find the other factor.

1. $32a^4b, \quad 2b$

2. $35x^3y^2, \quad 7x^3$

3. $44a^2b^2c, \quad 11b^2$

4. $50m^3n^5p^4q, \quad 10m^3q$

5. $51(a+1)^2(b+3)^4, \quad 3(a+1)$

6. $-26(x+2y)^3(x-y)^2, \quad -13(x-y)$

7. $-8x^5y^4(x+y)^4(x+3y)^3, \\ -2x(x+y)(x+3y)$

8. $-(6a-5b)^{10}(7a-b)^8(a+3b)^7, \\ -(6a-5b)^7(7a-b)^7(a+3b)^7$

9. $12x^{n+6}y^{2n-5}, \quad -3x^{n+1}y^{n+3}$

10. $-400a^{3n+10}b^{n-6}c^{4n+7}, \quad 20a^{2n+8}c^{2n-1}$

11. $16x - 32, \quad 16$

12. $35a - 45, \quad 513$

13. $24a^2 - 6a, \quad 6a$

14. $88x^4 - 33x^3 + 44x^2 + 55x, \quad 11x$

15. $9y^3 - 27y^2 + 36y, \quad -3y$

16. $4m^6 - 16m^4 + 16m^2, \quad 4m$

17. $-5x^4y^3 + 10x^3y^2 - 15x^2y^2, \quad -5x^2y^2$

18. $-21a^5b^6c^4(a+2)^3 + 35a^5bc^5(a+2)^4, \\ -7a^4b(a+2)^2$

19. $-x - 2y - c^2, \quad -1$

20. $a + 3b, \quad -1$

Sections 5.2 and 5.3

For problems 21–31, factor the polynomials.

21. $8a + 4$

22. $10x + 10$

23. $3y^2 + 27y$

24. $6a^2b^2 + 18a^2$

25. $21(x+5) + 9$

26. $14(2a+1) + 35$

27. $ma^3 - m$

28. $15y^3 - 24y + 24$

29. $r^2(r+1)^3 - 3r(r+1)^2 + r + 1$

30. $Pa + Pb + Pc$

31. $(10 - 3x)(2 + x) + 3(10 - 3x)(7 + x)$

Section 5.4

For problems 32–40, use the grouping method to factor the polynomials. Some may not be factorable.

32. $4ax + x + 4ay + y$

33. $xy + 4x - 3y - 12$

34. $2ab - 8b - 3ab - 12a$

35. $a^2 - 7a + ab - 7b$

36. $m^2 + 5m + nm + 5n$

37. $r^2 + rs - r - s$

38. $8a^2bc + 20a^2bc + 10a^3b^3c + 25a^3b^3$

39. $a(a+6) - (a+6) + a(a-4) - (a-4)$

40. $a(2x+7) - 4(2x+7) + a(x-10) - 4(x-10)$

Sections 5.5–5.7

For problems 41–75, factor the polynomials, if possible.

41. $m^2 - 36$

42. $r^2 - 81$

43. $a^2 + 8a + 16$

44. $c^2 + 10c + 25$

45. $m^2 + m + 1$

46. $r^2 - r - 6$

47. $a^2 + 9a + 20$

48. $s^2 + 9s + 18$

49. $x^2 + 14x + 40$

50. $a^2 - 12a + 36$

51. $n^2 - 14n + 49$

52. $a^2 + 6a + 5$

53. $a^2 - 9a + 20$

54. $6x^2 + 5x + 1$

55. $4a^2 - 9a - 9$

56. $4x^2 + 7x + 3$

57. $42a^2 + 5a - 2$

58. $30y^2 + 7y - 15$

59. $56m^2 + 26m + 6$

60. $27r^2 - 33r - 4$

61. $4x^2 + 4xy - 3y^2$

62. $25a^2 + 25ab + 6b^2$

63. $2x^2 + 6x - 20$

64. $-2y^2 + 4y + 48$

65. $x^3 + 3x^2 - 4x$

66. $3y^4 - 27y^3 + 24y^2$

67. $15a^2b^2 - ab - 2b$

68. $4x^3 - 16x^2 + 16x$

69. $18a^2 - 6a + \dfrac{1}{2}$

70. $a^4 + 16a^2b + 16b^2$

71. $4x^2 - 12xy + 9y^2$

72. $49b^4 + 84b^2 + 36$

73. $r^6s^8 + 6r^3s^4p^2q^6 + 9p^4q^{12}$

74. $a^4 - 2a^2b - 15b^2$

75. $81a^8b^{12}c^{10} - 25x^{20}y^{18}$

1. _____

1. (5.1) The product is $27a^3 + 9a^2 + 9a$ and a factor is $3a$. Find the other factor.

2. _____

2. (5.1) The product is $15x^{n+5}y^{3n-2}$ and a factor is $3x^5y^{n+1}$. Find the other factor.

3. _____

For problems 3–13, factor, if possible, the polynomials.

3. (5.3) $-14x^2y^4b - 28x^2y^3b - 42x^2y^2$

4. _____

4. (5.3) $(y+2)a + (y+2)c$

5. _____

5. (5.4) $6x^2y^2z + 5x^2y^3 - 12xyz - 10xy^2$

6. _____

6. (5.5) $4a^2 - 16c^2$

7. _____

7. (5.5) $m^4 - n^4$

8. _____

8. (5.5) $b^2 + 8b + 16$

9. _____

9. (5.5) $9y^2 - 30y + 25$

10. _____

10. (5.6) $x^2 + 5x - 15$

11. _____

11. (5.6) $x^2 - x - 30$

12. _____ 12. **(5.7)** $4x^6 - 36x^4 + 80x^2$

13. _____ 13. **(5.7)** $9x^2 + 25x - 6$

6

Graphing Linear Equations and Inequalities in One and Two Variables

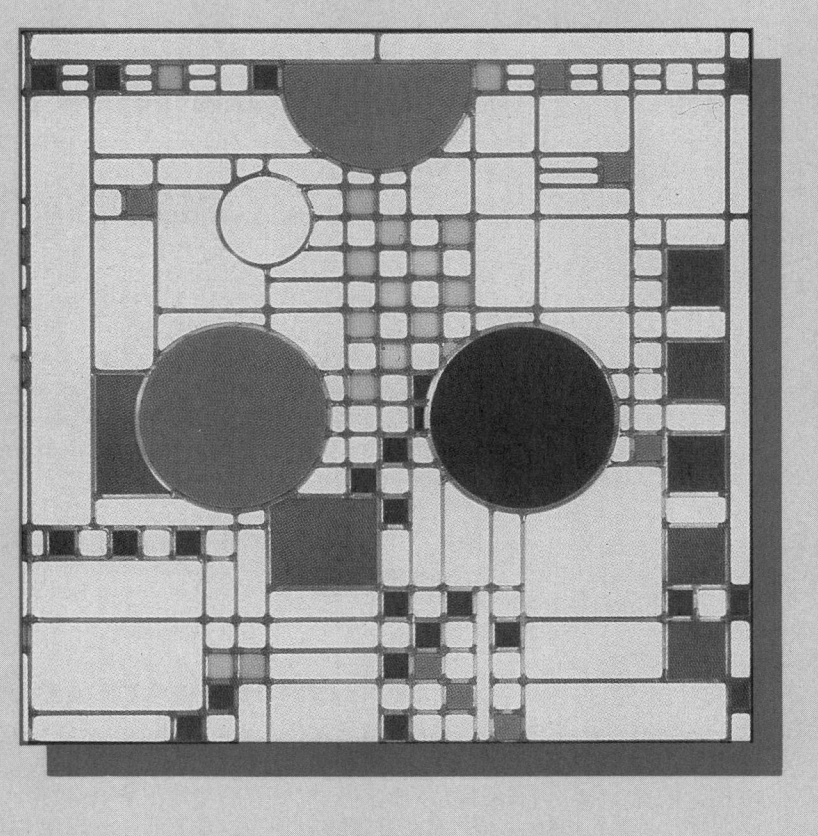

After completing this chapter, you should

Section 6.1 Graphing Linear Equations and Inequalities in One Variable
- understand the concept of a graph and the relationship between axes, coordinate systems, and dimension
- be able to construct one-dimensional graphs

Section 6.2 Plotting Points in the Plane
- be familiar with the plane
- know what is meant by the coordinates of a point
- be able to plot points in the plane

Section 6.3 Graphing Linear Equations in Two Variables
- be able to relate solutions to a linear equation to lines
- know the general form of a linear equation
- be able to construct the graph of a line using the intercept method
- be able to distinguish, by their equations, slanted, horizontal, and vertical lines

Section 6.4 The Slope-Intercept Form of a Line
- be more familiar with the general form of a line
- be able to recognize the slope-intercept form of a line
- be able to interpret the slope and intercept of a line
- be able to use the slope formula to find the slope of a line

Section 6.5 Graphing Equations in Slope-Intercept Form
- be able to use the slope and intercept to construct the graph of a line

Section 6.6 Finding the Equation of a Line
- be able to find the equation of a line using either the slope-intercept form or the point-slope form of a line

Section 6.7 Graphing Linear Inequalities in Two Variables
- be able to locate solutions to linear inequalities in two variables using graphical techniques

6.1 Graphing Linear Equations and Inequalities in One Variable

Section Overview

❑ GRAPHS
❑ AXES, COORDINATE SYSTEMS, AND DIMENSION
❑ GRAPHING IN ONE DIMENSION

❑ GRAPHS

We have, thus far in our study of algebra, developed and used several methods for obtaining solutions to linear equations in both one and two variables. Quite often it is helpful to obtain a picture of the solutions to an equation. These pictures are called *graphs* and they can reveal information that may not be evident from the equation alone.

The Graph of an Equation

> The geometric representation (picture) of the solutions to an equation is called the **graph** of the equation.

❑ AXES, COORDINATE SYSTEMS, AND DIMENSION

Axis

The Number Line is an Axis

The basic structure of the graph is the *axis*. It is with respect to the axis that all solutions to an equation are located. The most fundamental type of axis is the *number line*.

This number line is an axis.

We have the following general rules regarding axes.

Number of Variables and Number of Axes

> An equation in one variable requires one axis.
>
> An equation in two variables requires two axes.
>
> An equation in three variables requires three axes.
>
> .
> .
> .
>
> An equation in *n* variables requires *n* axes.

We shall always draw an axis as a straight line, and if more than one axis is required, we shall draw them so they are all mutually perpendicular (the lines forming the axes will be at 90° angles to one another).

Coordinate System

> A system of axes constructed for graphing an equation is called a **coordinate system.**

The Phrase, Graphing an Equation

The phrase **graphing an equation** is used frequently and should be interpreted as meaning geometrically locating the solutions to an equation.

Relating the Number of Variables and the Number of Axes

We will not start actually graphing equations until Section 6.2, but in the following examples we will *relate* the number of variables in an equation to the number of axes in the coordinate system.

1. If we wish to graph the equation $5x + 2 = 17$, we would need to construct a coordinate system consisting of a single axis (a single number line) since the equation consists of only one variable. We label the axis with the variable that appears in the equation.

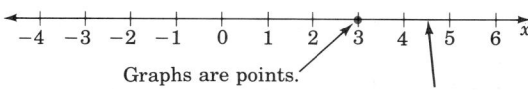

Graphs are points.

This axis is a line and lines are one-dimensional—length.

One-Dimensional Graphs

    We might interpret an equation in one variable as giving information in one-dimensional space. Since we live in three-dimensional space, one-dimensional space might be hard to imagine. Objects in one-dimensional space would have only length, no width or depth.

2. To graph an equation in two variables such as $y = 2x - 3$, we would need to construct a coordinate system consisting of two mutually perpendicular number lines **(axes).** We call the intersection of the two axes the **origin** and label it with a 0. The two axes are simply number lines; one drawn horizontally, one drawn vertically.

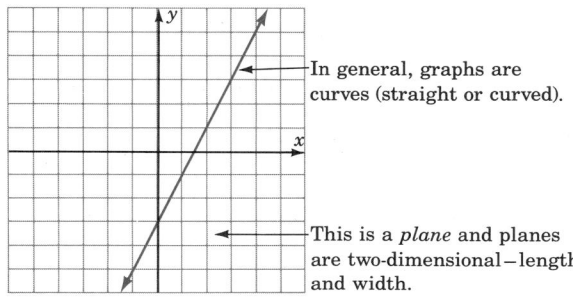

In general, graphs are curves (straight or curved).

This is a *plane* and planes are two-dimensional—length and width.

    Recall that an equation in two variables requires a solution to be a pair of numbers. The solutions can be written as ordered pairs (x, y). Since the equation $y = 2x - 3$ involves the variables x and y, we label one axis x and the other axis y. In mathematics it is customary to label the horizontal axis with the independent variable and the vertical axis with the dependent variable.

Two-Dimensional Graphs

    We might interpret equations in two variables as giving information in two-dimensional space. Objects in two-dimensional space would have length and width, but no depth.

3. An equation in three variables, such as $3x^2 - 4y^2 + 5z = 0$, requires three mutually perpendicular axes, one for each variable. We would construct the following coordinate system and graph.

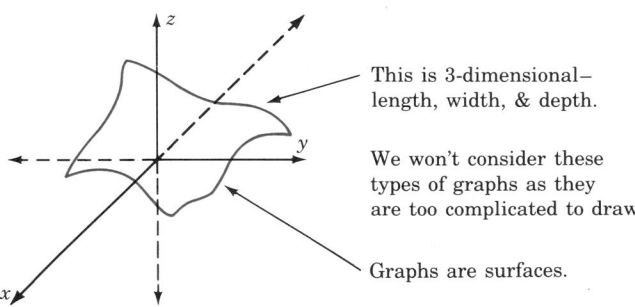

This is 3-dimensional—length, width, & depth.

We won't consider these types of graphs as they are too complicated to draw.

Graphs are surfaces.

Three-Dimensional Graphs

    We might interpret equations in three variables as giving information about three-dimensional space.

Four-Dimensional Graphs

Black Holes

4. To graph an equation in four variables, such as $3x - 2y + 8x - 5w = -7$, would require four mutually perpendicular number lines. These graphs are left to the imagination.

We might interpret equations in four variables as giving information in four-dimensional space. Four-dimensional objects would have length, width, depth, and some other dimension.

These other spaces are hard for us to imagine, but the existence of "black holes" makes the possibility of other universes of one-, two-, four-, or n-dimensions not entirely unlikely. Although it may be difficult for us "3-D" people to travel around in another dimensional space, at least we could be pretty sure that our mathematics would still work (since it is not restricted to only three dimensions)!

☐ GRAPHING IN ONE DIMENSION

Graphing a linear equation in one variable involves solving the equation, then locating the solution on the axis (number line), and marking a point at this location. We have observed that graphs may reveal information that may not be evident from the original equation. The graphs of linear equations in one variable do not yield much, if any, information, but they serve as a foundation to graphs of higher dimension (graphs of two variables and three variables).

1. Graph the equation $3x - 5 = 10$.

   Solve the equation for x and construct an axis. Since there is only one variable, we need only one axis. Label the axis x.

   $3x - 5 = 10$
   $3x = 15$
   $x = 5$

   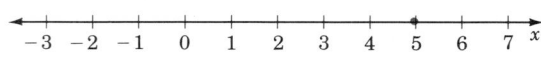

2. Graph the equation $3x + 4 + 7x - 1 + 8 = 31$.

   Solving the equation we get,

   $10x + 11 = 31$
   $10x = 20$
   $x = 2$

   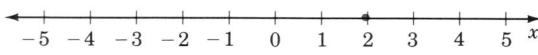

★ PRACTICE SET A

Graph the equation $4x + 1 = -7$.

☆ **SAMPLE SET B**

1. Graph the linear inequality $4x \geq 12$.

We proceed by solving the inequality.

$4x \geq 12$ **Divide each side by 4.**
$x \geq 3$

As we know, any value greater than or equal to 3 will satisfy the original inequality. Hence we have infinitely many solutions and, thus, infinitely many points to mark off on our graph.

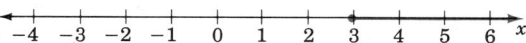

    The *closed circle* at 3 means that 3 is included as a solution. All the points beginning at 3 and in the direction of the arrow are solutions.

2. Graph the linear inequality $-2y - 1 > 3$.

We first solve the inequality.

$-2y - 1 > 3$
$-2y > 4$

$y < -2$ **The inequality symbol reversed direction because we divided by -2.**

Thus, all numbers strictly less than -2 will satisfy the inequality and are thus solutions.
    Since -2 itself is *not* to be included as a solution, we draw an *open circle* at -2. The solutions are to the left of -2 so we draw an arrow pointing to the left of -2 to denote the region of solutions.

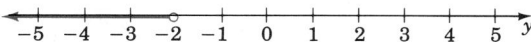

3. Graph the inequality $-2 \leq y + 1 < 1$.

We recognize this inequality as a *compound inequality* and solve it by subtracting 1 from all three parts.

$-2 \leq y + 1 < 1$
$-3 \leq y < 0$

Thus, the solution is all numbers between -3 and 0, more precisely, all numbers greater than or equal to -3 but strictly less than 0.

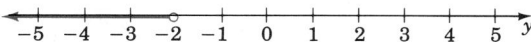

4. Graph the linear equation $5x = -125$.

The solution is $x = -25$. Scaling the axis by units of 5 rather than 1, we obtain

★ **PRACTICE SET B**

1. Graph the inequality $3x \leq 18$.

2. Graph the inequality $-3m + 1 < 13$.

3. Graph the inequality $-3 \leq x - 5 < 5$.

4. Graph the linear equation $-6y = 480$.

Answers to Practice Sets are on p. 245.

Section 6.1 EXERCISES

For problems 1–25, graph the linear equations and inequalities.

1. $4x + 7 = 19$

2. $8x - 1 = 7$

3. $2x + 3 = 4$

4. $x + 3 = 15$

5. $6y + 3 = y + 8$

6. $2x = 0$

7. $4 + 1 - 4 = 3z$

8. $x + \dfrac{1}{2} = \dfrac{4}{3}$

9. $7r = \dfrac{1}{4}$

10. $2x - 6 = \dfrac{2}{5}$

11. $x + 7 \leq 12$

12. $y - 5 < 3$

13. $x + 19 > 2$

14. $z + 5 > 11$

15. $3m - 7 \leq 8$

16. $-5t \geq 10$

17. $-8x - 6 \geq 34$

18. $\dfrac{x}{4} < 2$

19. $\dfrac{y}{7} \leq 3$

20. $\dfrac{2y}{9} \geq 4$

21. $\dfrac{-5y}{8} \leq 4$

22. $\dfrac{-6a}{7} < -4$ **23.** $-1 \le x - 3 < 0$ **24.** $6 \le x + 4 \le 7$

25. $-12 < -2x - 2 \le -8$

EXERCISES FOR REVIEW

(1.5)   **26.** Simplify $(3x^8y^2)^3$.

(3.1)   **27.** List, if any should appear, the common factors in the expression $10x^4 - 15x^2 + 5x^6$.

(4.6)   **28.** Solve the inequality $-4(x + 3) < -3x + 1$.

(4.7)   **29.** Solve the equation $y = -5x + 8$ if $x = -2$.

(4.7)   **30.** Solve the equation $2y = 5(3x + 7)$ if $x = -1$.

★ **Answers to Practice Sets (6.1)**

A. $x = -2$

B. 1. $x \le 6$ **2.** $m > -4$

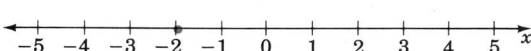  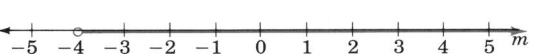

3. $2 \le x < 10$ **4.** $y = -80$

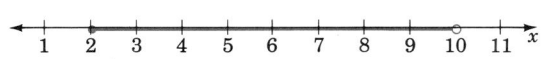 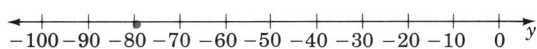

6.2 Plotting Points in the Plane

Section Overview

- ☐ **THE PLANE**
- ☐ **COORDINATES OF A POINT**
- ☐ **PLOTTING POINTS**

☐ THE PLANE

We are now interested in studying graphs of linear equations in two variables. We know that solutions to equations in two variables consist of a pair of values, one value for each variable. We have called these pairs of values **ordered pairs.** Since we have a pair of values to graph, we must have a pair of axes (number lines) upon which the values can be located. We draw the axes so they are perpendicular to each

Ordered Pairs

Origin

other and so that they intersect each other at their 0's. This point is called the **origin.**

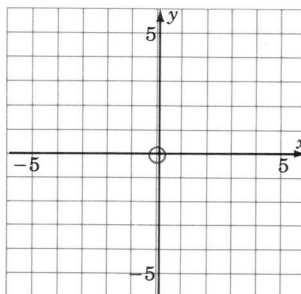

Rectangular Coordinate System

These two lines form what is called a **rectangular coordinate system.** They also determine a plane. A **plane** is a flat surface, and a result from geometry states that through any two intersecting lines (the axes) exactly one plane (flat surface) may be passed. If we are dealing with a linear equation in the two variables x and y, we sometimes say we are graphing the equation using a rectangular coordinate system, or that we are graphing the equation in the xy-plane.

xy-plane

Quadrant

Notice that the two intersecting coordinate axes divide the plane into four equal regions. Since there are four regions, we call each one a **quadrant** and number them counterclockwise using Roman numerals.

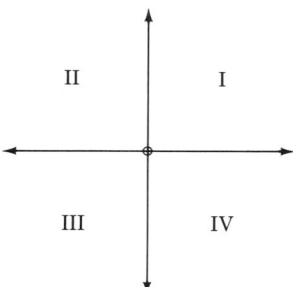

Recall that when we first studied the number line we observed the following:

For each real number there exists a unique point on the number line, and for each point on the number line we can associate a unique real number.

We have a similar situation for the plane.

For each ordered pair (a, b), there exists a unique point in the plane, and to each point in the plane we can associate a unique ordered pair (a, b) of real numbers.

☐ COORDINATES OF A POINT

The numbers in an ordered pair that are associated with a particular point are called the **coordinates of the point.** The *first number* in the ordered pair expresses the point's horizontal distance and direction (left or right) from the origin. The *second number* expresses the point's vertical distance and direction (up or down) from the origin. A *positive number* means a direction to the *right or up*. A *negative number* means a direction to the *left or down*.

Coordinates of a Point

The Coordinates Determine Distance and Direction

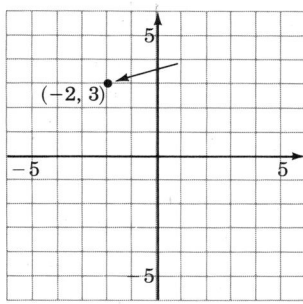

This point is located 2 units to the left of the origin and 3 units up from the origin.

❑ PLOTTING POINTS

Since points and ordered pairs are so closely related, the two terms are sometimes used interchangeably. The following two phrases have the same meaning:

1. Plot the point (a, b).
2. Plot the ordered pair (a, b).

Plotting a Point

Both phrases mean: Locate, in the plane, the point associated with the ordered pair (a, b) and draw a mark at that position.

☆ SAMPLE SET A

Plot the ordered pair $(2, 6)$.

We begin at the origin. The first number in the ordered pair, 2, tells us we move 2 units to the right ($+2$ means 2 units to the right). The second number in the ordered pair, 6, tells us we move 6 units up ($+6$ means 6 units up).

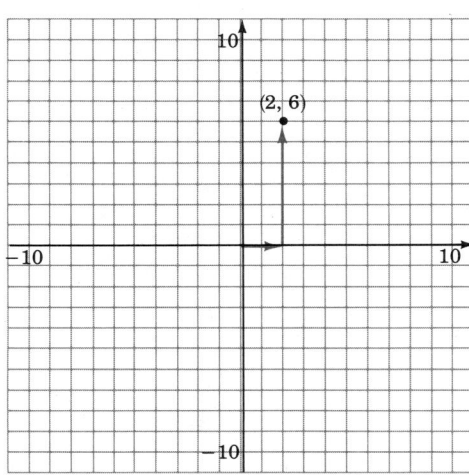

It is sometimes helpful to read $(2, 6)$ as "if $x = 2$, then $y = 6$".

★ **PRACTICE SET A**

Plot the ordered pairs.

1. $(1, 3)$, $(4, -5)$, $(0, 1)$, $(-4, 0)$.

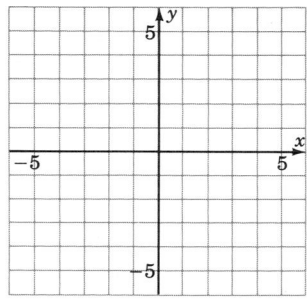

Answers to the Practice Set are on p. 251.

Section 6.2 EXERCISES

1. Plot the following ordered pairs. (Do not draw the arrows as in Practice Set A.)

$(8, 2)$, $(10, -3)$, $(-3, 10)$, $(0, 5)$, $(5, 0)$, $(0, 0)$,

$\left(-7, -\dfrac{3}{2}\right)$.

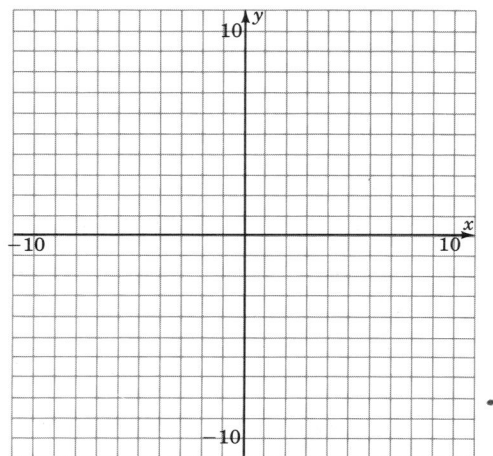

2. As accurately as possible, state the coordinates of the points that have been plotted on the following graph.

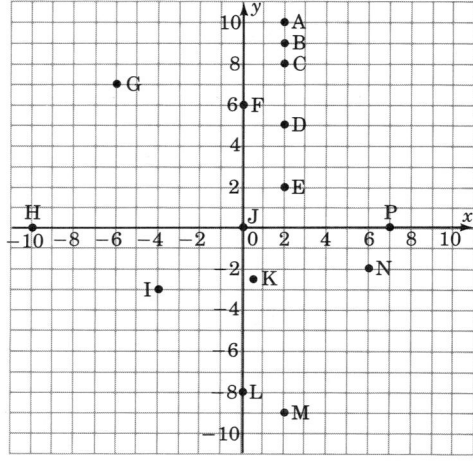

3. Using ordered pair notation, what are the coordinates of the origin?

4. We know that solutions to linear equations in two variables can be expressed as ordered pairs. Hence, the solutions can be represented as points in the plane. Consider the linear equation $y = 2x - 1$. Find at least ten solutions to this equation by choosing x-values between -4 and 5 and computing the corresponding y-values. Plot these solutions on the coordinate system below. Fill in the table to help you keep track of the ordered pairs.

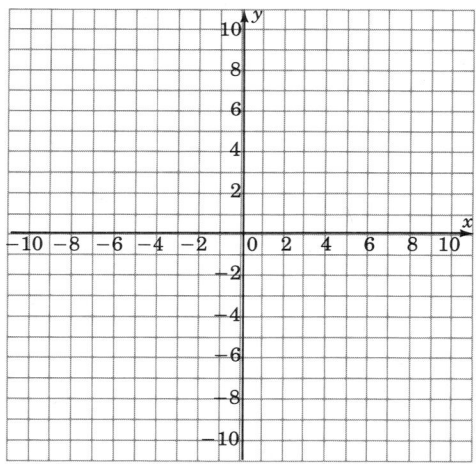

Keeping in mind that there are infinitely many ordered pair solutions to $y = 2x - 1$, speculate on the geometric structure of the graph of *all* the solutions. Complete the following statement:

The name of the type of geometric structure of the graph of all the solutions to the linear equation $y = 2x - 1$ seems to be _____.

Where does this figure cross the y-axis? Does this number appear in the equation $y = 2x - 1$?

Place your pencil at any point on the figure (you may have to connect the dots to see the figure clearly). Move your pencil exactly *one* unit to the right (horizontally). To get back onto the figure, you must move your pencil either up or down a particular number of units. How many units must you move vertically to get back onto the figure, and do you see this number in the equation $y = 2x - 1$?

5. Consider the xy-plane.

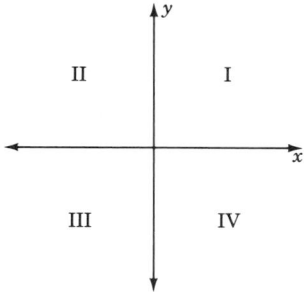

Complete the table by writing the appropriate inequalities.

I	II	III	IV
$x > 0$	$x < 0$	x	x
$y > 0$	y	y	y

In problems 6–9, the graphs of points are called *scatter diagrams* and are frequently used by statisticians to determine if there is a relationship between the two variables under consideration. The first component of the ordered pair is called the *input variable* and the second component is called the *output variable*. Construct the scatter diagrams. Determine if there appears to be a relationship between the two variables under consideration by making the following observations: A relationship may exist if

a. as one variable increases, the other variable increases
b. as one variable increases, the other variable decreases

6. A psychologist, studying the effects of a placebo on assembly line workers at a particular industrial site, noted the time it took to assemble a certain item before the subject was given the placebo, x, and the time it took to assemble a similar item after the subject was given the placebo, y. The psychologist's data are

x	y
10	8
12	9
11	9
10	7
14	11
15	12
13	10

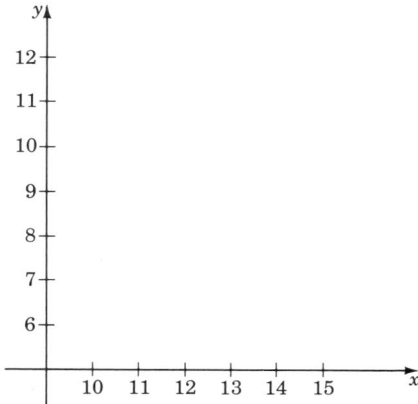

7. The following data were obtained in an engineer's study of the relationship between the amount of pressure used to form a piece of machinery, x, and the number of defective pieces of machinery produced, y.

x	y
50	0
60	1
65	2
70	3
80	4
70	5
90	5
100	5

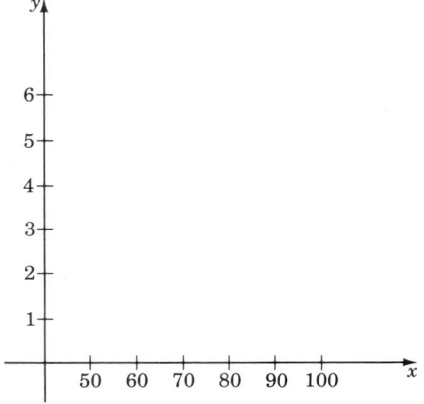

8. The following data represent the number of work days missed per year, x, by the employees of an insurance company and the number of minutes they arrive late from lunch, y.

x	y
1	3
6	4
2	2
2	3
3	1
1	4
4	4
6	3
5	2
6	1

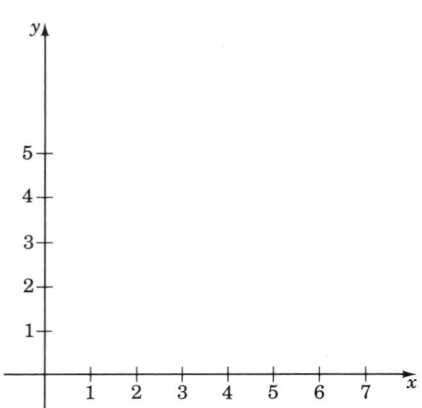

9. A manufacturer of dental equipment has the following data on the unit cost (in dollars), y, of a particular item and the number of units, x, manufactured for each order.

x	y
1	85
3	92
5	99
3	91
4	100
1	87
6	105
8	111
8	114

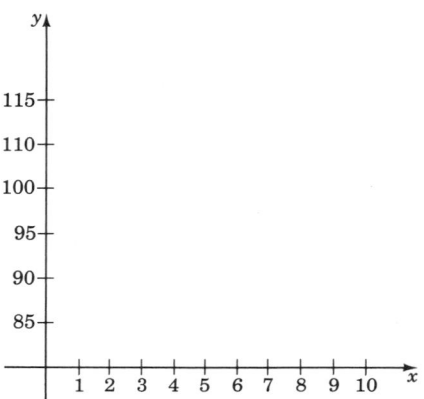

EXERCISES FOR REVIEW

(1.6) **10.** Simplify $\left(\dfrac{18x^5y^6}{9x^2y^4}\right)^5$.

(3.2) **11.** Supply the missing word. An _____ is a statement that two algebraic expressions are equal.

(3.3) **12.** Simplify the expression $5xy(xy - 2x + 3y) - 2xy(3xy - 4x) - 15xy^2$.

(4.1) **13.** Identify the equation $x + 2 = x + 1$ as an identity, a contradiction, or a conditional equation.

(6.1) **14.** Supply the missing phrase. A system of axes constructed for graphing an equation is called a _____.

★ **Answer to Practice Set (6.2)**

A. (Notice that the dotted lines on the graph are only for illustration and should not be included when plotting points.)

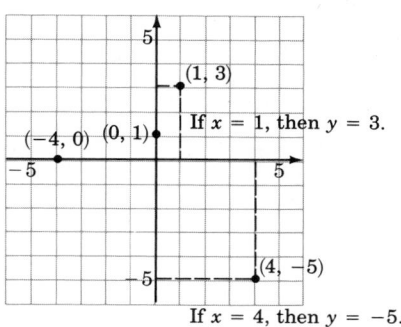

6.3 Graphing Linear Equations in Two Variables

Section Overview

☐ **SOLUTIONS AND LINES**
☐ **GENERAL FORM OF A LINEAR EQUATION**
☐ **THE INTERCEPT METHOD OF GRAPHING**
☐ **GRAPHING USING ANY TWO OR MORE POINTS**
☐ **SLANTED, HORIZONTAL, AND VERTICAL LINES**

☐ SOLUTIONS AND LINES

We know that solutions to linear equations in two variables can be expressed as ordered pairs. Hence, the solutions can be represented by points in the plane. We also know that the phrase "graph the equation" means to locate the solutions to the given equation in the plane. Consider the equation $y - 2x = -3$. We'll graph six solutions (ordered pairs) to this equation on the coordinate system below. We'll find the solutions by choosing x-values (from -1 to $+4$), substituting them into the equation $y - 2x = -3$, and then solving to obtain the corresponding y-values. We can keep track of the ordered pairs by using a table.

$y - 2x = -3$

If $x =$	Then $y =$	Ordered Pair
-1	-5	$(-1, -5)$
0	-3	$(0, -3)$
1	-1	$(1, -1)$
2	1	$(2, 1)$
3	3	$(3, 3)$
4	5	$(4, 5)$

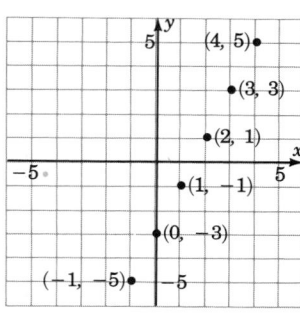

We have plotted only six solutions to the equation $y - 2x = -3$. There are, as we know, infinitely many solutions. By observing the six points we have plotted, we can speculate as to the location of all the other points. The six points we plotted seem to lie on a straight line. This would lead us to believe that all the other points (solutions) also lie on that same line. Indeed, this is true. In fact, this is precisely why first-degree equations are called *linear* equations.

Linear Equations Produce
Straight Lines

> Line
> ↕
> Linear

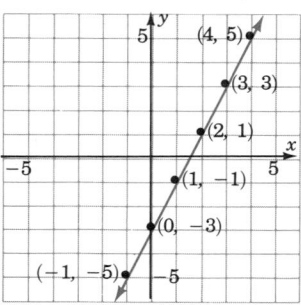

☐ GENERAL FORM OF A LINEAR EQUATION

General Form of a Linear
Equation in Two Variables

There is a standard form in which linear equations in two variables are written. Suppose that a, b, and c are any real numbers and that a and b cannot both be zero at the same time. Then, the linear equation in two variables

$ax + by = c$

is said to be in **general form.**

We must stipulate that a and b cannot both equal zero at the same time, for if they were we would have

$0x + 0y = c$
$0 = c$

This statement is true only if $c = 0$. If c were to be any other number, we would get a false statement.

Now, we have the following:

> The graphing of all ordered pairs that solve a linear equation in two variables produces a straight line.

This implies,

> The graph of a linear equation in two variables is a straight line.

From these statements we can conclude,

> If an ordered pair is a solution to a linear equation in two variables, then it lies on the graph of the equation.

Also,

> Any point (ordered pair) that lies on the graph of a linear equation in two variables is a solution to that equation.

☐ THE INTERCEPT METHOD OF GRAPHING

Intercepts

When we want to graph a linear equation, it is certainly impractical to graph infinitely many points. Since a straight line is determined by only two points, we need only find two solutions to the equation (although a third point is helpful as a check).

When a linear equation in two variables is given in general form, $ax + by = c$, often the two most convenient points (solutions) to find are called the **intercepts:** these are the points at which the line intercepts the coordinate axes. Of course, a horizontal or vertical line intercepts only one axis, so this method does not apply. Horizontal and vertical lines are easily recognized as they contain only *one* variable. (See Sample Set C.)

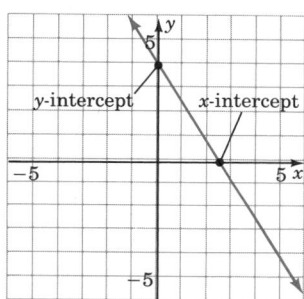

y-Intercept

x-Intercept

Intercept Method

The point at which the line crosses the y-axis is called the **y-intercept.** The x-value at this point is zero (since the point is neither to the left nor right of the origin). The point at which the line crosses the x-axis is called the **x-intercept** and the y-value at that point is zero. The y-intercept can be found by substituting the value 0 for x into the equation and solving for y. The x-intercept can be found by substituting the value 0 for y into the equation and solving for x. Since we are graphing an equation by finding the intercepts, we call this method the **intercept method.**

☆ SAMPLE SET A

Graph the following equations using the intercept method.

1. $y - 2x = -3$

To find the y-intercept, let $x = 0$ and $y = b$.

$b - 2(0) = -3$
$b - 0 = -3$
$b = -3$

Thus, we have the point $(0, -3)$. So, if $x = 0$, $y = b = -3$.

Continued

To find the x-intercept, let $y = 0$ and $x = a$.

$0 - 2a = -3$

$-2a = -3$ **Divide by -2.**

$a = \dfrac{-3}{-2}$

$a = \dfrac{3}{2}$

Thus, we have the point $\left(\dfrac{3}{2}, 0\right)$. So, if $x = a = \dfrac{3}{2}$, $y = 0$.

Construct a coordinate system, plot these two points, and draw a line through them. Keep in mind that every point on this line is a solution to the equation $y - 2x = -3$.

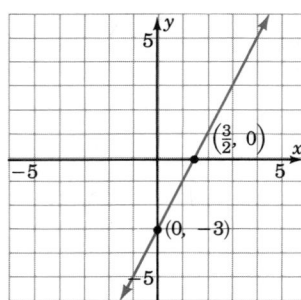

2. $-2x + 3y = 3$

To find the y-intercept, let $x = 0$ and $y = b$.

$-2(0) + 3b = 3$

$0 + 3b = 3$

$3b = 3$

$b = 1$

Thus, we have the point $(0, 1)$. So, if $x = 0$, $y = b = 1$.

To find the x-intercept, let $y = 0$ and $x = a$.

$-2a + 3(0) = 3$

$-2a + 0 = 3$

$-2a = 3$

$a = \dfrac{3}{-2}$

$a = -\dfrac{3}{2}$

Thus, we have the point $\left(-\dfrac{3}{2}, 0\right)$. So, if $x = a = -\dfrac{3}{2}$, $y = 0$.

Construct a coordinate system, plot these two points, and draw a line through them. Keep in mind that all the solutions to the equation $-2x + 3y = 3$ are precisely on this line.

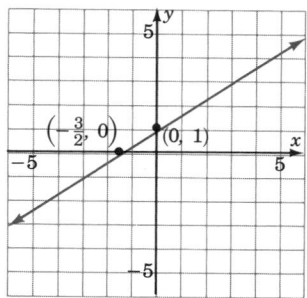

3. $4x + y = 5$

To find the y-intercept, let $x = 0$ and $y = b$.

$4(0) + b = 5$
$0 + b = 5$
$b = 5$

Thus, we have the point $(0, 5)$. So, if $x = 0$, $y = b = 5$.

To find the x-intercept, let $y = 0$ and $x = a$.

$4a + 0 = 5$
$4a = 5$

$a = \dfrac{5}{4}$

Thus, we have the point $\left(\dfrac{5}{4}, 0\right)$. So, if $x = a = \dfrac{5}{4}$, $y = 0$.

Construct a coordinate system, plot these two points, and draw a line through them.

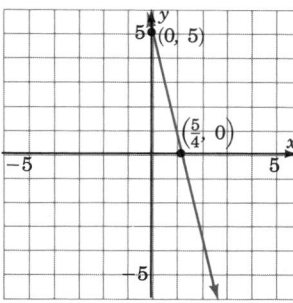

★ **PRACTICE SET A**

1. Graph $3x + y = 3$ using the intercept method.

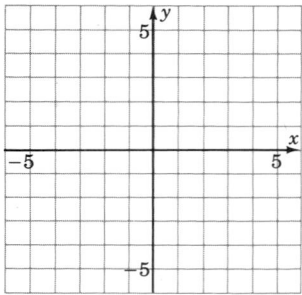

□ **GRAPHING USING ANY TWO OR MORE POINTS**

The graphs we have constructed so far have been done by finding two particular points, the intercepts. Actually, *any* two points will do. We chose to use the intercepts because they are usually the easiest to work with. In the next example, we will graph two equations using points other than the intercepts. We'll use three points, the extra point serving as a check.

☆ **SAMPLE SET B**

1. $x - 3y = -10$.   We can find three points by choosing three x-values and computing to find the corresponding y-values. We'll put our results in a table for ease of reading.

Since we are going to choose x-values and then compute to find the corresponding y-values, it will be to our advantage to solve the given equation for y.

$x - 3y = -10$ **Subtract x from both sides.**
$-3y = -x - 10$ **Divide both sides by -3.**
$y = \dfrac{1}{3}x + \dfrac{10}{3}$

x	y	(x, y)
1	If $x = 1$, then $y = \dfrac{1}{3}(1) + \dfrac{10}{3} = \dfrac{1}{3} + \dfrac{10}{3} = \dfrac{11}{3}$	$\left(1, \dfrac{11}{3}\right)$
-3	If $x = -3$, then $y = \dfrac{1}{3}(-3) + \dfrac{10}{3} = -1 + \dfrac{10}{3} = \dfrac{7}{3}$	$\left(-3, \dfrac{7}{3}\right)$
3	If $x = 3$, then $y = \dfrac{1}{3}(3) + \dfrac{10}{3} = 1 + \dfrac{10}{3} = \dfrac{13}{3}$	$\left(3, \dfrac{13}{3}\right)$

Thus, we have the three ordered pairs (points), $\left(1, \dfrac{11}{3}\right), \left(-3, \dfrac{7}{3}\right), \left(3, \dfrac{13}{3}\right)$. If we wish, we can

change the improper fractions to mixed numbers, $\left(1, 3\dfrac{2}{3}\right), \left(-3, 2\dfrac{1}{3}\right), \left(3, 4\dfrac{1}{3}\right)$.

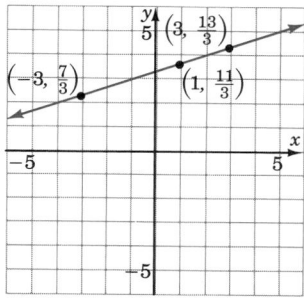

2. $4x + 4y = 0$

We solve for y.

$4y = -4x$
$y = -x$

x	y	(x, y)
0	0	(0, 0)
2	-2	(2, -2)
-3	3	(-3, 3)

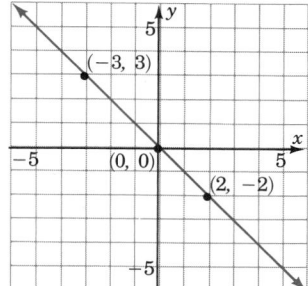

Notice that the x- and y-intercepts are the same point. Thus the intercept method does not provide enough information to construct this graph.

When an equation is given in the general form $ax + by = c$, usually the most efficient approach to constructing the graph is to use the intercept method, when it works.

★ **PRACTICE SET B**

Graph the following equations.

1. $x - 5y = 5$

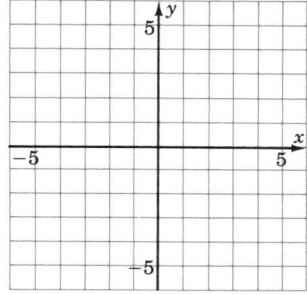

2. $x + 2y = 6$

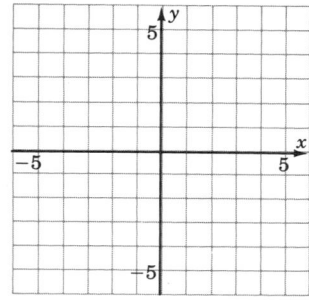

3. $2x + y = 1$

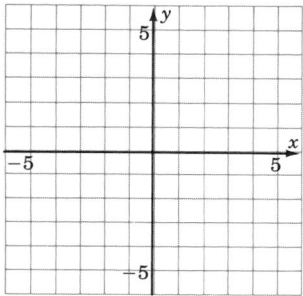

❏ SLANTED, HORIZONTAL, AND VERTICAL LINES

In all the graphs we have observed so far, the lines have been slanted. This will always be the case when *both* variables appear in the equation. If only one variable appears in the equation, then the line will be either vertical or horizontal. To see why, let's consider a specific case:

Using the general form of a line, $ax + by = c$, we can produce an equation with exactly one variable by choosing $a = 0$, $b = 5$, and $c = 15$. The equation $ax + by = c$ then becomes

$0x + 5y = 15$

Since $0 \cdot$ (any number) $= 0$, the term $0x$ is 0 for any number that is chosen for x. Thus,

$0x + 5y = 15$

becomes

$0 + 5y = 15$

But, 0 is the additive identity and $0 + 5y = 5y$.

$5y = 15$

Then, solving for y we get

$y = 3$

This is an equation in which exactly one variable appears.

This means that regardless of which number we choose for x, the corresponding y-value is 3. Since the y-value is always the same as we move from left-to-right through the x-values, the height of the line above the x-axis is always the same (in this case, 3 units). This type of line must be horizontal.

An argument similar to the one above will show that if the only variable that appears is x, we can expect to get a vertical line.

☆ **SAMPLE SET C**

---

1. Graph $y = 4$.   The only variable appearing is y. Regardless of which x-value we choose, the y-value is always 4. All points with a y-value of 4 satisfy the equation. Thus we get a horizontal line 4 units above the x-axis.

x	y	(x, y)
−3	4	(−3, 4)
−2	4	(−2, 4)
−1	4	(−1, 4)
0	4	(0, 4)
1	4	(1, 4)
2	4	(2, 4)
3	4	(3, 4)
4	4	(4, 4)

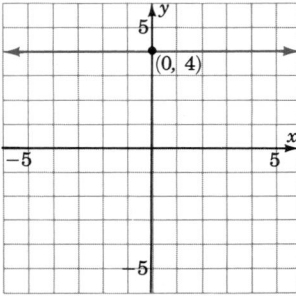

2. Graph $x = -2$. The only variable that appears is x. Regardless of which y-value we choose, the x-value will always be -2. Thus, we get a vertical line two units to the left of the y-axis.

x	y	(x, y)
-2	-4	$(-2, -4)$
-2	-3	$(-2, -3)$
-2	-2	$(-2, -2)$
-2	-1	$(-2, -1)$
-2	0	$(-2, 0)$
-2	1	$(-2, 1)$
-2	2	$(-2, 2)$
-2	3	$(-2, 3)$
-2	4	$(-2, 4)$

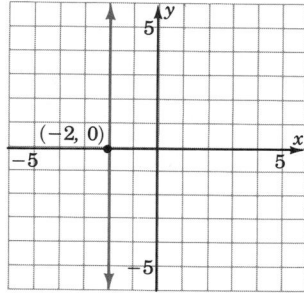

★ **PRACTICE SET C**

1. Graph $y = 2$.

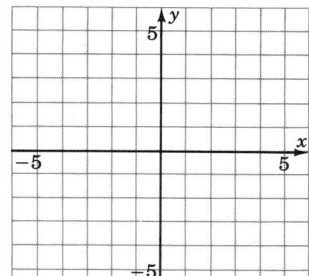

2. Graph $x = -4$.

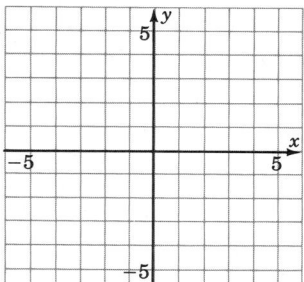

Answers to Practice Sets are on p. 263.

Summarizing our results we can make the following observations:

1. When a linear equation in two variables is written in the form $ax + by = c$, we say it is written in *general form*.
2. To graph an equation in general form it is sometimes convenient to use the intercept method.
3. A linear equation in which both variables appear will graph as a slanted line.
4. A linear equation in which only one variable appears will graph as either a vertical or horizontal line.

$x = a$ graphs as a vertical line passing through a on the x-axis.

$y = b$ graphs as a horizontal line passing through b on the y-axis.

Section 6.3 EXERCISES

For problems 1–30, graph the equations.

1. $-3x + y = -1$

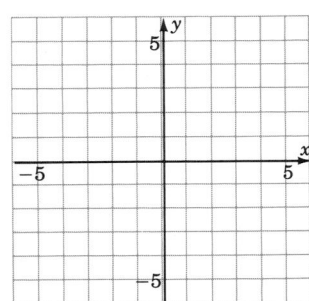

2. $3x - 2y = 6$

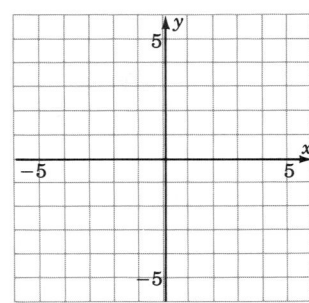

3. $-2x + y = 4$

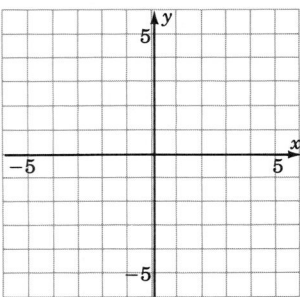

4. $x - 3y = 5$

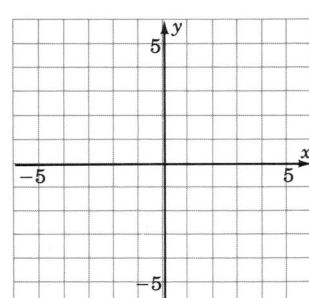

5. $2x - 3y = 6$

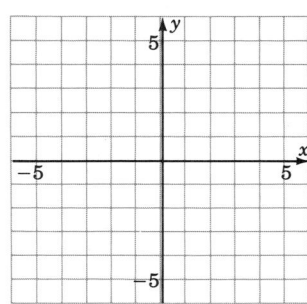

6. $2x + 5y = 10$

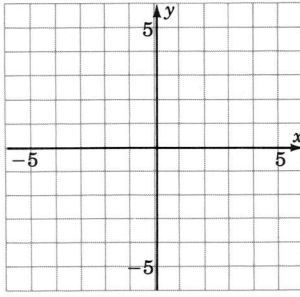

7. $3(x - y) = 9$

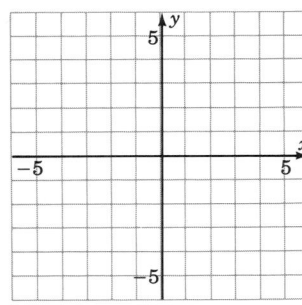

8. $-2x + 3y = -12$

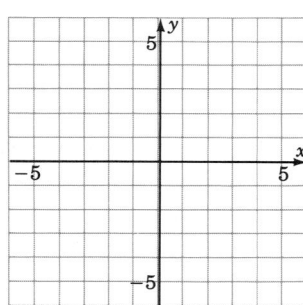

9. $y + x = 1$

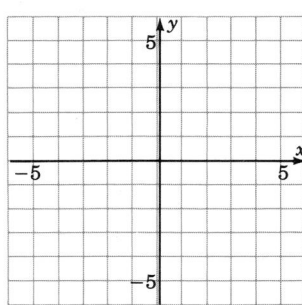

10. $4y - x - 12 = 0$

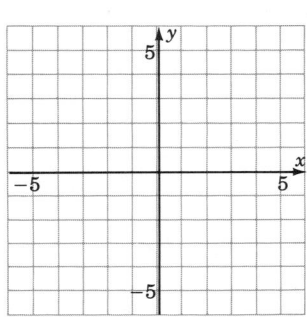

11. $2x - y + 4 = 0$

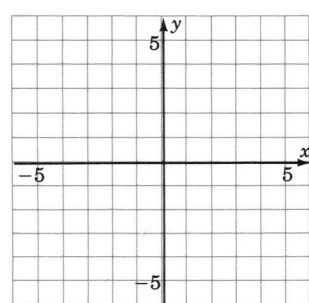

12. $-2x + 5y = 0$

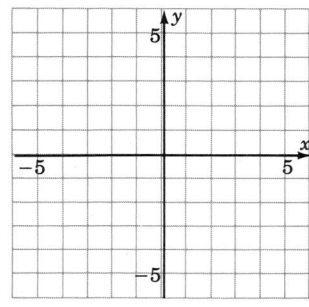

13. $y - 5x + 4 = 0$

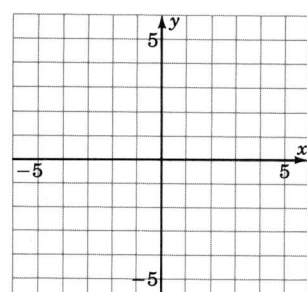

14. $0x + y = 3$

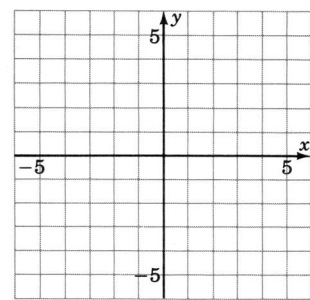

15. $0x + 2y = 2$

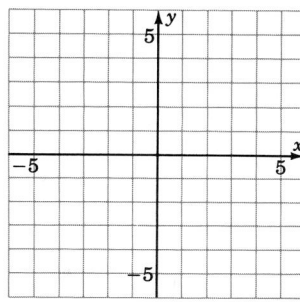

16. $0x + \dfrac{1}{4}y = 1$

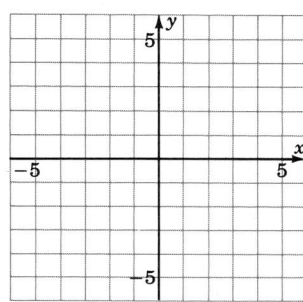

17. $4x + 0y = 16$

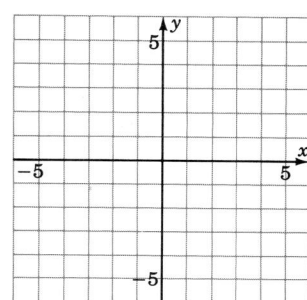

18. $\dfrac{1}{2}x + 0y = -1$

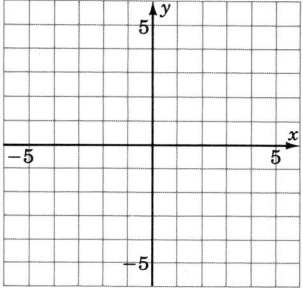

19. $\dfrac{2}{3}x + 0y = 1$

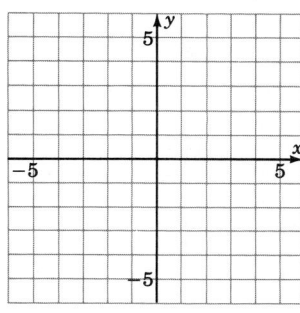

20. $y = 3$

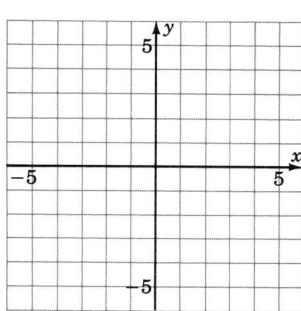

21. $y = -2$

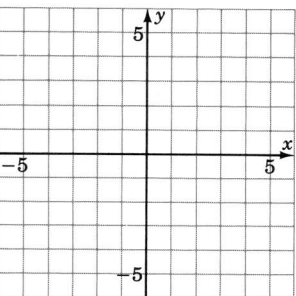

22. $-4y = 20$

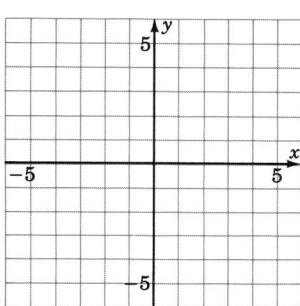

23. $x = -4$

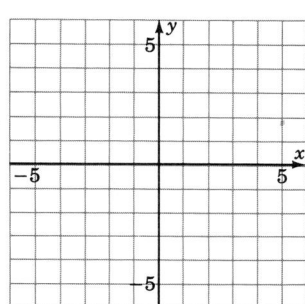

24. $-3x = -9$

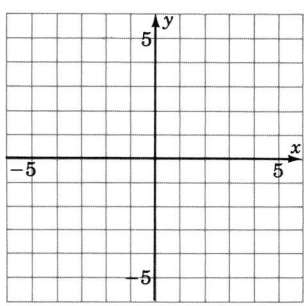

25. $-x + 4 = 0$

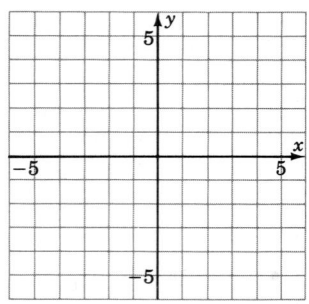

26. Construct the graph of all the points that have coordinates (a, a), that is, for each point, the x- and y- values are the same.

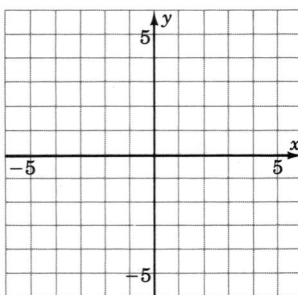

Calculator Problems

27. $2.53x + 4.77y = 8.45$

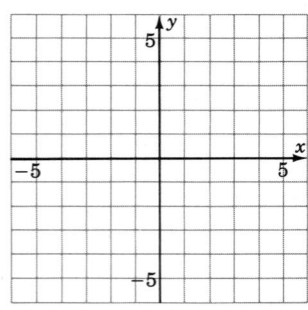

28. $1.96x + 2.05y = 6.55$

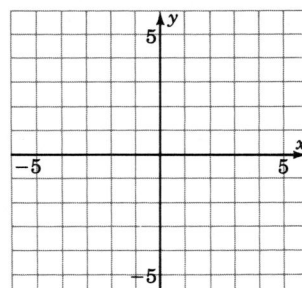

29. $4.1x - 6.6y = 15.5$

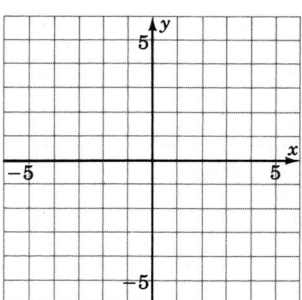

30. $626.01x - 506.73y = 2443.50$

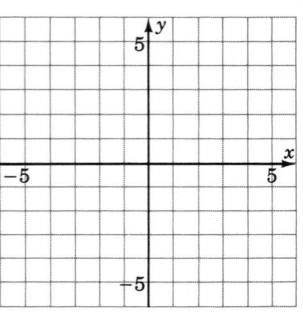

EXERCISES
FOR REVIEW

(1.2) **31.** Name the property of real numbers that makes $4 + x = x + 4$ a true statement.

(2.2) **32.** Supply the missing word. The absolute value of a number a, denoted $|a|$, is the _____ from a to 0 on the number line.

(3.5) **33.** Find the product $(3x + 2)(x - 7)$.

(4.3) **34.** Solve the equation $3[3(x - 2) + 4x] - 24 = 0$.

(6.2) **35.** Supply the missing word. The coordinate axes divide the plane into four equal regions called _____ .

★ Answers to Practice Sets (6.3)

A. **1.** When $x = 0$, $y = 3$; when $y = 0$, $x = 1$

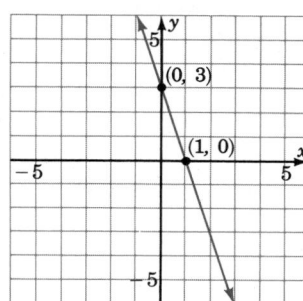

B. **1.**

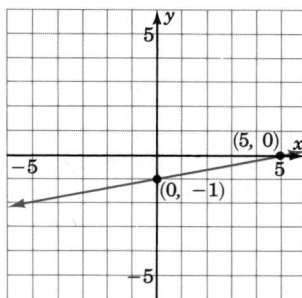

2.

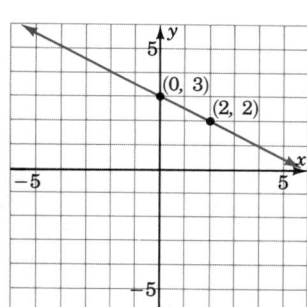

3.

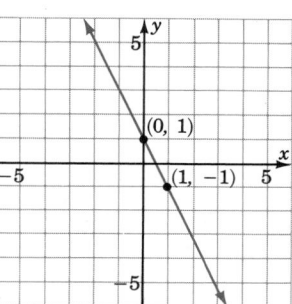

C. **1.**

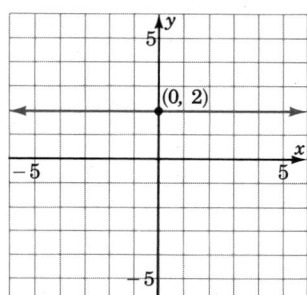

2.

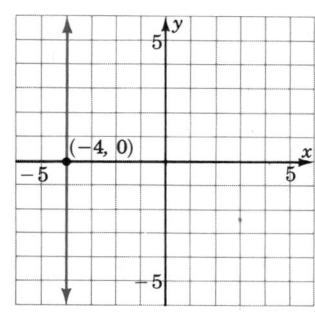

6.4 The Slope-Intercept Form of a Line

Section Overview

☐ THE GENERAL FORM OF A LINE
☐ THE SLOPE-INTERCEPT FORM OF A LINE
☐ SLOPE AND INTERCEPT
☐ THE FORMULA FOR THE SLOPE OF A LINE

☐ THE GENERAL FORM OF A LINE

We have seen that the general form of a linear equation in two variables is $ax + by = c$ (Section 6.3). When this equation is solved for y, the resulting form is called the slope-intercept form. Let's generate this new form.

$ax + by = c$ Subtract ax from *both* sides.
$by = -ax + c$ Divide *both* sides by b.

$$\frac{by}{b} = \frac{-ax}{b} + \frac{c}{b}$$

$$\frac{\cancel{b}y}{\cancel{b}} = \frac{-ax}{b} + \frac{c}{b}$$

$$y = \frac{-ax}{b} + \frac{c}{b}$$

$$y = \frac{-a}{b}x + \frac{c}{b}$$

This equation is of the form $y = mx + b$ if we replace $\frac{-a}{b}$ with m and the constant $\frac{c}{b}$ with b. (*Note:* The fact that we let $b = \frac{c}{b}$ is unfortunate and occurs because of the letters we have chosen to use in the general form. The letter b occurs on both sides of the equal sign and may not represent the same value at all. This problem is one of historical convention and, fortunately, does not occur very often.)

The following examples illustrate this procedure.

1. Solve $3x + 2y = 6$ for y.

   $3x + 2y = 6$ Subtract $3x$ from both sides.
   $2y = -3x + 6$ Divide both sides by 2.

   $$y = -\frac{3}{2}x + 3$$

   This equation is of the form $y = mx + b$. In this case, $m = -\frac{3}{2}$ and $b = 3$.

2. Solve $-15x + 5y = 20$ for y.

   $-15x + 5y = 20$
   $5y = 15x + 20$
   $y = 3x + 4$

   This equation is of the form $y = mx + b$. In this case, $m = 3$ and $b = 4$.

3. Solve $4x - y = 0$ for y.

$4x - y = 0$
$-y = -4x$
$y = 4x$

This equation is of the form $y = mx + b$. In this case, $m = 4$ and $b = 0$. Notice that we can write $y = 4x$ as $y = 4x + 0$.

❑ THE SLOPE-INTERCEPT FORM OF A LINE

The Slope-Intercept Form of
a Line $y = mx + b$

A linear equation in two variables written in the form $y = mx + b$ is said to be in **slope-intercept form.**

☆ SAMPLE SET A

The following equations *are* in slope-intercept form:

1. $y = 6x - 7$.   In this case $m = 6$ and $b = -7$.

2. $y = -2x + 9$.   In this case $m = -2$ and $b = 9$.

3. $y = \frac{1}{5}x + 4.8$ In this case $m = \frac{1}{5}$ and $b = 4.8$.

4. $y = 7x$.   In this case $m = 7$ and $b = 0$ since we can write $y = 7x$ as $y = 7x + 0$.

The following equations *are not* in slope-intercept form:

5. $2y = 4x - 1$.   The coefficient of y is 2. To be in slope-intercept form, the coefficient of y must be 1.

6. $y + 4x = 5$.   The equation is not solved for y. The x and y appear on the same side of the equal sign.

7. $y + 1 = 2x$.   The equation is not solved for y.

★ PRACTICE SET A

The following equations are in slope-intercept form. In each case, specify the slope and y-intercept.

1. $y = 2x + 7$; $m =$ $b =$ **2.** $y = -4x + 2$; $m =$ $b =$

3. $y = -5x - 1$; $m =$ $b =$ **4.** $y = \frac{2}{3}x - 10$; $m =$ $b =$

5. $y = \frac{-5}{8}x + \frac{1}{2}$; $m =$ $b =$ **6.** $y = -3x$; $m =$ $b =$

❑ SLOPE AND INTERCEPT

When the equation of a line is written in slope-intercept form, two important properties of the line can be seen: the *slope* and the *intercept*. Let's look at these two properties by graphing several lines and observing them carefully.

☆ **SAMPLE SET B**

1. Graph the line $y = x - 3$.

x	y	(x, y)
0	−3	(0, −3)
4	1	(4, 1)
−2	−5	(−2, −5)

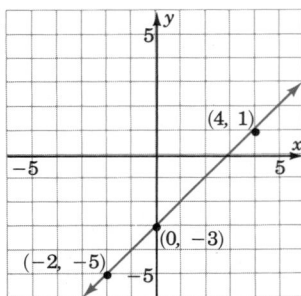

Looking carefully at this line, answer the following two questions.

1. At what number does the line cross the y-axis? Do you see this number in the equation?
2. Place your pencil at any point on the line. Move your pencil exactly *one* unit horizontally to the right. Now, how many units straight up or down must you move your pencil to get back on the line? Do you see this number in the equation?

Answers:

1. The line crosses the y-axis at −3.
2. After moving horizontally one unit to the right, we must move exactly one vertical unit up. This number is the coefficient of x.

2. Graph the line $y = \frac{2}{3}x + 1$.

x	y	(x, y)
0	1	(0, 1)
3	3	(3, 3)
−3	−1	(−3, −1)

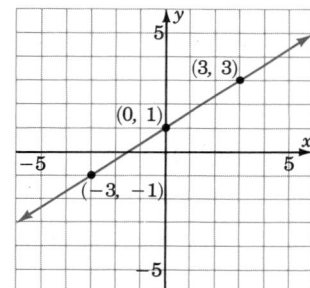

Looking carefully at this line, answer the following two questions.

1. At what number does this line cross the y-axis? Do you see this number in the equation?
2. Place your pencil at any point on the line. Move your pencil exactly *one* unit horizontally to the right. Now, how many units straight up or down must you move your pencil to get back on the line? Do you see this number in the equation?

Answers:

1. The line crosses the y-axis at +1.

2. After moving horizontally one unit to the right, we must move exactly $\frac{2}{3}$ unit upward. This number is the coefficient of x.

★ PRACTICE SET B

Graph the line $y = -3x + 4$.

x	y	(x, y)
0		
3		
2		

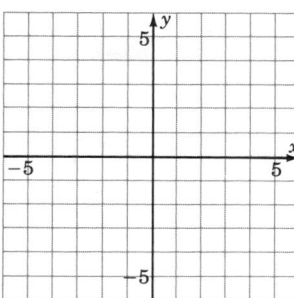

Looking carefully at this line, answer the following two questions.

1. At what number does the line cross the y-axis? Do you see this number in the equation?

2. Place your pencil at any point on the line. Move your pencil exactly *one* unit horizontally to the right. Now, how many units straight up or down must you move your pencil to get back on the line? Do you see this number in the equation?

In the graphs constructed in Sample Set B and Practice Set B, each equation had the form $y = mx + b$. We can answer the same questions by using this form of the equation (shown in the diagram).

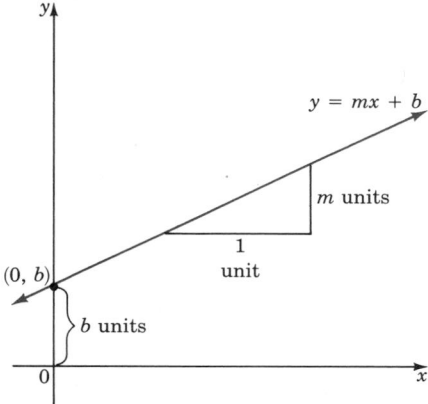

y-Intercept

Question 1: At what number does the line cross the y-axis? Do you see this number in the equation?

Answer: In each case, the line crosses the y-axis at the constant b. The number b is the number at which the line crosses the y-axis, and it is called the *y-intercept*. The ordered pair corresponding to the y-intercept is $(0, b)$.

Question 2: Place your pencil at any point on the line. Move your pencil exactly *one* unit horizontally to the right. Now, how many units straight up or down must you move your pencil to get back on the line? Do you see this number in the equation?

Slope

Answer: To get back on the line, we must move our pencil exactly m vertical units. The number m is the coefficient of the variable x. The number m is called the **slope** of the line and it is the number of units that y changes when x is increased by 1 unit. Thus, if x changes by 1 unit, y changes by m units.

Since the equation $y = mx + b$ contains both the slope of the line and the y-intercept, we call the form $y = mx + b$ the *slope-intercept* form.

The Slope-Intercept Form of the Equation of a Line

The slope-intercept form of a straight line is

$y = mx + b$

The slope of the line is m, and the y-intercept is the point $(0, b)$.

The Slope is a Measure of the Steepness of a Line

The word *slope* is really quite appropriate. It gives us a measure of the steepness of the line. Consider two lines, one with slope $\frac{1}{2}$ and the other with slope 3. The line with slope 3 is steeper than is the line with slope $\frac{1}{2}$. Imagine your pencil being placed at any point on the lines. We make a 1-unit increase in the x-value by moving the pencil *one* unit to the right. To get back to one line we need only move vertically $\frac{1}{2}$ unit, whereas to get back onto the other line we need to move vertically 3 units.

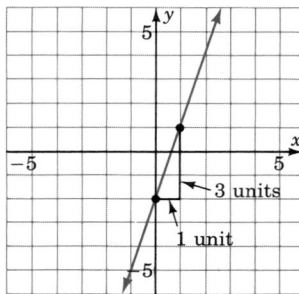 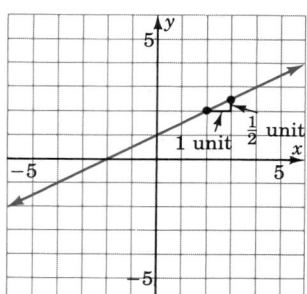

☆ **SAMPLE SET C**

Find the slope and the y-intercept of the following lines.

1. $y = 2x + 7$.   The line is in the slope-intercept form $y = mx + b$. The slope is m, the coefficient of x. Therefore, $m = 2$. The y-intercept is the point $(0, b)$. Since $b = 7$, the y-intercept is $(0, 7)$.

Slope: 2
y-intercept: $(0, 7)$

2. $y = -4x + 1$.   The line is in slope-intercept form $y = mx + b$. The slope is m, the coefficient of x. So, $m = -4$. The y-intercept is the point $(0, b)$. Since $b = 1$, the y-intercept is $(0, 1)$.

Slope: -4
y-intercept: $(0, 1)$

3. $3x + 2y = 5$.   The equation is written in general form. We can put the equation in slope-intercept form by solving for y.

$$3x + 2y = 5$$
$$2y = -3x + 5$$
$$y = -\frac{3}{2}x + \frac{5}{2}$$

Now the equation is in slope-intercept form.

$$\text{Slope: } -\frac{3}{2}$$

$$y\text{-intercept: } \left(0, \frac{5}{2}\right)$$

★ **PRACTICE SET C**

Find the slope and y-intercept of the line $2x + 5y = 15$.

☐ THE FORMULA FOR THE SLOPE OF A LINE

We have observed that the slope is a measure of the steepness of a line. We wish to develop a formula for measuring this steepness.

It seems reasonable to develop a slope formula that produces the following results:

Steepness of line 1 > steepness of line 2.

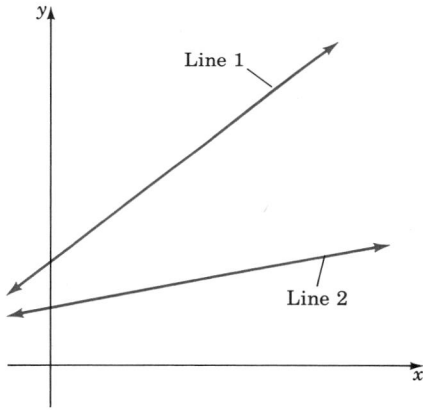

Consider a line on which we select any two points. We'll denote these points with the ordered pairs (x_1, y_1) and (x_2, y_2). The subscripts help us to identify the points.

(x_1, y_1) is the first point.   Subscript 1 indicates the first point.
(x_2, y_2) is the second point.   Subscript 2 indicates the second point.

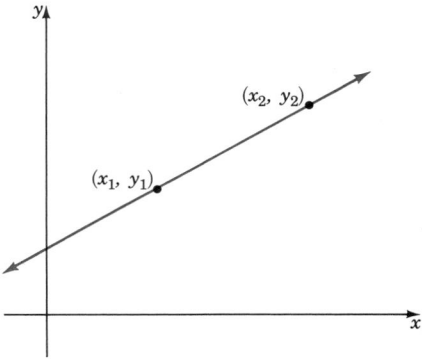

The difference in x values $(x_2 - x_1)$ gives us the horizontal change, and the difference in y values $(y_2 - y_1)$ gives us the vertical change. If the line is very steep, then when going from the first point to the second point, we would expect a large vertical change compared to the horizontal change. If the line is not very steep, then when going from the first point to the second point, we would expect a small vertical change compared to the horizontal change.

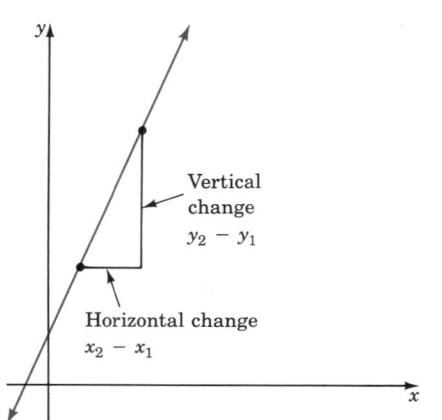   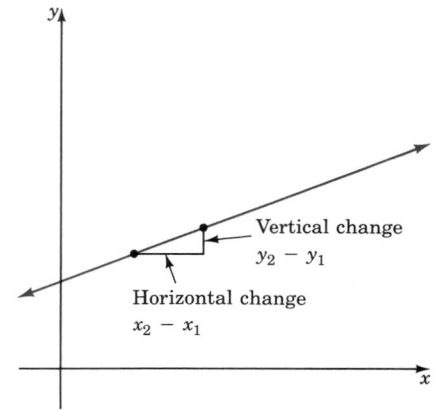

We are comparing changes. We see that we are comparing

The vertical change	to	the horizontal change
The change in y	to	the change in x
$y_2 - y_1$	to	$x_2 - x_1$

This is a comparison and is therefore a **ratio.** Ratios can be expressed as fractions. Thus, a measure of the steepness of a line can be expressed as a ratio.

The slope of a line is defined as the ratio

$$\text{Slope} = \frac{\text{change in } y}{\text{change in } x}$$

Mathematically, we can write these changes as

$$\text{Slope} = \frac{y_2 - y_1}{x_2 - x_1}$$

Finding the Slope of a Line

> The slope of a nonvertical line passing through the points (x_1, y_1) and (x_2, y_2) is found by the formula
>
> $$m = \frac{y_2 - y_1}{x_2 - x_1}$$

☆ SAMPLE SET D

For the two given points, find the slope of the line that passes through them.

1. $(0, 1)$ and $(1, 3)$.   Looking left to right on the line we can choose (x_1, y_1) to be $(0, 1)$, and (x_2, y_2) to be $(1, 3)$. Then,

$$m = \frac{y_2 - y_1}{x_2 - x_1} = \frac{3 - 1}{1 - 0} = \frac{2}{1} = 2$$

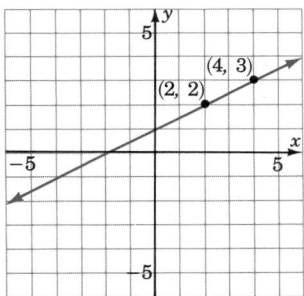

This line has slope 2. It appears fairly steep. When the slope is written in fraction form, $2 = \frac{2}{1}$, we can see, by recalling the slope formula, that as x changes 1 unit to the right (because of the $+1$), y changes 2 units upward (because of the $+2$).

$$m = \frac{\text{change in } y}{\text{change in } x} = \frac{2}{1}$$

Notice that as we look left to right, the line rises.

2. $(2, 2)$ and $(4, 3)$.   Looking left to right on the line we can choose (x_1, y_1) to be $(2, 2)$ and (x_2, y_2) to be $(4, 3)$. Then,

$$m = \frac{y_2 - y_1}{x_2 - x_1} = \frac{3 - 2}{4 - 2} = \frac{1}{2}$$

This line has slope $\frac{1}{2}$. Thus, as x changes 2 units to the right (because of the $+2$), y changes 1 unit upward (because of the $+1$).

$$m = \frac{\text{change in } y}{\text{change in } x} = \frac{1}{2}$$

Notice that in examples 1 and 2, both lines have positive slopes, $+2$ and $+\frac{1}{2}$, and both lines *rise* as we look left to right.

3. $(-2, 4)$ and $(1, 1)$.   Looking left to right on the line we can choose (x_1, y_1) to be $(-2, 4)$ and (x_2, y_2) to be $(1, 1)$. Then,

$$m = \frac{y_2 - y_1}{x_2 - x_1} = \frac{1 - 4}{1 - (-2)} = \frac{-3}{1 + 2} = \frac{-3}{3} = -1$$

Continued

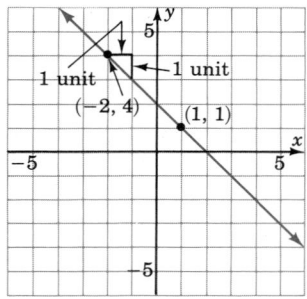

This line has slope -1.

When the slope is written in fraction form, $m = -1 = \dfrac{-1}{+1}$, we can see that as x changes 1 unit to the right (because of the $+1$), y changes 1 unit downward (because of the -1).

Notice also that this line has a negative slope and declines as we look left to right.

4. $(1, 3)$ and $(5, 3)$.

$$m = \frac{y_2 - y_1}{x_2 - x_1} = \frac{3 - 3}{5 - 1} = \frac{0}{4} = 0$$

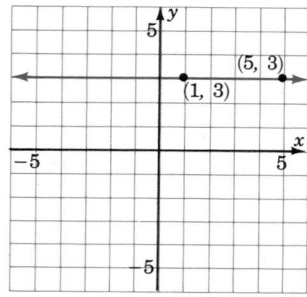

This line has 0 slope. This means it has *no* rise and, therefore, is a horizontal line. This does not mean that the line has no slope, however.

5. $(4, 4)$ and $(4, 0)$. This problem shows why the slope formula is valid only for nonvertical lines.

$$m = \frac{y_2 - y_1}{x_2 - x_1} = \frac{0 - 4}{4 - 4} = \frac{-4}{0}$$

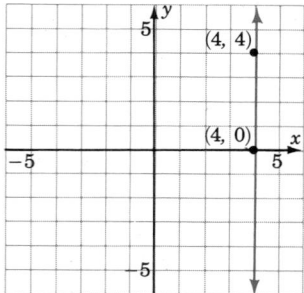

Since division by 0 is undefined, we say that vertical lines have undefined slope. Since there is no real number to represent the slope of this line, we sometimes say that vertical lines have *undefined slope,* or *no slope.*

★ **PRACTICE SET D**

1. Find the slope of the line passing through (2, 1) and (6, 3). Graph this line on the graph of problem 2 below.
2. Find the slope of the line passing through (3, 4) and (5, 5). Graph this line.

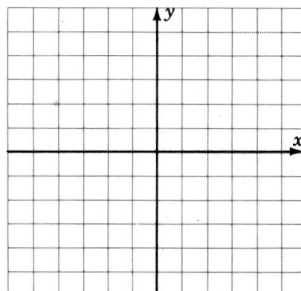

3. Compare the lines of problems 1 and 2. Do the lines appear to cross? What is it called when lines do not meet (parallel or intersecting)? Compare their slopes. Make a statement about the condition of these lines and their slopes.

Answers to Practice Sets are on p. 277.

Before trying some problems, let's summarize what we have observed.

1. The equation $y = mx + b$ is called the slope-intercept form of the equation of a line. The number m is the slope of the line and the point $(0, b)$ is the y-intercept.
2. The slope, m, of a line is defined as the steepness of the line, and it is the number of units that y changes when x changes 1 unit.
3. The formula for finding the slope of a line through any two given points (x_1, y_1) and (x_2, y_2) is

$$m = \frac{y_2 - y_1}{x_2 - x_1}$$

4. The fraction $\dfrac{y_2 - y_1}{x_2 - x_1}$ represents the $\dfrac{\text{change in } y}{\text{change in } x}$.
5. As we look at a graph from left to right, lines with positive slope rise and lines with negative slope decline.
6. Parallel lines have the same slope.
7. Horizontal lines have 0 slope.
8. Vertical lines have undefined slope (or no slope).

Section 6.4 EXERCISES

For problems 1–25, determine the slope and y-intercept of the lines.

1. $y = 3x + 4$

2. $y = 2x + 9$

3. $y = 9x + 1$

4. $y = 7x + 10$

5. $y = -4x + 5$

6. $y = -2x + 8$

7. $y = -6x - 1$

8. $y = -x - 6$

9. $y = -x + 2$

10. $2y = 4x + 8$

For problems 26–47, find the slope of the line through the pairs of points.

26. $(1, 6)$, $(4, 9)$

11. $4y = 16x + 20$

12. $-5y = 15x + 55$

27. $(1, 3)$, $(4, 7)$

13. $-3y = 12x - 27$

14. $y = \dfrac{3}{5}x - 8$

28. $(3, 5)$, $(4, 7)$

15. $y = \dfrac{2}{7}x - 12$

16. $y = \dfrac{-1}{8}x + \dfrac{2}{3}$

29. $(6, 1)$, $(2, 8)$

17. $y = \dfrac{-4}{5}x - \dfrac{4}{7}$

18. $-3y = 5x + 8$

30. $(0, 5)$, $(2, -6)$

19. $-10y = -12x + 1$

31. $(-2, 1)$, $(0, 5)$

20. $-y = x + 1$

21. $-y = -x + 3$

32. $(3, -9)$, $(5, 1)$

22. $3x - y = 7$

23. $5x + 3y = 6$

24. $-6x - 7y = -12$

25. $-x + 4y = -1$

33. $(4, -6)$, $(-2, 1)$

34. $(-5, 4)$, $(-1, 0)$

35. $(-3, 2)$, $(-4, 6)$

36. $(9, 12)$, $(6, 0)$

37. $(0, 0)$, $(6, 6)$

38. $(-2, -6)$, $(-4, -1)$

39. $(-1, -7)$, $(-2, -9)$

40. $(-6, -6)$, $(-5, -4)$

41. $(-1, 0)$, $(-2, -2)$

42. $(-4, -2)$, $(0, 0)$

43. $(2, 3)$, $(10, 3)$

44. $(4, -2)$, $(4, 7)$

45. $(8, -1)$, $(8, 3)$

46. $(4, 2)$, $(6, 2)$

47. $(5, -6)$, $(9, -6)$

48. Do lines with a positive slope rise or decline as we look left to right?

49. Do lines with a negative slope rise or decline as we look left to right?

50. Make a statement about the slopes of parallel lines.

Calculator Problems

For problems 51–54, determine the slope and y-intercept of the lines. Round to two decimal places.

51. $3.8x + 12.1y = 4.26$

52. $8.09x + 5.57y = -1.42$

53. $10.813x - 17.0y = -45.99$

54. $-6.003x - 92.388y = 0.008$

For problems 55 – 60, find the slope of the line through the pairs of points. Round to two decimal places.

55. $(5.56, 9.37), \quad (2.16, 4.90)$

56. $(33.1, 8.9), \quad (42.7, -1.06)$

57. $(155.89, 227.61), \quad (157.04, 227.61)$

58. $(0.00426, -0.00404), \quad (-0.00191, -0.00404)$

59. $(88.81, -23.19), \quad (88.81, -26.87)$

60. $(-0.0000567, -0.0000567), \quad (-0.00765, 0.00764)$

EXERCISES FOR REVIEW

(1.6)   **61.** Simplify $(x^2y^3w^4)^0$.

(4.3)   **62.** Solve the equation $3x - 4(2 - x) - 3(x - 2) + 4 = 0$.

(4.5)   **63.** When four times a number is divided by five, and that result is decreased by eight, the result is zero. What is the original number?

(4.7)   **64.** Solve $-3y + 10 = x + 2$ if $x = -4$.

(6.3)   **65.** Graph the linear equation $x + y = 3$.

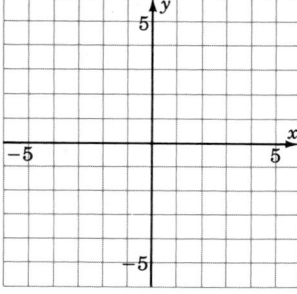

★ **Answers to Practice Sets (6.4)**

A. **1.** $m = 2, b = 7$ **2.** $m = -4, b = 2$ **3.** $m = -5, b = -1$ **4.** $m = \frac{2}{3}, b = -10$

5. $m = \frac{-5}{8}, b = \frac{1}{2}$ **6.** $m = -3, b = 0$

B. The line crosses the y-axis at $+4$. After moving horizontally 1 unit to the right, we must move exactly 3 units downward.

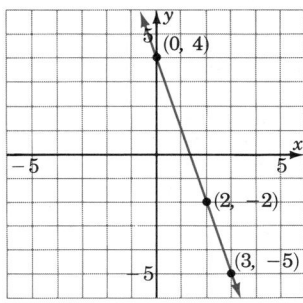

C. Solving for y we get $y = \frac{-2}{5}x + 3$. Now, $m = \frac{-2}{5}$ and $b = 3$.

D. **1.** $m = \frac{3-1}{6-2} = \frac{2}{4} = \frac{1}{2}$. **2.** The line has slope $\frac{1}{2}$. **3.** The lines appear to be parallel. Parallel lines have the same slope, and lines that have the same slope are parallel.

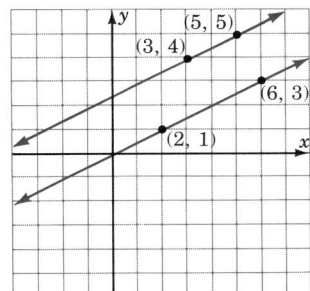

6.5 Graphing Equations in Slope-Intercept Form

Section Overview

❏ **USING THE SLOPE AND INTERCEPT TO GRAPH A LINE**

❏ **USING THE SLOPE AND INTERCEPT TO GRAPH A LINE**

When a linear equation is given in the *general form*, $ax + by = c$, we observed that an efficient graphical approach was the intercept method. We let $x = 0$ and computed the corresponding value of y, then let $y = 0$ and computed the corresponding value of x.

When an equation is written in the *slope-intercept form*, $y = mx + b$, there are also efficient ways of constructing the graph. One way, but less efficient, is to choose two or three x-values and compute to find the corresponding y-values. However, computations are tedious, time consuming, and can lead to errors. Another way, the method listed below, makes use of the slope and the y-intercept for graphing the line. It is quick, simple, and involves no computations.

GRAPHING METHOD

1. Plot the y-intercept $(0, b)$.
2. Determine another point by using the slope m.
3. Draw a line through the two points.

Recall that we defined the slope m as the ratio $\dfrac{y_2 - y_1}{x_2 - x_1}$. The numerator $y_2 - y_1$ represents the number of units that y changes and the denominator $x_2 - x_1$ represents the number of units that x changes. Suppose $m = \dfrac{p}{q}$. Then p is the number of units that y changes and q is the number of units that x changes. Since these changes occur simultaneously, start with your pencil at the y-intercept, move p units in the appropriate vertical direction, and then move q units in the appropriate horizontal direction. Mark a point at this location.

☆ **SAMPLE SET A**

Graph the following lines.

1. $y = \dfrac{3}{4}x + 2$

Step 1:   The y-intercept is the point $(0, 2)$. Thus the line crosses the y-axis 2 units above the origin. Mark a point at $(0, 2)$.

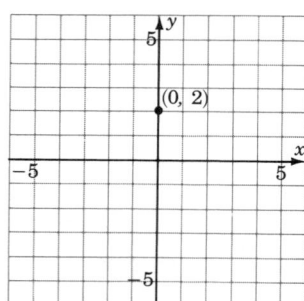

Step 2:   The slope, m, is $\dfrac{3}{4}$. This means that if we start at any point on the line and move our pencil 3 units up and then 4 units to the right, we'll be back on the line. Start at a known point, the y-intercept $(0, 2)$. Move up 3 units, then move 4 units to the right. Mark a point at this location. (Note also that $\dfrac{3}{4} = \dfrac{-3}{-4}$. This means that if we start at any point on the line and move our pencil 3 units *down* and 4 units to the *left*, we'll be back on the line. Note also that $\dfrac{3}{4} = \dfrac{\frac{3}{4}}{1}$.

This means that if we start at any point on the line and move to the right 1 unit, we'll have to move up 3/4 unit to get back on the line.)

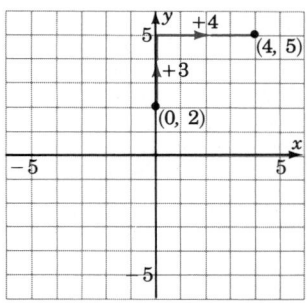

Step 3: Draw a line through both points.

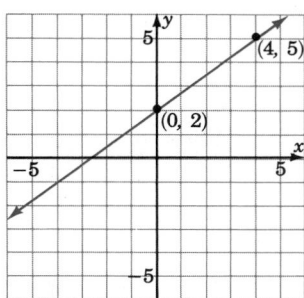

2. $y = -\dfrac{1}{2}x + \dfrac{7}{2}$

Step 1: The y-intercept is the point $\left(0, \dfrac{7}{2}\right)$. Thus the line crosses the y-axis $\dfrac{7}{2}$ units above the origin. Mark a point at $\left(0, \dfrac{7}{2}\right)$, or $\left(0, 3\dfrac{1}{2}\right)$.

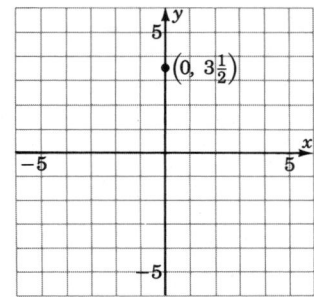

Continued

Step 2:   The slope, m, is $-\dfrac{1}{2}$. We can write $-\dfrac{1}{2}$ as $\dfrac{-1}{2}$. Thus, we start at a known point, the y-intercept $\left(0, 3\dfrac{1}{2}\right)$, move *down* one unit (because of the -1), then move right 2 units. Mark a point at this location.

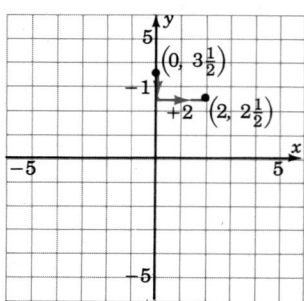

Step 3:   Draw a line through both points.

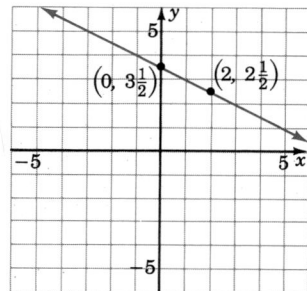

3. $y = \dfrac{2}{5}x$

Step 1:   We can put this equation into explicit slope-intercept by writing it as

$$y = \frac{2}{5}x + 0$$

The y-intercept is the point $(0, 0)$, the origin. This line goes right through the origin.

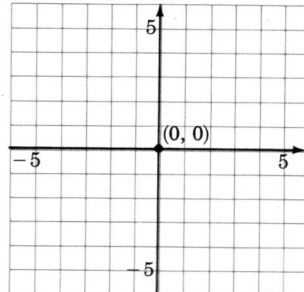

Step 2: The slope, m, is $\frac{2}{5}$. Starting at the origin, we move up 2 units, then move to the right 5 units. Mark a point at this location.

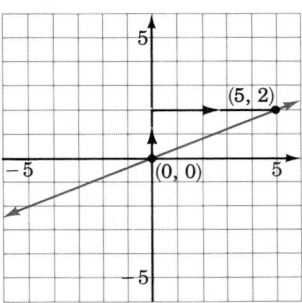

Step 3: Draw a line through the two points.

4. $y = 2x - 4$

Step 1: The y-intercept is the point $(0, -4)$. Thus the line crosses the y-axis 4 units below the origin. Mark a point at $(0, -4)$.

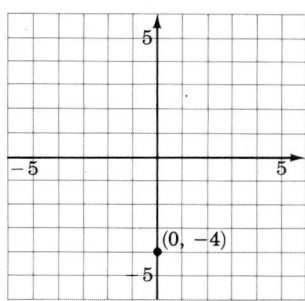

Step 2: The slope, m, is 2. If we write the slope as a fraction, $2 = \frac{2}{1}$, we can read how to make the changes. Start at the known point $(0, -4)$, move up 2 units, then move right 1 unit. Mark a point at this location.

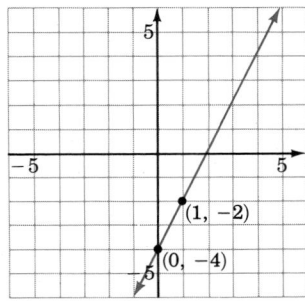

Step 3: Draw a line through the two points.

★ **PRACTICE SET A**

Use the *y*-intercept and the slope to graph each line.

1. $y = \dfrac{-2}{3}x + 4$

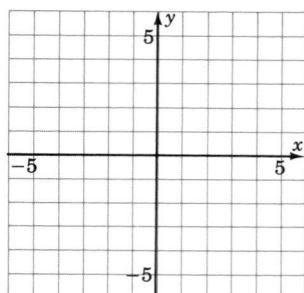

2. $y = \dfrac{3}{4}x$

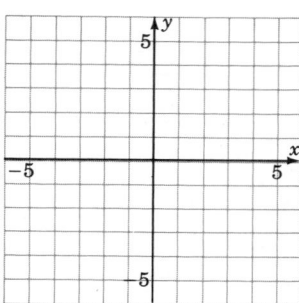

Answers to the Practice Set are on p. 284.

Section 6.5 EXERCISES

For problems 1–18, graph the equations.

1. $y = \dfrac{2}{3}x + 1$

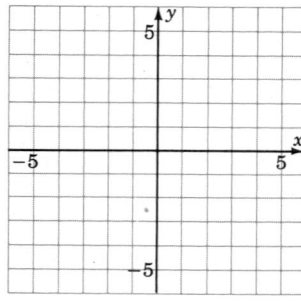

2. $y = \dfrac{1}{4}x - 2$

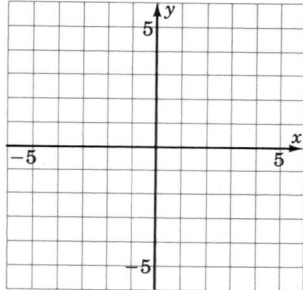

3. $y = 5x - 4$

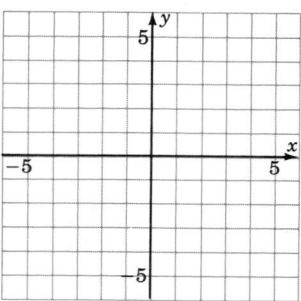

4. $y = -\dfrac{6}{5}x - 3$

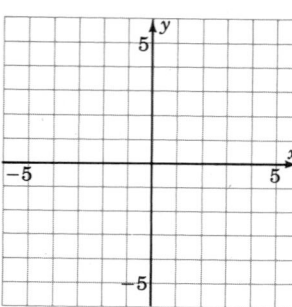

5. $y = \dfrac{3}{2}x - 5$

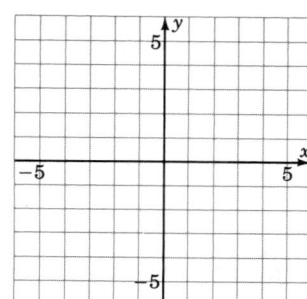

6. $y = \dfrac{1}{5}x + 2$

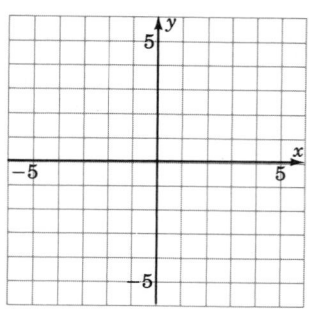

7. $y = -\dfrac{8}{3}x + 4$

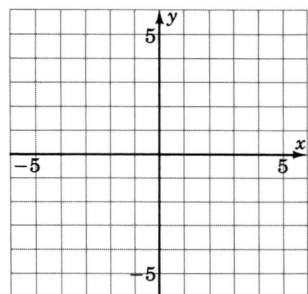

8. $y = -\dfrac{10}{3}x + 6$

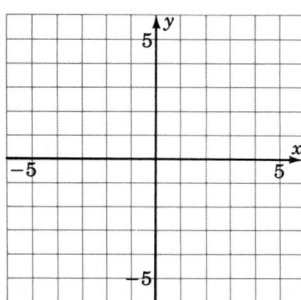

9. $y = 1x - 4$

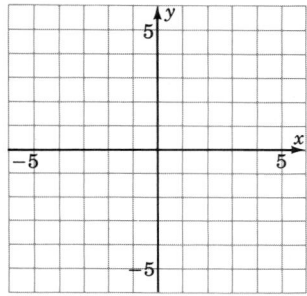

10. $y = -2x + 1$

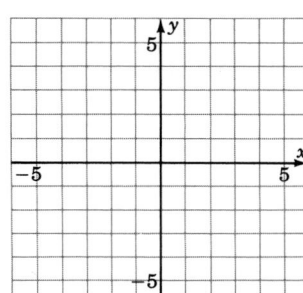

11. $y = x + 2$

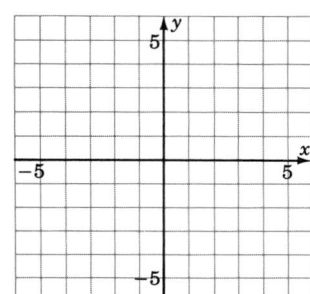

12. $y = \dfrac{3}{5}x$

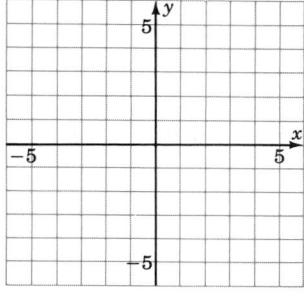

13. $y = -\dfrac{4}{3}x$

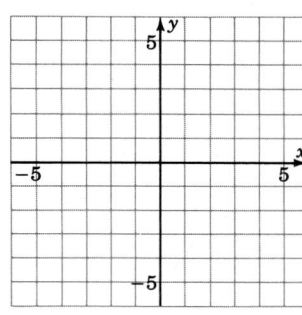

14. $y = x$

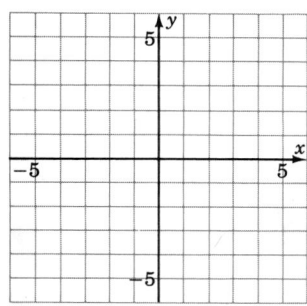

15. $y = -x$

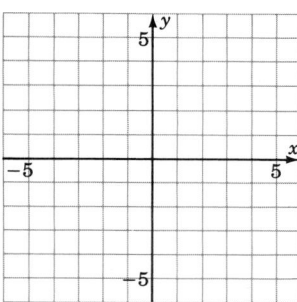

16. $3y - 2x = -3$

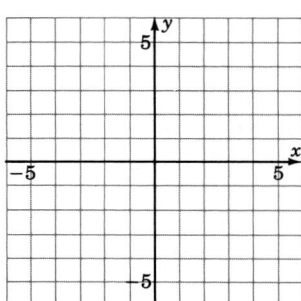

17. $6x + 10y = 30$

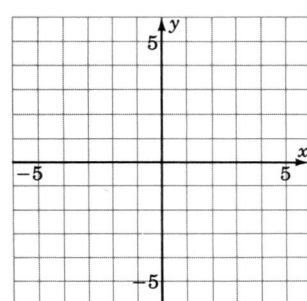

18. $x + y = 0$

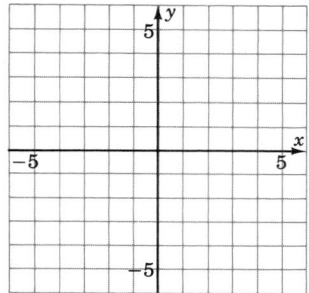

EXERCISES FOR REVIEW

(4.6)   **19.** Solve the inequality $2 - 4x \geq x - 3$.

(6.1)   **20.** Graph the inequality $y + 3 > 1$.

$$\longleftarrow\qquad\qquad\longrightarrow$$

(6.3)   **21.** Graph the equation $y = -2$.

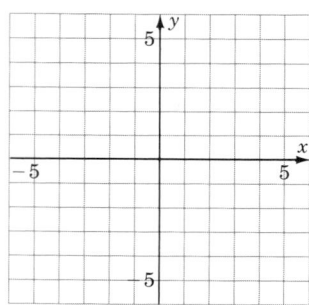

(6.4)   **22.** Determine the slope and y-intercept of the line $-4y - 3x = 16$.

(6.4)   **23.** Find the slope of the line passing through the points $(-1, 5)$ and $(2, 3)$.

★ Answers to Practice Set (6.5)

A. **1.**

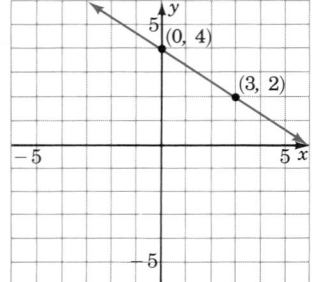

2.

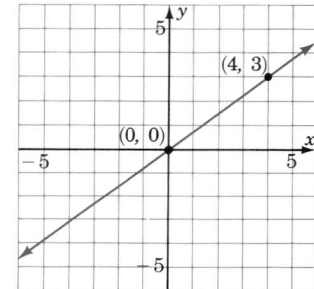

6.6 Finding the Equation of a Line

Section Overview

❑ **THE SLOPE-INTERCEPT AND POINT-SLOPE FORMS**

❑ THE SLOPE-INTERCEPT AND POINT-SLOPE FORMS

In the previous sections we have been given an equation and have constructed the line to which it corresponds. Now, however, suppose we're given some geometric information about the line and we wish to construct the corresponding equation. We wish to find the equation of a line.

We know that the formula for the slope of a line is $m = \dfrac{y_2 - y_1}{x_2 - x_1}$. We can find the equation of a line using the slope formula in either of two ways:

1. If we're given the slope, m, and *any* point (x_1, y_1) on the line, we can substitute this information into the formula for slope.

Let (x_1, y_1) be the known point on the line and let (x, y) be any other point on the line. Then

$$m = \frac{y - y_1}{x - x_1}$$

Multiply both sides by $x - x_1$.

$$m(x - x_1) = (x - x_1) \cdot \frac{y - y_1}{x - x_1}$$

$$m(x - x_1) = y - y_1$$

For convenience, we'll rewrite the equation.

$$y - y_1 = m(x - x_1)$$

Since this equation was derived using a point and the slope of a line, it is called the **point-slope** form of a line.

2. If we are given the slope, m, and the y-intercept, $(0, b)$, we can substitute this information into the formula for slope.

Let $(0, b)$ be the y-intercept and (x, y) be any other point on the line. Then,

$$m = \frac{y - b}{x - 0}$$

$$m = \frac{y - b}{x}$$

Multiply both sides by x.

$$m \cdot x = x \cdot \frac{y - b}{x}$$

$$mx = y - b$$

Solve for y.

$$mx + b = y$$

For convenience, we'll rewrite this equation.

$$y = mx + b$$

Since this equation was derived using the slope and the intercept, it was called the **slope-intercept** form of a line.

We summarize these two derivations as follows.

Forms of the Equation of a Line

> We can find the equation of a line if we're given either of the following sets of information:
>
> 1. The slope, m, and the y-intercept, $(0, b)$, by substituting these values into
>
> $$y = mx + b$$
>
> This is the slope-intercept form.
> 2. The slope, m, and *any* point, (x_1, y_1), by substituting these values into
>
> $$y - y_1 = m(x - x_1)$$
>
> This is the point-slope form.

Notice that both forms rely on knowing the slope. If we are given two points on the line we may still find the equation of the line passing through them by first finding the slope of the line, then using the point-slope form.

It is customary to use either the slope-intercept form or the general form for the final form of the line. We will use the slope-intercept form as the final form.

Find the equation of the line using the given information.

1. $m = 6$, y-intercept $(0, 4)$

Since we're given the slope and the y-intercept, we'll use the slope-intercept form. $m = 6, b = 4$.

$y = mx + b$
$y = 6x + 4$

2. $m = -\dfrac{3}{4}$, y-intercept $\left(0, \dfrac{1}{8}\right)$

Since we're given the slope and the y-intercept, we'll use the slope-intercept form. $m = \dfrac{-3}{4}$, $b = \dfrac{1}{8}$.

$y = mx + b$
$y = -\dfrac{3}{4}x + \dfrac{1}{8}$

3. $m = 2$,   the point $(4, 3)$.   Write the equation in slope-intercept form.

Since we're given the slope and some point, we'll use the point-slope form.

$y - y_1 = m(x - x_1)$ Let (x_1, y_1) be $(4, 3)$.
$y - 3 = 2(x - 4)$ **Put this equation in slope-intercept form by solving for y.**
$y - 3 = 2x - 8$
$y = 2x - 5$

4. $m = -5$,   the point $(-3, 0)$.   Write the equation in slope-intercept form.

Since we're given the slope and some point, we'll use the point-slope form.

$y - y_1 = m(x - x_1)$ Let (x_1, y_1) be $(-3, 0)$.
$y - 0 = -5[x - (-3)]$
$y = -5(x + 3)$ **Solve for y.**
$y = -5x - 15$

5. $m = -1$,   the point $(0, 7)$.   Write the equation in slope-intercept form.

We're given the slope and a point, but careful observation reveals that this point is actually the y-intercept. Thus, we'll use the slope-intercept form. If we had not seen that this point was the y-intercept we would have proceeded with the point-slope form. This would create slightly more work, but still give the same result.

Slope-intercept form   Point-slope form

$y = mx + b$ $y - y_1 = m(x - x_1)$
$y = -1x + 7$ $y - 7 = -1(x - 0)$
$y = -x + 7$ $y - 7 = -x$
          $y = -x + 7$

6. The two points $(4, 1)$ and $(3, 5)$.   Write the equation in slope-intercept form.

Since we're given two points, we'll find the slope first.

$$m = \frac{y_2 - y_1}{x_2 - x_1} = \frac{5 - 1}{3 - 4} = \frac{4}{-1} = -4$$

Now, we have the slope and two points. We can use either point and the point-slope form.

Using (4, 1)	**Using (3, 5)**
$y - y_1 = m(x - x_1)$	$y - y_1 = m(x - x_1)$
$y - 1 = -4(x - 4)$	$y - 5 = -4(x - 3)$
$y - 1 = -4x + 16$	$y - 5 = -4x + 12$
$y = -4x + 17$	$y = -4x + 17$

We can see that the use of either point gives the same result.

★ **PRACTICE SET A**

Find the equation of each line given the following information. Use the slope-intercept form as the final form of the equation.

1. $m = 5$, y-intercept $(0, 8)$.

2. $m = -8$, y-intercept $(0, 3)$.

3. $m = 2$, y-intercept $(0, -7)$.

4. $m = 1$, y-intercept $(0, -1)$.

5. $m = -1$, y-intercept $(0, -10)$.

6. $m = 4$,   the point $(5, 2)$.

7. $m = -6$,   the point $(-1, 0)$.

8. $m = -1$,   the point $(-5, -5)$.

9. The two points $(4, 1)$ and $(6, 5)$.

10. The two points $(-7, -1)$ and $(-4, 8)$.

☆ SAMPLE SET B

1. Find the equation of the line passing through the point $(4, -7)$ having slope 0.

We're given the slope and some point, so we'll use the point-slope form. With $m = 0$ and (x_1, y_1) as $(4, -7)$, we have

$y - y_1 = m(x - x_1)$
$y - (-7) = 0(x - 4)$
$y + 7 = 0$
$y = -7$

This is a horizontal line.

2. Find the equation of the line passing through the point $(1, 3)$ given that the line is vertical.

Since the line is vertical, the slope does not exist. Thus, we cannot use either the slope-intercept form or the point-slope form. We must recall what we know about vertical lines. The equation of this line is simply $x = 1$.

★ PRACTICE SET B

1. Find the equation of the line passing through the point $(-2, 9)$ having slope 0.

2. Find the equation of the line passing through the point $(-1, 6)$ given that the line is vertical.

☆ SAMPLE SET C

Reading only from the graph, determine the equation of the line.

The slope of the line is $\dfrac{2}{3}$, and the line crosses the y-axis at the point $(0, -3)$. Using the slope-intercept form we get

$y = \dfrac{2}{3}x - 3$

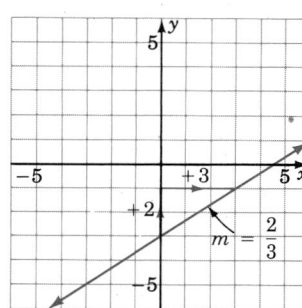

★ **PRACTICE SET C**

1. Reading only from the graph, determine the equation of the line.

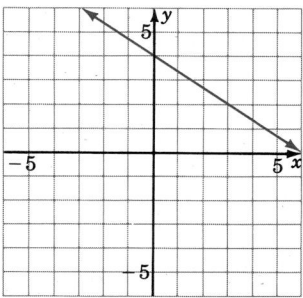

Answers to Practice Sets are on p. 291.

Section 6.6 EXERCISES

For problems 1–30, write the equation of the line using the given information in slope-intercept form.

1. $m = 3$, y-intercept $(0, 4)$

2. $m = 2$, y-intercept $(0, 5)$

3. $m = 8$, y-intercept $(0, 1)$

4. $m = 5$, y-intercept $(0, -3)$

5. $m = -6$, y-intercept $(0, -1)$

6. $m = -4$, y-intercept $(0, 0)$

7. $m = -\dfrac{3}{2}$, y-intercept $(0, 0)$

8. $m = 3$, $(1, 4)$

9. $m = 1$, $(3, 8)$

10. $m = 2$, $(1, 4)$

11. $m = 8$, $(4, 0)$

12. $m = -3$, $(3, 0)$

13. $m = -1$, $(6, 0)$

14. $m = -6$, $(0, 0)$

15. $m = -2$, $(0, 1)$

16. $(0, 0)$, $(3, 2)$

17. $(0, 0)$, $(5, 8)$

18. $(4, 1)$, $(6, 3)$

19. $(2, 5)$, $(1, 4)$

20. $(5, -3)$, $(6, 2)$

21. $(2, 3)$, $(5, 3)$

22. $(-1, 5)$, $(4, 5)$

23. $(4, 1)$, $(4, 2)$

24. $(2, 7)$, $(2, 8)$

25. $(3, 3)$, $(5, 5)$

26. $(0, 0)$, $(1, 1)$

27. $(-2, 4)$, $(3, -5)$

28. $(1, 6)$, $(-1, -6)$

29. $(14, 12)$, $(-9, -11)$

30. $(0, -4)$, $(5, 0)$

For problems 31–37, read only from the graph and determine the equation of the lines.

31.

32.

33.

34.

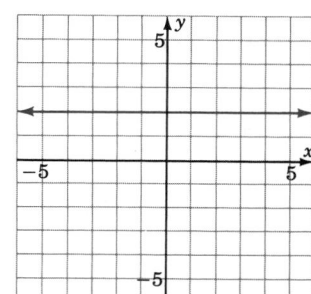

35.

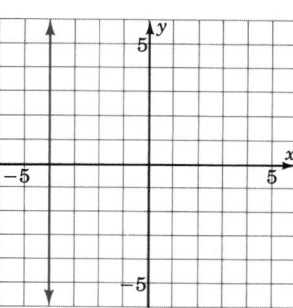

36.

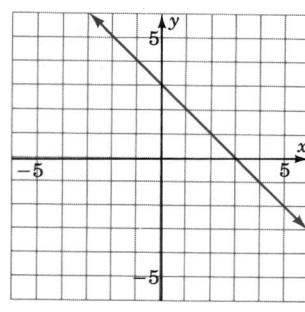

37.

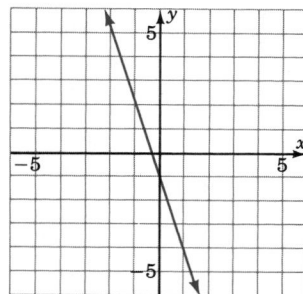

EXERCISES FOR REVIEW

(6.1) **38.** Graph the equation $x - 3 = 0$.

⟵————————————————⟶

(6.3) **39.** Supply the missing word. The point at which a line crosses the y-axis is called the _____ .

(6.4) **40.** Supply the missing word. The _____ of a line is a measure of the steepness of the line.

(6.4) **41.** Find the slope of the line that passes through the points $(4, 0)$ and $(-2, -6)$.

(6.5) **42.** Graph the equation $3y = 2x + 3$.

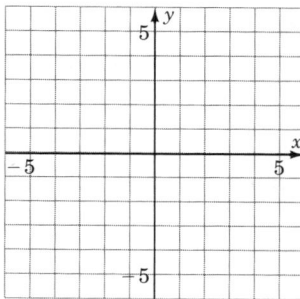

★ Answers to Practice Sets (6.6)

A. **1.** $y = 5x + 8$ **2.** $y = -8x + 3$ **3.** $y = 2x - 7$ **4.** $y = x - 1$ **5.** $y = -x - 10$
  6. $y = 4x - 18$ **7.** $y = -6x - 6$ **8.** $y = -x - 10$ **9.** $y = 2x - 7$ **10.** $y = 3x + 20$

B. **1.** $y = 9$ **2.** $x = -1$

C. **1.** $y = \dfrac{-2}{3}x + 4$

6.7 Graphing Linear Inequalities in Two Variables

Section Overview

□ **LOCATION OF SOLUTIONS**
□ **METHOD OF GRAPHING**

□ LOCATION OF SOLUTIONS

In our study of linear equations in two variables, we observed that *all* the solutions to the equation, and only the solutions to the equation, were located on the graph of the equation. We now wish to determine the location of the solutions to linear inequalities in two variables. Linear inequalities in two variables are inequalities of the forms:

$$ax + by \leq c \qquad ax + by \geq c$$
$$ax + by < c \qquad ax + by > c$$

Half-Planes

Boundary Line

> A straight line drawn through the plane divides the plane into two **half-planes.**
>
> The straight line is called the **boundary line.**

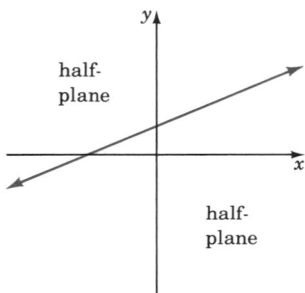

Recall that when working with linear equations in two variables, we observed that ordered pairs that produced true statements when substituted into an equation were called solutions to that equation. We can make a similar statement for inequalities in two variables. We say that an inequality in two variables has a solution when a pair of values has been found such that when these values are substituted into the inequality a true statement results.

Solution to an Inequality in Two Variables

The Location of Solutions in the Plane

As with equations, solutions to linear inequalities have particular locations in the plane. All solutions to a linear inequality in two variables are located in one and only in one entire half-plane. For example, consider the inequality

$$2x + 3y \leq 6$$

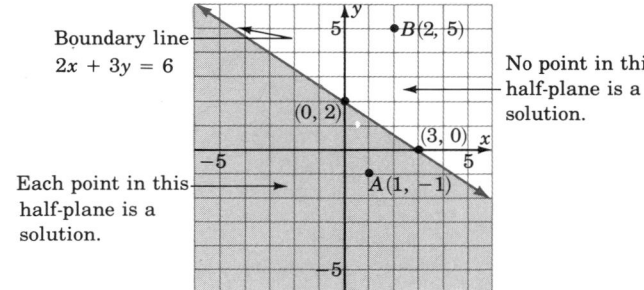

All the solutions to the inequality $2x + 3y \leq 6$ lie in the shaded half-plane.

1. Point A $(1, -1)$ is a solution since

$$2x + 3y \leq 6$$
$$2(1) + 3(-1) \leq 6?$$
$$2 - 3 \leq 6?$$
$$-1 \leq 6. \qquad \text{True}$$

2. Point $B(2, 5)$ is *not* a solution since

$$2x + 3y \leq 6$$
$$2(2) + 3(5) \leq 6?$$
$$4 + 15 \leq 6?$$
$$19 \leq 6. \qquad \text{False}$$

☐ METHOD OF GRAPHING

The method of graphing linear inequalities in two variables is as follows:

1. Graph the boundary line (consider the inequality as an equation, that is, replace the inequality sign with an equal sign).

   a. If the inequality is $\leq$ or $\geq$, draw the boundary line *solid*. This means that points on the line are solutions and are part of the graph.
   b. If the inequality is $<$ or $>$, draw the boundary line *dotted*. This means that points on the line are *not* solutions and are *not* part of the graph.

2. Determine which half-plane to shade by choosing a test point.

   a. If, when substituted, the test point yields a true statement, shade the half-plane containing it.
   b. If, when substituted, the test point yields a false statement, shade the half-plane on the opposite side of the boundary line.

☆ **SAMPLE SET A**

---

1. Graph $3x - 2y \geq -4$.

1. Graph the boundary line. The inequality is $\geq$ so we'll draw the line *solid*. Consider the inequality as an equation.

$$3x - 2y = -4$$

x	y	(x, y)
0	2	$(0, 2)$
$\dfrac{-4}{3}$	0	$\left(\dfrac{-4}{3}, 0\right)$

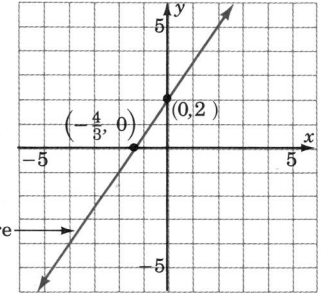

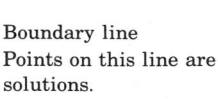
Boundary line
Points on this line are
solutions.

Continued

2. Choose a test point. The easiest one is (0, 0). Substitute (0, 0) into the original inequality.

$3x - 2y \geq -4$
$3(0) - 2(0) \geq -4?$
$0 - 0 \geq -4?$
$0 \geq -4.$ True

Shade the half-plane containing (0, 0).

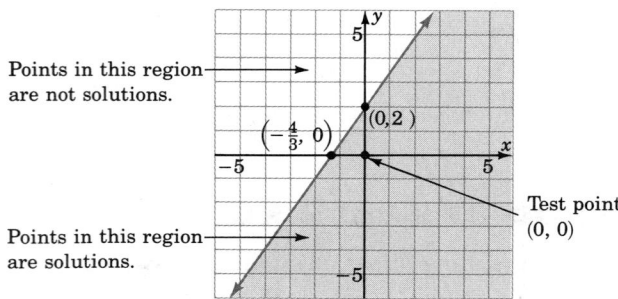

2. Graph $x + y - 3 < 0$.

1. Graph the boundary line: $x + y - 3 = 0$. The inequality is $<$ so we'll draw the line *dotted*.

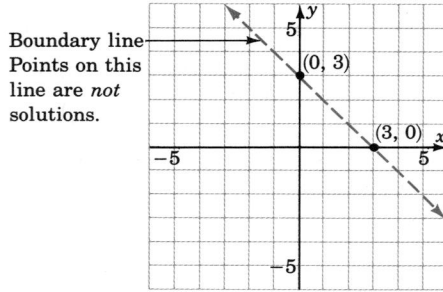

2. Choose a test point, say (0, 0).

$x + y - 3 < 0$
$0 + 0 - 3 < 0?$
$-3 < 0.$ True

Shade the half-plane containing (0, 0).

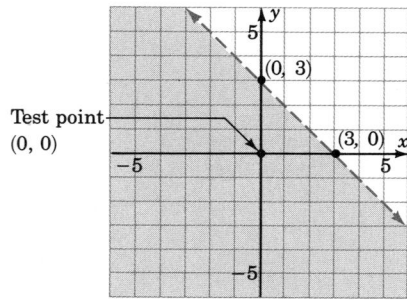

3. Graph $y \leq 2x$.

1. Graph the boundary line $y = 2x$. The inequality is $\leq$, so we'll draw the line *solid*.

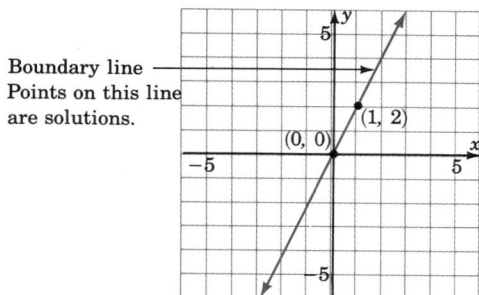

2. Choose a test point, say (0, 0).

$y \leq 2x$
$0 \leq 2(0)$?
$0 \leq 0.$ True

Shade the half-plane containing (0, 0). We can't! (0, 0) is right on the line! Pick another test point, say (1, 6).

$y \leq 2x$
$6 \leq 2(1)$?
$6 \leq 2.$ False

Shade the half-plane on the opposite side of the boundary line.

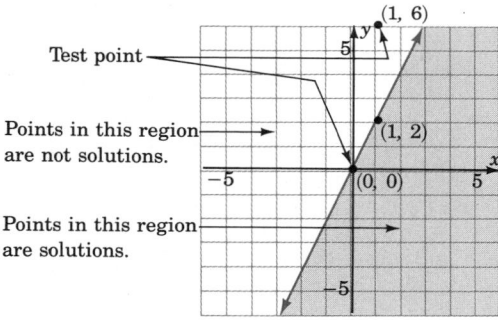

4. Graph $y > 2$.

1. Graph the boundary line $y = 2$. The inequality is $>$, so we'll draw the line *dotted*.

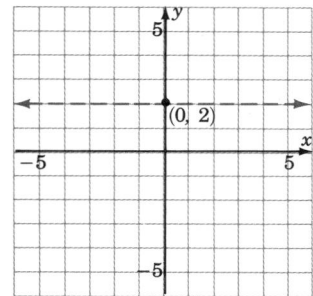

Continued

2. We don't really need a test point. Where is $y > 2$? *Above* the line $y = 2$! Any point above the line clearly has a y-coordinate greater than 2.

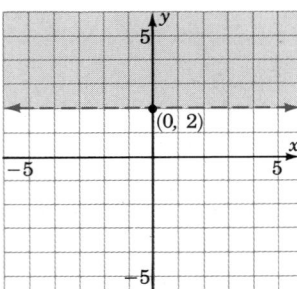

★ **PRACTICE SET A**

Solve the following inequalities by graphing.

1. $-3x + 2y \leq 4$

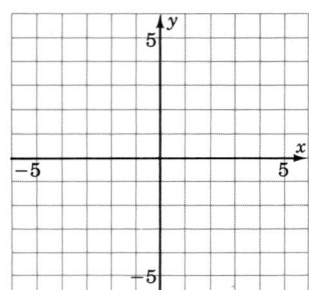

2. $x - 4y < 4$

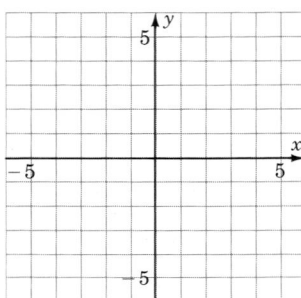

3. $3x + y > 0$

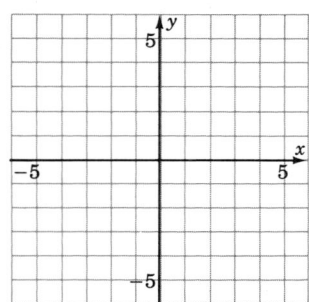

4. $x \geq 1$

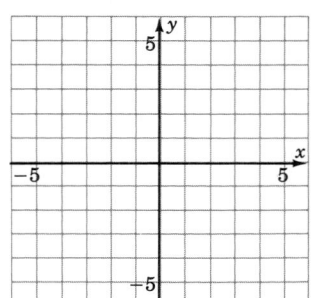

Answers to the Practice Set are on p. 298.

Section 6.7 EXERCISES

Solve the inequalities by graphing.

1. $y < x + 1$

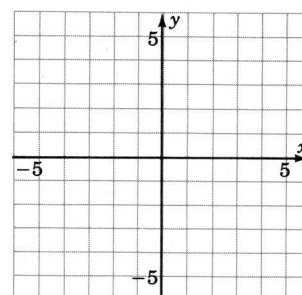

2. $x + y \leq 1$

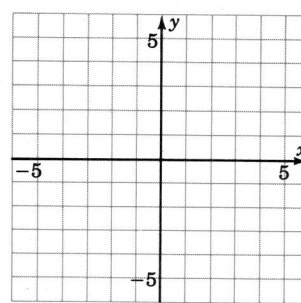

3. $-x + 2y + 4 \geq 0$

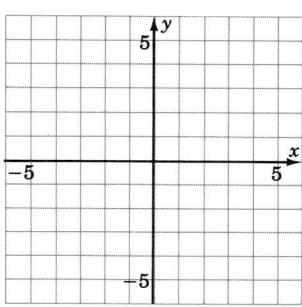

4. $-x + 5y - 10 < 0$

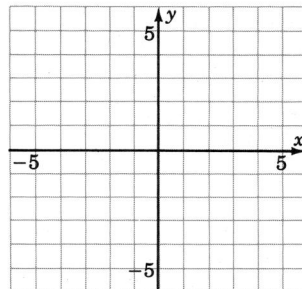

5. $-3x + 4y > -12$

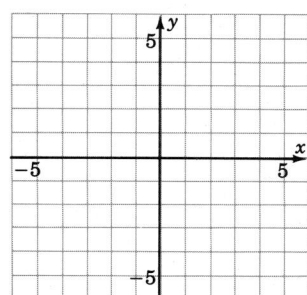

6. $2x + 5y - 15 \geq 0$

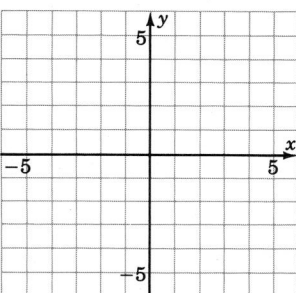

7. $y \leq 4$

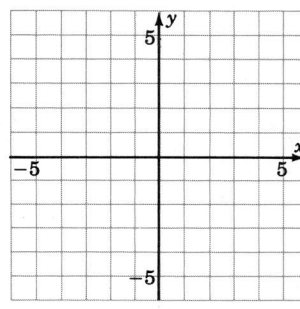

8. $x \geq 2$

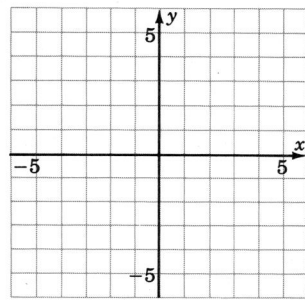

9. $x \leq 0$

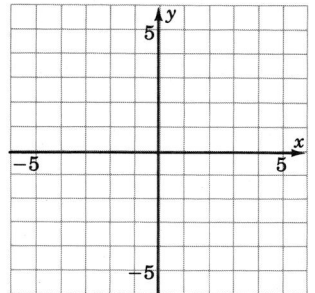

10. $x - y < 0$

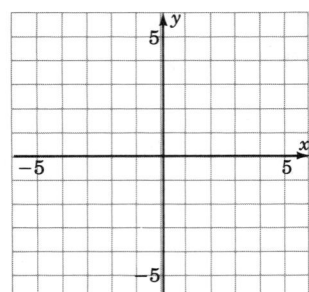

11. $x + 3y \geq 0$

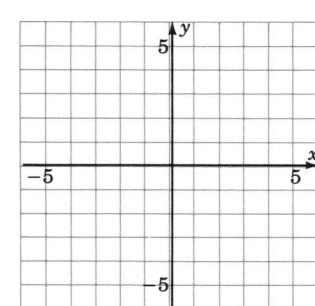

12. $-2x + 4y > 0$

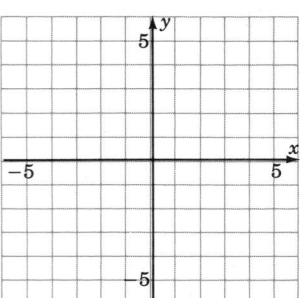

EXERCISES FOR REVIEW

(6.1)   **13.** Graph the inequality $-3x + 5 \geq -1$.

$$\longleftrightarrow$$

(6.1)   **14.** Supply the missing word. The geometric representation (picture) of the solutions to an equation is called the _____ of the equation.

(6.4)   **15.** Supply the denominator: $m = \dfrac{y_2 - y_1}{?}$.

(6.5)   **16.** Graph the equation $y = -3x + 2$.

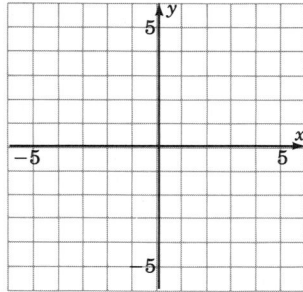

(6.6)   **17.** Write the equation of the line that has slope 4 and passes through the point $(-1, 2)$.

★ **Answers to Practice Set (6.7)**

A. 1.

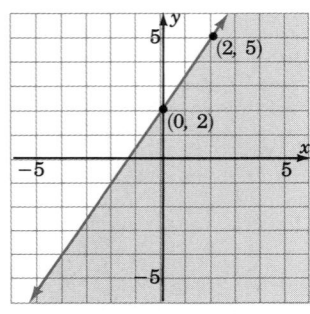

2.

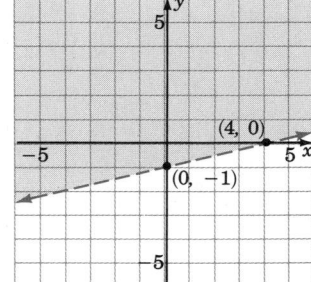

3.

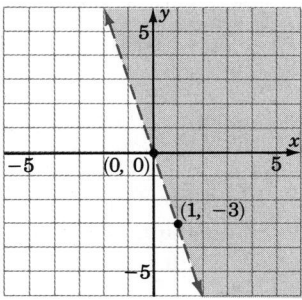

4.

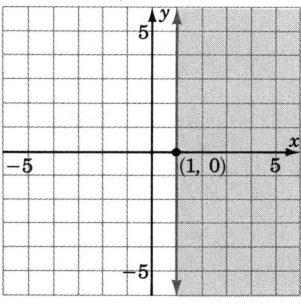

Chapter 6 SUMMARY OF KEY CONCEPTS

Graph of a Function (6.1)

The geometric representation (picture) of the solutions to an equation is called the *graph* of the equation.

Axis (6.1)

An *axis* is the most basic structure of a graph. In mathematics, the number line is used as an axis.

Number of Variables and the Number of Axes (6.1)

An equation in one variable requires one axis.   One-dimension.
An equation in two variables requires two axes.   Two-dimensions.
An equation in three variables requires three axes.   Three-dimensions.

．
．
．

An equation in n variables requires n axes. n-dimensions.

Coordinate System (6.1)

A system of axes that is constructed for graphing an equation is called a *coordinate system.*

Graphing an Equation (6.1)

The phrase *graphing an equation* is interpreted as meaning geometrically locating the solutions to that equation.

Uses of a Graph (6.1)

A graph may reveal information that may not be evident from the equation.

Rectangular Coordinate System *xy*-Plane (6.2)

A *rectangular coordinate system* is constructed by placing two number lines at 90° angles. These lines form a plane that is referred to as the *xy*-plane.

Ordered Pairs and Points (6.2)

For each ordered pair (a, b), there exists a unique point in the plane, and for each point in the plane we can associate a unique ordered pair (a, b) of real numbers.

Graphs of Linear Equations (6.3)

When graphed, a linear equation produces a straight line.

General Form of a Linear Equation in Two Variables (6.3)

The *general form* of a linear equation in two variables is $ax + by = c$, where a and b are not both 0.

Graphs, Ordered Pairs, Solutions, and Lines (6.3)

The graphing of all ordered pairs that solve a linear equation in two variables produces a straight line.
The graph of a linear equation in two variables is a straight line.
If an ordered pair is a solution to a linear equation in two variables, then it lies on the graph of the equation.
Any point (ordered pair) that lies on the graph of a linear equation in two variables is a solution to that equation.

Intercept (6.3)

An *intercept* is a point where a line intercepts a coordinate axis.

Intercept Method (6.3)

The *intercept method* is a method of graphing a linear equation in two variables by finding the intercepts, that is, by finding the points where the line crosses the x-axis and the y-axis.

Slanted, Vertical, and Horizontal Lines (6.3)

An equation in which both variables appear will graph as a *slanted* line.
A linear equation in which only one variable appears will graph as either a *vertical* or *horizontal* line.

$x = a$ graphs as a vertical line passing through a on the x-axis.

$y = b$ graphs as a horizontal line passing through b on the y-axis.

Slope of a Line (6.4)

The *slope* of a line is a measure of the line's steepness. If (x_1, y_1) and (x_2, y_2) are any two points on a line, the slope of the line passing through these points can be found using the slope formula.

$$m = \frac{y_2 - y_1}{x_2 - x_1} = \frac{\text{vertical change}}{\text{horizontal change}}$$

300

Slope and Rise and Decline **(6.4)**	Moving left to right, lines with positive slope rise, and lines with negative slope decline.
Graphing an Equation Given in Slope-Intercept Form **(6.5)**	An equation written in slope intercept form can be graphed by 1. Plotting the y-intercept $(0, b)$. 2. Determining another point using the slope, m. 3. Drawing a line through these two points.

Forms of Equations of Lines **(5.6)**

General form	Slope-intercept form	Point-slope form
$ax + by = c$	$y = mx + b$	$y - y_1 = m(x - x_1)$
	To use this form, the slope and y-intercept are needed.	To use this form, the slope and one point, or two points, are needed.

Half-Planes and Boundary Lines **(6.7)**	A straight line drawn through the plane divides the plane into two *half-planes*. The straight line is called a *boundary line*.
Solution to an Inequality in Two Variables **(6.7)**	A solution to an inequality in two variables is a pair of values that produce a true statement when substituted into the inequality.
Location of Solutions to Inequalities in Two Variables **(6.7)**	All solutions to a linear inequality in two variables are located in one, and only one, half-plane.

EXERCISE SUPPLEMENT

Section 6.1

For problems 1–10, graph the equations and inequalities.

1. $6x - 18 = 6$

2. $4x - 3 = -7$

3. $5x - 1 = 2$

4. $10x - 16 < 4$

5. $-2y + 1 \leq 5$

6. $\dfrac{-7a}{12} \geq 2$

7. $3x + 4 \leq 12$

8. $-16 \leq 5x - 1 \leq -11$

9. $0 < -3y + 9 \leq 9$

10. $\dfrac{-5c}{2} + 1 = 7$

Section 6.2

11. Draw a coordinate system and plot the following ordered pairs.

$(3, 1), (4, -2), (-1, -3), (0, 3), (3, 0), \left(5, -\dfrac{2}{3}\right)$

12. As accurately as possible, state the coordinates of the points that have been plotted on the graph.

302

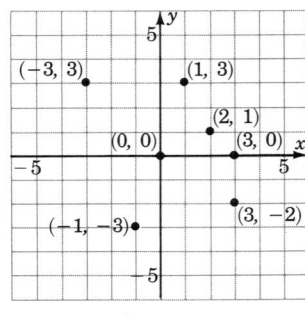

Section 6.3

13. What is the geometric structure of the graph of all the solutions to the linear equation $y = 4x - 9$?

Sections 6.3–6.5

For problems 14–25, graph the equations.

14. $y - x = 2$ **15.** $y + x - 3 = 0$

16. $-2x + 3y = -6$ **17.** $2y + x - 8 = 0$

18. $4(x - y) = 12$ **19.** $3y - 4x + 12 = 0$

20. $y = -3$ **21.** $y - 2 = 0$

22. $x = 4$ **23.** $x + 1 = 0$

24. $x = 0$ **25.** $y = 0$

Section 6.4

26. Write the slope-intercept form of a straight line.

27. The slope of a straight line is a _____ of the steepness of the line.

28. Write the formula for the slope of a line that passes through the points (x_1, y_1) and (x_2, y_2).

For problems 29–45, determine the slope and y-intercept of the lines.

29. $y = 4x + 10$ **30.** $y = 3x - 11$

31. $y = 9x - 1$ **32.** $y = -x + 2$

33. $y = -5x - 4$ **34.** $y = x$

35. $y = -6x$ **36.** $3y = 4x + 9$

37. $4y = 5x + 1$ **38.** $2y = 9x$

39. $5y + 4x = 6$ **40.** $7y + 3x = 10$

41. $6y - 12x = 24$

42. $5y - 10x - 15 = 0$

43. $3y + 3x = 1$ **44.** $7y + 2x = 0$

45. $y = 4$

For problems 46–55, find the slope, if it exists, of the line through the given pairs of points.

46. $(5, 2)$, $(6, 3)$

47. $(8, -2)$, $(10, -6)$

48. $(0, 5)$, $(3, 4)$

49. $(1, -4)$, $(3, 3)$

50. $(0, 0)$, $(-8, -5)$

51. $(-6, 1)$, $(-2, 7)$

52. $(-3, -2)$, $(-4, -5)$

53. $(4, 7)$, $(4, -2)$

54. $(-3, 1)$, $(4, 1)$

55. $\left(\dfrac{1}{3}, \dfrac{3}{4}\right)$, $\left(\dfrac{2}{9}, -\dfrac{5}{6}\right)$

56. Moving left to right, lines with _____ slope rise while lines with _____ slope decline.

57. Compare the slopes of parallel lines.

Section 6.6

For problems 58–85, write the equation of the line using the given information. Write the equation in slope-intercept form.

58. Slope $= 4$, y-intercept $= 5$

59. Slope $= 3$, y-intercept $= -6$

60. Slope $= 1$, y-intercept $= 8$

61. Slope $= 1$, y-intercept $= -2$

62. Slope $= -5$, y-intercept $= 1$

63. Slope $= -11$, y-intercept $= -4$

64. Slope $= 2$, y-intercept $= 0$

65. Slope $= -1$, y-intercept $= 0$

66. $m = 3$, $(4, 1)$

67. $m = 2$, $(1, 5)$

68. $m = 6$, $(5, -2)$

69. $m = -5$, $(2, -3)$

70. $m = -9$, $(-4, -7)$

71. $m = -2$, $(0, 2)$

72. $m = -1$, $(2, 0)$

73. $(2, 3)$, $(3, 5)$

74. $(4, 4)$, $(5, 1)$

75. $(6, 1)$, $(5, 3)$

76. $(8, 6)$, $(7, 2)$

77. $(-3, 1)$, $(2, 3)$

78. $(-1, 4)$, $(-2, -4)$

79. $(0, -5)$, $(6, -1)$

80. $(2, 1)$, $(6, 1)$

81. $(-5, 7)$, $(-2, 7)$

82. $(4, 1)$, $(4, 3)$

83. $(-1, -1)$, $(-1, 5)$

84. $(0, 4)$, $(0, -3)$

85. $(0, 2)$, $(1, 0)$

For problems 86–91, reading only from the graph, determine the equation of the line.

86.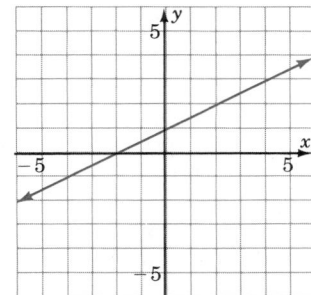

87.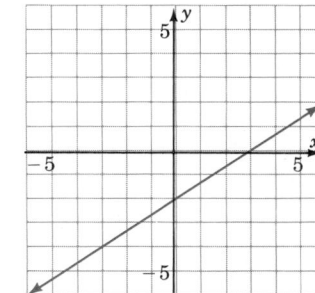

88.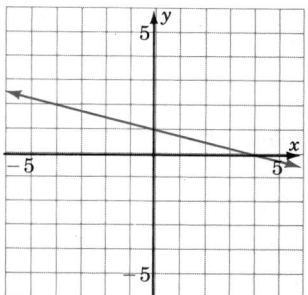

89.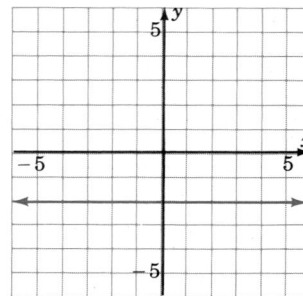

90.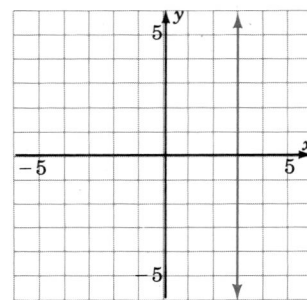

91.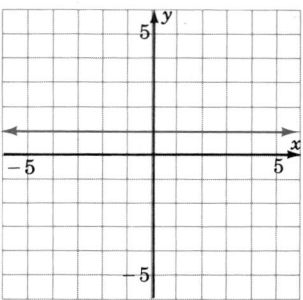

Section 6.7

For problems 92–100, graph the inequalities.

92. $y \le x + 2$

93. $y < -\dfrac{1}{2}x + 3$

94. $y > \dfrac{1}{3}x - 3$

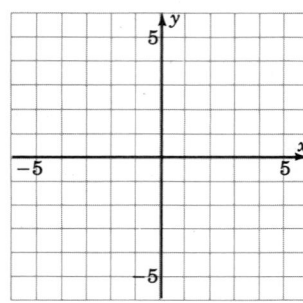

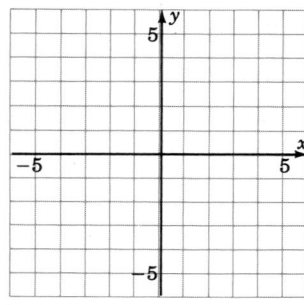

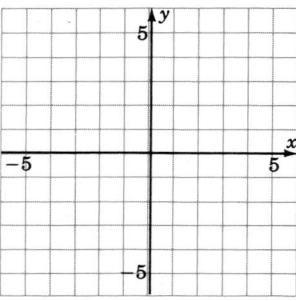

95. $-2x + 3y \le -6$

96. $2x + 5y \ge 20$

97. $4x - y + 12 > 0$

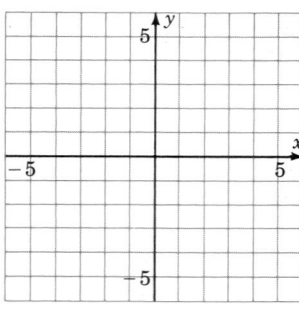

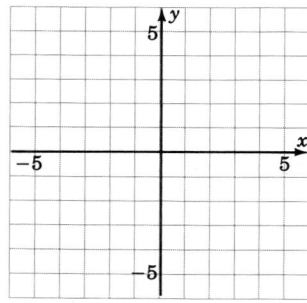

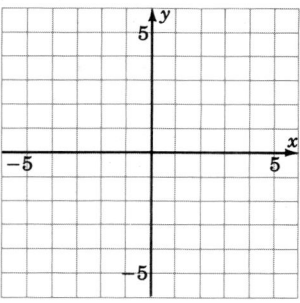

98. $y \ge -2$

99. $x < 3$

100. $y \le 0$

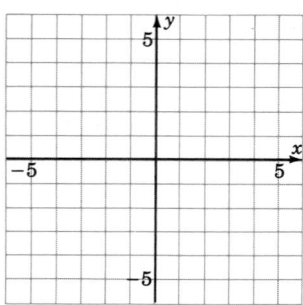

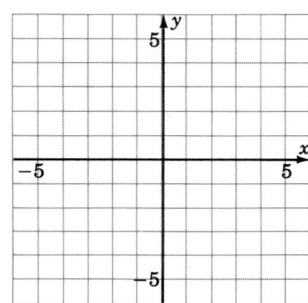

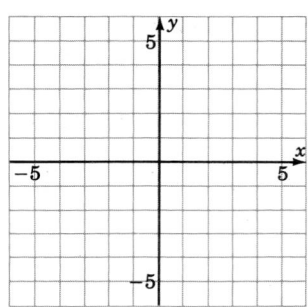

For problems 1 and 2, construct a coordinate system and graph the inequality.

1. _____

1. (6.1) $-6x + 4 > -14$

2. (6.1) $-8 < x + 6 \leq -4$

2. _____

3. (6.1) Plot the ordered pairs $(3, 1)$, $(-2, 4)$, $(0, 5)$, $(-2, -2)$.

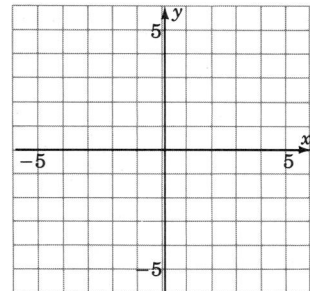

3. _____

4. _____

4. (6.2) As accurately as possible, label the coordinates of the points that have been plotted on the graph.

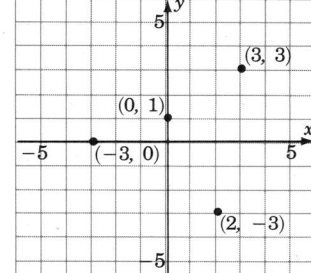

5. _____

6. _____

5. (6.3) What is the geometric structure of the graph of all the solutions to the equation $2y + 3x = -4$?

6. (6.3) In what form is the linear equation in two variables $ax + by = c$?

7. _____

7. (6.4) In what form is the linear equation in two variables $y = mx + b$?

8. (6.3) If an ordered pair is a solution to a linear equation in two variables, where does it lie geometrically?

8. _____

9. (6.4) Consider the graph of $y = \dfrac{2}{7} x + 16$. If we were to place our pencil at any point on the line and then move it horizontally 7 units to the right, how many units and in what direction would we have to move our pencil to get back on the line?

9. _____

For problems 10 and 11, find the slope, if it exists, of the line containing the following points.

10. (6.4) $(-6, -1)$ and $(0, 8)$

11. (6.4) $(-2, -8)$ and $(-2, 10)$

12. (6.4) Determine the slope and y-intercept of the line $3y + 2x + 1 = 0$.

13. (6.4) As we look at a graph left to right, do lines with a positive slope rise or decline?

For problems 14–18, find the equation of the line using the information provided. Write the equation in slope-intercept form.

14. (6.6) Slope $= 4$, y-intercept $= -3$.

15. (6.6) Slope $= -\dfrac{3}{2}$, y-intercept $= \dfrac{4}{3}$.

16. (6.6) Slope $= \dfrac{2}{3}$, passes through $(-1, 2)$.

17. (6.6) Slope $= 7$, passes through $(0, 0)$.

18. (6.6) Passes through the points $(5, 2)$ and $(2, 1)$.

For problems 19–24, graph the equation or inequality.

19. (6.3 – 6.5) $y = \dfrac{1}{3}x - 2$

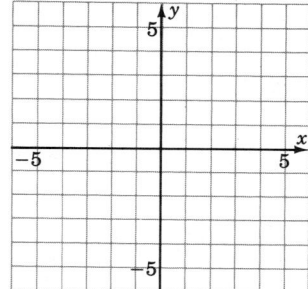

10. _____

11. _____

12. _____

13. _____

14. _____

15. _____

16. _____

17. _____

18. _____

20. (6.3 – 6.5) $5y - 2x + 15 = 0$

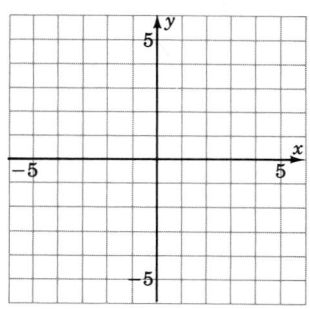

21. (6.3 – 6.5) $4(x + y) = 8$

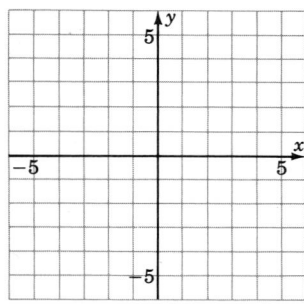

22. (6.3 – 6.5) $\dfrac{3}{2}y + 2 = 0$

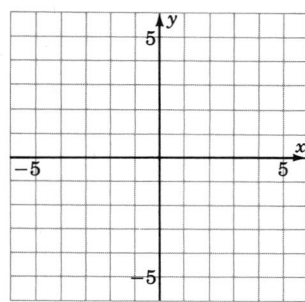

23. (6.3 – 6.5) $x = -2$

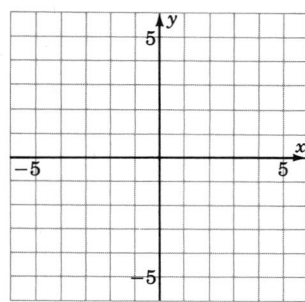

24. (6.8) $2x + 3y > 6$

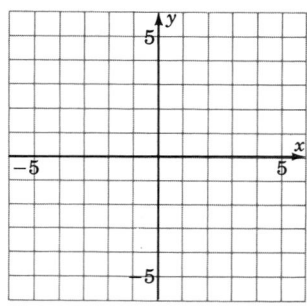

25. (6.6) Reading only from the graph, determine the equation of the line.

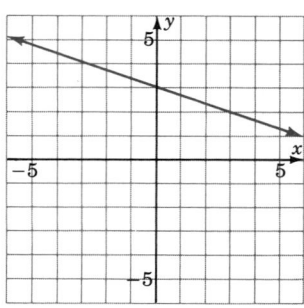

7

Rational Expres- sions

After completing this chapter, you should

Section 7.1 Rational Expressions
- be able to recognize a rational expression
- be familiar with the equality and negative properties of fractions

Section 7.2 Reducing Rational Expressions
- understand and be able to use the process of reducing rational expressions

Section 7.3 Multiplying and Dividing Rational Expressions
- be able to multiply and divide rational expressions

Section 7.4 Building Rational Expressions and the LCD
- understand and be able to use the process of building rational expressions and know why it is often necessary to build them
- be able to find the LCD of one or more expressions

Section 7.5 Adding and Subtracting Rational Expressions
- be familiar with the basic rule for adding and subtracting rational expressions
- be able to add and subtract fractions with the same and with different denominators

Section 7.6 Rational Equations
- be able to identify rational equations
- understand and be able to use the method of solving rational expressions
- be able to recognize extraneous solutions

Section 7.7 Applications
- be able to use the five-step method to solve various applied problems

Section 7.8 Complex Rational Expressions
- be able to distinguish between simple and complex fractions
- be able to simplify complex fractions using the combine-divide and the LCD-multiply-divide method

Section 7.9 Dividing Polynomials
- be able to divide a polynomial by a monomial
- understand the process and be able to divide a polynomial by a polynomial

7.1 Rational Expressions

Section
Overview

☐ RATIONAL EXPRESSIONS
☐ ZERO-FACTOR PROPERTY
☐ THE EQUALITY PROPERTY OF FRACTIONS
☐ THE NEGATIVE PROPERTY OF FRACTIONS

☐ RATIONAL EXPRESSIONS

In arithmetic it is noted that a fraction is a quotient of two whole numbers. The expression $\frac{a}{b}$, where a and b are any two whole numbers and $b \neq 0$, is called a fraction. The top number, a, is called the numerator, and the bottom number, b, is called the denominator.

Simple Algebraic Fraction

We define a simple algebraic fraction in a similar manner. Rather than restricting ourselves only to numbers, we use polynomials for the numerator and denominator. Another term for a simple algebraic fraction is a *rational expression*. A rational expression is an expression of the form $\frac{P}{Q}$, where P and Q are both polynomials and Q never represents the zero polynomial.

Rational Expression

> A **rational expression** is an algebraic expression that can be written as the quotient of two polynomials.

Examples 1–4 are rational expressions:

1. $\frac{x+9}{x-7}$ is a rational expression: P is $x + 9$ and Q is $x - 7$.

2. $\frac{x^3 + 5x^2 - 12x + 1}{x^4 - 10}$ is a rational expression: P is $x^3 + 5x^2 - 12x + 1$ and Q is $x^4 - 10$.

3. $\frac{3}{8}$ is a rational expression: P is 3 and Q is 8.

4. $4x - 5$ is a rational expression since $4x - 5$ can be written as $\frac{4x-5}{1}$: P is $4x - 5$ and Q is 1.

5. $\frac{\sqrt{5x^2-8}}{2x-1}$ is *not* a rational expression since $\sqrt{5x^2 - 8}$ is not a polynomial.

In the rational expression $\frac{P}{Q}$, P is called the numerator and Q is called the denominator.

Domain of a Rational Expression

Finding the Domain of a Rational Expression

Since division by zero is not defined, we must be careful to note the values for which the rational expression is valid. The collection of values for which the rational expression is defined is called the **domain** of the rational expression. (Recall our study of the domain of an equation in Section 3.7.)

To find the domain of a rational expression we must ask, "What values, if any, of the variable will make the denominator zero?" To find these values, we set the denominator equal to zero and solve. If any zero-producing values are obtained, they are not included in the domain. All other real numbers are included in the domain (unless some have been excluded for particular situational reasons).

☐ ZERO-FACTOR PROPERTY

Sometimes to find the domain of a rational expression, it is necessary to factor the denominator and use the *zero-factor property* of real numbers.

Zero-factor Property

> If two real numbers a and b are multiplied together and the resulting product is 0, then at least one of the factors must be zero, that is, either $a = 0$, $b = 0$, or both $a = 0$ and $b = 0$.

The following examples illustrate the use of the zero-factor property.

1. What value will produce zero in the expression $4x$? By the zero-factor property, if $4x = 0$, then $x = 0$.

2. What value will produce zero in the expression $8(x - 6)$? By the zero-factor property, if $8(x - 6) = 0$, then

   $x - 6 = 0$
   $x = 6$

   Thus, $8(x - 6) = 0$ when $x = 6$.

3. What value(s) will produce zero in the expression $(x - 3)(x + 5)$? By the zero-factor property, if $(x - 3)(x + 5) = 0$, then

   $x - 3 = 0$ or $x + 5 = 0$
   $x = 3$ $x = -5$

   Thus, $(x - 3)(x + 5) = 0$ when $x = 3$ or $x = -5$.

4. What value(s) will produce zero in the expression $x^2 + 6x + 8$? We must factor $x^2 + 6x + 8$ to put it into the zero-factor property form.

   $x^2 + 6x + 8 = (x + 2)(x + 4)$

   Now, $(x + 2)(x + 4) = 0$ when

   $x + 2 = 0$ or $x + 4 = 0$
   $x = -2$ $x = -4$

   Thus, $x^2 + 6x + 8 = 0$ when $x = -2$ or $x = -4$.

5. What value(s) will produce zero in the expression $6x^2 - 19x - 7$? We must factor $6x^2 - 19x - 7$ to put it into the zero-factor property form.

   $6x^2 - 19x - 7 = (3x + 1)(2x - 7)$

   Now, $(3x + 1)(2x - 7) = 0$ when

   $3x + 1 = 0$ or $2x - 7 = 0$
   $3x = -1$ $2x = 7$

   $x = \dfrac{-1}{3}$ $x = \dfrac{7}{2}$

   Thus, $6x^2 - 19x - 7 = 0$ when $x = \dfrac{-1}{3}$ or $\dfrac{7}{2}$.

☆ SAMPLE SET A

Find the domain of the following expressions.

1. $\dfrac{5}{x-1}$. The domain is the collection of all real numbers except 1. One is not included, for if $x = 1$, division by zero results.

2. $\dfrac{3a}{2a-8}$. If we set $2a - 8$ equal to zero, we find that $a = 4$.

$$2a - 8 = 0$$
$$2a = 8$$
$$a = 4$$

Thus 4 must be excluded from the domain since it will produce division by zero. The domain is the collection of all real numbers except 4.

3. $\dfrac{5x-1}{(x+2)(x-6)}$. Setting $(x + 2)(x - 6) = 0$, we find that $x = -2$ and $x = 6$. Both these values produce division by zero and must be excluded from the domain. The domain is the collection of all real numbers except -2 and 6.

4. $\dfrac{9}{x^2-2x-15}$. Setting $x^2 - 2x - 15 = 0$, we get

$$(x + 3)(x - 5) = 0$$
$$x = -3, \quad 5$$

Thus, $x = -3$ and $x = 5$ produce division by zero and must be excluded from the domain. The domain is the collection of all real numbers except -3 and 5.

5. $\dfrac{2x^2+x-7}{x(x-1)(x-3)(x+10)}$. Setting $x(x - 1)(x - 3)(x + 10) = 0$, we get $x = 0, 1, 3, -10$. These numbers must be excluded from the domain. The domain is the collection of all real numbers except $0, 1, 3, -10$.

6. $\dfrac{8b+7}{(2b+1)(3b-2)}$. Setting $(2b + 1)(3b - 2) = 0$, we get $b = -\dfrac{1}{2}, \dfrac{2}{3}$. The domain is the collection of all real numbers except $-\dfrac{1}{2}$ and $\dfrac{2}{3}$.

7. $\dfrac{4x-5}{x^2+1}$. No value of x is excluded since for any choice of x, the denominator is never zero. The domain is the collection of all real numbers.

8. $\dfrac{x-9}{6}$. No value of x is excluded since for any choice of x, the denominator is never zero. The domain is the collection of all real numbers.

★ PRACTICE SET A

Find the domain of each of the following rational expressions.

1. $\dfrac{2}{x-7}$

2. $\dfrac{5x}{x(x+4)}$

3. $\dfrac{2x+1}{(x+2)(1-x)}$

4. $\dfrac{5a+2}{a^2+6a+8}$

5. $\dfrac{12y}{3y^2 - 2y - 8}$ **6.** $\dfrac{2m - 5}{m^2 + 3}$ **7.** $\dfrac{k^2 - 4}{5}$

❑ THE EQUALITY PROPERTY OF FRACTIONS

From our experience with arithmetic we may recall the equality property of fractions. Let a, b, c, d be real numbers such that $b \neq 0$ and $d \neq 0$.

Equality Property of Fractions

> 1. If $\dfrac{a}{b} = \dfrac{c}{d}$, then $ad = bc$.
>
> 2. If $ad = bc$, then $\dfrac{a}{b} = \dfrac{c}{d}$.

Two fractions are equal when their cross-products are equal.
We see this property in the following examples:

1. $\dfrac{2}{3} = \dfrac{8}{12}$, since $2 \cdot 12 = 3 \cdot 8$.

2. $\dfrac{5y}{2} = \dfrac{15y^2}{6y}$, since $5y \cdot 6y = 2 \cdot 15y^2$ and $30y^2 = 30y^2$.

3. Since $9a \cdot 4 = 18a \cdot 2$, $\dfrac{9a}{18a} = \dfrac{2}{4}$.

❑ THE NEGATIVE PROPERTY OF FRACTIONS

A useful property of fractions is the *negative property of fractions*.

Negative Property of Fractions

> The negative sign of a fraction may be placed
>
> 1. in front of the fraction, $-\dfrac{a}{b}$,
>
> 2. in the numerator of the fraction, $\dfrac{-a}{b}$,
>
> 3. in the denominator of the fraction, $\dfrac{a}{-b}$.
>
> All three fractions will have the same value, that is,
>
> $$-\frac{a}{b} = \frac{-a}{b} = \frac{a}{-b}$$

The negative property of fractions is illustrated by the fractions

$$-\frac{3}{4} = \frac{-3}{4} = \frac{3}{-4}$$

To see this, consider $-\dfrac{3}{4} \overset{?}{=} \dfrac{-3}{4}$. By the equality property of fractions, $-(3 \cdot 4) = -12$ and $-3 \cdot 4 = -12$. Thus, $-\dfrac{3}{4} = \dfrac{-3}{4}$. Convince yourself that the other two fractions are equal as well.

This same property holds for rational expressions and negative signs. This property is often quite helpful in simplifying a rational expression (as we shall need to do in subsequent sections).

If either the numerator or denominator of a fraction or a fraction itself is immediately preceded by a negative sign, it is usually most convenient to place the negative sign in the numerator for later operations.

☆ SAMPLE SET B

1. $\dfrac{x}{-4}$ is best written as $\dfrac{-x}{4}$.

2. $-\dfrac{y}{9}$ is best written as $\dfrac{-y}{9}$.

3. $-\dfrac{x-4}{2x-5}$ could be written as $\dfrac{-(x-4)}{2x-5}$, which would then yield $\dfrac{-x+4}{2x-5}$.

4. $\dfrac{-5}{-10-x}$.   Factor out -1 from the denominator.

$\dfrac{-5}{-(10+x)}$ A negative divided by a negative is a positive.

$\dfrac{5}{10+x}$

5. $-\dfrac{3}{7-x}$.   Rewrite this.

$\dfrac{-3}{7-x}$ Factor out -1 from the denominator.

$\dfrac{-3}{-(-7+x)}$ A negative divided by a negative is positive.

$\dfrac{3}{-7+x}$ Rewrite.

$\dfrac{3}{x-7}$

This expression seems less cumbersome than does the original (fewer minus signs).

★ PRACTICE SET B

Fill in the missing term.

1. $-\dfrac{5}{y-2} = \dfrac{}{y-2}$ **2.** $-\dfrac{a+2}{-a+3} = \dfrac{}{a-3}$ **3.** $-\dfrac{8}{5-y} = \dfrac{}{y-5}$

Answers to Practice Sets are on p. 316.

Section 7.1 EXERCISES

For problems 1–20, find the domain of each of the rational expressions.

1. $\dfrac{6}{x-4}$

2. $\dfrac{-3}{x-8}$

3. $\dfrac{-11x}{x+1}$

4. $\dfrac{x+10}{x+4}$

5. $\dfrac{x-1}{x^2-4}$

6. $\dfrac{x+7}{x^2-9}$

7. $\dfrac{-x+4}{x^2-36}$

8. $\dfrac{-a+5}{a(a-5)}$

9. $\dfrac{2b}{b(b+6)}$

10. $\dfrac{3b+1}{b(b-4)(b+5)}$

11. $\dfrac{3x+4}{x(x-10)(x+1)}$

12. $\dfrac{-2x}{x^2(4-x)}$

13. $\dfrac{6a}{a^3(a-5)(7-a)}$

14. $\dfrac{-5}{a^2+6a+8}$

15. $\dfrac{-8}{b^2-4b+3}$

16. $\dfrac{x-1}{x^2-9x+2}$

17. $\dfrac{y-9}{y^2-y-20}$

18. $\dfrac{y-6}{2y^2-3y-2}$

19. $\dfrac{2x+7}{6x^3+x^2-2x}$

20. $\dfrac{-x+4}{x^3-8x^2+12x}$

For problems 21–25, show that the fractions are equivalent.

21. $\dfrac{-3}{5}$ and $-\dfrac{3}{5}$

22. $\dfrac{-2}{7}$ and $-\dfrac{2}{7}$

23. $-\dfrac{1}{4}$ and $\dfrac{-1}{4}$

24. $\dfrac{-2}{3}$ and $-\dfrac{2}{3}$

25. $\dfrac{-9}{10}$ and $\dfrac{9}{-10}$

27. $-\dfrac{2}{x+7} = \dfrac{}{x+7}$

28. $-\dfrac{3x+4}{2x-1} = \dfrac{}{2x-1}$

29. $-\dfrac{2x+7}{5x-1} = \dfrac{}{5x-1}$

30. $-\dfrac{x-2}{6x-1} = \dfrac{}{6x-1}$

31. $-\dfrac{x-4}{2x-3} = \dfrac{}{2x-3}$

32. $-\dfrac{x+5}{-x-3} = \dfrac{}{x+3}$

33. $-\dfrac{a+1}{-a-6} = \dfrac{}{a+6}$

34. $\dfrac{x-7}{-x+2} = \dfrac{}{x-2}$

For problems 26–35, fill in the missing term.

26. $-\dfrac{4}{x-1} = \dfrac{}{x-1}$

35. $\dfrac{y+10}{-y-6} = \dfrac{}{y+6}$

EXERCISES FOR REVIEW

(2.6) **36.** Write $\left(\dfrac{15x^{-3}y^4}{5x^2y^{-7}}\right)^{-2}$ so that only positive exponents appear.

(4.6) **37.** Solve the compound inequality $1 \le 6x - 5 < 13$.

(5.7) **38.** Factor $8x^2 - 18x - 5$.

(5.7) **39.** Factor $x^2 - 12x + 36$.

(6.1) **40.** Supply the missing word. The phrase "graphing an equation" is interpreted as meaning "geometrically locate the _____ to an equation."

★ Answers to Practice Sets (7.1)

A. For problems 1–5, all real numbers except **1.** 7   **2.** $0, -4$ **3.** $-2, 1$ **4.** $-2, -4$ **5.** $-\dfrac{4}{3}, 2$

   6. All real numbers comprise the domain.   **7.** All real numbers comprise the domain.

B. **1.** -5 **2.** $a+2$ **3.** 8

7.2 Reducing Rational Expressions

Section Overview

☐ **THE LOGIC BEHIND THE PROCESS**
☐ **THE PROCESS**

☐ THE LOGIC BEHIND THE PROCESS

When working with rational expressions, it is often best to write them in the simplest possible form. For example, the rational expression

$$\frac{x^2 - 4}{x^2 - 6x + 8}$$

can be reduced to the simpler expression $\frac{x + 2}{x - 4}$ for all x except $x = 2, 4$.

From our discussion of equality of fractions in Section 7.1, we know that $\frac{a}{b} = \frac{c}{d}$

when $ad = bc$. This fact allows us to deduce that, if $k \neq 0$, $\frac{ak}{bk} = \frac{a}{b}$, since $akb = abk$

(recall the commutative property of multiplication). But this fact means that if a factor (in this case, k) is common to both the numerator and denominator of a fraction, we may remove it without changing the value of the fraction.

$$\frac{ak}{bk} = \frac{a\cancel{k}}{b\cancel{k}} = \frac{a}{b}$$

Cancelling

The process of removing common factors is commonly called **cancelling.**

1. $\frac{16}{40}$ can be reduced to $\frac{2}{5}$.   Process:

$$\frac{16}{40} = \frac{2 \cdot 2 \cdot 2 \cdot 2}{2 \cdot 2 \cdot 2 \cdot 5}$$

Remove the three factors of 1; $\frac{2}{2} \cdot \frac{2}{2} \cdot \frac{2}{2}$.

$$\frac{\cancel{2} \cdot \cancel{2} \cdot \cancel{2} \cdot 2}{\cancel{2} \cdot \cancel{2} \cdot \cancel{2} \cdot 5} = \frac{2}{5}$$

Notice that in $\frac{2}{5}$, there is no factor common to the numerator and denominator.

2. $\frac{111}{148}$ can be reduced to $\frac{3}{4}$.   Process:

$$\frac{111}{148} = \frac{3 \cdot 37}{4 \cdot 37}$$

Remove the factor of 1; $\frac{37}{37}$.

$$\frac{3 \cdot \cancel{37}}{4 \cdot \cancel{37}}$$

$$\frac{3}{4}$$

Notice that in $\frac{3}{4}$, there is no factor common to the numerator and denominator.

3. $\dfrac{3}{9}$ can be reduced to $\dfrac{1}{3}$.   Process:

$$\frac{3}{9} = \frac{3 \cdot 1}{3 \cdot 3}$$

Remove the factor of 1; $\dfrac{3}{3}$.

$$\frac{\cancel{3} \cdot 1}{\cancel{3} \cdot 3} = \frac{1}{3}$$

Notice that in $\dfrac{1}{3}$ there is no factor common to the numerator and denominator.

4. $\dfrac{5}{7}$ cannot be reduced since there are no factors common to the numerator and denominator.

Problems 1, 2, and 3 shown above could all be reduced. The process in each reduction included the following steps:

1. Both the numerator and denominator were factored.
2. Factors that were common to both the numerator and denominator were noted and removed by dividing them out.

We know that we can divide both sides of an equation by the same nonzero number, but why should we be able to divide both the numerator and denominator of a fraction by the same nonzero number? The reason is that any nonzero number divided by itself is 1, and that if a number is multiplied by 1, it is left unchanged. Consider the fraction $\dfrac{6}{24}$. Multiply this fraction by 1. This is written $\dfrac{6}{24} \cdot 1$. But 1 can be rewritten as $\dfrac{\frac{1}{6}}{\frac{1}{6}}$.

$$\frac{6}{24} \cdot \frac{\frac{1}{6}}{\frac{1}{6}} = \frac{6 \cdot \frac{1}{6}}{24 \cdot \frac{1}{6}} = \frac{1}{4}$$

The answer, $\dfrac{1}{4}$, is the reduced form. Notice that in $\dfrac{1}{4}$ there is no factor common to both the numerator and denominator. This reasoning provides justification for the following rule.

> Multiplying or dividing the numerator and denominator by the same nonzero number does not change the value of a fraction.

❑ THE PROCESS

We can now state a process for reducing a rational expression.

Reducing a Rational Expression

> 1. Factor the numerator and denominator completely.
> 2. Divide the numerator and denominator by all factors they have in common, that is, remove all factors of 1.

Reduced to Lowest Terms   A rational expression is said to be **reduced to lowest terms** when the numerator and denominator have *no* factors in common.

☆ SAMPLE SET A

Reduce the following rational expressions.

1. $\dfrac{15x}{20x}.$ **Factor.**

$\dfrac{15x}{20x} = \dfrac{5 \cdot 3 \cdot x}{5 \cdot 2 \cdot 2 \cdot x}$ **The factors that are common to both the numerator and denominator are 5 and x. Divide each by $5x$.**

$\dfrac{\cancel{5} \cdot 3 \cdot \cancel{x}}{\cancel{5} \cdot 2 \cdot 2 \cdot \cancel{x}} = \dfrac{3}{4}, \quad x \neq 0$

It is helpful to draw a line through the divided-out factors.

2. $\dfrac{x^2 - 4}{x^2 - 6x + 8}.$ **Factor.**

$\dfrac{(x + 2)(x - 2)}{(x - 2)(x - 4)}$ **The factor that is common to both the numerator and denominator is $x - 2$. Divide each by $x - 2$.**

$\dfrac{(x + 2)\cancel{(x - 2)}}{\cancel{(x - 2)}(x - 4)} = \dfrac{x + 2}{x - 4}, \quad x \neq 2, 4$

The expression $\dfrac{x - 2}{x - 4}$ is the reduced form since there are no *factors* common to both the numerator and denominator. Although there is an x in both, it is a *common term*, not a *common factor*, and therefore cannot be divided out.

CAUTION — This is a common error: $\dfrac{x - 2}{x - 4} = \dfrac{\cancel{x} - 2}{\cancel{x} - 4} = \dfrac{2}{4}$ is *incorrect!*

3. $\dfrac{a + 2b}{6a + 12b}.$ **Factor.**

$\dfrac{a + 2b}{6(a + 2b)} = \dfrac{\cancel{a + 2b}}{6\cancel{(a + 2b)}} = \dfrac{1}{6}, \quad a \neq -2b$

Since $a + 2b$ is a common factor to both the numerator and denominator, we divide both by $a + 2b$. Since $\dfrac{(a + 2b)}{(a + 2b)} = 1$, we get 1 in the numerator.

Sometimes we may reduce a rational expression by using the division rule of exponents.

4. $\dfrac{8x^2 y^5}{4xy^2}.$ **Factor and use the rule $\dfrac{a^n}{a^m} = a^{n-m}$.**

$\dfrac{8x^2 y^5}{4xy^2} = \dfrac{2 \cdot 2 \cdot 2}{2 \cdot 2} x^{2-1} y^{5-2}$

$= 2xy^3, \quad x \neq 0, y \neq 0$

Continued

5. $\dfrac{-10x^3a(x^2-36)}{2x^3-10x^2-12x}.$ Factor.

$$\dfrac{-10x^3a(x^2-36)}{2x^3-10x^2-12x}=\dfrac{-5\cdot 2x^3a(x+6)(x-6)}{2x(x^2-5x-6)}$$

$$=\dfrac{-5\cdot 2x^3a(x+6)(x-6)}{2x(x-6)(x+1)}$$

$$=\dfrac{-5\cdot \cancel{2}x^{\cancel{3}2}a(x+6)\cancel{(x-6)}}{\cancel{2x}\cancel{(x-6)}(x+1)}$$

$$=\dfrac{-5x^2a(x+6)}{x+1},\quad x\neq -1,6$$

6. $\dfrac{x^2-x-12}{-x^2+2x+8}.$ Since it is most convenient to have the leading terms of a polynomial positive, factor out -1 from the denominator.

$\dfrac{x^2-x-12}{-(x^2-2x-8)}$ Rewrite this.

$-\dfrac{x^2-x-12}{x^2-2x-8}$ Factor.

$-\dfrac{\cancel{(x-4)}(x+3)}{\cancel{(x-4)}(x+2)}$

$-\dfrac{x+3}{x+2}=\dfrac{-(x+3)}{x+2}=\dfrac{-x-3}{x+2},\quad x\neq -2,4$

7. $\dfrac{a-b}{b-a}.$ The numerator and denominator have the same terms but they occur with opposite signs. Factor -1 from the denominator.

$$\dfrac{a-b}{-(-b+a)}=\dfrac{a-b}{-(a-b)}=-\dfrac{\cancel{a-b}}{\cancel{a-b}}=-1,\quad a\neq b$$

★ **PRACTICE SET A**

Reduce each of the following fractions to lowest terms.

1. $\dfrac{30y}{35y}$ **2.** $\dfrac{x^2-9}{x^2+5x+6}$ **3.** $\dfrac{x+2b}{4x+8b}$ **4.** $\dfrac{18a^3b^5c^7}{3ab^3c^5}$

5. $\dfrac{-3a^4+75a^2}{2a^3-16a^2+30a}$ **6.** $\dfrac{x^2-5x+4}{-x^2+12x-32}$ **7.** $\dfrac{2x-y}{y-2x}$

Answers to the Practice Set are on p. 323.

Section 7.2 EXERCISES

For problems 1–42, reduce each rational expression to lowest terms.

1. $\dfrac{6}{3x-12}$

2. $\dfrac{8}{4a-16}$

3. $\dfrac{9}{3y-21}$

4. $\dfrac{10}{5x-5}$

5. $\dfrac{7}{7x-14}$

6. $\dfrac{6}{6x-18}$

7. $\dfrac{2y^2}{8y}$

8. $\dfrac{4x^3}{2x}$

9. $\dfrac{16a^2b^3}{2ab^2}$

10. $\dfrac{20a^4b^4}{4ab^2}$

11. $\dfrac{(x+3)(x-2)}{(x+3)(x+5)}$

12. $\dfrac{(y-1)(y-7)}{(y-1)(y+6)}$

13. $\dfrac{(a+6)(a-5)}{(a-5)(a+2)}$

14. $\dfrac{(m-3)(m-1)}{(m-1)(m+4)}$

15. $\dfrac{(y-2)(y-3)}{(y-3)(y-2)}$

16. $\dfrac{(x+7)(x+8)}{(x+8)(x+7)}$

17. $\dfrac{-12x^2(x+4)}{4x}$

18. $\dfrac{-3a^4(a-1)(a+5)}{-2a^3(a-1)(a+9)}$

19. $\dfrac{6x^2y^5(x-1)(x+4)}{-2xy(x+4)}$

20. $\dfrac{22a^4b^6c^7(a+2)(a-7)}{4c(a+2)(a-5)}$

21. $\dfrac{(x+10)^3}{x+10}$

22. $\dfrac{(y-6)^7}{y-6}$

23. $\dfrac{(x-8)^2(x+6)^4}{(x-8)(x+6)}$

24. $\dfrac{(a+1)^5(a-1)^7}{(a+1)^3(a-1)^4}$

25. $\dfrac{(y-2)^6(y-1)^4}{(y-2)^3(y-1)^2}$

26. $\dfrac{(x+10)^5(x-6)^3}{(x-6)(x+10)^2}$

27. $\dfrac{(a+6)^2(a-7)^6}{(a+6)^5(a-7)^2}$

28. $\dfrac{(m+7)^4(m-8)^5}{(m+7)^7(m-8)^2}$

29. $\dfrac{(a+2)(a-1)^3}{(a+1)(a-1)}$

30. $\dfrac{(b+6)(b-2)^4}{(b-1)(b-2)}$

31. $\dfrac{8(x+2)^3(x-5)^6}{2(x+2)(x-5)^2}$

32. $\dfrac{14(x-4)^3(x-10)^6}{-7(x-4)^2(x-10)^2}$

33. $\dfrac{x^2+x-12}{x^2-4x+3}$

34. $\dfrac{x^2+3x-10}{x^2+2x-15}$

35. $\dfrac{x^2-10x+21}{x^2-6x-7}$

36. $\dfrac{x^2+10x+24}{x^2+6x}$

37. $\dfrac{x^2+9x+14}{x^2+7x}$

38. $\dfrac{6b^2-b}{6b^2+11b-2}$

39. $\dfrac{3b^2+10b+3}{3b^2+7b+2}$

40. $\dfrac{4b^2-1}{2b^2+5b-3}$

41. $\dfrac{16a^2-9}{4a^2-a-3}$

42. $\dfrac{20x^2+28xy+9y^2}{4x^2+4xy+y^2}$

For problems 43–62, reduce each rational expression if possible. If not possible, state the answer in lowest terms.

43. $\dfrac{x+3}{x+4}$

44. $\dfrac{a+7}{a-1}$

45. $\dfrac{3a+6}{3}$

46. $\dfrac{4x+12}{4}$

47. $\dfrac{5a - 5}{-5}$ **48.** $\dfrac{6b - 6}{-3}$ **55.** $\dfrac{y^4 - y}{y}$ **56.** $\dfrac{a^5 - a^2}{a}$

49. $\dfrac{8x - 16}{-4}$ **50.** $\dfrac{4x - 7}{-7}$ **57.** $\dfrac{a^6 - a^4}{a^3}$ **58.** $\dfrac{4b^2 + 3b}{b}$

51. $\dfrac{-3x + 10}{10}$ **52.** $\dfrac{x - 2}{2 - x}$ **59.** $\dfrac{2a^3 + 5a}{a}$ **60.** $\dfrac{a}{a^3 + a}$

53. $\dfrac{a - 3}{3 - a}$ **54.** $\dfrac{x^3 - x}{x}$ **61.** $\dfrac{x^4}{x^5 - 3x}$ **62.** $\dfrac{-a}{-a^2 - a}$

EXERCISES FOR REVIEW

(2.6)   **63.** Write $\left(\dfrac{4^4 a^8 b^{10}}{4^2 a^6 b^2}\right)^{-1}$ so that only positive exponents appear.

(5.5)   **64.** Factor $y^4 - 16$.

(5.7)   **65.** Factor $10x^2 - 17x + 3$.

(6.3)   **66.** Supply the missing word. An equation expressed in the form $ax + by = c$ is said to be expressed in _____ form.

(7.1)   **67.** Find the domain of the rational expression $\dfrac{2}{x^2 - 3x - 18}$.

★ **Answers to Practice Set (7.2)**

A. 1. $\dfrac{6}{7}$ **2.** $\dfrac{x - 3}{x + 2}$ **3.** $\dfrac{1}{4}$ **4.** $6a^2 b^2 c^2$ **5.** $\dfrac{-3a(a + 5)}{2(a - 3)}$ **6.** $\dfrac{-x + 1}{x - 8}$ **7.** -1

7.3 Multiplying and Dividing Rational Expressions

Section Overview
- ☐ **MULTIPLICATION OF RATIONAL EXPRESSIONS**
- ☐ **DIVISION OF RATIONAL EXPRESSIONS**

☐ MULTIPLICATION OF RATIONAL EXPRESSIONS

Rational expressions are multiplied together in much the same way that arithmetic fractions are multiplied together. To multiply rational numbers, we do the following:

Method for Multiplying
Rational Numbers

1. Reduce each fraction to lowest terms.
2. Multiply the numerators together.
3. Multiply the denominators together.

Rational expressions are multiplied together using exactly the same three steps. Since rational expressions tend to be longer than arithmetic fractions, we can simplify the multiplication process by adding one more step.

Method for Multiplying
Rational Expressions

1. Factor all numerators and denominators.
2. Reduce to lowest terms first by dividing out all common factors. (It is perfectly legitimate to cancel the numerator of one fraction with the denominator of another.)
3. Multiply numerators together.
4. Multiply denominators. It is often convenient, but not necessary, to leave denominators in factored form.

☆ SAMPLE SET A

Perform the following multiplications.

1. $\dfrac{3}{4} \cdot \dfrac{1}{2} = \dfrac{3 \cdot 1}{4 \cdot 2} = \dfrac{3}{8}$

2. $\dfrac{8}{9} \cdot \dfrac{1}{6} = \dfrac{\overset{4}{\cancel{8}}}{9} \cdot \dfrac{1}{\underset{3}{\cancel{6}}} = \dfrac{4 \cdot 1}{9 \cdot 3} = \dfrac{4}{27}$

3. $\dfrac{3x}{5y} \cdot \dfrac{7}{12y} = \dfrac{\overset{1}{\cancel{3}}x}{5y} \cdot \dfrac{7}{\underset{4}{\cancel{12}}y} = \dfrac{x \cdot 7}{5y \cdot 4y} = \dfrac{7x}{20y^2}$

4. $\dfrac{x+4}{x-2} \cdot \dfrac{x+7}{x+4}$ Divide out the common factor $x+4$.

$\dfrac{\cancel{x+4}}{x-2} \cdot \dfrac{x+7}{\cancel{x+4}}$ Multiply numerators and denominators together.

$\dfrac{x+7}{x-2}$

5. $\dfrac{x^2+x-6}{x^2-4x+3} \cdot \dfrac{x^2-2x-3}{x^2+4x-12}.$ Factor.

$\dfrac{(x+3)(x-2)}{(x-3)(x-1)} \cdot \dfrac{(x-3)(x+1)}{(x+6)(x-2)}$ Divide out the common factors $x-2$ and $x-3$.

$\dfrac{(x+3)\cancel{(x-2)}}{\cancel{(x-3)}(x-1)} \cdot \dfrac{\cancel{(x-3)}(x+1)}{(x+6)\cancel{(x-2)}}$ Multiply.

$\dfrac{(x+3)(x+1)}{(x-1)(x+6)}$ or $\dfrac{x^2+4x+3}{(x-1)(x+6)}$ or $\dfrac{x^2+4x+3}{x^2+5x-6}$

Each of these three forms is an acceptable form of the same answer.

6. $\dfrac{2x+6}{8x-16} \cdot \dfrac{x^2-4}{x^2-x-12}.$ Factor.

$\dfrac{2(x+3)}{8(x-2)} \cdot \dfrac{(x+2)(x-2)}{(x-4)(x+3)}$ Divide out the common factors 2, $x+3$, and $x-2$.

$\dfrac{\overset{1}{\cancel{2}(\cancel{x+3})}}{\underset{4}{\cancel{8}(\cancel{x-2})}} \cdot \dfrac{(x+2)(\cancel{x-2})}{(\cancel{x+3})(x-4)}$ Multiply.

$\dfrac{x+2}{4(x-4)}$ or $\dfrac{x+2}{4x-16}$

Both these forms are acceptable forms of the same answer.

7. $3x^2 \cdot \dfrac{x+7}{x-5}.$ Rewrite $3x^2$ as $\dfrac{3x^2}{1}$.

$\dfrac{3x^2}{1} \cdot \dfrac{x+7}{x-5}$ Multiply.

$\dfrac{3x^2(x+7)}{x-5}$

8. $(x-3) \cdot \dfrac{4x-9}{x^2-6x+9}.$

$\dfrac{(\cancel{x-3})}{1} \cdot \dfrac{4x-9}{(\cancel{x-3})(x-3)}$

$\dfrac{4x-9}{x-3}$

9. $\dfrac{-x^2-3x-2}{x^2+8x+15} \cdot \dfrac{4x+20}{x^2+2x}.$ Factor -1 from the first numerator.

$\dfrac{-(x^2+3x+2)}{x^2+8x+15} \cdot \dfrac{4x+20}{x^2+2x}$ Factor.

$\dfrac{-(x+1)(\cancel{x+2})}{(x+3)(\cancel{x+5})} \cdot \dfrac{4(\cancel{x+5})}{x(\cancel{x+2})}$ Multiply.

$\dfrac{-4(x+1)}{x(x+3)} = \dfrac{-4x-1}{x(x+3)}$ or $\dfrac{-4x-1}{x^2+3x}$

★ PRACTICE SET A

Perform each multiplication.

1. $\dfrac{5}{3} \cdot \dfrac{6}{7}$ **2.** $\dfrac{a^3}{b^2c^2} \cdot \dfrac{c^5}{a^5}$ **3.** $\dfrac{y-1}{y^2+1} \cdot \dfrac{y+1}{y^2-1}$

4. $\dfrac{x^2 - x - 12}{x^2 + 7x + 6} \cdot \dfrac{x^2 - 4x - 5}{x^2 - 9x + 20}$ **5.** $\dfrac{x^2 + 6x + 8}{x^2 - 6x + 8} \cdot \dfrac{x^2 - 2x - 8}{x^2 + 2x - 8}$

☐ DIVISION OF RATIONAL EXPRESSIONS

To divide one rational expression by another, we first invert the divisor then multiply the two expressions. Symbolically, if we let P, Q, R, and S represent polynomials, we can write

$$\frac{P}{Q} \div \frac{R}{S} = \frac{P}{Q} \cdot \frac{S}{R} = \frac{P \cdot S}{Q \cdot R}$$

☆ **SAMPLE SET B**

Perform the following divisions.

1. $\dfrac{6x^2}{5a} \div \dfrac{2x}{10a^3}$.   Invert the divisor and multiply.

$$\frac{\overset{3}{\cancel{6}x^{\cancel{2}}}}{\cancel{5a}} \cdot \frac{\overset{2}{\cancel{10}}a^{\overset{2}{\cancel{3}}}}{\cancel{2x}} = \frac{3x \cdot 2a^2}{1} = 6a^2x$$

2. $\dfrac{x^2 + 3x - 10}{2x - 2} \div \dfrac{x^2 + 9x + 20}{x^2 + 3x - 4}$ Invert and multiply.

$$\frac{x^2 + 3x - 10}{2x - 2} \cdot \frac{x^2 + 3x - 4}{x^2 + 9x + 20}$$ Factor.

$$\frac{\cancel{(x+5)}(x - 2)}{2\cancel{(x-1)}} \cdot \frac{(x+4)\cancel{(x-1)}}{\cancel{(x+5)}(x+4)}$$

$$\frac{x - 2}{2}$$

3. $(4x + 7) \div \dfrac{12x + 21}{x - 2}$.   Write $4x + 7$ as $\dfrac{4x + 7}{1}$.

$$\frac{4x + 7}{1} \div \frac{12x + 21}{x - 2}$$ Invert and multiply.

$$\frac{4x + 7}{1} \cdot \frac{x - 2}{12x + 21}$$ Factor.

$$\frac{\cancel{4x+7}}{1} \cdot \frac{x - 2}{3\cancel{(4x+7)}} = \frac{x - 2}{3}$$

★ **PRACTICE SET B**

Perform each division.

1. $\dfrac{8m^2n}{3a^5b^2} \div \dfrac{2m}{15a^7b^2}$

2. $\dfrac{x^2-4}{x^2+x-6} \div \dfrac{x^2+x-2}{x^2+4x+3}$

3. $\dfrac{6a^2+17a+12}{3a+2} \div (2a+3)$

Answers to Practice Sets are on p. 333.

Section 7.3 EXERCISES

For problems 1–77, perform the multiplications and divisions.

1. $\dfrac{4a^3}{5b} \cdot \dfrac{3b}{2a}$

2. $\dfrac{9x^4}{4y^3} \cdot \dfrac{10y}{x^2}$

3. $\dfrac{a}{b} \cdot \dfrac{b}{a}$

4. $\dfrac{2x}{5y} \cdot \dfrac{5y}{2x}$

5. $\dfrac{12a^3}{7} \cdot \dfrac{28}{15a}$

6. $\dfrac{39m^4}{16} \cdot \dfrac{4}{13m^2}$

7. $\dfrac{18x^6}{7} \cdot \dfrac{1}{4x^2}$

8. $\dfrac{34a^6}{21} \cdot \dfrac{42}{17a^5}$

9. $\dfrac{16x^6y^3}{15x^2} \cdot \dfrac{25x}{4y}$

10. $\dfrac{27a^7b^4}{39b} \cdot \dfrac{13a^4b^2}{16a^5}$

11. $\dfrac{10x^2y^3}{7y^5} \cdot \dfrac{49y}{15x^6}$

12. $\dfrac{22m^3n^4}{11m^6n} \cdot \dfrac{33mn}{4mn^3}$

13. $\dfrac{-10p^2q}{7a^3b^2} \cdot \dfrac{21a^5b^3}{2p}$

14. $\dfrac{-25m^4n^3}{14r^3s^3} \cdot \dfrac{21rs^4}{10mn}$

15. $\dfrac{9}{a} \div \dfrac{3}{a^2}$

16. $\dfrac{10}{b^2} \div \dfrac{4}{b^3}$

23. $\dfrac{x+8}{x+1} \cdot \dfrac{x+2}{x+8}$

17. $\dfrac{21a^4}{5b^2} \div \dfrac{14a}{15b^3}$

24. $\dfrac{x+10}{x-4} \cdot \dfrac{x-4}{x-1}$

18. $\dfrac{42x^5}{16y^4} \div \dfrac{21x^4}{8y^3}$

25. $\dfrac{2x+5}{x+8} \cdot \dfrac{x+8}{x-2}$

19. $\dfrac{39x^2y^2}{55p^2} \div \dfrac{13x^3y}{15p^6}$

26. $\dfrac{y+2}{2y-1} \cdot \dfrac{2y-1}{y-2}$

20. $\dfrac{14mn^3}{25n^6} \div \dfrac{32m}{20m^2n^3}$

27. $\dfrac{x-5}{x-1} \div \dfrac{x-5}{4}$

21. $\dfrac{12a^2b^3}{-5xy^4} \div \dfrac{6a^2}{15x^2}$

28. $\dfrac{x}{x-4} \div \dfrac{2x}{5x+1}$

22. $\dfrac{24p^3q}{9mn^3} \div \dfrac{10pq}{-21n^2}$

29. $\dfrac{a+2b}{a-1} \div \dfrac{4a+8b}{3a-3}$

30. $\dfrac{6m + 2}{m - 1} \div \dfrac{4m - 4}{m - 1}$

37. $(x - 2) \cdot \dfrac{x - 1}{x - 2}$

31. $x^3 \cdot \dfrac{4ab}{x}$

38. $(a - 6)^3 \cdot \dfrac{(a + 2)^2}{a - 6}$

32. $y^4 \cdot \dfrac{3x^2}{y^2}$

39. $(b + 1)^4 \cdot \dfrac{(b - 7)^3}{b + 1}$

33. $2a^5 \div \dfrac{6a^2}{4b}$

40. $(b^2 + 2)^3 \cdot \dfrac{b - 3}{(b^2 + 2)^2}$

34. $16x^2 y^3 \div \dfrac{10xy}{3}$

41. $(x^3 - 7)^4 \cdot \dfrac{x^2 - 1}{(x^3 - 7)^2}$

35. $21m^4 n^2 \div \dfrac{3mn^2}{7n}$

42. $(x - 5) \div \dfrac{x - 5}{x - 2}$

36. $(x + 8) \cdot \dfrac{x + 2}{x + 8}$

43. $(y - 2) \div \dfrac{y - 2}{y - 1}$

44. $(y+6)^3 \div \dfrac{(y+6)^2}{y-6}$

45. $(a-2b)^4 \div \dfrac{(a-2b)^2}{a+b}$

46. $\dfrac{x^2+3x+2}{x^2-4x+3} \cdot \dfrac{x^2-2x-3}{2x+2}$

47. $\dfrac{6x-42}{x^2-2x-3} \cdot \dfrac{x^2-1}{x-7}$

48. $\dfrac{3a+3b}{a^2-4a-5} \div \dfrac{9a+9b}{a^2-3a-10}$

49. $\dfrac{a^2-4a-12}{a^2-9} \div \dfrac{a^2-5a-6}{a^2+6a+9}$

50. $\dfrac{b^2-5b+6}{b^2-b-2} \cdot \dfrac{b^2-2b-3}{b^2-9b+20}$

51. $\dfrac{m^2-4m+3}{m^2+5m-6} \cdot \dfrac{m^2+4m-12}{m^2-5m+6}$

52. $\dfrac{r^2+7r+10}{r^2-2r-8} \div \dfrac{r^2+6r+5}{r^2-3r-4}$

53. $\dfrac{2a^2+7a+3}{3a^2-5a-2} \cdot \dfrac{a^2-5a+6}{a^2+2a-3}$

54. $\dfrac{6x^2+x-2}{2x^2+7x-4} \cdot \dfrac{x^2+2x-12}{3x^2-4x-4}$

55. $\dfrac{x^3y-x^2y^2}{x^2y-y^2} \cdot \dfrac{x^2-y}{x-xy}$

56. $\dfrac{4a^3b - 4a^2b^2}{15a - 10} \cdot \dfrac{3a - 2}{4ab - 2b^2}$

61. $\dfrac{-3a^2}{4b} \cdot \dfrac{-8b^3}{15a}$

57. $\dfrac{x+3}{x-4} \cdot \dfrac{x-4}{x+1} \cdot \dfrac{x-2}{x+3}$

62. $\dfrac{-6x^3}{5y^2} \cdot \dfrac{20y}{-2x}$

58. $\dfrac{x-7}{x+8} \cdot \dfrac{x+1}{x-7} \cdot \dfrac{x+8}{x-2}$

63. $\dfrac{-8x^2y^3}{-5x} \div \dfrac{4}{-15xy}$

59. $\dfrac{2a-b}{a+b} \cdot \dfrac{a+3b}{a-5b} \cdot \dfrac{a-5b}{2a-b}$

64. $\dfrac{-4a^3}{3b} \div \dfrac{2a}{6b^2}$

60. $\dfrac{3a(a+1)^2}{a-5} \cdot \dfrac{6(a-5)^2}{5a+5} \cdot \dfrac{15a+30}{4a-20}$

65. $\dfrac{-3a-3}{2a+2} \cdot \dfrac{a^2-3a+2}{a^2-5a-6}$

66. $\dfrac{x^2 - x - 2}{x^2 - 3x - 4} \cdot \dfrac{-x^2 + 2x + 3}{-4x - 8}$

71. $\dfrac{4x + 12}{x - 7} \cdot \dfrac{7 - x}{2x + 2}$

67. $\dfrac{-5x - 10}{x^2 - 4x + 3} \cdot \dfrac{x^2 + 4x + 1}{x^2 + x - 2}$

72. $\dfrac{-2b - 2}{b^2 + b - 6} \cdot \dfrac{-b + 2}{b + 5}$

68. $\dfrac{-a^2 - 2a + 15}{-6a - 12} \div \dfrac{a^2 - 2a - 8}{-2a - 10}$

73. $\dfrac{3x^2 - 6x - 9}{2x^2 - 6x - 4} \div \dfrac{3x^2 - 5x - 2}{6x^2 - 7x - 3}$

69. $\dfrac{-b^2 - 5b + 14}{3b - 6} \div \dfrac{-b^2 - 9b - 14}{-b + 8}$

74. $\dfrac{-2b^2 - 2b + 4}{8b^2 - 28b - 16} \div \dfrac{b^2 - 2b + 1}{2b^2 - 5b - 3}$

70. $\dfrac{3a + 6}{4a - 24} \cdot \dfrac{6 - a}{3a + 15}$

75. $\dfrac{x^2 + 4x + 3}{x^2 + 5x + 4} \div (x + 3)$

76. $\dfrac{x^2 - 3x + 2}{x^2 - 4x + 3} \div (x - 3)$

77. $\dfrac{3x^2 - 21x + 18}{x^2 + 5x + 6} \div (x + 2)$

EXERCISES FOR REVIEW

(2.2) **78.** If $a < 0$, then $|a| = $ _____ .

(3.3) **79.** Classify the polynomial $4xy + 2y$ as a monomial, binomial, or trinomial. State its degree and write the numerical coefficient of each term.

(3.4) **80.** Find the product: $y^2(2y - 1)(2y + 1)$.

(4.4) **81.** Translate the sentence "four less than twice some number is two more than the number" into an equation.

(7.2) **82.** Reduce the fraction $\dfrac{x^2 - 4x + 4}{x^2 - 4}$.

★ **Answers to Practice Sets (7.3)**

A. 1. $\dfrac{10}{7}$ **2.** $\dfrac{c^3}{a^2 b^2}$ **3.** $\dfrac{1}{y^2 + 1}$ **4.** $\dfrac{x + 3}{x + 6}$ **5.** $\dfrac{(x + 2)^2}{(x - 2)^2}$

B. 1. $20a^2 mn$ **2.** $\dfrac{x + 1}{x - 1}$ **3.** $\dfrac{3a + 4}{3a + 2}$

7.4 Building Rational Expressions and the LCD

Section Overview

☐ **THE PROCESS**
☐ **THE REASON FOR BUILDING RATIONAL EXPRESSIONS**
☐ **THE LEAST COMMON DENOMINATOR (LCD)**

☐ THE PROCESS

Recall, from Section 7.1, the equality property of fractions.

EQUALITY PROPERTY OF FRACTIONS

If $\dfrac{a}{b} = \dfrac{c}{d}$, then $ad = bc$.

Using the fact that $1 = \dfrac{b}{b}$, $b \neq 0$, and that 1 is the multiplicative identity, it follows that if $\dfrac{P}{Q}$ is a rational expression, then

$$\frac{P}{Q} \cdot \frac{b}{b} = \frac{Pb}{Qb}, \qquad b \neq 0$$

This equation asserts that a rational expression can be transformed into an equivalent rational expression by multiplying both the numerator and denominator by the same nonzero number.

Process of Building Rational Expressions

This process is known as the process of **building rational expressions** and it is exactly the opposite of reducing rational expressions. The process is shown in these examples:

1. $\dfrac{3}{4}$ can be built to $\dfrac{12}{16}$ since

$$\frac{3}{4} \cdot 1 = \frac{3}{4} \cdot \frac{4}{4} = \frac{3 \cdot 4}{4 \cdot 4} = \frac{12}{16}$$

2. $\dfrac{-4}{5}$ can be built to $\dfrac{-8}{10}$ since

$$\frac{-4}{5} \cdot 1 = \frac{-4}{5} \cdot \frac{2}{2} = \frac{-4 \cdot 2}{5 \cdot 2} = \frac{-8}{10}$$

3. $\dfrac{3}{7}$ can be built to $\dfrac{3xy}{7xy}$ since

$$\frac{3}{7} \cdot 1 = \frac{3}{7} \cdot \frac{xy}{xy} = \frac{3xy}{7xy}$$

4. $\dfrac{4a}{3b}$ can be built to $\dfrac{4a^2(a+1)}{3ab(a+1)}$ since

$$\frac{4a}{3b} \cdot 1 = \frac{4a}{3b} \cdot \frac{a(a+1)}{a(a+1)} = \frac{4a^2(a+1)}{3ab(a+1)}$$

Suppose we're given a rational expression $\dfrac{P}{Q}$ and wish to build it into a rational expression with denominator Qb^2, that is,

$$\frac{P}{Q} \rightarrow \frac{?}{Qb^2}$$

Since we changed the denominator, we must certainly change the numerator in the same way. To determine how to change the numerator we need to know how the denominator was changed. Since one rational expression is built into another equivalent expression by multiplication by 1, the first denominator must have been multiplied by some quantity. Observation of

$$\frac{P}{Q} \rightarrow \frac{?}{Qb^2}$$

tells us that Q was multiplied by b^2. Hence, we must multiply the numerator P by b^2. Thus,

$$\frac{P}{Q} = \frac{Pb^2}{Qb^2}$$

Quite often a simple comparison of the original denominator with the new denominator will tell us the factor being used. However, there will be times when the factor is unclear by simple observation. We need a method for finding the factor.

Observe the following examples; then try to speculate on the method.

1. $\dfrac{3}{4} = \dfrac{?}{20}$.   The original denominator 4 was multiplied by 5 to yield 20. What arithmetic process will yield 5 using 4 and 20?

2. $\dfrac{9}{10} = \dfrac{?}{10y}$.   The original denominator 10 was multiplied by y to yield $10y$.

3. $\dfrac{-6xy}{2a^3b} = \dfrac{?}{16a^5b^3}$.   The original denominator $2a^3b$ was multiplied by $8a^2b^2$ to yield $16a^5b^3$.

4. $\dfrac{5ax}{(a+1)^2} = \dfrac{?}{4(a+1)^2(a-2)}$.   The original denominator $(a+1)^2$ was multiplied by $4(a-2)$ to yield $4(a+1)^2(a-2)$.

> To determine the quantity that the original denominator was multiplied by to yield the new denominator, we ask, "What did I multiply the original denominator by to get the new denominator?" We find this factor by dividing the original denominator into the new denominator.

It is precisely this quantity that we multiply the numerator by to build the rational expression.

☆ SAMPLE SET A

Determine N in each of the following problems.

1. $\dfrac{8}{3} = \dfrac{N}{15}$.

   The original denominator is 3 and the new denominator is 15. Divide the original denominator into the new denominator and multiply the numerator 8 by this result.

   $15 \div 3 = 5$

   Then, $8 \cdot 5 = 40$. So,

   $\dfrac{8}{3} = \dfrac{40}{15}$ and $N = 40$.

   Check by reducing $\dfrac{40}{15}$.

2. $\dfrac{2x}{5b^2y} = \dfrac{N}{20b^5y^4}$.

   The original denominator is $5b^2y$ and the new denominator is $20b^5y^4$. Divide the original denominator into the new denominator and multiply the numerator $2x$ by this result.

   $\dfrac{20b^5y^4}{5b^2y} = 4b^3y^3$

   So, $2x \cdot 4b^3y^3 = 8b^3xy^3$. Thus,

   $\dfrac{2x}{5b^2y} = \dfrac{8b^3xy^3}{20b^5y^4}$ and $N = 8b^3xy^3$.

Continued

3. $\dfrac{-6a}{a+2} = \dfrac{N}{(a+2)(a-7)}$.

The new denominator divided by the original denominator is

$$\dfrac{(a+2)(a-7)}{a+2} = a - 7$$

Multiply $-6a$ by $a-7$.

$$-6a(a-7) = -6a^2 + 42a$$

$\dfrac{-6a}{a+2} = \dfrac{-6a^2 + 42a}{(a+2)(a-7)}$ and $N = -6a^2 + 42a$.

4. $\dfrac{-3(a-1)}{a-4} = \dfrac{N}{a^2 - 16}$.

The new denominator divided by the original denominator is

$$\dfrac{a^2 - 16}{a - 4} = \dfrac{(a+4)(a-4)}{a-4}$$

$$= a + 4$$

Multiply $-3(a-1)$ by $a + 4$.

$$-3(a-1)(a+4) = -3(a^2 + 3a - 4)$$
$$= -3a^2 - 9a + 12$$

$\dfrac{-3(a-1)}{a-4} = \dfrac{-3a^2 - 9a + 12}{a^2 - 16}$ and $N = -3a^2 - 9a + 12$.

5. $7x = \dfrac{N}{x^2 y^3}$.

Write $7x$ as $\dfrac{7x}{1}$.

$\dfrac{7x}{1} = \dfrac{N}{x^2 y^3}$

Now we can see clearly that the original denominator 1 was multiplied by $x^2 y^3$. We need to multiply the numerator $7x$ by $x^2 y^3$.

$7x = \dfrac{7x \cdot x^2 y^3}{x^2 y^3}$

$7x = \dfrac{7x^3 y^3}{x^2 y^3}$ and $N = 7x^3 y^3$.

6. $\dfrac{5x}{x+3} = \dfrac{5x^2 - 20x}{N}$.

The same process works in this case. Divide the original numerator $5x$ into the new numerator $5x^2 - 20x$.

$$\dfrac{5x^2 - 20x}{5x} = \dfrac{5x(x-4)}{5x}$$

$$= x - 4$$

Multiply the denominator by $x - 4$.

$$(x+3)(x-4)$$

$\dfrac{5x}{x+3} = \dfrac{5x^2 - 20}{(x+3)(x-4)}$ and $N = 5x^2 - 20$.

7. $\dfrac{4x}{3-x} = \dfrac{N}{x-3}$.

The two denominators have nearly the same terms; each has the opposite sign. Factor -1 from the original denominator.

$$3 - x = -1(-3 + x)$$
$$= -(x - 3)$$

$$\dfrac{4x}{3-x} = \dfrac{4x}{-(x-3)} = \dfrac{-4x}{x-3} \text{ and } N = -4x.$$

It is important to note that we *factored* -1 from the original denominator. We *did not* multiply it by -1. Had we multiplied only the denominator by -1 we would have had to multiply the numerator by -1 also.

★ **PRACTICE SET A**

Determine N.

1. $\dfrac{3}{8} = \dfrac{N}{48}$

2. $\dfrac{9a}{5b} = \dfrac{N}{35b^2x^3}$

3. $\dfrac{-2y}{y-1} = \dfrac{N}{y^2-1}$

4. $\dfrac{a+7}{a-5} = \dfrac{N}{a^2-3a-10}$

5. $4a = \dfrac{N}{6a^3(a-1)}$

6. $-2x = \dfrac{N}{8x^3y^3z^5}$

7. $\dfrac{6ab}{b+3} = \dfrac{N}{b^2+6b+9}$

8. $\dfrac{3m}{m+5} = \dfrac{3m^2-18m}{N}$

9. $\dfrac{-2r^2}{r-3} = \dfrac{-2r^3+8r^2}{N}$

10. $\dfrac{-8ab^2}{a-4} = \dfrac{N}{4-a}$

❏ THE REASON FOR BUILDING RATIONAL EXPRESSIONS

Building Rational Expressions

Normally, when we write a rational expression, we write it in reduced form. The reason for building rational expressions is to make addition and subtraction of rational expressions convenient (simpler).

> To add or subtract two or more rational expressions they *must* have the *same denominator*.

Building rational expressions allows us to transform fractions into fractions with the same denominators (which we can then add or subtract). The most convenient new denominator is the *least common denominator* (LCD) of the given fractions.

❏ THE LEAST COMMON DENOMINATOR (LCD)

In arithmetic, the **least common denominator** is the smallest (least) quantity that each of the given denominators will divide into without a remainder. For algebraic expressions, the LCD is the polynomial of *least degree* divisible by each denominator. Some examples are shown below.

1. $\dfrac{3}{4}$, $\dfrac{1}{6}$, $\dfrac{5}{12}$.   The LCD is 12 since 12 is the smallest number that 4, 6, and 12 will divide into without a remainder.

2. $\dfrac{1}{3}$, $\dfrac{5}{6}$, $\dfrac{5}{8}$, $\dfrac{7}{12}$.   The LCD is 24 since 24 is the smallest number that 3, 6, 8, and 12 will divide into without a remainder.

3. $\dfrac{2}{x}$, $\dfrac{3}{x^2}$.   The LCD is x^2 since x^2 is the smallest quantity that x and x^2 will divide into without a remainder.

4. $\dfrac{5a}{6a^2b}$, $\dfrac{3a}{8ab^3}$.   The LCD is $24a^2b^3$ since $24a^2b^3$ is the smallest quantity that $6a^2b$ and $8ab^3$ will divide into without a remainder.

5. $\dfrac{2y}{y-6}$, $\dfrac{4y^2}{(y-6)^3}$, $\dfrac{y}{y-1}$.   The LCD is $(y-6)^3(y-1)$ since $(y-6)^3 \cdot (y-1)$ is the smallest quantity that $y-6$, $(y-6)^3$ and $y-1$ will divide into without a remainder.

We'll now propose and demonstrate a method for obtaining the LCD:

Finding the LCD

METHOD FOR OBTAINING THE LCD

1. Factor each denominator. Use exponents for repeated factors. It is usually not necessary to factor numerical quantities.
2. Write down each *different* factor that appears. If a factor appears more than once, use only the factor with the highest exponent.
3. The LCD is the product of the factors written in step 2.

☆ SAMPLE SET B

Find the LCD.

1. $\dfrac{1}{x}$, $\dfrac{3}{x^3}$, $\dfrac{2}{4y}$

   1. The denominators are already factored.
   2. Note that x appears as x and x^3. Use only the x with the higher exponent, x^3. The term $4y$ appears, so we must also use $4y$.
   3. The LCD is $4x^3y$.

2. $\dfrac{5}{(x-1)^2}$, $\dfrac{2x}{(x-1)(x-4)}$, $\dfrac{-5x}{x^2-3x+2}$

   1. Only the third denominator needs to be factored.

      $x^2 - 3x + 2 = (x-2)(x-1)$

      Now the three denominators are $(x-1)^2$, $(x-1)(x-4)$, and $(x-2)(x-1)$.
   2. Note that $x-1$ appears as $(x-1)^2$, $x-1$, and $x-1$. Use only the $x-1$ with the highest exponent, $(x-1)^2$. Also appearing are $x-4$ and $x-2$.
   3. The LCD is $(x-1)^2(x-4)(x-2)$.

3. $\dfrac{-1}{6a^4}, \quad \dfrac{3}{4a^3 b}, \quad \dfrac{1}{3a^3(b+5)}$

1. The denominators are already factored.
2. We can see that the LCD of the numbers 6, 4, and 3 is 12. We also need a^4, b, and $b+5$.
3. The LCD is $12a^4 b(b+5)$.

4. $\dfrac{9}{x}, \quad \dfrac{4}{8y}$

1. The denominators are already factored.
2. x, $8y$.
3. The LCD is $8xy$.

★ PRACTICE SET B

Find the LCD.

1. $\dfrac{3}{x^2}, \quad \dfrac{4}{x^5}, \quad \dfrac{-6}{xy}$ **2.** $\dfrac{x+1}{x-4}, \quad \dfrac{x-7}{(x-4)^2}, \quad \dfrac{-6}{x+1}$

3. $\dfrac{2}{m-6}, \quad \dfrac{-5m}{(m+1)^2(m-2)}, \quad \dfrac{12m^2}{(m-2)^3(m-6)}$ **4.** $\dfrac{1}{x^2-1}, \quad \dfrac{2}{x^2-2x-3}, \quad \dfrac{-3x}{x^2-6x+9}$

5. $\dfrac{3}{4y^2-8y}, \quad \dfrac{8}{y^2-4y+4}, \quad \dfrac{10y-1}{3y^3-6y^2}$

☆ SAMPLE SET C

Change the given rational expressions into rational expressions having the same denominator.

1. $\dfrac{3}{x^2}, \quad \dfrac{4}{x}$.

The LCD, by inspection, is x^2. Rewrite each expression with x^2 as the new denominator.

$\dfrac{}{x^2}, \quad \dfrac{}{x^2}$

Determine the numerators. In $\dfrac{3}{x^2}$, the denominator was not changed so we need not change the numerator.

$\dfrac{3}{x^2}, \quad \dfrac{}{x^2}$

In the second fraction, the original denominator was x. We can see that x must be multiplied by x to build it to x^2. So we must also multiply the numerator 4 by x. Thus, $4 \cdot x = 4x$.

$\dfrac{3}{x^2}, \quad \dfrac{4x}{x^2}$

Continued

2. $\dfrac{4b}{b-1}, \quad \dfrac{-2b}{b+3}.$

By inspection, the LCD is

$(b-1)(b+3).$

Rewrite each fraction with new denominator $(b-1)(b+3)$.

$\dfrac{}{(b-1)(b+3)}, \quad \dfrac{}{(b-1)(b+3)}$

The denominator of the first rational expression has been multiplied by $b+3$, so the numerator $4b$ must be multiplied by $b+3$.

$4b(b+3) = 4b^2 + 12b$

$\dfrac{4b^2+12b}{(b-1)(b+3)}, \quad \dfrac{}{(b-1)(b+3)}$

The denominator of the second rational expression has been multiplied by $b-1$, so the numerator $-2b$ must be multiplied by $b-1$.

$-2b(b-1) = -2b^2 + 2b$

$\dfrac{4b^2+12b}{(b-1)(b+3)}, \quad \dfrac{-2b^2+2b}{(b-1)(b+3)}$

3. $\dfrac{6x}{x^2-8x+15}, \quad \dfrac{-2x^2}{x^2-7x+12}.$

We first find the LCD. Factor.

$\dfrac{6x}{(x-3)(x-5)}, \quad \dfrac{-2x^2}{(x-3)(x-4)}$

The LCD is $(x-3)(x-5)(x-4)$. Rewrite each of these fractions with new denominator $(x-3)(x-5)(x-4)$.

$\dfrac{}{(x-3)(x-5)(x-4)}, \quad \dfrac{}{(x-3)(x-5)(x-4)}$

By comparing the denominator of the first fraction with the LCD we see that we must multiply the numerator $6x$ by $x-4$.

$6x(x-4) = 6x^2 - 24x$

$\dfrac{6x^2-24x}{(x-3)(x-5)(x-4)}, \quad \dfrac{}{(x-3)(x-5)(x-4)}$

By comparing the denominator of the second fraction with the LCD, we see that we must multiply the numerator $-2x^2$ by $x-5$.

$-2x^2(x-5) = -2x^3 + 10x^2$

$\dfrac{6x^2-24x}{(x-3)(x-5)(x-4)}, \quad \dfrac{-2x^3+10x^2}{(x-3)(x-5)(x-4)}$

These examples have been done step-by-step and include explanations. This makes the process seem fairly long. In practice, however, the process is much quicker.

4. $\dfrac{6ab}{a^2-5a+4}, \quad \dfrac{a+b}{a^2-8a+16}$

$\dfrac{6ab}{(a-1)(a-4)}, \quad \dfrac{a+b}{(a-4)^2}$

LCD $= (a-1)(a-4)^2$.

$\dfrac{6ab(a-4)}{(a-1)(a-4)^2}, \quad \dfrac{(a+b)(a-1)}{(a-1)(a-4)^2}$

5. $\dfrac{x+1}{x^3+3x^2}$, $\dfrac{2x}{x^3-4x}$, $\dfrac{x-4}{x^2-4x+4}$

$\dfrac{x+1}{x^2(x+3)}$, $\dfrac{2x}{x(x+2)(x-2)}$, $\dfrac{x-4}{(x-2)^2}$ $\qquad$ $\mathbf{LCD} = x^2(x+3)(x+2)(x-2)^2.$

$\dfrac{(x+1)(x+2)(x-2)^2}{x^2(x+3)(x+2)(x-2)^2}$, $\dfrac{2x^2(x+3)(x-2)}{x^2(x+3)(x+2)(x-2)^2}$, $\dfrac{x^2(x+3)(x+2)(x-4)}{x^2(x+3)(x+2)(x-2)^2}$

★ **PRACTICE SET C**

Change the given rational expressions into rational expressions with the same denominators.

1. $\dfrac{4}{x^3}$, $\dfrac{7}{x^5}$ $\qquad$ **2.** $\dfrac{2x}{x+6}$, $\dfrac{x}{x-1}$ $\qquad$ **3.** $\dfrac{-3}{b^2-b}$, $\dfrac{4b}{b^2-1}$ $\qquad$ **4.** $\dfrac{8}{x^2-x-6}$, $\dfrac{-1}{x^2+x-2}$

5. $\dfrac{10x}{x^2+8x+16}$, $\dfrac{5x}{x^2-16}$ $\qquad$ **6.** $\dfrac{-2ab^2}{a^3-6a^2}$, $\dfrac{6b}{a^4-2a^3}$, $\dfrac{-2a}{a^2-4a+4}$

Answers to Practice Sets are on p. 346.

Section 7.4 EXERCISES

For problems 1–49, replace N with the proper quantity.

1. $\dfrac{3}{x} = \dfrac{N}{x^3}$ $\qquad$ **2.** $\dfrac{4}{a} = \dfrac{N}{a^2}$

8. $\dfrac{b^3}{6a} = \dfrac{N}{18a^5}$

3. $\dfrac{-2}{x} = \dfrac{N}{xy}$ $\qquad$ **4.** $\dfrac{-7}{m} = \dfrac{N}{ms}$

9. $\dfrac{-4a}{5x^2y} = \dfrac{N}{15x^3y^3}$

5. $\dfrac{6a}{5} = \dfrac{N}{10b}$ $\qquad$ **6.** $\dfrac{a}{3z} = \dfrac{N}{12z}$

10. $\dfrac{-10z}{7a^3b} = \dfrac{N}{21a^4b^5}$

7. $\dfrac{x^2}{4y^2} = \dfrac{N}{20y^4}$

11. $\dfrac{8x^2y}{5a^3} = \dfrac{N}{25a^3x^2}$

12. $\dfrac{2}{a^2} = \dfrac{N}{a^2(a-1)}$

13. $\dfrac{5}{x^3} = \dfrac{N}{x^3(x-2)}$

14. $\dfrac{2a}{b^2} = \dfrac{N}{b^3-b}$

15. $\dfrac{4x}{a} = \dfrac{N}{a^4-4a^2}$

16. $\dfrac{6b^3}{5a} = \dfrac{N}{10a^2-30a}$

17. $\dfrac{4x}{3b} = \dfrac{N}{3b^5-15b}$

18. $\dfrac{2m}{m-1} = \dfrac{N}{(m-1)(m+2)}$

19. $\dfrac{3s}{s+12} = \dfrac{N}{(s+12)(s-7)}$

20. $\dfrac{a+1}{a-3} = \dfrac{N}{(a-3)(a-4)}$

21. $\dfrac{a+2}{a-2} = \dfrac{N}{(a-2)(a-4)}$

22. $\dfrac{b+7}{b-6} = \dfrac{N}{(b-6)(b+6)}$

23. $\dfrac{5m}{2m+1} = \dfrac{N}{(2m+1)(m-2)}$

24. $\dfrac{4}{a+6} = \dfrac{N}{a^2+5a-6}$

25. $\dfrac{9}{b-2} = \dfrac{N}{b^2-6b+8}$

26. $\dfrac{3b}{b-3} = \dfrac{N}{b^2-11b+24}$

27. $\dfrac{-2x}{x-7} = \dfrac{N}{x^2-4x-21}$

28. $\dfrac{-6m}{m+6} = \dfrac{N}{m^2+10m+24}$

29. $\dfrac{4y}{y+1} = \dfrac{N}{y^2 + 9y + 8}$

36. $\dfrac{a+2}{2a-1} = \dfrac{N}{2a^2 + 9a - 5}$

30. $\dfrac{x+2}{x-2} = \dfrac{N}{x^2 - 4}$

37. $\dfrac{-3}{4x+3} = \dfrac{N}{4x^2 - 13x - 12}$

31. $\dfrac{y-3}{y+3} = \dfrac{N}{y^2 - 9}$

38. $\dfrac{b+2}{3b-1} = \dfrac{N}{6b^2 + 7b - 3}$

32. $\dfrac{a+5}{a-5} = \dfrac{N}{a^2 - 25}$

39. $\dfrac{x-1}{4x-5} = \dfrac{N}{12x^2 - 11x - 5}$

33. $\dfrac{z-4}{z+4} = \dfrac{N}{z^2 - 16}$

40. $\dfrac{3}{x+2} = \dfrac{3x-21}{N}$

34. $\dfrac{4}{2a+1} = \dfrac{N}{2a^2 - 5a - 3}$

41. $\dfrac{4}{y+6} = \dfrac{4y+8}{N}$

35. $\dfrac{1}{3b-1} = \dfrac{N}{3b^2 + 11b - 4}$

42. $\dfrac{-6}{a-1} = \dfrac{-6a-18}{N}$

43. $\dfrac{-8a}{a+3} = \dfrac{-8a^2 - 40a}{N}$

44. $\dfrac{y+1}{y-8} = \dfrac{y^2 - 2y - 3}{N}$

45. $\dfrac{x-4}{x+9} = \dfrac{x^2 + x - 20}{N}$

46. $\dfrac{3x}{2-x} = \dfrac{N}{x-2}$

47. $\dfrac{7a}{5-a} = \dfrac{N}{a-5}$

48. $\dfrac{-m+1}{3-m} = \dfrac{N}{m-3}$

49. $\dfrac{k+6}{10-k} = \dfrac{N}{k-10}$

For problems 50–75, convert the given rational expressions to rational expressions having the same denominators.

50. $\dfrac{2}{a}$, $\dfrac{3}{a^4}$

51. $\dfrac{5}{b^2}$, $\dfrac{4}{b^3}$

52. $\dfrac{8}{z}$, $\dfrac{3}{4z^3}$

53. $\dfrac{9}{x^2}$, $\dfrac{1}{4x}$

54. $\dfrac{2}{a+3}$, $\dfrac{4}{a+1}$

55. $\dfrac{2}{x+5}$, $\dfrac{4}{x-5}$

56. $\dfrac{1}{x-7}$, $\dfrac{4}{x-1}$

57. $\dfrac{10}{y+2}$, $\dfrac{1}{y+8}$

58. $\dfrac{4}{a^2}$, $\dfrac{a}{a+4}$

59. $\dfrac{-3}{b^2}$, $\dfrac{b^2}{b+5}$

60. $\dfrac{-6}{b-1}, \quad \dfrac{5b}{4b}$

61. $\dfrac{10a}{a-6}, \quad \dfrac{2}{a^2-6a}$

62. $\dfrac{4}{x^2+2x}, \quad \dfrac{1}{x^2-4}$

63. $\dfrac{x+1}{x^2-x-6}, \quad \dfrac{x+4}{x^2+x-2}$

64. $\dfrac{x-5}{x^2-9x+20}, \quad \dfrac{4}{x^2-3x-10}$

65. $\dfrac{-4}{b^2+5b-6}, \quad \dfrac{b+6}{b^2-1}$

66. $\dfrac{b+2}{b^2+6b+8}, \quad \dfrac{b-1}{b^2+8b+12}$

67. $\dfrac{x+7}{x^2-2x-3}, \quad \dfrac{x+3}{x^2-6x-7}$

68. $\dfrac{2}{a^2+a}, \quad \dfrac{a+3}{a^2-1}$

69. $\dfrac{x-2}{x^2+7x+6}, \quad \dfrac{2x}{x^2+4x-12}$

70. $\dfrac{x-2}{2x^2+5x-3}, \quad \dfrac{x-1}{5x^2+16x+3}$

71. $\dfrac{2}{x-5}$, $\dfrac{-3}{5-x}$

74. $\dfrac{k}{5-k}$, $\dfrac{3k}{k-5}$

72. $\dfrac{4}{a-6}$, $\dfrac{-5}{6-a}$

75. $\dfrac{2m}{m-8}$, $\dfrac{7}{8-m}$

73. $\dfrac{6}{2-x}$, $\dfrac{5}{x-2}$

EXERCISES FOR REVIEW

(5.3) **76.** Factor $m^2x^3 + mx^2 + mx$.

(5.6) **77.** Factor $y^2 - 10y + 21$.

(6.6) **78.** Write the equation of the line that passes through the points $(1, 1)$ and $(4, -2)$. Express the equation in slope-intercept form.

(7.2) **79.** Reduce $\dfrac{y^2 - y - 6}{y - 3}$.

(7.3) **80.** Find the quotient: $\dfrac{x^2 - 6x + 9}{x^2 - x - 6} \div \dfrac{x^2 + 2x - 15}{x^2 + 2x}$.

★ **Answers to Practice Sets (7.4)**

A. **1.** $N = 18$ **2.** $N = 63abx^3$ **3.** $N = -2y^2 - 2y$ **4.** $N = a^2 + 9a + 14$ **5.** $N = 24a^4(a - 1)$
   6. $N = -16x^4y^3z^5$ **7.** $N = 6ab^2 + 18ab$ **8.** $N = m^2 - m - 30$ **9.** $N = r^2 - 7r + 12$
   10. $N = 8ab^2$

B. **1.** x^5y **2.** $(x - 4)^2(x + 1)$ **3.** $(m - 6)(m + 1)^2(m - 2)^3$ **4.** $(x + 1)(x - 1)(x - 3)^2$
   5. $12y^2(y - 2)^2$

C. **1.** $\dfrac{4x^2}{x^5}$, $\dfrac{7}{x^5}$ **2.** $\dfrac{2x(x - 1)}{(x + 6)(x - 1)}$, $\dfrac{x(x + 6)}{(x + 6)(x - 1)}$ **3.** $\dfrac{-3(b + 1)}{b(b - 1)(b + 1)}$, $\dfrac{4b^2}{b(b - 1)(b + 1)}$

4. $\dfrac{8(x-1)}{(x-3)(x+2)(x-1)}$, $\dfrac{-1(x-3)}{(x-3)(x+2)(x-1)}$ **5.** $\dfrac{10x(x-4)}{(x+4)^2(x-4)}$, $\dfrac{5x(x+4)}{(x+4)^2(x-4)}$

6. $\dfrac{-2a^2b^2(a-2)^2}{a^3(a-6)(a-2)^2}$, $\dfrac{6b(a-6)(a-2)}{a^3(a-6)(a-2)^2}$, $\dfrac{-2a^4(a-6)}{a^3(a-6)(a-2)^2}$

7.5 Adding and Subtracting Rational Expressions

Section Overview

- ☐ **BASIC RULE**
- ☐ **FRACTIONS WITH THE SAME DENOMINATOR**
- ☐ **FRACTIONS WITH DIFFERENT DENOMINATORS**

☐ BASIC RULE

We are now in a position to study the process of adding and subtracting rational expressions. There is a most basic rule to which we must strictly adhere if we wish to conveniently add or subtract rational expressions.

> To add or subtract rational expressions conveniently, they should have the same denominators.

Thus, to add or subtract two or more rational expressions conveniently, we must ensure that they all have the same denominator. The denominator that is most convenient is the LCD.

☐ FRACTIONS WITH THE SAME DENOMINATOR

The Rule for Adding and Subtracting Rational Expressions

> To add (or subtract) two or more rational expressions with the same denominators, add (or subtract) the numerators and place the result over the LCD. Reduce if necessary. Symbolically,
>
> $$\frac{a}{c} + \frac{b}{c} = \frac{a+b}{c}$$
>
> $$\frac{a}{c} - \frac{b}{c} = \frac{a-b}{c}$$
>
> Note that we combine *only* the numerators.

☆ SAMPLE SET A

Add or subtract the following rational expressions.

1. $\dfrac{1}{6} + \dfrac{3}{6}$ The denominators are the same. Add the numerators.

$\dfrac{1}{6} + \dfrac{3}{6} = \dfrac{1+3}{6} = \dfrac{4}{6}$ Reduce.

$\dfrac{1}{6} + \dfrac{3}{6} = \dfrac{2}{3}$

Continued

2. $\dfrac{5}{x} + \dfrac{8}{x}$

The denominators are the same. Add the numerators.

$$\dfrac{5}{x} + \dfrac{8}{x} = \dfrac{5+8}{x} = \dfrac{13}{x}$$

3. $\dfrac{2ab}{y^2 w} - \dfrac{5b}{y^2 w}$

The denominators are the same. Subtract the numerators.

$$\dfrac{2ab}{y^2 w} - \dfrac{5b}{y^2 w} = \dfrac{2ab - 5b}{y^2 w}$$

4. $\dfrac{3x^2 + x + 2}{x - 7} + \dfrac{x^2 - 4x + 1}{x - 7}$

The denominators are the same. Add the numerators.

$$\dfrac{3x^2 + x + 2}{x - 7} + \dfrac{x^2 - 4x + 1}{x - 7} = \dfrac{3x^2 + x + 2 + x^2 - 4x + 1}{x - 7}$$
$$= \dfrac{4x^2 - 3x + 3}{x - 7}$$

5. $\dfrac{5y + 3}{2y - 5} - \dfrac{2y + 4}{2y - 5}$

The denominators are the same. Subtract the numerators. But *be careful* to subtract the *entire* numerator. Use parentheses!

$$\dfrac{5y + 3}{2y - 5} - \dfrac{2y + 4}{2y - 5} = \dfrac{5y + 3 - (2y + 4)}{2y - 5}$$
$$= \dfrac{5y + 3 - 2y - 4}{2y - 5}$$
$$= \dfrac{3y - 1}{2y - 5}$$

Note: $\dfrac{5y + 3}{2y - 5} - \underbrace{\dfrac{2y + 4}{2y - 5}}.$

Observe this part

The term $-\dfrac{2y + 4}{2y - 5}$ could be written as

$$+\dfrac{-(2y + 4)}{2y - 5} = \dfrac{-2y - 4}{2y - 5}$$

A common mistake is to write

$$-\dfrac{2y + 4}{2y - 5} \text{ as } \dfrac{-2y + 4}{2y - 5}$$

This is *not* correct, as the negative sign is not being applied to the entire numerator.

6. $\dfrac{3x^2 + 4x + 5}{(x+6)(x-2)} + \dfrac{2x^2 + x + 6}{x^2 + 4x - 12} - \dfrac{x^2 - 4x - 6}{x^2 + 4x - 12}$

Factor the denominators to determine if they're the same.

$$\dfrac{3x^2 + 4x + 5}{(x+6)(x-2)} + \dfrac{2x^2 + x + 6}{(x+6)(x-2)} - \dfrac{x^2 - 4x - 6}{(x+6)(x-2)}$$

The denominators are the same. Combine the numerators being careful to note the negative sign.

$$\dfrac{3x^2 + 4x + 5 + 2x^2 + x + 6 - (x^2 - 4x - 6)}{(x+6)(x-2)}$$

$$\dfrac{3x^2 + 4x + 5 + 2x^2 + x + 6 - x^2 + 4x + 6}{(x+6)(x-2)}$$

$$\dfrac{4x^2 + 9x + 17}{(x+6)(x-2)}$$

★ **PRACTICE SET A**

Add or subtract the following rational expressions.

1. $\dfrac{4}{9} + \dfrac{2}{9}$

2. $\dfrac{3}{b} + \dfrac{2}{b}$

3. $\dfrac{5x}{2y^2} - \dfrac{3x}{2y^2}$

4. $\dfrac{x+y}{x-y} + \dfrac{2x+3y}{x-y}$

5. $\dfrac{4x^2 - x + 4}{3x + 10} - \dfrac{x^2 + 2x + 5}{3x + 10}$

6. $\dfrac{x(x+1)}{x(2x+3)} + \dfrac{3x^2 - x + 7}{2x^2 + 3x}$

7. $\dfrac{4x+3}{x^2 - x - 6} - \dfrac{8x - 4}{(x+2)(x-3)}$

8. $\dfrac{5a^2 + a - 4}{2a(a-6)} + \dfrac{2a^2 + 3a + 4}{2a^2 - 12a} + \dfrac{a^2 + 2}{2a^2 - 12a}$

9. $\dfrac{8x^2 + x - 1}{x^2 - 6x + 8} + \dfrac{2x^2 + 3x}{x^2 - 6x + 8} - \dfrac{5x^2 + 3x - 4}{(x-4)(x-2)}$

☐ FRACTIONS WITH DIFFERENT DENOMINATORS

☆ **SAMPLE SET B**

Add or subtract the following rational expressions.

1. $\dfrac{4a}{3y} + \dfrac{2a}{9y^2}$.

The denominators are *not* the same. Find the LCD. By inspection, the LCD is $9y^2$.

$\dfrac{}{9y^2} + \dfrac{2a}{9y^2}$

The denominator of the first rational expression has been multiplied by $3y$, so the numerator must be multiplied by $3y$.

$4a \cdot 3y = 12ay$

$\dfrac{12ay}{9y^2} + \dfrac{2a}{9y^2}$

The denominators are now the same. Add the numerators.

$\dfrac{12ay + 2a}{9y^2}$

2. $\dfrac{3b}{b+2} + \dfrac{5b}{b-3}$.

The denominators are *not* the same. The LCD is $(b+2)(b-3)$.

$\dfrac{}{(b+2)(b-3)} + \dfrac{}{(b+2)(b-3)}$

The denominator of the first rational expression has been multiplied by $b-3$, so the numerator must be multiplied by $b-3$. $3b(b-3)$

$\dfrac{3b(b-3)}{(b+2)(b-3)} + \dfrac{}{(b+2)(b-3)}$

The denominator of the second rational expression has been multiplied by $b+2$, so the numerator must be multiplied by $b+2$. $5b(b+2)$

$\dfrac{3b(b-3)}{(b+2)(b-3)} + \dfrac{5b(b+2)}{(b+2)(b-3)}$

The denominators are now the same. Add the numerators.

$$\dfrac{3b(b-3) + 5b(b+2)}{(b-3)(b+2)} = \dfrac{3b^2 - 9b + 5b^2 + 10b}{(b-3)(b+2)}$$

$$= \dfrac{8b^2 + b}{(b-3)(b-2)}$$

3. $\dfrac{x+3}{x-1} + \dfrac{x-2}{4x+4}$.

The denominators are *not* the same. Find the LCD.

$\dfrac{x+3}{x-1} + \dfrac{x-2}{4(x+1)}$

The LCD is $4(x+1)(x-1)$.

$\dfrac{}{4(x+1)(x-1)} + \dfrac{}{4(x+1)(x-1)}$

The denominator of the first rational expression has been multiplied by $4(x+1)$ so the numerator must be multiplied by $4(x+1)$. $4(x+3)(x+1)$

$$\frac{4(x+3)(x+1)}{4(x+1)(x-1)} + \frac{}{4(x+1)(x-1)}$$

The denominator of the second rational expression has been multiplied by $x - 1$ so the numerator must be multiplied by $x - 1$. $(x-1)(x-2)$

$$\frac{4(x+3)(x+1)}{4(x+1)(x-1)} + \frac{(x-1)(x-2)}{4(x+1)(x-1)}$$

The denominators are now the same. Add the numerators.

$$\frac{4(x+3)(x+1) + (x-1)(x-2)}{4(x+1)(x-1)}$$

$$\frac{4(x^2+4x+3) + x^2 - 3x + 2}{4(x+1)(x-1)}$$

$$\frac{4x^2 + 16x + 12 + x^2 - 3x + 2}{4(x+1)(x-1)} = \frac{5x^2 + 13x + 14}{4(x+1)(x-1)}$$

4. $\dfrac{x+5}{x^2-7x+12} + \dfrac{3x-1}{x^2-2x-3}$

Determine the LCD.

$$\frac{x+5}{(x-4)(x-3)} + \frac{3x-1}{(x-3)(x+1)}$$

The LCD is $(x-4)(x-3)(x+1)$.

$$\frac{}{(x-4)(x-3)(x+1)} + \frac{}{(x-4)(x-3)(x+1)}$$

The first numerator must be multiplied by $x+1$ and the second by $x-4$.

$$\frac{(x+5)(x+1)}{(x-4)(x-3)(x+1)} + \frac{(3x-1)(x-4)}{(x-4)(x-3)(x+1)}$$

The denominators are now the same. Add the numerators.

$$\frac{(x+5)(x+1) + (3x-1)(x-4)}{(x-4)(x-3)(x+1)}$$

$$\frac{x^2 + 6x + 5 + 3x^2 - 13x + 4}{(x-4)(x-3)(x+1)}$$

$$\frac{4x^2 - 7x + 9}{(x-4)(x-3)(x+1)}$$

5. $\dfrac{a+4}{a^2+5a+6} - \dfrac{a-4}{a^2-5a-24}$

Determine the LCD.

$$\frac{a+4}{(a+3)(a+2)} - \frac{a-4}{(a+3)(a-8)}$$

The LCD is $(a+3)(a+2)(a-8)$.

$$\frac{}{(a+3)(a+2)(a-8)} - \frac{}{(a+3)(a+2)(a-8)}$$

The first numerator must be multiplied by $a-8$ and the second by $a+2$.

$$\frac{(a+4)(a-8)}{(a+3)(a+2)(a-8)} - \frac{(a-4)(a+2)}{(a+3)(a+2)(a-8)}$$

The denominators are now the same. Subtract the numerators.

$$\frac{(a+4)(a-8) - (a-4)(a+2)}{(a+3)(a+2)(a-8)}$$

$$\frac{a^2 - 4a - 32 - (a^2 - 2a - 8)}{(a+3)(a+2)(a-8)}$$

$$\frac{a^2 - 4a - 32 - a^2 + 2a + 8}{(a+3)(a+2)(a-8)}$$

Continued

$$\frac{-2a - 24}{(a + 3)(a + 2)(a - 8)}$$

Factor -2 from the numerator.

$$\frac{-2(a + 12)}{(a + 3)(a + 2)(a - 8)}$$

6. $\dfrac{3x}{7 - x} + \dfrac{5x}{x - 7}.$

The denominators are *nearly* the same. They differ only in sign. Our technique is to factor -1 from one of them.

Factor -1 from the first term.

$$\frac{3x}{7 - x} = \frac{3x}{-(x - 7)} = \frac{-3x}{x - 7}$$

$$\frac{3x}{7 - x} + \frac{5x}{x - 7} = \frac{-3x}{x - 7} + \frac{5x}{x - 7}$$

$$= \frac{-3x + 5x}{x - 7}$$

$$= \frac{2x}{x - 7}$$

★ PRACTICE SET B

Add or subtract the following rational expressions.

1. $\dfrac{3x}{4a^2} + \dfrac{5x}{12a^3}$ **2.** $\dfrac{5b}{b + 1} + \dfrac{3b}{b - 2}$ **3.** $\dfrac{a - 7}{a + 2} + \dfrac{a - 2}{a + 3}$

4. $\dfrac{4x + 1}{x + 3} - \dfrac{x + 5}{x - 3}$ **5.** $\dfrac{2y - 3}{y} + \dfrac{3y + 1}{y + 4}$ **6.** $\dfrac{a - 7}{a^2 - 3a + 2} + \dfrac{a + 2}{a^2 - 6a + 8}$

7. $\dfrac{6}{b^2 + 6b + 9} - \dfrac{2}{b^2 + 4b + 4}$ **8.** $\dfrac{x}{x + 4} - \dfrac{x - 2}{3x - 3}$ **9.** $\dfrac{5x}{4 - x} + \dfrac{7x}{x - 4}$

☆ **SAMPLE SET C**

Combine the following rational expressions.

1. $3 + \dfrac{7}{x-1}$.   **Rewrite the expression.**

$\dfrac{3}{1} + \dfrac{7}{x-1}$ **The LCD is $x - 1$.**

$\dfrac{3(x-1)}{x-1} + \dfrac{7}{x-1} = \dfrac{3x-3}{x-1} + \dfrac{7}{x-1} = \dfrac{3x-3+7}{x-1}$

$\qquad\qquad\qquad\qquad\qquad\qquad = \dfrac{3x+4}{x-1}$

2. $3y + 4 - \dfrac{y^2 - y + 3}{y-6}$.   **Rewrite the expression.**

$\dfrac{3y+4}{1} - \dfrac{y^2-y+3}{y-6}$ **The LCD is $y - 6$.**

$\dfrac{(3y+4)(y-6)}{y-6} - \dfrac{y^2-y+3}{y-6} = \dfrac{(3y+4)(y-6)-(y^2-y+3)}{y-6}$

$\qquad\qquad\qquad\qquad\qquad\quad = \dfrac{3y^2 - 14y - 24 - y^2 + y - 3}{y-6}$

$\qquad\qquad\qquad\qquad\qquad\quad = \dfrac{2y^2 - 13y - 27}{y-6}$

★ **PRACTICE SET C**

1. Simplify $8 + \dfrac{3}{x-6}$.   **2.** Simplify $2a - 5 - \dfrac{a^2 + 2a - 1}{a+3}$.

Answers to Practice Sets are on p. 360.

Section 7.5 EXERCISES

For problems 1–78, add or subtract the rational expressions.

1. $\dfrac{3}{8} + \dfrac{1}{8}$ **2.** $\dfrac{1}{9} + \dfrac{4}{9}$ **3.** $\dfrac{7}{10} - \dfrac{2}{5}$ **4.** $\dfrac{3}{4} - \dfrac{5}{12}$

5. $\dfrac{3}{4x} + \dfrac{5}{4x}$

6. $\dfrac{2}{7y} + \dfrac{3}{7y}$

19. $\dfrac{b+7}{b-6} - \dfrac{b-1}{b-6}$

7. $\dfrac{6y}{5x} + \dfrac{8y}{5x}$

8. $\dfrac{9a}{7b} + \dfrac{3a}{7b}$

20. $\dfrac{2b+3}{b+1} - \dfrac{b-4}{b+1}$

9. $\dfrac{15n}{2m} - \dfrac{6n}{2m}$

10. $\dfrac{8p}{11q} - \dfrac{3p}{11q}$

21. $\dfrac{3y+4}{y+8} - \dfrac{2y-5}{y+8}$

11. $\dfrac{y+4}{y-6} + \dfrac{y+8}{y-6}$

12. $\dfrac{y-1}{y+4} + \dfrac{y+7}{y+4}$

22. $\dfrac{2a-7}{a-9} + \dfrac{3a+5}{a-9}$

13. $\dfrac{a+6}{a-1} + \dfrac{3a+5}{a-1}$

14. $\dfrac{5a+1}{a+7} + \dfrac{2a-6}{a+7}$

23. $\dfrac{8x-1}{x+2} - \dfrac{15x+7}{x+2}$

15. $\dfrac{x+1}{5x} + \dfrac{x+3}{5x}$

16. $\dfrac{a-6}{a+2} + \dfrac{a-2}{a+2}$

24. $\dfrac{7}{2x^2} + \dfrac{1}{6x^3}$

17. $\dfrac{b+1}{b-3} + \dfrac{b+2}{b-3}$

18. $\dfrac{a+2}{a-5} - \dfrac{a+3}{a-5}$

25. $\dfrac{2}{3x} + \dfrac{4}{6x^2}$

26. $\dfrac{5}{6y^3} - \dfrac{2}{18y^5}$

33. $\dfrac{x+2}{x-5} + \dfrac{x-1}{x+2}$

27. $\dfrac{2}{5a^2} - \dfrac{1}{10a^3}$

34. $\dfrac{a+3}{a-3} - \dfrac{a+2}{a-2}$

28. $\dfrac{3}{x+1} + \dfrac{5}{x-2}$

35. $\dfrac{y+1}{y-1} - \dfrac{y+4}{y-4}$

29. $\dfrac{4}{x-6} + \dfrac{1}{x-1}$

36. $\dfrac{x-1}{(x+2)(x-3)} + \dfrac{x+4}{x-3}$

30. $\dfrac{2a}{a+1} - \dfrac{3a}{a+4}$

37. $\dfrac{y+2}{(y+1)(y+6)} + \dfrac{y-2}{y+6}$

31. $\dfrac{6y}{y+4} + \dfrac{2y}{y+3}$

32. $\dfrac{x-1}{x-3} + \dfrac{x+4}{x-4}$

38. $\dfrac{2a+1}{(a+3)(a-3)} - \dfrac{a+2}{a+3}$

39. $\dfrac{3a+5}{(a+4)(a-1)} - \dfrac{2a-1}{a-1}$

44. $\dfrac{a-4}{a^2+2a-3} + \dfrac{a+2}{a^2+3a-4}$

40. $\dfrac{2x}{x^2-3x+2} + \dfrac{3}{x-2}$

45. $\dfrac{b-3}{b^2+9b+20} + \dfrac{b+4}{b^2+b-12}$

41. $\dfrac{4a}{a^2-2a-3} + \dfrac{3}{a+1}$

46. $\dfrac{y-1}{y^2+4y-12} - \dfrac{y+3}{y^2+6y-16}$

42. $\dfrac{3y}{y^2-7y+12} - \dfrac{y^2}{y-3}$

47. $\dfrac{x+3}{x^2+9x+14} - \dfrac{x-5}{x^2-4}$

43. $\dfrac{x-1}{x^2+6x+8} + \dfrac{x+3}{x^2+2x-8}$

48. $\dfrac{x-1}{x^2-4x+3} + \dfrac{x+3}{x^2-5x+6} + \dfrac{2x}{x^2-3x+2}$

49. $\dfrac{4x}{x^2 + 6x + 8} + \dfrac{3}{x^2 + x - 6} + \dfrac{x - 1}{x^2 + x - 12}$

54. $\dfrac{4}{3b^2 - 12b} - \dfrac{2}{6b^2 - 6b}$

50. $\dfrac{y + 2}{y^2 - 1} + \dfrac{y - 3}{y^2 - 3y - 4} - \dfrac{y + 3}{y^2 - 5y + 4}$

55. $\dfrac{3}{2x^5 - 4x^4} + \dfrac{-2}{8x^3 + 24x^2}$

51. $\dfrac{a - 2}{a^2 - 9a + 18} + \dfrac{a - 2}{a^2 - 4a - 12} - \dfrac{a - 2}{a^2 - a - 6}$

56. $\dfrac{x + 2}{12x^3} + \dfrac{x + 1}{4x^2 + 8x - 12} - \dfrac{x + 3}{16x^2 - 32x + 16}$

52. $\dfrac{y - 2}{y^2 + 6y} + \dfrac{y + 4}{y^2 + 5y - 6}$

57. $\dfrac{2x}{x^2 - 9} - \dfrac{x + 1}{4x^2 - 12x} - \dfrac{x - 4}{8x^3}$

53. $\dfrac{a + 1}{a^3 + 3a^2} - \dfrac{a + 6}{a^2 - a}$

58. $4 + \dfrac{3}{x + 2}$

59. $8 + \dfrac{2}{x+6}$

64. $6 - \dfrac{4y}{y+2}$

60. $1 + \dfrac{4}{x-7}$

65. $2x + \dfrac{x^2-4}{x+1}$

61. $3 + \dfrac{5}{x-6}$

66. $-3y + \dfrac{4y^2+2y-5}{y+3}$

62. $-2 + \dfrac{4x}{x+5}$

67. $x + 2 + \dfrac{x^2+4}{x-1}$

63. $-1 + \dfrac{3a}{a-1}$

68. $b + 6 + \dfrac{2b+5}{b-2}$

69. $\dfrac{3x-1}{x-4}-8$

74. $\dfrac{2x}{1-x}+\dfrac{6x}{x-1}$

70. $\dfrac{4y+5}{y+1}-9$

75. $\dfrac{5m}{6-m}+\dfrac{3m}{m-6}$

71. $\dfrac{2y^2+11y-1}{y+4}-3y$

76. $\dfrac{-a+7}{8-3a}+\dfrac{2a+1}{3a-8}$

72. $\dfrac{5y^2-2y+1}{y^2+y-6}-2$

77. $\dfrac{-2y+4}{4-5y}-\dfrac{9}{5y-4}$

73. $\dfrac{4a^3+2a^2+a-1}{a^2+11a+28}+3a$

78. $\dfrac{m-1}{1-m}-\dfrac{2}{m-1}$

EXERCISES FOR REVIEW

(1.6) 79. Simplify $(x^3 y^2 z^5)^6 (x^2 yz)^2$.

(2.6) 80. Write $6a^{-3}b^4 c^{-2} a^{-1} b^{-5} c^3$ so that only positive exponents appear.

(6.5) 81. Construct the graph of $y = -2x + 4$.

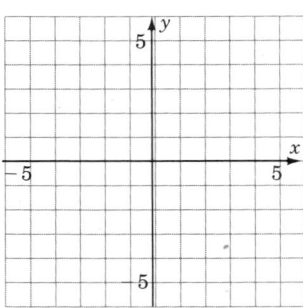

(7.3) 82. Find the product: $\dfrac{x^2 - 3x - 4}{x^2 + 6x + 5} \cdot \dfrac{x^2 + 5x + 6}{x^2 - 2x - 8}$.

(7.4) 83. Replace N with the proper quantity: $\dfrac{x+3}{x-5} = \dfrac{N}{x^2 - 7x + 10}$.

★ **Answers to Practice Sets (7.5)**

A. 1. $\dfrac{2}{3}$ 2. $\dfrac{5}{b}$ 3. $\dfrac{x}{y^2}$ 4. $\dfrac{3x + 4y}{x - y}$ 5. $\dfrac{3x^2 - 3x - 1}{3x + 10}$ 6. $\dfrac{4x^2 + 7}{x(2x + 3)}$ 7. $\dfrac{-4x + 7}{(x + 2)(x - 3)}$

     8. $\dfrac{4a^2 + 2a + 1}{a(a - 6)}$ 9. $\dfrac{5x^2 + x + 3}{(x - 4)(x - 2)}$

B. 1. $\dfrac{9ax + 5x}{12a^3}$ 2. $\dfrac{8b^2 - 7b}{(b + 1)(b - 2)}$ 3. $\dfrac{2a^2 - 4a - 25}{(a + 2)(a + 3)}$ 4. $\dfrac{3x^2 - 19x - 18}{(x + 3)(x - 3)}$ 5. $\dfrac{5y^2 + 6y - 12}{y(y + 4)}$

     6. $\dfrac{2a^2 - 10a + 26}{(a - 2)(a - 1)(a - 4)}$ 7. $\dfrac{4b^2 + 12b + 6}{(b + 3)^2 (b + 2)^2}$ 8. $\dfrac{2x^2 - 5x + 8}{3(x + 4)(x - 1)}$ 9. $\dfrac{2x}{x - 4}$

C. 1. $\dfrac{8x - 45}{x - 6}$ 2. $\dfrac{a^2 - a - 14}{a + 3}$

7.6 Rational Equations

Section Overview

☐ **RATIONAL EQUATIONS**
☐ **THE LOGIC BEHIND THE PROCESS**
☐ **THE PROCESS**
☐ **EXTRANEOUS SOLUTIONS**

☐ RATIONAL EQUATIONS

Rational Equation

When one rational expression is set equal to another rational expression, a **rational equation** results.

     Some examples of rational equations are the following (except for number 5):

1. $\dfrac{3x}{4} = \dfrac{15}{2}$

2. $\dfrac{x+1}{x-2} = \dfrac{x-7}{x-3}$

3. $\dfrac{5a}{2} = 10$

4. $\dfrac{3}{x} + \dfrac{x-3}{x+1} = \dfrac{6}{5x}$

5. $\dfrac{x-6}{x+1}$ is a rational *expression,* not a rational equation.

❏ THE LOGIC BEHIND THE PROCESS

It seems most reasonable that an equation without any fractions would be easier to solve than an equation with fractions. Our goal, then, is to convert any rational equation to an equation that contains no fractions. This is easily done.

To develop this method, let's consider the rational equation

$$\dfrac{1}{6} + \dfrac{x}{4} = \dfrac{17}{12}$$

The LCD is 12. We know that we can multiply both sides of an equation by the same nonzero quantity, so we'll multiply both sides by the LCD, 12.

$$12\left(\dfrac{1}{6} + \dfrac{x}{4}\right) = 12 \cdot \dfrac{17}{12}$$

Now distribute 12 to each term on the left side using the distributive property.

$$12 \cdot \dfrac{1}{6} + 12 \cdot \dfrac{x}{4} = 12 \cdot \dfrac{17}{12}$$

Now divide to eliminate all denominators.

$2 \cdot 1 + 3 \cdot x = 17$
$2 + 3x = 17$

Now there are no more fractions, and we can solve this equation using our previous techniques to obtain 5 as the solution.

❏ THE PROCESS

We have cleared the equation of fractions by multiplying both sides by the LCD. This development generates the following rule.

Clearing an Equation of
Fractions

> To clear an equation of fractions, multiply both sides of the equation by the LCD.

When multiplying both sides of the equation by the LCD, we use the distributive property to distribute the LCD to each term. This means we can simplify the above rule.

Clearing an Equation of
Fractions

> To clear an equation of fractions, multiply *every* term on both sides of the equation by the LCD.

The complete method for solving a rational equation is

1. Determine all the values that must be excluded from consideration by finding the values that will produce zero in the denominator (and thus, division by zero). These excluded values are not in the domain of the equation and are called nondomain values.
2. Clear the equation of fractions by multiplying every term by the LCD.
3. Solve this nonfractional equation for the variable. Check to see if any of these potential solutions are excluded values.
4. Check the solution by substitution.

❑ EXTRANEOUS SOLUTIONS

Extraneous Solutions

Potential solutions that have been excluded because they make an expression undefined (or produce a false statement for an equation) are called **extraneous solutions.** Extraneous solutions are discarded. If there are no other potential solutions, the equation has no solution.

☆ SAMPLE SET A

Solve the following rational equations.

1. $\dfrac{3x}{4} = \dfrac{15}{2}$.

Since the denominators are constants, there are no excluded values. No values must be excluded. The LCD is 4. Multiply each term by 4.

$$4 \cdot \dfrac{3x}{4} = 4 \cdot \dfrac{15}{2}$$

$$\cancel{4} \cdot \dfrac{3x}{\cancel{4}} = \overset{2}{\cancel{4}} \cdot \dfrac{15}{\cancel{2}}$$

$$3x = 2 \cdot 15$$
$$3x = 30$$

$$x = 10$$

10 is not an excluded value. Check it as a solution.

Check: $\dfrac{3x}{4} = \dfrac{15}{2}$

$$\dfrac{3(10)}{4} \overset{?}{=} \dfrac{15}{2}$$

$$\dfrac{30}{4} \overset{?}{=} \dfrac{15}{2}$$

$$\dfrac{15}{2} \overset{\checkmark}{=} \dfrac{15}{2}$$

10 is the solution.

2. $\dfrac{4}{x-1} = \dfrac{2}{x+6}$.

1 and -6 are nondomain values. Exclude them from consideration.

The LCD is $(x-1)(x+6)$. Multiply every term by $(x-1)(x+6)$.

$$(x-1)(x+6) \cdot \dfrac{4}{x-1} = (x-1)(x+6) \cdot \dfrac{2}{x+6}$$

$$(x-1)(x+6) \cdot \frac{4}{x-1} = (x-1)(x+6) \cdot \frac{2}{x+6}$$

$4(x+6) = 2(x-1)$ **Solve this nonfractional equation.**

$4x + 24 = 2x - 2$
$2x = -26$

$x = -13$ **-13 is not an excluded value. Check it as a solution.**

Check: $\dfrac{4}{x-1} = \dfrac{2}{x+6}$

$$\frac{4}{-13-1} \overset{?}{=} \frac{2}{-13+6}$$

$$\frac{4}{-14} \overset{?}{=} \frac{2}{-7}$$

$$\frac{2}{-7} \overset{\checkmark}{=} \frac{2}{-7}$$

-13 is the solution.

3. $\dfrac{4a}{a-4} = 2 + \dfrac{16}{a-4}$.   **4 is a nondomain value. Exclude it from consideration. The LCD is $a - 4$. Multiply every term by $a - 4$.**

$$(a-4) \cdot \frac{4a}{a-4} = 2(a-4) + (a-4) \cdot \frac{16}{a-4}$$

$$(a-4) \cdot \frac{4a}{a-4} = 2(a-4) + (a-4) \cdot \frac{16}{a-4}$$

$4a = 2(a-4) + 16$ **Solve this nonfractional equation.**

$4a = 2a - 8 + 16$
$4a = 2a + 8$
$2a = 8$
$a = 4$

This value, $a = 4$, has been excluded from consideration. It is not to be considered as a solution. It is extraneous. As there are no other potential solutions to consider, we conclude that this equation has *no solution*.

★ PRACTICE SET A

Solve the following rational equations.

1. $\dfrac{2x}{5} = \dfrac{x-14}{6}$ **2.** $\dfrac{3a}{a-1} = \dfrac{3a+8}{a+3}$ **3.** $\dfrac{3}{y-3} + 2 = \dfrac{y}{y-3}$

☆ **SAMPLE SET B**

Solve the following rational equations.

1. $\dfrac{3}{x} + \dfrac{4x}{x-1} = \dfrac{4x^2 + x + 5}{x^2 - x}.$ **Factor all denominators to find any excluded values and the LCD.**

$\dfrac{3}{x} + \dfrac{4x}{x-1} = \dfrac{4x^2 + x + 5}{x(x-1)}$ **Nondomain values are 0 and 1. Exclude them from consideration.**
The LCD is $x(x-1)$. Multiply each term by $x(x-1)$ and simplify.

$$\cancel{x}(x-1) \cdot \dfrac{3}{\cancel{x}} + x\cancel{(x-1)} \cdot \dfrac{4x}{\cancel{x-1}} = \cancel{x(x-1)} \cdot \dfrac{4x^2 + x + 5}{\cancel{x(x-1)}}$$

$3(x-1) + 4x \cdot x = 4x^2 + x + 5$ **Solve this nonfractional equation to obtain the potential solutions.**

$3x - 3 + 4x^2 = 4x^2 + x + 5$
$3x - 3 = x + 5$
$2x = 8$

$x = 4$ **4 is not an excluded value. Check it as a solution.**

Check: $\dfrac{3}{x} + \dfrac{4x}{x-1} = \dfrac{4x^2 + x + 5}{x^2 - x}$

$\dfrac{3}{4} + \dfrac{4 \cdot 4}{4 - 1} \overset{?}{=} \dfrac{4 \cdot 4^2 + 4 + 5}{16 - 4}$

$\dfrac{3}{4} + \dfrac{16}{3} \overset{?}{=} \dfrac{64 + 4 + 5}{12}$

$\dfrac{9}{12} + \dfrac{64}{12} \overset{?}{=} \dfrac{73}{12}$

$\dfrac{73}{12} \overset{\checkmark}{=} \dfrac{73}{12}$

4 is the solution.

The zero-factor property can be used to solve certain types of rational equations. We studied the zero-factor property in Section 7.1, and you may remember that it states that if a and b are real numbers and that $a \cdot b = 0$, then either or both $a = 0$ or $b = 0$. The zero-factor property is useful in solving the following rational equation.

2. $\dfrac{3}{a^2} - \dfrac{2}{a} = 1.$ **Zero is an excluded value.**
The LCD is a^2. Multiply each term by a^2 and simplify.

$$\cancel{a^2} \cdot \dfrac{3}{\cancel{a^2}} - a^{\cancel{2}} \cdot \dfrac{2}{\cancel{a}} = 1 \cdot a^2$$

$3 - 2a = a^2$ **Solve this nonfractional quadratic equation. Set it equal to zero.**

$0 = a^2 + 2a - 3$
$0 = (a + 3)(a - 1)$

$a = -3, \qquad a = 1$ **Check these as solutions.**

Check:

If $a = -3$: $\qquad \dfrac{3}{(-3)^2} - \dfrac{2}{-3} \overset{?}{=} 1$

$\qquad\qquad\qquad \dfrac{3}{9} + \dfrac{2}{3} \overset{?}{=} 1$

$\qquad\qquad\qquad \dfrac{1}{3} + \dfrac{2}{3} \overset{?}{=} 1$

$\qquad\qquad\qquad 1 \overset{\checkmark}{=} 1$

$\qquad\qquad a = -3$ checks and is a solution.

If $a = 1$: $\qquad \dfrac{3}{(1)^2} - \dfrac{2}{1} \overset{?}{=} 1$

$\qquad\qquad\qquad \dfrac{3}{1} - \dfrac{2}{1} \overset{?}{=} 1$

$\qquad\qquad\qquad 1 \overset{\checkmark}{=} 1$

$\qquad\qquad a = 1$ checks and is a solution.

-3 and 1 are the solutions.

★ **PRACTICE SET B**

1. Solve the equation $\dfrac{a+3}{a-2} = \dfrac{a+1}{a-1}$.

2. Solve the equation $\dfrac{1}{x-1} - \dfrac{1}{x+1} = \dfrac{2x}{x^2-1}$.

Answers to Practice Sets are on p. 370.

Section 7.6 EXERCISES

For problems 1–55, solve the rational equations.

1. $\dfrac{32}{x} = \dfrac{16}{3}$

2. $\dfrac{54}{y} = \dfrac{27}{4}$

5. $\dfrac{x+1}{4} = \dfrac{x-3}{2}$

3. $\dfrac{8}{y} = \dfrac{2}{3}$

4. $\dfrac{x}{28} = \dfrac{3}{7}$

6. $\dfrac{a+3}{6} = \dfrac{a-1}{4}$

7. $\dfrac{y-3}{6} = \dfrac{y+1}{4}$

14. $\dfrac{b+11}{3} + 8 = 6$

8. $\dfrac{x-7}{8} = \dfrac{x+5}{6}$

15. $\dfrac{y-1}{y+2} = \dfrac{y+3}{y-2}$

9. $\dfrac{a+6}{9} - \dfrac{a-1}{6} = 0$

16. $\dfrac{x+2}{x-6} = \dfrac{x-1}{x+2}$

10. $\dfrac{y+11}{4} = \dfrac{y+8}{10}$

17. $\dfrac{3m+1}{2m} = \dfrac{4}{3}$

11. $\dfrac{b+1}{2} + 6 = \dfrac{b-4}{3}$

18. $\dfrac{2k+7}{3k} = \dfrac{5}{4}$

12. $\dfrac{m+3}{2} + 1 = \dfrac{m-4}{5}$

19. $\dfrac{4}{x+2} = 1$

13. $\dfrac{a-6}{2} + 4 = -1$

20. $\dfrac{-6}{x-3} = 1$

21. $\dfrac{a}{3} + \dfrac{10 + a}{4} = 6$

27. $\dfrac{4a}{a + 2} - \dfrac{3a}{a - 1} = \dfrac{a^2 - 8a - 4}{a^2 + a - 2}$

22. $\dfrac{k + 17}{5} - \dfrac{k}{2} = 2k$

28. $\dfrac{3a - 7}{a - 3} = \dfrac{4a - 10}{a - 3}$

23. $\dfrac{2b + 1}{3b - 5} = \dfrac{1}{4}$

29. $\dfrac{2x - 5}{x - 6} = \dfrac{x + 1}{x - 6}$

24. $\dfrac{-3a + 4}{2a - 7} = \dfrac{-7}{9}$

25. $\dfrac{x}{x + 3} - \dfrac{x}{x - 2} = \dfrac{10}{x^2 + x - 6}$

30. $\dfrac{3}{x + 4} + \dfrac{5}{x + 4} = \dfrac{3}{x - 1}$

26. $\dfrac{3y}{y - 1} + \dfrac{2y}{y - 6} = \dfrac{5y^2 - 15y + 20}{y^2 - 7y + 6}$

31. $\dfrac{2}{y + 2} + \dfrac{8}{y + 2} = \dfrac{9}{y + 3}$

32. $\dfrac{4}{a^2 + 2a} = \dfrac{3}{a^2 + a - 2}$

37. $\dfrac{-1}{x + 4} - \dfrac{2}{x + 1} = \dfrac{4x + 19}{x^2 + 5x + 4}$

38. $\dfrac{2}{x^2} + \dfrac{1}{x} = 1$ **39.** $\dfrac{6}{y^2} - \dfrac{5}{y} = 1$

33. $\dfrac{2}{b(b + 2)} = \dfrac{3}{b^2 + 6b + 8}$

40. $\dfrac{12}{a^2} - \dfrac{4}{a} = 1$ **41.** $\dfrac{20}{x^2} - \dfrac{1}{x} = 1$

34. $\dfrac{x}{x - 1} + \dfrac{3x}{x - 4} = \dfrac{4x^2 - 8x + 1}{x^2 - 5x + 4}$

42. $\dfrac{12}{y} + \dfrac{12}{y^2} = -3$ **43.** $\dfrac{16}{b^2} + \dfrac{12}{b} = 4$

35. $\dfrac{4x}{x + 2} - \dfrac{x}{x + 1} = \dfrac{3x^2 + 4x + 4}{x^2 + 3x + 2}$

44. $\dfrac{1}{x^2} = 1$ **45.** $\dfrac{16}{y^2} = 1$

36. $\dfrac{2}{a - 5} - \dfrac{4a - 2}{a^2 - 6a + 5} = \dfrac{-3}{a - 1}$

46. $\dfrac{25}{a^2} = 1$ **47.** $\dfrac{36}{y^2} = 1$

48. $\dfrac{2}{x^2} + \dfrac{3}{x} = 2$

49. $\dfrac{2}{a^2} - \dfrac{5}{a} = 3$

59. $P = 2(1 + w)$ for w.

50. $\dfrac{2}{x^2} + \dfrac{7}{x} = -6$

51. $\dfrac{4}{a^2} + \dfrac{9}{a} = 9$

60. $A = \dfrac{1}{2}h(b + B)$ for B.

61. $A = P(1 + rt)$ for r.

52. $\dfrac{2}{x} = \dfrac{3}{x + 2} + 1$

53. $\dfrac{1}{x} = \dfrac{2}{x + 4} - \dfrac{3}{2}$

62. $z = \dfrac{x - \bar{x}}{s}$ for $\bar{x}$.

54. $\dfrac{4}{m} - \dfrac{5}{m - 3} = 7$

55. $\dfrac{6}{a + 1} - \dfrac{2}{a - 2} = 5$

63. $F = \dfrac{S_x{}^2}{S_y{}^2}$ for $S_y{}^2$.

For problems 56–68, solve each literal equation for
the designated letter.

56. $V = \dfrac{GMm}{D}$ for D.

64. $\dfrac{1}{R} = \dfrac{1}{E} + \dfrac{1}{F}$ for F.

57. $PV = nrt$ for n.

65. $K = \dfrac{1}{2}h(s_1 + s_2)$ for s_2.

58. $E = mc^2$ for m.

66. $Q = \dfrac{2mn}{s + t}$ for s.

67. $V = \frac{1}{6}\pi(3a^2 + h^2)$ for h^2.

68. $I = \frac{E}{R + r}$ for R.

EXERCISES FOR REVIEW

(2.6) **69.** Write $(4x^3y^{-4})^{-2}$ so that only positive exponents appear.

(5.5) **70.** Factor $x^4 - 16$.

(6.4) **71.** Supply the missing word. The slope of a line is a measure of the _____ of the line.

(7.2) **72.** Find the product. $\dfrac{x^2 - 3x + 2}{x^2 - x - 12} \cdot \dfrac{x^2 + 6x + 9}{x^2 + x - 2} \cdot \dfrac{x^2 - 6x + 8}{x^2 + x - 6}$.

(7.5) **73.** Find the sum. $\dfrac{2x}{x + 1} + \dfrac{1}{x - 3}$.

★ Answers to Practice Sets (7.6)

A. **1.** $x = -10$ **2.** $a = -2$ **3.** $y = 3$ is extraneous, so no solution.

B. **1.** $a = \dfrac{1}{3}$ **2.** This equation has no solution. $x = 1$ is extraneous.

7.7 Applications

Section Overview	☐ THE FIVE-STEP METHOD

☐ THE FIVE-STEP METHOD

We are now in a position to study some applications of rational equations. Some of these problems will have practical applications while others are intended as logic developers.

   We will apply the five-step method for solving word problems.

Five-Step Method

1. Represent all unknown quantities in terms of x or some other letter.
2. Translate the verbal phrases to mathematical symbols and form an equation.
3. Solve this equation.
4. Check the solution by substituting the result into the original statement of the problem.
5. Write the conclusion.

Remember, step 1 is very important: always

Introduce a variable.

☆ SAMPLE SET A

When the same number is added to the numerator and denominator of the fraction $\frac{3}{5}$, the result is $\frac{7}{9}$. What is the number that is added?

Step 1:   Let $x =$ the number being added.

Step 2: $\dfrac{3 + x}{5 + x} = \dfrac{7}{9}$.

Step 3: $\dfrac{3 + x}{5 + x} = \dfrac{7}{9}$.   An excluded value is -5. The LCD is $9(5 + x)$. Multiply each term by $9(5 + x)$.

$$9(5 + x) \cdot \dfrac{3 + x}{5 + x} = 9(5 + x) \cdot \dfrac{7}{9}$$

$$9(3 + x) = 7(5 + x)$$
$$27 + 9x = 35 + 7x$$
$$2x = 8$$

$$x = 4$$ Check this potential solution.

Step 4: $\dfrac{3 + 4}{5 + 4} \overset{\checkmark}{=} \dfrac{7}{9}$.

Step 5:   The number added is 4.

★ PRACTICE SET A

The same number is added to the numerator and denominator of the fraction $\frac{4}{9}$. The result is $\frac{2}{3}$. What is the number that is added?

Step 1:   Let $x =$

Step 2:

Step 3:

Step 4:

Step 5:   The number added is _____ .

☆ SAMPLE SET B

Two thirds of a number added to the reciprocal of the number yields $\dfrac{25}{6}$. What is the number?

Step 1:   Let $x =$ the number.

Step 2:   Recall that the reciprocal of a number x is the number $\dfrac{1}{x}$.

$$\frac{2}{3} \cdot x + \frac{1}{x} = \frac{25}{6}$$

Step 3: $\dfrac{2}{3} \cdot x + \dfrac{1}{x} = \dfrac{25}{6}$ **The LCD is $6x$. Multiply each term by $6x$.**

$$6x \cdot \frac{2}{3} x + 6x \cdot \frac{1}{x} = 6x \cdot \frac{25}{6}$$

$$4x^2 + 6 = 25x$$ **Solve this nonfractional quadratic equation to obtain the potential solutions. (Use the zero-factor property.)**

$$4x^2 - 25x + 6 = 0$$
$$(4x - 1)(x - 6) = 0$$

$$x = \frac{1}{4}, \quad 6$$ **Check these potential solutions.**

Step 4:   Substituting into the original equation, it can be seen that both solutions check.

Step 5:   There are two solutions: $\dfrac{1}{4}$ and 6.

★ PRACTICE SET B

Seven halves of a number added to the reciprocal of the number yields $\dfrac{23}{6}$. What is the number?

Step 1:   Let $x =$

Step 2:

Step 3:

Step 4:

Step 5:   The number is _____ .

☆ **SAMPLE SET C**

Person A, working alone, can pour a concrete walkway in 6 hours. Person B, working alone, can pour the same walkway in 4 hours. How long will it take both people to pour the concrete walkway working together?

Step 1: Let x = the number of hours to pour the concrete walkway working together (since this is what we're looking for).

Step 2: If person A can complete the job in 6 hours, A can complete $\frac{1}{6}$ of the job in 1 hour.

If person B can complete the job in 4 hours, B can complete $\frac{1}{4}$ of the job in 1 hour.

If A and B, working together, can complete the job in x hours, they can complete $\frac{1}{x}$ of the job in 1 hour. Putting these three facts into equation form, we have

$$\frac{1}{6} + \frac{1}{4} = \frac{1}{x}$$

Step 3: $\frac{1}{6} + \frac{1}{4} = \frac{1}{x}$. An excluded value is 0. The LCD is $12x$. Multiply each term by $12x$.

$$12x \cdot \frac{1}{6} + 12x \cdot \frac{1}{4} = 12x \cdot \frac{1}{x}$$

$2x + 3x = 12$ Solve this nonfractional equation to obtain the potential solutions.

$5x = 12$

$x = \frac{12}{5}$ or $x = 2\frac{2}{5}$ Check this potential solution.

Step 4: $\frac{1}{6} + \frac{1}{4} = \frac{1}{x}$

$$\frac{1}{6} + \frac{1}{4} \overset{?}{=} \frac{\frac{1}{12}}{5}.$$

$\frac{1}{6} + \frac{1}{4} \overset{?}{=} \frac{5}{12}$ The LCD is 12.

$\frac{2}{12} + \frac{3}{12} \overset{?}{=} \frac{5}{12}$

$\frac{5}{12} \overset{\checkmark}{=} \frac{5}{12}$

Step 5: Working together, A and B can pour the concrete walkway in $2\frac{2}{5}$ hours.

★ **PRACTICE SET C**

Person A, working alone, can pour a concrete walkway in 9 hours. Person B, working alone, can pour the same walkway in 6 hours. How long will it take both people to pour the concrete walkway working together?

Step 1:

Step 2:

Step 3:

Step 4:

Step 5:   Working together, A and B _____ .

☆ **SAMPLE SET D**

An inlet pipe can fill a water tank in 12 hours. An outlet pipe can drain the tank in 20 hours. If both pipes are open, how long will it take to fill the tank?

Step 1:   Let $x =$ the number of hours required to fill the tank.

Step 2:   If the inlet pipe can fill the tank in 12 hours, it can fill $\dfrac{1}{12}$ of the tank in 1 hour.

If the outlet pipe can drain the tank in 20 hours, it can drain $\dfrac{1}{20}$ of the tank in 1 hour.

If both pipes are open, it takes x hours to fill the tank. So $\dfrac{1}{x}$ of the tank will be filled in 1 hour.

Since water is being added (inlet pipe) and subtracted (outlet pipe) we get

$$\frac{1}{12} - \frac{1}{20} = \frac{1}{x}$$

Step 3: $\dfrac{1}{12} - \dfrac{1}{20} = \dfrac{1}{x}.$ An excluded value is 0. The LCD is $60x$. Multiply each term by $60x$.

$$60x \cdot \frac{1}{12} - 60x \cdot \frac{1}{20} = 60x \cdot \frac{1}{x}$$

$$5x - 3x = 60$$ Solve this nonfractional equation to obtain the potential solutions.

$$2x = 60$$

$$x = 30$$ Check this potential solution.

Step 4: $\dfrac{1}{12} - \dfrac{1}{20} = \dfrac{1}{x}$

$$\frac{1}{12} - \frac{1}{20} \overset{?}{=} \frac{1}{30}.$$ The LCD is 60.

$$\frac{5}{60} - \frac{3}{60} \overset{?}{=} \frac{1}{30}$$

$$\frac{1}{30} \overset{\checkmark}{=} \frac{1}{30}$$

Step 5:   With both pipes open, it will take 30 hours to fill the water tank.

★ PRACTICE SET D

An inlet pipe can fill a water tank in 8 hours and an outlet pipe can drain the tank in 10 hours. If both pipes are open, how long will it take to fill the tank?

Step 1:

Step 2:

Step 3:

Step 4:

Step 5:

☆ SAMPLE SET E

It takes person A 3 hours longer than person B to complete a certain job. Working together, both can complete the job in 2 hours. How long does it take each person to complete the job working alone?

Step 1:   Let x = time required for B to complete the job working alone. Then, $(x + 3)$ = time required for A to complete the job working alone.

Step 2: $\dfrac{1}{x} + \dfrac{1}{x + 3} = \dfrac{1}{2}$.

Step 3: $\dfrac{1}{x} + \dfrac{1}{x + 3} = \dfrac{1}{2}$. **The two excluded values are 0 and -3. The LCD is $2x(x + 3)$.**

$$2x(x + 3) \cdot \frac{1}{x} + 2x(x + 3) \cdot \frac{1}{x + 3} = 2x(x + 3) \cdot \frac{1}{2}$$

$$2(x + 3) + 2x = x(x + 3)$$

Continued

$$2x + 6 + 2x = x^2 + 3x$$

This is a quadratic equation that can be solved using the zero-factor property.

$$4x + 6 = x^2 + 3x$$
$$x^2 - x - 6 = 0$$
$$(x - 3)(x + 2) = 0$$

$$x = 3, \quad -2$$

Check these potential solutions.

Step 4: If $x = -2$, the equation checks, but does not even make physical sense.

If $x = 3$, the equation checks.

$$x = 3 \quad \text{and} \quad x + 3 = 6$$

Step 5: Person B can do the job in 3 hours and person A can do the job in 6 hours.

★ **PRACTICE SET E**

It takes person A 4 hours less than person B to complete a certain task. Working together, both can complete the task in $\frac{8}{3}$ hours. How long does it take each person to complete the task working alone?

Step 1:

Step 2:

Step 3:

Step 4:

Step 5:

☆ **SAMPLE SET F**

The width of a rectangle is $\frac{1}{3}$ its length. Find the dimensions (length and width) if the perimeter is 16 cm.

Step 1: Let x = length. Then, $\frac{x}{3}$ = width.

Step 2: Make a sketch of the rectangle.

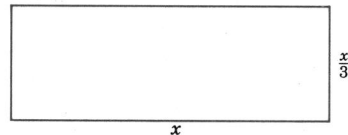

The perimeter of a figure is the total length around the figure.

$$x + \frac{x}{3} + x + \frac{x}{3} = 16$$

$$2x + \frac{2x}{3} = 16$$

Step 3: $2x + \frac{2x}{3} = 16.$ **The LCD is 3.**

$$3 \cdot 2x + 3 \cdot \frac{2x}{3} = 3 \cdot 16$$

$$6x + 2x = 48$$
$$8x = 48$$

$$x = 6$$ **Check this potential solution.**

Step 4: $6 + \frac{6}{3} + 6 + \frac{6}{3} \overset{?}{=} 16$

$6 + 2 + 6 + 2 \overset{?}{=} 16$

$16 \overset{\checkmark}{=} 16$

Since $x = 6$, $\frac{x}{3} = \frac{6}{3} = 2$

Step 5: The length = 6 cm and the width = 2 cm.

★ **PRACTICE SET F**

The width of a rectangle is $\frac{1}{12}$ its length. Find the dimensions (length and width) if the perimeter is 78 feet.

Step 1:

Step 2:

Step 3:

Step 4:

Step 5:

Answers to Practice Sets are on p. 383.

Section 7.7 EXERCISES

For problems 1–31, solve using the five-step method.

1. When the same number is added to both the numerator and denominator of the fraction $\frac{3}{7}$, the result is $\frac{2}{3}$. What is the number?

3. When the same number is added to both the numerator and denominator of the fraction $\frac{3}{8}$, the result is $\frac{1}{6}$. What is the number?

2. When the same number is added to both the numerator and denominator of the fraction $\frac{5}{8}$, the result is $\frac{3}{4}$. What is the number?

4. When the same number is added to both the numerator and denominator of the fraction $\frac{7}{9}$, the result is $\frac{2}{3}$. What is the number?

5. When the same number is subtracted from both the numerator and denominator of $\frac{1}{10}$, the result is $\frac{2}{3}$. What is the number?

6. When the same number is subtracted from both the numerator and denominator of $\frac{3}{4}$, the result is $\frac{5}{6}$. What is the number?

7. One third of a number added to the reciprocal of the number yields $\frac{13}{6}$. What is the number?

8. Four fifths of a number added to the reciprocal of the number yields $\frac{81}{10}$. What is the number?

9. One half of a number added to twice the reciprocal of the number yields 2. What is the number?

10. One fourth of a number added to four times the reciprocal of the number yields $\frac{-10}{3}$. What is the number?

11. One inlet pipe can fill a tank in 8 hours. Another inlet pipe can fill the tank in 5 hours. How long does it take both pipes working together to fill the tank?

12. One pipe can drain a pool in 12 hours. Another pipe can drain the pool in 15 hours. How long does it take both pipes working together to drain the pool?

13. A faucet can fill a bathroom sink in 1 minute. The drain can empty the sink in 2 minutes. If both the faucet and drain are open, how long will it take to fill the sink?

14. A faucet can fill a bathtub in $6\frac{1}{2}$ minutes. The drain can empty the tub in $8\frac{1}{3}$ minutes. If both the faucet and drain are open, how long will it take to fill the bathtub?

15. An inlet pipe can fill a tank in 5 hours. An outlet pipe can empty the tank in 4 hours. If both pipes are open, can the tank be filled? Explain.

16. An inlet pipe can fill a tank in a units of time. An outlet pipe can empty the tank in b units of time. If both pipes are open, how many units of time are required to fill the tank? Are there any restrictions on a and b?

17. A delivery boy, working alone, can deliver all his goods in 6 hours. Another delivery boy, working alone, can deliver the same goods in 5 hours. How long will it take the boys to deliver all the goods working together?

18. A Space Shuttle astronaut can perform a certain experiment in 2 hours. Another Space Shuttle astronaut who is not as familiar with the experiment can perform it in $2\frac{1}{2}$ hours. Working together, how long will it take both astronauts to perform the experiment?

19. One person can complete a task 8 hours sooner than another person. Working together, both people can perform the task in 3 hours. How many hours does it take each person to complete the task working alone?

20. Find two consecutive integers such that two thirds of the smaller number added to the other yields 11.

21. Find two consecutive integers such that three fourths of the smaller number added to the other yields 29.

22. The width of a rectangle is $\frac{2}{5}$ its length. Find the dimensions if the perimeter is 42 meters.

23. The width of a rectangle is $\frac{3}{7}$ the length. Find the dimensions if the perimeter is 60 feet.

24. Two sides of a triangle have the same length. The third side is twice as long as either of the other two sides. The perimeter of the triangle is 56 inches. What is the length of each side?

25. In a triangle, the second side is 3 inches longer than first side. The third side is $\frac{3}{4}$ the length of the second side. If the perimeter is 30 inches, how long is each side?

26. The pressure due to surface tension in a spherical drop of liquid is given by $P = \dfrac{2T}{r}$, where T is the surface tension of the liquid and r is the radius of the drop. If the liquid is a bubble, it has two surfaces and the surface tension is given by

$$P = \frac{2T}{r} + \frac{2T}{r} = \frac{4T}{r}$$

(a) Determine the pressure due to surface tension within a soap bubble of radius 2 inches and surface tension 28.
(b) Determine the radius of a bubble if the pressure due to surface tension is 52 and the surface tension is 39.

27. The equation $\dfrac{1}{p} + \dfrac{1}{q} = \dfrac{1}{f}$ relates the distance p of an object from a lens and the image distance q from the lens to the focal length f of the lens.

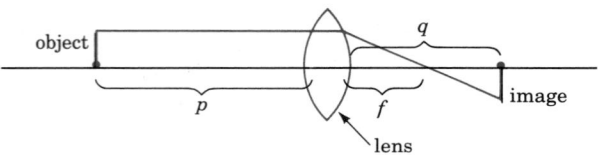

(a) Determine the focal length of a lens in which an object 10 feet away produces an image 6 feet away.
(b) Determine how far an object is from a lens if the focal length of the lens is 6 inches and the image distance is 10 inches.
(c) Determine how far an image will be from a lens that has a focal length of $4\,\dfrac{4}{5}$ cm and the object is 12 cm away from the lens.

28. Person A can complete a task in 4 hours, person B can complete the task in 6 hours, and person C can complete the task in 3 hours. If all three people are working together, how long will it take to complete the task?

29. Three inlet pipes can fill a storage tank in 4, 6, and 8 hours, respectively. How long will it take all three pipes to fill the tank?

30. An inlet pipe can fill a tank in 10 hours. The tank has two drain pipes, each of which can empty the tank in 30 hours. If all three pipes are open, can the tank be filled? If so, how long will it take?

31. An inlet pipe can fill a tank in 4 hours. The tank has three drain pipes. Two of the drain pipes can empty the tank in 12 hours, and the third can empty the tank in 20 hours. If all four pipes are open, can the tank be filled? If so, how long will it take?

EXERCISES FOR REVIEW

(5.7)   **32.** Factor $12a^2 + 13a - 4$.

(6.4)   **33.** Find the slope of the line passing through the points $(4, -3)$ and $(1, -6)$.

(7.3)   **34.** Find the quotient: $\dfrac{2x^2 - 11x - 6}{x^2 - 2x - 24} \div \dfrac{2x^2 - 3x - 2}{x^2 + 2x - 8}$.

(7.5)   **35.** Find the difference: $\dfrac{x + 2}{x^2 + 5x + 6} - \dfrac{x + 1}{x^2 + 4x + 3}$.

(7.6)   **36.** Solve the equation $\dfrac{9}{2m - 5} = -2$.

★ **Answers to Practice Sets (7.7)**

A. The number added is 6.

B. There are two numbers: $\dfrac{3}{7}, \dfrac{2}{3}$.

C. Working together, A and B can pour the concrete walkway in $3\dfrac{3}{5}$ hr.

D. It will take 40 hr to fill the tank.

E. Person A, 4 hr to complete the task; person B, 8 hr to complete the task.

F. length $= 36$ ft, width $= 3$ ft.

7.8 Complex Rational Expressions

Section Overview

- ☐ **SIMPLE AND COMPLEX FRACTIONS**
- ☐ **THE COMBINE-DIVIDE METHOD**
- ☐ **THE LCD-MULTIPLY-DIVIDE METHOD**

☐ SIMPLE AND COMPLEX FRACTIONS

Simple Fraction

In Section 7.1 we saw that a **simple fraction** was a fraction of the form $\dfrac{P}{Q}$, where P and Q are polynomials and $Q \neq 0$.

Complex Fraction

A **complex fraction** is a fraction in which the numerator or denominator, or both, is a fraction. The fractions

$$\frac{\dfrac{8}{15}}{\dfrac{2}{3}} \quad \text{and} \quad \frac{1 - \dfrac{1}{x}}{1 - \dfrac{1}{x^2}}$$

are examples of complex fractions, or more generally, complex rational expressions.

There are two methods for simplifying complex rational expressions: the combine-divide method and the LCD-multiply-divide method.

❏ THE COMBINE-DIVIDE METHOD

1. If necessary, combine the terms of the numerator together.
2. If necessary, combine the terms of the denominator together.
3. Divide the numerator by the denominator.

☆ **SAMPLE SET A**

Simplify each complex rational expression.

1. $\dfrac{\dfrac{x^3}{8}}{\dfrac{x^5}{12}}$

Steps 1 and 2 are not necessary so we proceed with step 3.

$$\frac{\dfrac{x^3}{8}}{\dfrac{x^5}{12}} = \frac{x^3}{8} \cdot \frac{12}{x^5} = \frac{\cancel{x^3}}{\cancel{8}_2} \cdot \frac{\cancel{12}^3}{\cancel{x^5}^2} = \frac{3}{2x^2}$$

2. $\dfrac{1 - \dfrac{1}{x}}{1 - \dfrac{1}{x^2}}$

Step 1: Combine the terms of the numerator: LCD $= x$.

$$1 - \frac{1}{x} = \frac{x}{x} - \frac{1}{x} = \frac{x-1}{x}$$

Step 2: Combine the terms of the denominator: LCD $= x^2$

$$1 - \frac{1}{x^2} = \frac{x^2}{x^2} - \frac{1}{x^2} = \frac{x^2-1}{x^2}$$

Step 3: Divide the numerator by the denominator.

$$\frac{\dfrac{x-1}{x}}{\dfrac{x^2-1}{x^2}} = \frac{x-1}{x} \cdot \frac{x^2}{x^2-1}$$

$$= \frac{\cancel{x-1}}{\cancel{x}} \cdot \frac{x^{\cancel{2}}}{(x+1)\cancel{(x-1)}}$$

$$= \frac{x}{x+1}$$

Thus,

$$\frac{1 - \dfrac{1}{x}}{1 - \dfrac{1}{x^2}} = \frac{x}{x+1}$$

3. $\dfrac{2 - \dfrac{13}{m} - \dfrac{7}{m^2}}{2 + \dfrac{3}{m} + \dfrac{1}{m^2}}$

Step 1:   Combine the terms of the numerator: LCD $= m^2$.

$$2 - \frac{13}{m} - \frac{7}{m^2} = \frac{2m^2}{m^2} - \frac{13m}{m^2} - \frac{7}{m^2} = \frac{2m^2 - 13m - 7}{m^2}$$

Step 2:   Combine the terms of the denominator: LCD $= m^2$

$$2 + \frac{3}{m} + \frac{1}{m^2} = \frac{2m^2}{m^2} + \frac{3m}{m^2} + \frac{1}{m^2} = \frac{2m^2 + 3m + 1}{m^2}$$

Step 3:   Divide the numerator by the denominator.

$$\frac{\dfrac{2m^2 - 13m - 7}{m^2}}{\dfrac{2m^2 + 3m - 1}{m^2}} = \frac{2m^2 - 13m - 7}{m^2} \cdot \frac{m^2}{2m^2 + 3m + 1}$$

$$= \frac{(2m + 1)(m - 7)}{m^2} \cdot \frac{m^2}{(2m + 1)(m + 1)}$$

$$= \frac{m - 7}{m + 1}$$

Thus,

$$\frac{2 - \dfrac{13}{m} - \dfrac{7}{m^2}}{2 + \dfrac{3}{m} + \dfrac{1}{m^2}} = \frac{m - 7}{m + 1}$$

★ **PRACTICE SET A**

Use the combine-divide method to simplify each expression.

1. $\dfrac{\dfrac{27x^2}{6}}{\dfrac{15x^3}{8}}$ **2.** $\dfrac{3 - \dfrac{1}{x}}{3 + \dfrac{1}{x}}$ **3.** $\dfrac{1 + \dfrac{x}{y}}{x - \dfrac{y^2}{x}}$ **4.** $\dfrac{m - 3 + \dfrac{2}{m}}{m - 4 + \dfrac{3}{m}}$ **5.** $\dfrac{1 + \dfrac{1}{x - 1}}{1 - \dfrac{1}{x - 1}}$

☐ THE LCD-MULTIPLY-DIVIDE METHOD

1. Find the LCD of all the terms.
2. Multiply the numerator and denominator by the LCD.
3. Reduce if necessary.

☆ **SAMPLE SET B**

Simplify each complex fraction.

1. $\dfrac{1 - \dfrac{4}{a^2}}{1 + \dfrac{2}{a}}$

Step 1: The LCD = a^2.
Step 2: Multiply both the numerator and denominator by a^2.

$$\frac{a^2 \left(1 - \dfrac{4}{a^2}\right)}{a^2 \left(1 + \dfrac{2}{a}\right)} = \frac{a^2 \cdot 1 - a^2 \cdot \dfrac{4}{a^2}}{a^2 \cdot 1 + a^2 \cdot \dfrac{2}{a}}$$

$$= \frac{a^2 - 4}{a^2 + 2a}$$

Step 3: Reduce.

$$\frac{a^2 - 4}{a^2 + 2a} = \frac{(a + 2)(a - 2)}{a(a + 2)}$$

$$= \frac{a - 2}{a}$$

Thus,

$$\frac{1 - \dfrac{4}{a^2}}{1 + \dfrac{2}{a}} = \frac{a - 2}{a}$$

2. $\dfrac{1 - \dfrac{5}{x} - \dfrac{6}{x^2}}{1 + \dfrac{6}{x} + \dfrac{5}{x^2}}$

Step 1: The LCD is x^2.
Step 2: Multiply the numerator and denominator by x^2.

$$\frac{x^2 \left(1 - \dfrac{5}{x} - \dfrac{6}{x^2}\right)}{x^2 \left(1 + \dfrac{6}{x} + \dfrac{5}{x^2}\right)} = \frac{x^2 \cdot 1 - x^2 \cdot \dfrac{5}{x} - x^2 \cdot \dfrac{6}{x^2}}{x^2 \cdot 1 + x^2 \cdot \dfrac{6}{x} + x^2 \cdot \dfrac{5}{x^2}}$$

$$= \frac{x^2 - 5x - 6}{x^2 + 6x + 5}$$

Step 3:   Reduce.

$$\frac{x^2 - 5x - 6}{x^2 + 6x + 5} = \frac{(x-6)(x+1)}{(x+5)(x+1)}$$

$$= \frac{x-6}{x+5}$$

Thus,

$$\frac{1 - \dfrac{5}{x} - \dfrac{6}{x^2}}{1 + \dfrac{6}{x} + \dfrac{5}{x^2}} = \frac{x-6}{x+5}$$

★ PRACTICE SET B

The following problems are the same problems as the problems in Practice Set A. Simplify these expressions using the LCD-multiply-divide method. Compare the answers to the answers produced in Practice Set A.

1. $\dfrac{\dfrac{27x^2}{6}}{\dfrac{15x^3}{8}}$

2. $\dfrac{3 - \dfrac{1}{x}}{3 + \dfrac{1}{x}}$

3. $\dfrac{1 + \dfrac{x}{y}}{x - \dfrac{y^2}{x}}$

4. $\dfrac{m - 3 + \dfrac{2}{m}}{m - 4 + \dfrac{3}{m}}$

5. $\dfrac{1 + \dfrac{1}{x-1}}{1 - \dfrac{1}{x-1}}$

Answers to Practice Sets are on p. 390.

Section 7.8 EXERCISES

For problems 1–27, simplify each complex rational expression.

1. $\dfrac{1 + \dfrac{1}{4}}{1 - \dfrac{1}{4}}$

2. $\dfrac{1 - \dfrac{1}{3}}{1 + \dfrac{1}{3}}$

3. $\dfrac{1 - \dfrac{1}{y}}{1 + \dfrac{1}{y}}$

4. $\dfrac{a + \dfrac{1}{x}}{a - \dfrac{1}{x}}$

5. $\dfrac{\dfrac{a}{b} + \dfrac{c}{b}}{\dfrac{a}{b} - \dfrac{c}{b}}$

6. $\dfrac{\dfrac{5}{m} + \dfrac{4}{m}}{\dfrac{5}{m} - \dfrac{4}{m}}$

13. $\dfrac{k - \dfrac{1}{k}}{\dfrac{k+1}{k}}$

14. $\dfrac{\dfrac{m}{m+1} - 1}{\dfrac{m+1}{2}}$

7. $\dfrac{3 + \dfrac{1}{x}}{\dfrac{3x+1}{x^2}}$

8. $\dfrac{1 + \dfrac{x}{x+y}}{1 - \dfrac{x}{x+y}}$

15. $\dfrac{\dfrac{2xy}{2x-y} - y}{\dfrac{2x-y}{3}}$

16. $\dfrac{\dfrac{1}{a+b} - \dfrac{1}{a-b}}{\dfrac{1}{a+b} + \dfrac{1}{a-b}}$

9. $\dfrac{2 + \dfrac{5}{a+1}}{2 - \dfrac{5}{a+1}}$

10. $\dfrac{1 - \dfrac{1}{a-1}}{1 + \dfrac{1}{a-1}}$

17. $\dfrac{\dfrac{5}{x+3} - \dfrac{5}{x-3}}{\dfrac{5}{x+3} + \dfrac{5}{x-3}}$

18. $\dfrac{2 + \dfrac{1}{y+1}}{\dfrac{1}{y} + \dfrac{2}{3}}$

11. $\dfrac{4 - \dfrac{1}{m^2}}{2 + \dfrac{1}{m}}$

12. $\dfrac{9 - \dfrac{1}{x^2}}{3 - \dfrac{1}{x}}$

19. $\dfrac{\dfrac{1}{x^2} - \dfrac{1}{y^2}}{\dfrac{1}{x} + \dfrac{1}{y}}$

20. $\dfrac{1 + \dfrac{5}{x} + \dfrac{6}{x^2}}{1 - \dfrac{1}{x} - \dfrac{12}{x^2}}$

21. $\dfrac{1 + \dfrac{1}{y} - \dfrac{2}{y^2}}{1 + \dfrac{7}{y} + \dfrac{10}{y^2}}$

22. $\dfrac{\dfrac{3n}{m} - 2 - \dfrac{m}{n}}{\dfrac{3n}{m} + 4 + \dfrac{m}{n}}$

27. $\dfrac{x - \dfrac{1}{1 - \dfrac{1}{x}}}{x + \dfrac{1}{1 + \dfrac{1}{x}}}$

23. $\dfrac{x - \dfrac{4}{3x - 1}}{1 - \dfrac{2x - 2}{3x - 1}}$

24. $\dfrac{\dfrac{y}{x + y} - \dfrac{x}{x - y}}{\dfrac{x}{x + y} + \dfrac{y}{x - y}}$

28. In electricity theory, when two resistors of resistance R_1 and R_2 ohms are connected in parallel, the total resistance R is

$$R = \dfrac{1}{\dfrac{1}{R_1} + \dfrac{1}{R_2}}$$

Write this complex fraction as a simple fraction.

25. $\dfrac{\dfrac{a}{a - 2} - \dfrac{a}{a + 2}}{\dfrac{2a}{a - 2} + \dfrac{a^2}{a + 2}}$

29. According to Einstein's theory of relativity, two velocities v_1 and v_2 are not added according to $v = v_1 + v_2$, but rather by

$$v = \dfrac{v_1 + v_2}{1 + \dfrac{v_1 v_2}{c^2}}$$

Write this complex fraction as a simple fraction.

Einstein's formula is really only applicable for velocities near the speed of light ($c = 186,000$ miles per second). At very much lower velocities, such as 500 miles per hour, the formula $v = v_1 + v_2$ provides an extremely good approximation.

26. $3 - \dfrac{2}{1 - \dfrac{1}{m + 1}}$

EXERCISES FOR REVIEW

(2.2) 30. Supply the missing word. Absolute value speaks to the question of how _____ and not "which way."

(3.6) 31. Find the product. $(3x + 4)^2$.

(5.5) 32. Factor $x^4 - y^4$.

(7.6) 33. Solve the equation $\dfrac{3}{x-1} - \dfrac{5}{x+3} = 0$.

(7.7) 34. One inlet pipe can fill a tank in 10 minutes. Another inlet pipe can fill the same tank in 4 minutes. How long does it take both pipes working together to fill the tank?

★ **Answers to Practice Sets (7.8)**

A. 1. $\dfrac{12}{5x}$ 2. $\dfrac{3x-1}{3x+1}$ 3. $\dfrac{x}{y(x-y)}$ 4. $\dfrac{m-2}{m-3}$ 5. $\dfrac{x}{x-2}$

B. 1. $\dfrac{12}{5x}$ 2. $\dfrac{3x-1}{3x+1}$ 3. $\dfrac{x}{y(x-y)}$ 4. $\dfrac{m-2}{m-3}$ 5. $\dfrac{x}{x-2}$

7.9 Dividing Polynomials

Section Overview

- ☐ **DIVIDING A POLYNOMIAL BY A MONOMIAL**
- ☐ **THE PROCESS OF DIVISION**
- ☐ **REVIEW OF SUBTRACTION OF POLYNOMIALS**
- ☐ **DIVIDING A POLYNOMIAL BY A POLYNOMIAL**

☐ DIVIDING A POLYNOMIAL BY A MONOMIAL

The following examples illustrate how to divide a polynomial by a monomial. The division process is quite simple and is based on addition of rational expressions.

$$\frac{a}{c} + \frac{b}{c} = \frac{a+b}{c}$$

Turning this equation around we get

$$\frac{a+b}{c} = \frac{a}{c} + \frac{b}{c}$$

Now we simply divide c into a, and c into b. This should suggest a rule.

Dividing a Polynomial by a Monomial

> To divide a polynomial by a monomial, divide every term of the polynomial by the monomial.

☆ **SAMPLE SET A**

1. $\dfrac{3x^2 + x - 11}{x}$.   **Divide every term of $3x^2 + x - 11$ by x.**

$$\frac{3x^2}{x} + \frac{x}{x} - \frac{11}{x} = 3x + 1 - \frac{11}{x}$$

2. $\dfrac{8a^3 + 4a^2 - 16a + 9}{2a^2}$.   Divide every term of $8a^3 + 4a^2 - 16a + 9$ by $2a^2$.

$\dfrac{8a^3}{2a^2} + \dfrac{4a^2}{2a^2} - \dfrac{16a}{2a^2} + \dfrac{9}{2a^2} = 4a + 2 - \dfrac{8}{a} + \dfrac{9}{2a^2}$

3. $\dfrac{4b^6 - 9b^4 - 2b + 5}{-4b^2}$.   Divide every term of $4b^6 - 9b^4 - 2b + 5$ by $-4b^2$.

$\dfrac{4b^6}{-4b^2} - \dfrac{9b^4}{-4b^2} - \dfrac{2b}{-4b^2} + \dfrac{5}{-4b^2} = -b^4 + \dfrac{9}{4}b^2 + \dfrac{1}{2b} - \dfrac{5}{4b^2}$

★ **PRACTICE SET A**

Perform the following divisions.

1. $\dfrac{2x^2 + x - 1}{x}$

2. $\dfrac{3x^3 + 4x^2 + 10x - 4}{x^2}$

3. $\dfrac{a^2b + 3ab^2 + 2b}{ab}$

4. $\dfrac{14x^2y^2 - 7xy}{7xy}$

5. $\dfrac{10m^3n^2 + 15m^2n^3 - 20mn}{-5m}$

❏ THE PROCESS OF DIVISION

In Section 7.2 we studied the method of reducing rational expressions. For example, we observed how to reduce an expression such as

$\dfrac{x^2 - 2x - 8}{x^2 - 3x - 4}$

Our method was to factor both the numerator and denominator, then divide out common factors.

$\dfrac{(x - 4)(x + 2)}{(x - 4)(x + 1)}$

$\dfrac{(\cancel{x - 4})(x + 2)}{(\cancel{x - 4})(x + 1)}$

$\dfrac{x + 2}{x + 1}$

When the numerator and denominator have no factors in common, the division may still occur, but the process is a little more involved than merely factoring. The method of dividing one polynomial by another is much the same as that of dividing one number by another. First, we'll review the steps in dividing numbers.

1. $\dfrac{35}{8}$.   We are to divide 35 by 8.

2. $8\overline{)35}$.   We try 4, since 32 divided by 8 is 4.

3. $\begin{array}{r} 4 \\ 8\overline{)35} \end{array}$ Multiply 4 and 8.

4. $\begin{array}{r} 4 \\ 8\overline{)35} \\ 32 \end{array}$ Subtract 32 from 35.

5. $\begin{array}{r} 4 \\ 8\overline{)35} \\ 32 \\ \hline 3 \end{array}$ Since the remainder 3 is less than the divisor 8, we are done with the division.

6. $4\dfrac{3}{8}$.   The quotient is expressed as a mixed number.

The process was to divide, multiply, and subtract.

❑ REVIEW OF SUBTRACTION OF POLYNOMIALS

A very important step in the process of dividing one polynomial by another is subtraction of polynomials. Let's review the process of subtraction by observing a few examples.

1. Subtract $x - 2$ from $x - 5$; that is, find $(x - 5) - (x - 2)$.

    Since $x - 2$ is preceded by a minus sign, remove the parentheses, change the sign of each term, then add.

$$\begin{array}{r} x - 5 \\ -(x - 2) \end{array} = \begin{array}{r} x - 5 \\ -x + 2 \\ \hline -3 \end{array}$$

The result is -3.

2. Subtract $x^3 + 3x^2$ from $x^3 + 4x^2 + x - 1$.

    Since $x^3 + 3x^2$ is preceded by a minus sign, remove the parentheses, change the sign of each term, then add.

$$\begin{array}{r} x^3 + 4x^2 + x - 1 \\ -(x^3 + 3x^2) \end{array} = \begin{array}{r} x^3 + 4x^2 + x - 1 \\ -x^3 - 3x^2 \\ \hline x^2 + x - 1 \end{array}$$

The result is $x^2 + x - 1$.

3. Subtract $x^2 + 3x$ from $x^2 + 1$.

    We can write $x^2 + 1$ as $x^2 + 0x + 1$.

$$\begin{array}{r} x^2 + 1 \\ -(x^2 + 3x) \end{array} = \begin{array}{r} x^2 + 0x + 1 \\ -(x^2 + 3x) \end{array} = \begin{array}{r} x^2 + 0x + 1 \\ -x^2 - 3x \\ \hline -3x + 1 \end{array}$$

❑ DIVIDING A POLYNOMIAL BY A POLYNOMIAL

Now we'll observe some examples of dividing one polynomial by another. The process is the same as the process used with whole numbers: divide, multiply, subtract, divide, multiply, subtract, ...

The division, multiplication, and subtraction take place one term at a time. The process is concluded when the polynomial remainder is of lesser degree than the polynomial divisor.

☆ **SAMPLE SET B**

Perform the division.

1. $\dfrac{x-5}{x-2}$. $\qquad$ We are to divide $x - 5$ by $x - 2$.

$x-2\overline{)x-5}$ $\qquad$ Divide x into x.

$\boxed{x}-2\overline{)x-5}$ $\overset{1}{}$ $\qquad$ Multiply 1 and $x - 2$.

$\begin{array}{r} 1 \\ x-2\overline{)x-5} \\ x-2 \end{array}$ $\qquad$ Multiply 1 and $x - 2$.
Subtract $x - 2$ from $x - 5$.

$\begin{array}{r} 1 \\ x-2\overline{)x-5} \\ \underline{x \not{+} 2} \\ -3 \end{array}$ $\qquad$ We write the quotient as

$1 - \dfrac{3}{x-2}$

Thus,

$\dfrac{x-5}{x-2} = 1 - \dfrac{3}{x-2}$

2. $\dfrac{x^3 + 4x^2 + x - 1}{x+3}$. $\qquad$ We are to divide $x^3 + 4x^2 + x - 1$ by $x + 3$.

$x+3\overline{)x^3 + 4x^2 + x - 1}$ $\qquad$ Divide x into x^3.

$\boxed{x}+3\overline{)x^3 + 4x^2 + x - 1}$ $\overset{x^2}{}$ $\qquad$ Multiply x^2 and $x + 3$.

$\begin{array}{r} x^2 \\ x+3\overline{)x^3 + 4x^2 + x - 1} \\ x^3 + 3x^2 \end{array}$ $\qquad$ Multiply x^2 and $x + 3$.
Subtract $x^3 + 3x^2$ from $x^3 + 4x^2 + x - 1$.

$\begin{array}{r} x^2 \\ x+3\overline{)x^3 + 4x^2 + x - 1} \\ \underline{x^3 \not{+} 3x^2} \\ x^2 + x - 1 \end{array}$ $\qquad$ Now divide x into x^2.

$\begin{array}{r} x^2 + x \\ \boxed{x}+3\overline{)x^3 + 4x^2 + x - 1} \\ x^3 + 3x^2 \\ \rightarrow x^2 + x - 1 \end{array}$ $\qquad$ Multiply x and $x + 3$.

Continued

$$\frac{x^2 + x}{x+3)\overline{x^3 + 4x^2 + x - 1}}$$

$$\underline{x^3 + 3x^3}$$

$$x^2 + x - 1 \qquad \text{Multiply } x \text{ and } x + 3.$$

$$x^2 + 3x \qquad \text{Subtract } x^2 + 3x \text{ from } x^2 + x - 1.$$

$$\frac{x^2 + x}{x+3)\overline{x^3 + 4x^2 + x - 1}}$$

$$\underline{x^3 + 3x^3}$$

$$x^2 + x - 1$$

$$\underline{-x^2 \cancel{+} 3x}$$

$$-2x - 1 \qquad \text{Divide } x \text{ into } -2x.$$

$$\frac{x^2 + x - 2}{\boxed{x} + 3)\overline{x^3 + 4x^2 + x - 1}}$$

$$\underline{x^3 + 3x^2}$$

$$x^2 + x - 1$$

$$\underline{x^2 + 3x}$$

$$\longrightarrow -2x - 1$$

$$\frac{x^2 + x - 2}{x+3)\overline{x^3 + 4x^2 + x - 1}}$$

$$\underline{x^3 + 3x^2}$$

$$x^2 + x - 1$$

$$\underline{x^2 + 3x}$$

$$-2x - 1 \qquad \text{Multiply } -2 \text{ and } x + 3.$$

$$-2x - 6 \qquad \text{Subtract } -2x - 6 \text{ from } -2x - 1.$$

$$\frac{x^2 + x - 2}{x+3)\overline{x^3 + 4x^2 + x - 1}}$$

$$\underline{x^3 + 3x^2}$$

$$x^2 + x - 1$$

$$\underline{x^2 + 3x}$$

$$-2x - 1$$

$$\underline{+-2x \pm 6}$$

$$5$$

Since the polynomial 5 is of lesser degree than $x + 3$ $(0 < 1)$, we are finished. We write the quotient as

$$x^2 + x - 2 + \frac{5}{x+3}$$

Thus,

$$\frac{x^3 + 4x^2 + x - 1}{x+3} = x^2 + x - 2 + \frac{5}{x+3}$$

★ **PRACTICE SET B**

Perform the following divisions.

1. $\dfrac{x+6}{x-1}$
2. $\dfrac{x^2 + 2x + 5}{x+3}$
3. $\dfrac{x^3 + x^2 - x - 2}{x+8}$
4. $\dfrac{x^3 + x^2 - 3x + 1}{x^2 + 4x - 5}$

☆ SAMPLE SET C

Divide $2x^3 - 4x + 1$ by $x + 6$.

$$\frac{2x^3 - 4x + 1}{x + 6}$$

Notice that the x^2 term in the numerator is missing. We can avoid any confusion by writing

$$\frac{2x^3 + 0x^2 - 4x + 1}{x + 6}$$

Divide, multiply, and subtract.

$$\begin{array}{r} 2x^2 \\ \boxed{x} + 6 \overline{)\, 2x^3 + 0x^2 - 4x + 1} \\ 2x^3 + 12x^2 \\ \hline -12x^2 - 4x + 1 \end{array}$$

$$\begin{array}{r} 2x^2 - 12x \\ \boxed{x} + 6 \overline{)\, 2x^3 + 0x^2 - 4x + 1} \\ 2x^3 + 12x^2 \\ \hline -12x^2 - 4x + 1 \\ -12x^2 - 72x \\ \hline 68x + 1 \end{array}$$

Divide, multiply, and subtract.

$$\begin{array}{r} 2x^2 - 12x + 68 \\ \boxed{x} + 6 \overline{)\, 2x^3 + 0x^2 - 4x + 1} \\ 2x^3 + 12x^2 \\ \hline -12x^2 - 4x + 1 \\ -12x^2 - 72x + 1 \\ \hline 68x + 1 \\ 68x + 408 \\ \hline -407 \end{array}$$

$$\frac{2x^3 - 4x + 1}{x + 6} = 2x^2 - 12x + 68 - \frac{407}{x + 6}$$

★ PRACTICE SET C

Perform the following divisions.

1. $\dfrac{x^2 - 3}{x + 2}$ 2. $\dfrac{4x^2 - 1}{x - 3}$ 3. $\dfrac{x^3 + 2x + 2}{x - 2}$ 4. $\dfrac{6x^3 + 5x^2 - 1}{2x + 3}$

Answers to Practice Sets are on p. 400.

Section 7.9 EXERCISES

For problems 1–60, perform the divisions.

1. $\dfrac{6a + 12}{2}$

2. $\dfrac{12b - 6}{3}$

3. $\dfrac{8y - 4}{-4}$

4. $\dfrac{21a - 9}{-3}$

5. $\dfrac{3x^2 - 6x}{-3}$

6. $\dfrac{4y^2 - 2y}{2y}$

7. $\dfrac{9a^2 + 3a}{3a}$

8. $\dfrac{20x^2 + 10x}{5x}$

9. $\dfrac{6x^3 + 2x^2 + 8x}{2x}$

10. $\dfrac{26y^3 + 13y^2 + 39y}{13y}$

11. $\dfrac{a^2b^2 + 4a^2b + 6ab^2 - 10ab}{ab}$

12. $\dfrac{7x^3y + 8x^2y^3 + 3xy^4 - 4xy}{xy}$

13. $\dfrac{5x^3y^3 - 15x^2y^2 + 20xy}{-5xy}$

14. $\dfrac{4a^2b^3 - 8ab^4 + 12ab^2}{-2ab^2}$

15. $\dfrac{6a^2y^2 + 12a^2y + 18a^2}{24a^2}$

16. $\dfrac{3c^3y^3 + 99c^3y^4 - 12c^3y^5}{3c^3y^3}$

17. $\dfrac{16ax^2 - 20ax^3 + 24ax^4}{6a^4}$

18. $\dfrac{21ay^3 - 18ay^2 - 15ay}{6ay^2}$

19. $\dfrac{-14b^2c^2 + 21b^3c^3 - 28c^3}{-7a^2c^3}$

20. $\dfrac{-30a^2b^4 - 35a^2b^3 - 25a^2}{-5b^3}$

21. $\dfrac{x + 6}{x - 2}$

22. $\dfrac{y + 7}{y + 1}$

23. $\dfrac{x^2 - x + 4}{x + 2}$

24. $\dfrac{x^2 + 2x - 1}{x + 1}$

25. $\dfrac{x^2 - x + 3}{x + 1}$

26. $\dfrac{x^2 + 5x + 5}{x + 5}$

27. $\dfrac{x^2 - 2}{x + 1}$

28. $\dfrac{a^2 - 6}{a + 2}$

29. $\dfrac{y^2 + 4}{y + 2}$

30. $\dfrac{x^2 + 36}{x + 6}$

31. $\dfrac{x^3 - 1}{x + 1}$

32. $\dfrac{a^3 - 8}{a + 2}$

33. $\dfrac{x^3 - 1}{x - 1}$

34. $\dfrac{a^3 - 8}{a - 2}$

35. $\dfrac{x^3 + 3x^2 + x - 2}{x - 2}$

36. $\dfrac{a^3 + 2a^2 - a + 1}{a - 3}$

37. $\dfrac{a^3 + a + 6}{a - 1}$

38. $\dfrac{x^3 + 2x + 1}{x - 3}$

39. $\dfrac{y^3 + 3y^2 + 4}{y + 2}$

40. $\dfrac{y^3 + 5y^2 - 3}{y - 1}$

49. $\dfrac{6x^2 + 8x - 1}{3x + 4}$

50. $\dfrac{20y^2 + 15y - 4}{4y + 3}$

41. $\dfrac{x^3 + 3x^2}{x + 3}$

42. $\dfrac{a^2 + 2a}{a + 2}$

51. $\dfrac{4x^3 + 4x^2 - 3x - 2}{2x - 1}$

43. $\dfrac{x^2 - x - 6}{x^2 - 2x - 3}$

44. $\dfrac{a^2 + 5a + 4}{a^2 - a - 2}$

52. $\dfrac{9a^3 - 18a^2 + 8a - 1}{3a - 2}$

45. $\dfrac{2y^2 + 5y + 3}{y^2 - 3y - 4}$

46. $\dfrac{3a^2 + 4a - 4}{a^2 + 3a + 3}$

53. $\dfrac{4x^4 - 4x^3 + 2x^2 - 2x - 1}{x - 1}$

47. $\dfrac{2x^2 - x + 4}{2x - 1}$

48. $\dfrac{3a^2 + 4a + 2}{3a + 4}$

54. $\dfrac{3y^4 + 9y^3 - 2y^2 - 6y + 4}{y + 3}$

55. $\dfrac{3y^2 + 3y + 5}{y^2 + y + 1}$

58. $\dfrac{9a^7 + 15a^6 + 4a^5 - 3a^4 - a^3 + 12a^2 + a - 5}{3a + 1}$

56. $\dfrac{2a^2 + 4a + 1}{a^2 + 2a + 3}$

59. $(2x^5 + 5x^4 - 1) \div (2x + 5)$

57. $\dfrac{8z^6 - 4z^5 - 8z^4 + 8z^3 + 3z^2 - 14z}{2z - 3}$

60. $(6a^4 - 2a^3 - 3a^2 + a + 4) \div (3a - 1)$

EXERCISES FOR REVIEW

(7.3) **61.** Find the product. $\dfrac{x^2 + 2x - 8}{x^2 - 9} \cdot \dfrac{2x + 6}{4x - 8}$.

(7.5) **62.** Find the sum. $\dfrac{x - 7}{x + 5} + \dfrac{x + 4}{x - 2}$.

(7.6) **63.** Solve the equation $\dfrac{1}{x + 3} + \dfrac{1}{x - 3} = \dfrac{1}{x^2 - 9}$.

(7.7) **64.** When the same number is subtracted from both the numerator and denominator of $\dfrac{3}{10}$, the result is $\dfrac{1}{8}$. What is the number that is subtracted?

(7.8) **65.** Simplify $\dfrac{\dfrac{1}{x + 5}}{\dfrac{4}{x^2 - 25}}$.

★ **Answers to Practice Sets (7.9)**

A. **1.** $2x + 1 - \dfrac{1}{x}$ **2.** $3x + 4 + \dfrac{10}{x} - \dfrac{4}{x^2}$ **3.** $a + 3b + \dfrac{2}{a}$ **4.** $2xy - 1$ **5.** $-2m^2n^2 - 3mn^3 + 4n$

B. **1.** $1 + \dfrac{7}{x-1}$ **2.** $x - 1 + \dfrac{8}{x+3}$ **3.** $x^2 - 7x + 55 - \dfrac{442}{x+8}$ **4.** $x - 3 + \dfrac{14x - 14}{x^2 + 4x - 5} = x - 3 + \dfrac{14}{x+5}$

C. **1.** $x - 2 + \dfrac{1}{x+2}$ **2.** $4x + 12 + \dfrac{35}{x-3}$ **3.** $x^2 + 2x + 6 + \dfrac{14}{x-2}$ **4.** $3x^2 - 2x + 3 - \dfrac{10}{2x+3}$

Rational Expression (7.1)

A *rational expression* is an algebraic expression that can be written as the quotient of two polynomials. An example of a rational expression is

$$\frac{x^2 + 3x - 1}{7x - 4}$$

Domain of a Rational Expression (7.1)

The *domain* of a rational expression is the collection of values for which the rational expression is defined. These values can be found by determining the values that will not produce zero in the denominator of the expression.

The domain of $\frac{x + 6}{x + 8}$ is the collection of all numbers except -8.

Equality Property of Fractions (7.1)

1. If $\frac{a}{b} = \frac{c}{d}$, then $ad = bc$.

2. If $ad = bc$, then $\frac{a}{b} = \frac{c}{d}$.

Negative Property of Fractions (7.1)

$$\frac{-a}{b} = \frac{a}{-b} = -\frac{a}{b}$$

Reducing a Rational Expression (7.2)

1. Factor the numerator and denominator completely.
2. Divide the numerator and denominator by any factors they have in common.

Common Cancelling Error (7.2)

$$\frac{x + 4}{x + 7} \neq \frac{\cancel{x} + 4}{\cancel{x} + 7} \neq \frac{4}{7}$$

Since x is not a common factor, it cannot be cancelled.

Multiplying Rational Expressions (7.3)

1. Factor all numerators and denominators.
2. Reduce to lowest terms first by dividing out all common factors.
3. Multiply numerators together.
4. Multiply denominators together.

It will be more convenient to leave the denominator in factored form.

Division of Rational Expressions (7.3)

$$\frac{P}{Q} \div \frac{R}{S} = \frac{P}{Q} \cdot \frac{S}{R} = \frac{P \cdot S}{Q \cdot R}$$

Building Rational Expressions (7.4)

$$\frac{P}{Q} \cdot \frac{b}{b} = \frac{Pb}{Qb}$$

Building rational expressions is exactly the opposite of reducing rational expressions. It is often useful in adding or subtracting rational expressions.

The building factor may be determined by dividing the original denominator into the new denominator. The quotient will be the building factor. It is this factor that will multiply the original numerator.

Least Common Denominator LCD (7.4)

The LCD is the polynomial of least degree divisible by each denominator. It is found as follows:

1. Factor each denominator. Use exponents for repeated factors.
2. Write each *different* factor that appears. If a factor appears more than once, use only the factor with the highest exponent.
3. The LCD is the product of the factors written in step 2.

Fundamental Rule for Adding or Subtracting Rational Expressions (7.5)

To add or subtract rational expressions conveniently, they should have the same denominator.

Adding and Subtracting Rational Expressions **(7.5)**

$$\frac{a}{c} + \frac{b}{c} = \frac{a+b}{c} \qquad \text{and} \qquad \frac{a}{c} - \frac{b}{c} = \frac{a-b}{c}$$

Note that we combine *only* the numerators.

Rational Equation **(7.6)**

A *rational equation* is a statement that two rational expressions are equal.

Clearing an Equation of Fractions **(7.6)**

To clear an equation of fractions, multiply both sides of the equation by the LCD. This amounts to multiplying every term by the LCD.

Solving a Rational Equation **(7.6)**

1. Determine all values that must be excluded as solutions by finding the values that produce zero in the denominator.
2. Clear the equation of fractions by multiplying every term by the LCD.
3. Solve this nonfractional equation for the variable. Check to see if any of these potential solutions are excluded values.
4. Check the solution by substitution.

Extraneous Solution **(7.6)**

A potential solution that has been excluded because it creates an undefined expression (perhaps, division by zero) is called an *extraneous solution*.

EXERCISE SUPPLEMENT

Section 7.1

For problems 1–10, find the domain of each rational expression.

1. $\dfrac{9}{x+4}$

2. $\dfrac{10x}{x+6}$

3. $\dfrac{x+1}{2x-5}$

4. $\dfrac{2a+3}{7a+5}$

5. $\dfrac{3m}{2m(m-1)}$

6. $\dfrac{5r+6}{9r(2r+1)}$

7. $\dfrac{s}{s(s+8)(4s+7)}$

8. $\dfrac{-11x}{x^2-9x+18}$

9. $\dfrac{-y+5}{12y^2+28y-5}$

10. $\dfrac{16}{12a^3+21a^2-6a}$

For problems 11–13, show that the fractions are equivalent.

11. $\dfrac{-4}{5}, \quad -\dfrac{4}{5}$

12. $\dfrac{-3}{8}, \quad -\dfrac{3}{8}$

13. $\dfrac{-7}{10}, \quad -\dfrac{7}{10}$

For problems 14–20, fill in the missing term.

14. $-\dfrac{3}{y-5}=\dfrac{}{y-5}$

15. $-\dfrac{6a}{2a+1}=\dfrac{}{2a+1}$

16. $-\dfrac{x+1}{x-3}=\dfrac{}{x-3}$

17. $-\dfrac{9}{-a+4}=\dfrac{}{a-4}$

18. $\dfrac{y+3}{-y-5}=\dfrac{}{y+5}$

19. $\dfrac{-6m-7}{-5m-1}=\dfrac{6m+7}{}$

20. $-\dfrac{2r-5}{7r+1}=\dfrac{2r-5}{}$

Section 7.2

For problems 21–42, reduce the rational expressions to lowest terms.

21. $\dfrac{12}{6x+24}$

22. $\dfrac{16}{4y-16}$

23. $\dfrac{5m+25}{10m^2+15m}$

24. $\dfrac{7+21r}{7r^2+28r}$

25. $\dfrac{3a^2+4a}{5a^3+6a^2}$

26. $\dfrac{4x-4}{x^2+2x-3}$

27. $\dfrac{5y+20}{y^2-16}$

28. $\dfrac{4y^3-12}{y^4-2y^2-3}$

29. $\dfrac{6a^9-12a^7}{2a^7-14a^5}$

30. $\dfrac{8x^4y^8+24x^3y^9}{4x^2y^5-12x^3y^6}$

31. $\dfrac{21y^8z^{10}w^2}{-7y^7w^2}$

32. $\dfrac{-35a^5b^2c^4d^8}{-5abc^3d^6}$

33. $\dfrac{x^2+9x+18}{x^3+3x^2}$

34. $\dfrac{a^2-12a+35}{2a^4-14a^3}$

35. $\dfrac{y^2-7y+12}{y^2-4y+3}$

36. $\dfrac{m^2-6m-16}{m^2-9m-22}$

37. $\dfrac{12r^2-7r-10}{4r^2-13r+10}$

38. $\dfrac{14a^2 - 5a - 1}{6a^2 + 9a - 6}$

39. $\dfrac{4a^4 - 8a^3}{4a^2}$

40. $\dfrac{5m^2}{10m^3 + 5m^2}$

41. $\dfrac{-6a - 1}{-5a - 2}$

42. $\dfrac{-r}{-5r - 1}$

Sections 7.3, 7.5

For problems 43–75, perform the indicated operations.

43. $\dfrac{x^2}{18} \cdot \dfrac{3}{x^3}$

44. $\dfrac{4a^2b^3}{15x^4y^5} \cdot \dfrac{10x^6y^3}{ab^2}$

45. $\dfrac{x + 6}{x - 1} \cdot \dfrac{x + 7}{x + 6}$

46. $\dfrac{8a - 12}{3a + 3} \div \dfrac{(a + 1)^2}{4a - 6}$

47. $\dfrac{10m^4 - 5m^2}{4r^7 + 20r^3} \div \dfrac{m}{16r^8 + 80r^4}$

48. $\dfrac{5}{r + 7} - \dfrac{3}{r + 7}$

49. $\dfrac{2a}{3a - 1} - \dfrac{9a}{3a - 1}$

50. $\dfrac{9x + 7}{4x - 6} + \dfrac{3x + 2}{4x - 6}$

51. $\dfrac{15y - 4}{8y + 1} - \dfrac{2y + 1}{8y + 1}$

52. $\dfrac{4}{a + 3} + \dfrac{6}{a - 5}$

53. $\dfrac{7a}{a + 6} + \dfrac{5a}{a - 8}$

54. $\dfrac{x + 4}{x - 2} + \dfrac{x + 7}{x - 1}$

55. $\dfrac{2y + 1}{y + 4} - \dfrac{y + 6}{y + 1}$

56. $\dfrac{x - 3}{(x + 2)(x + 4)} + \dfrac{2x - 1}{x + 4}$

57. $\dfrac{6a + 5}{(2a + 1)(4a - 3)} + \dfrac{4a + 1}{2a + 1}$

58. $\dfrac{4}{x^2 + 3x + 2} + \dfrac{9}{x^2 + 6x + 8}$

59. $\dfrac{6r}{r^2 + 7r - 18} - \dfrac{-3r}{r^2 - 3r + 2}$

60. $\dfrac{y + 3}{y^2 - 11y + 10} - \dfrac{y + 1}{y^2 + 3y - 4}$

61. $\dfrac{2a + 5}{16a^2 - 1} - \dfrac{6a + 7}{16a^2 - 12a + 2}$

62. $\dfrac{7y + 4}{6y^2 - 32y + 32} + \dfrac{6y - 10}{2y^2 - 18y + 40}$

63. $\dfrac{x^2 - x - 12}{x^2 - 3x + 2} \cdot \dfrac{x^2 + 3x - 4}{x^2 - 3x - 18}$

64. $\dfrac{y^2 - 1}{y^2 + 9y + 20} \div \dfrac{y^2 + 5y - 6}{y^2 - 16}$

65. $(r + 3)^4 \cdot \dfrac{r + 4}{(r + 3)^3}$

66. $(b + 5)^3 \cdot \dfrac{(b + 1)^2}{(b + 5)^2}$

67. $(x - 7)^4 \div \dfrac{(x - 7)^3}{x + 1}$

68. $(4x + 9)^6 \div \dfrac{(4x + 9)^2}{(3x + 1)^4}$

69. $5x + \dfrac{2x^2 + 1}{x - 4}$

70. $2y + \dfrac{4y^2 + 5}{y - 1}$

71. $\dfrac{y^2 + 4y + 4}{y^2 + 10y + 21} \div (y + 2)$

72. $2x - 3 + \dfrac{4x^2 + x - 1}{x - 1}$

73. $\dfrac{3x + 1}{x^2 + 3x + 2} + \dfrac{5x + 6}{x^2 + 6x + 5} - \dfrac{3x - 7}{x^2 - 2x - 35}$

74. $\dfrac{5a + 3b}{8a^2 + 2ab - b^2} - \dfrac{3a - b}{4a^2 - 9ab + 2b^2} - \dfrac{a + 5b}{4a^2 + 3ab - b^2}$

75. $\dfrac{3x^2 + 6x + 10}{10x^2 + 11x - 6} + \dfrac{2x^2 - 4x + 15}{2x^2 - 11x - 21}$

Section 7.6

For problems 76–90, solve the rational equations.

76. $\dfrac{4x}{5} + \dfrac{3x-1}{15} = \dfrac{29}{25}$

77. $\dfrac{6a}{7} + \dfrac{2a-3}{21} = \dfrac{77}{21}$

78. $\dfrac{5x-1}{6} + \dfrac{3x+4}{9} = \dfrac{-8}{9}$

79. $\dfrac{4y-5}{4} + \dfrac{8y+1}{6} = \dfrac{-69}{12}$

80. $\dfrac{4}{x-1} + \dfrac{7}{x+2} = \dfrac{43}{x^2+x-2}$

81. $\dfrac{5}{a+3} + \dfrac{6}{a-4} = \dfrac{9}{a^2-a-12}$

82. $\dfrac{-5}{y-3} + \dfrac{2}{y-3} = \dfrac{3}{y-3}$

83. $\dfrac{2m+5}{m-8} + \dfrac{9}{m-8} = \dfrac{30}{m-8}$

84. $\dfrac{r+6}{r-1} - \dfrac{3r+2}{r-1} = \dfrac{-6}{r-1}$

85. $\dfrac{8b+1}{b-7} - \dfrac{b+5}{b-7} = \dfrac{45}{b-7}$

86. Solve $z = \dfrac{x-\bar{x}}{s}$ for s.

87. Solve $A = P(1+rt)$ for t.

88. Solve $\dfrac{1}{R} = \dfrac{1}{E} + \dfrac{1}{F}$ for E.

89. Solve $Q = \dfrac{2mn}{s+t}$ for t.

90. Solve $I = \dfrac{E}{R+r}$ for r.

Section 7.7

For problems 91–100, find the solution.

91. When the same number is subtracted from both terms of the fraction $\dfrac{7}{12}$, the result is $\dfrac{1}{2}$. What is the number?

92. When the same number is added to both terms of the fraction $\dfrac{13}{15}$, the result is $\dfrac{8}{9}$. What is the number?

93. When three fourths of a number is added to the reciprocal of the number, the result is $\dfrac{173}{16}$. What is the number?

94. When one third of a number is added to the reciprocal of the number, the result is $\dfrac{-127}{90}$. What is the number?

95. Person A working alone can complete a job in 9 hours. Person B working alone can complete the same job in 7 hours. How long will it take both people to complete the job working together?

96. Debbie can complete an algebra assignment in $\dfrac{3}{4}$ of an hour. Sandi, who plays her radio while working, can complete the same assignment in $1\dfrac{1}{4}$ hours. If Debbie and Sandi work together, how long will it take them to complete the assignment?

97. An inlet pipe can fill a tank in 6 hours and an outlet pipe can drain the tank in 8 hours. If both pipes are open, how long will it take to fill the tank?

98. Two pipes can fill a tank in 4 and 5 hours, respectively. How long will it take both pipes to fill the tank?

99. The pressure due to surface tension in a spherical bubble is given by $P = \dfrac{4T}{r}$, where T is the surface tension of the liquid, and r is the radius of the bubble.
(a) Determine the pressure due to surface tension within a soap bubble of radius $\dfrac{1}{2}$ inch and surface tension 22.
(b) Determine the radius of a bubble if the pressure due to surface tension is 57.6 and the surface tension is 18.

100. The equation $\dfrac{1}{p} + \dfrac{1}{q} = \dfrac{1}{f}$ relates an objects distance p from a lens and the image distance q from the lens to the focal length f of the lens.
(a) Determine the focal length of a lens in which an object 8 feet away produces an image 6 feet away.
(b) Determine how far an object is from a lens if the focal length of the lens is 10 inches and the image distance is 10 inches.

(c) Determine how far an object will be from a lens that has a focal length of $1\frac{7}{8}$ cm and the object distance is 3 cm away from the lens.

Section 7.9

For problems 101–115, divide the polynomials.

101. $a^2 + 9a + 18$ by $a + 3$

102. $c^2 + 3c - 88$ by $c - 8$

103. $x^3 + 9x^2 + 18x + 28$ by $x + 7$

104. $y^3 - 2y^2 - 49y - 6$ by $y + 6$

105. $m^4 + 2m^3 - 8m^2 - m + 2$ by $m - 2$

106. $3r^2 - 17r - 27$ by $r - 7$

107. $a^3 - 3a^2 - 56a + 10$ by $a - 9$

108. $x^3 - x + 1$ by $x + 3$

109. $y^3 + y^2 - y$ by $y + 4$

110. $5x^6 + 5x^5 - 2x^4 + 5x^3 - 7x^2 - 8x + 6$ by $x^2 + x - 1$

111. $y^{10} - y^7 + 3y^4 - 3y$ by $y^4 - y$

112. $-4b^7 - 3b^6 - 22b^5 - 19b^4 + 12b^3 - 6b^2 + b + 4$ by $b^2 + 6$

113. $x^3 + 1$ by $x + 1$

114. $a^4 + 6a^3 + 4a^2 + 12a + 8$ by $a^2 + 3a + 2$

115. $y^{10} + 6y^5 + 9$ by $y^5 + 3$

1. _____

1. (7.1) Find the domain of $\dfrac{5a + 1}{a^2 - 5a - 24}$.

For problems 2 and 3, fill in the missing term.

2. _____

2. (7.1) $-\dfrac{3}{x + 4} = \dfrac{}{x + 4}$

3. _____

3. (7.1) $\dfrac{2x + 5}{-x + 1} = \dfrac{}{x - 1}$

For problems 4–6, reduce to lowest terms.

4. _____

4. (7.2) $\dfrac{30x^6 y^3 (x - 3)^2 (x + 5)^2}{6xy^3 (x + 5)}$

5. _____

5. (7.2) $\dfrac{x^2 + 10x + 24}{x^2 + x - 30}$

6. _____

6. (7.2) $\dfrac{8x^2 + 2x - 3}{4x^2 + 12x - 7}$

7. (7.4) Replace N with the proper quantity.

7. _____

$\dfrac{x + 2}{x - 1} = \dfrac{N}{x^2 - 4x + 3}$

8. _____

8. (7.4) Assume that $a^2 + a - 6$, $a^2 - a - 12$, and $a^2 - 2a - 8$ are denominators of rational expressions. Find the LCD.

For problems 9–16, perform the operations.

9. _____

9. (7.5) $\dfrac{3a + 4}{a + 6} - \dfrac{2a - 1}{a + 6}$

10. _____

10. (7.3) $\dfrac{18x^3 y}{5a^2} \cdot \dfrac{15a^3 b}{6x^2 y}$

11. _____

11. (7.3) $\dfrac{y^2 - y - 12}{y^2 + 3y + 2} \cdot \dfrac{y^2 + 10y + 16}{y^2 - 7y + 12}$

12. _____

12. (7.5) $\dfrac{y - 2}{y^2 - 11y + 24} + \dfrac{y + 4}{y^2 + 3y - 18}$

407

13. _____

13. (7.5) $\dfrac{9}{2x+7} + \dfrac{4}{6x-1}$

14. _____

14. (7.3) $\dfrac{16x^5(x^2-1)}{9x-9} \div \dfrac{2x^2-2x}{3}$

15. _____

15. (7.3) $(m+3) \div \dfrac{2m+6}{5m+1}$

16. _____

16. (7.5) $\dfrac{3y+10}{8y^2+10y-3} - \dfrac{5y-1}{4y^2+23y-6}$

17. _____

17. (7.6) Solve $\dfrac{1}{x+3} + \dfrac{3}{x-3} = \dfrac{x}{x^2-9}$.

18. _____

18. (7.6) Solve $\dfrac{12}{m-4} + 5 = \dfrac{3m}{m-4}$.

19. (7.7) When the same number is added to both the numerator and denominator of the fraction $\dfrac{5}{3}$, the result is $\dfrac{6}{5}$. What is the number that is added?

19. _____

20. (7.7) Person A, working alone, can complete a job in 20 hours. Person B, working alone, can complete the same job in 30 hours. How long will it take both people, working together, to complete the job?

20. _____

21. (7.7) The width of a rectangle is 1 foot longer than one half the length. Find the dimensions (lengh and width) of the rectangle if the perimeter is 44 feet.

21. _____

22. (7.8) Simplify the complex fraction $\dfrac{4 - \dfrac{3}{x}}{4 + \dfrac{3}{x}}$.

22. _____

23. (7.8) Simplify the complex fraction $\dfrac{1 - \dfrac{5}{x} - \dfrac{6}{x^2}}{1 + \dfrac{6}{x} + \dfrac{5}{x^2}}$.

23. _____

24. (7.9) Perform the division: $\dfrac{x^3 + 10x^2 + 21x - 18}{x+6}$.

24. _____

25. (7.9) Perform the division: $\dfrac{2x^3 + 5x - 1}{x-2}$.

25. _____

8

Roots, Radicals, and Square Root Equations

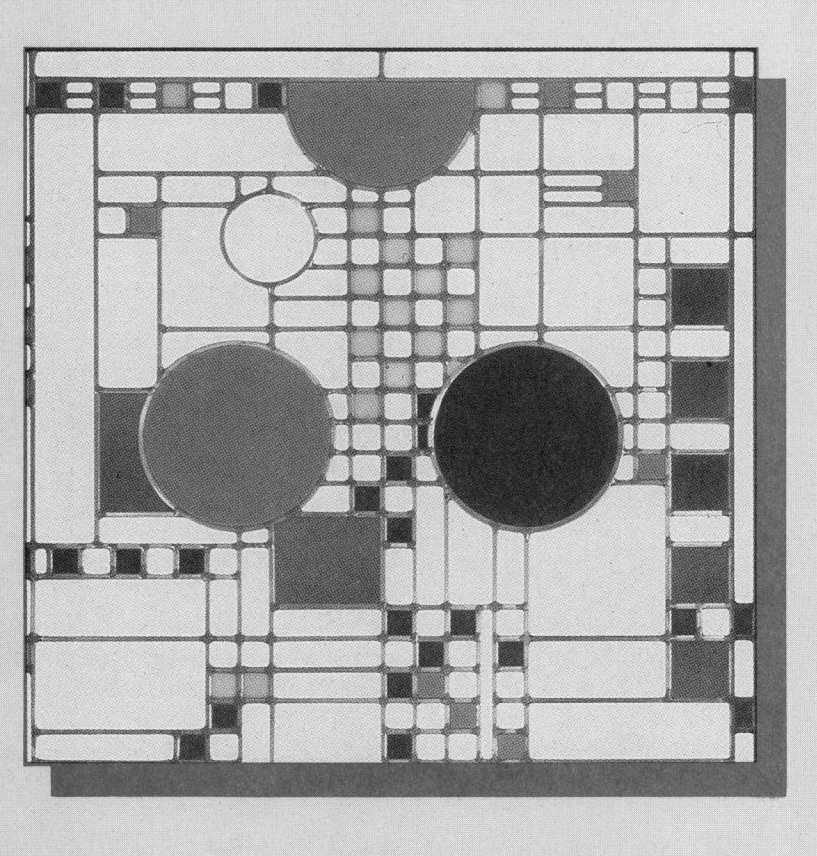

After completing this chapter, you should

Section 8.1 Square Root Expressions
- understand the concept of square root
- be able to distinguish between the principal and secondary square roots of a number
- be able to relate square roots and meaningful expressions and to simplify a square root expression

Section 8.2 Simplifying Square Root Expressions
- be able to identify a perfect square
- be familiar with the product and quotient properties of square roots
- be able to simplify square roots involving and not involving fractions

Section 8.3 Multiplication of Square Root Expressions
- be able to use the product property of square roots to multiply square roots

Section 8.4 Division of Square Root Expressions
- be able to use the division property of square roots, the method of rationalizing the denominator, and conjugates to divide square roots

Section 8.5 Addition and Subtraction of Square Root Expressions
- understand the process used in adding and subtracting square roots
- be able to add and subtract square roots

Section 8.6 Square Root Equations with Applications
- be able to recognize square root equations and extraneous solutions
- be able to solve square root equations

8.1 Square Root Expressions

Section Overview

- ☐ SQUARE ROOTS
- ☐ PRINCIPAL AND SECONDARY SQUARE ROOTS
- ☐ MEANINGFUL EXPRESSIONS
- ☐ SIMPLIFYING SQUARE ROOTS

☐ SQUARE ROOTS

When we studied exponents in Section 1.4, we noted that $4^2 = 16$ and $(-4)^2 = 16$.

We can see that 16 is the square of both 4 and -4. Since 16 comes from squaring 4 or -4, 4 and -4 are called the *square roots* of 16. Thus 16 has two square roots, 4 and -4. Notice that these two square roots are opposites of each other.

We can say that

Square Root

> The square root of a positive number x is a number such that when it is squared the number x results.

Every positive number has two square roots, one positive square root and one negative square root. Furthermore, the two square roots of a positive number are opposites of each other. The square root of 0 is 0.

☆ SAMPLE SET A

1. The two square roots of 49 are 7 and -7 since

   $$7^2 = 49 \qquad \text{and} \qquad (-7)^2 = 49$$

2. The two square roots of $\dfrac{49}{64}$ are $\dfrac{7}{8}$ and $\dfrac{-7}{8}$ since

   $$\left(\frac{7}{8}\right)^2 = \frac{7}{8} \cdot \frac{7}{8} = \frac{49}{64} \qquad \text{and} \qquad \left(\frac{-7}{8}\right)^2 = \frac{-7}{8} \cdot \frac{-7}{8} = \frac{49}{64}$$

★ PRACTICE SET A

Name both square roots of each of the following numbers.

1. 36   **2.** 25   **3.** 100   **4.** 64   **5.** 1

6. $\dfrac{1}{4}$ **7.** $\dfrac{9}{16}$ **8.** 0.01   **9.** 0.09

☐ PRINCIPAL AND SECONDARY SQUARE ROOTS

There is a notation for distinguishing the positive square root of a number x from the negative square root of x.

> If x is a positive real number, then
>
> $\sqrt{x}$ represents the positive square root of x. The positive square root of a number is called the **principal square root** of the number.

Principal Square Root: $\sqrt{x}$

Secondary Square Root: $-\sqrt{x}$

$-\sqrt{x}$ represents the negative square root of x. The negative square root of a number is called the **secondary square root** of the number.
$-\sqrt{x}$ indicates the secondary square root of x.

Radical Sign
Radicand
Radical

In the expression $\sqrt{x}$,
$\sqrt{}$ is called a **radical sign.**
x is called the **radicand.**
$\sqrt{x}$ is called a **radical.**

The horizontal bar that appears attached to the radical sign, $\sqrt{}$, is a grouping symbol that specifies the radicand.

Because $\sqrt{x}$ and $-\sqrt{x}$ are the two square root of x,

$$(\sqrt{x})(\sqrt{x}) = x \qquad \text{and} \qquad (-\sqrt{x})(-\sqrt{x}) = x$$

☆ **SAMPLE SET B**

Write the principal and secondary square roots of each number.

1. 9.   Principal square root is $\sqrt{9} = 3$.
           Secondary square root is $-\sqrt{9} = -3$.

2. 15.   Principal square root is $\sqrt{15}$.
           Secondary square root is $-\sqrt{15}$.

3. Use a calculator to obtain a decimal approximation for the two square roots of 34. Round to two decimal places.

On the Calculator

Type	34
Press	$\boxed{\sqrt{x}}$
Display reads:	5.8309519
Round to 5.83.	

Notice that the square root symbol on the calculator is $\sqrt{}$. This means, of course, that a calculator will produce only the positive square root. We must supply the negative square root ourselves.

$$\sqrt{34} \approx 5.83 \qquad \text{and} \qquad -\sqrt{34} \approx -5.83$$

Note: The symbol $\approx$ means "approximately equal to."

4. The number $\sqrt{50}$ is between what two whole numbers?

Since $7^2 = 49$, $\sqrt{49} = 7$.

Since $8^2 = 64$, $\sqrt{64} = 8$. Thus,

$$7 < \sqrt{50} < 8$$

Thus, $\sqrt{50}$ is a number between 7 and 8.

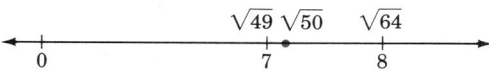

★ PRACTICE SET B

Write the principal and secondary square roots of each number.

1. 100   **2.** 121   **3.** 35

4. Use a calculator to obtain a decimal approximation for the two square roots of 35. Round to two decimal places.

❑ MEANINGFUL EXPRESSIONS

Since we know that the square of any real number is a positive number or zero, we can see that expressions such as $\sqrt{-16}$ do not describe real numbers. There is no real number that can be squared that will produce -16. For $\sqrt{x}$ to be a real number, we must have $x \geq 0$. In our study of algebra, we will assume that all variables and all expressions in radicands represent nonnegative numbers (numbers greater than or equal to zero).

☆ SAMPLE SET C

Write the proper restrictions that must be placed on the variable so that each expression represents a real number.

1. For $\sqrt{x-3}$ to be a real number, we must have

$$x - 3 \geq 0 \quad \text{or} \quad x \geq 3$$

2. For $\sqrt{2m+7}$ to be a real number, we must have

$$2m + 7 \geq 0 \quad \text{or} \quad 2m \geq -7 \quad \text{or} \quad m \geq \frac{-7}{2}$$

★ PRACTICE SET C

Write the proper restrictions that must be placed on the variable so that each expression represents a real number.

1. $\sqrt{x+5}$ **2.** $\sqrt{y-8}$ **3.** $\sqrt{3a+2}$ **4.** $\sqrt{5m-6}$

❑ SIMPLIFYING SQUARE ROOTS

When variables occur in the radicand, we can often simplify the expression by removing the radical sign. We can do so by keeping in mind that the radicand is the square of some other expression. We can simplify a radical by seeking an expression whose square is the radicand. The following observations will help us find the square root of a variable quantity.

Since $(x^3)^2 = x^{3 \cdot 2} = x^6$, x^3 is a square root of x^6. Also

$$6 \div 2 = 3$$

exponent   exponent
of x^6 of $\sqrt{x^6}$

$$\sqrt{x^6} = x^3$$

Since $(x^4)^2 = x^{4 \cdot 2} = x^8$, x^4 is a square root of x^8. Also

$$8 \div 2 = 4$$

exponent   exponent
of x^8 of $\sqrt{x^8}$

$$\sqrt{x^8} = x^4$$

Since $(x^6)^2 = x^{6 \cdot 2} = x^{12}$, x^6 is a square root of x^{12}. Also

$$12 \div 2 = 6$$

exponent   exponent
of x^{12} of $\sqrt{x^{12}}$

$$\sqrt{x^{12}} = x^6$$

These examples suggest the following rule:

> If a variable has an even exponent, its square root can be found by dividing that exponent by 2.

The examples of Sample Set B illustrate the use of this rule.

☆ SAMPLE SET D

Simplify each expression by removing the radical sign. Assume each variable is **nonnegative.**

1. $\sqrt{a^2}$.   We seek an expression whose square is a^2. Since $(a)^2 = a^2$,

   $\sqrt{a^2} = a$ Notice that $2 \div 2 = 1$.

2. $\sqrt{y^8}$.   We seek an expression whose square is y^8. Since $(y^4)^2 = y^8$,

   $\sqrt{y^8} = y^4$ Notice that $8 \div 2 = 4$.

3. $\sqrt{25m^2n^6}$.   We seek an expression whose square is $25m^2n^6$. Since $(5mn^3)^2 = 25m^2n^6$,

   $\sqrt{25m^2n^6} = 5mn^3$ Notice that $2 \div 2 = 1$ and $6 \div 2 = 3$.

4. $-\sqrt{121a^{10}(b-1)^4}$.   We seek an expression whose square is $121a^{10}(b-1)^4$. Since $[11a^5(b-1)^2]^2 = 121a^{10}(b-1)^4$,

   $\sqrt{121a^{10}(b-1)^4} = 11a^5(b-1)^2$

   Then, $-\sqrt{121a^{10}(b-1)^4} = -11a^5(b-1)^2$ Notice that $10 \div 2 = 5$ and $4 \div 2 = 2$.

★ PRACTICE SET D

Simplify each expression by removing the radical sign. Assume each variable is nonnegative.

1. $\sqrt{y^8}$ 2. $\sqrt{16a^4}$ 3. $\sqrt{49x^4y^6}$ 4. $-\sqrt{100x^8y^{12}z^2}$ 5. $-\sqrt{36(a+5)^4}$

6. $\sqrt{225w^4(z^2-1)^2}$ 7. $\sqrt{0.25y^6z^{14}}$ 8. $\sqrt{x^{2n}}$, where n is a natural number.

9. $\sqrt{x^{4n}}$, where n is a natural number.

Answers to Practice Sets are on p. 416.

Section 8.1 EXERCISES

1. How many square roots does every positive real number have?

2. The symbol $\sqrt{}$ represents which square root of a number?

3. The symbol $-\sqrt{}$ represents which square root of a number?

For problems 4–17, find the two square roots of the given number.

4. 64

5. 81

6. 25

7. 121

8. 144

9. 225

10. 10,000

11. $\dfrac{1}{16}$

12. $\dfrac{1}{49}$

13. $\dfrac{25}{36}$

14. $\dfrac{121}{225}$

15. 0.04

16. 0.16

17. 1.21

For problems 18–30, evaluate each expression. If the expression does not represent a real number, write "not a real number."

18. $\sqrt{49}$

19. $\sqrt{64}$

20. $-\sqrt{36}$

21. $-\sqrt{100}$

22. $-\sqrt{169}$

23. $-\sqrt{\dfrac{36}{81}}$

24. $-\sqrt{\dfrac{121}{169}}$

25. $\sqrt{-225}$

26. $\sqrt{-36}$

27. $-\sqrt{-1}$

28. $-\sqrt{-5}$

29. $-(-\sqrt{9})$

30. $-(-\sqrt{0.81})$

For problems 31–38, write the proper restrictions that must be placed on the variable so that the expression represents a real number.

31. $\sqrt{y + 10}$

32. $\sqrt{x + 4}$

33. $\sqrt{a - 16}$

34. $\sqrt{h - 11}$

35. $\sqrt{2k - 1}$

36. $\sqrt{7x + 8}$

37. $\sqrt{-2x - 8}$

38. $\sqrt{-5y + 15}$

For problems 39–66, simplify each expression by removing the radical sign.

39. $\sqrt{m^6}$

40. $\sqrt{k^{10}}$

41. $\sqrt{a^8}$

42. $\sqrt{h^{16}}$

57. $\sqrt{(f-2)^2(g+6)^4}$

43. $\sqrt{x^4y^{10}}$

44. $\sqrt{a^6b^{20}}$

58. $\sqrt{(2c-3)^6(5c+1)^2}$

45. $\sqrt{a^4b^6}$

46. $\sqrt{x^8y^{14}}$

59. $-\sqrt{64r^4s^{22}}$

47. $\sqrt{81a^2b^2}$

48. $\sqrt{49x^6y^4}$

60. $-\sqrt{121a^6(a-4)^8}$

49. $\sqrt{100m^8n^2}$

50. $\sqrt{225p^{14}r^{16}}$

61. $-[-\sqrt{(w+6)^2}]$

51. $\sqrt{36x^{22}y^{44}}$

52. $\sqrt{169w^4z^6(m-1)^2}$

62. $-[-\sqrt{4a^2b^2(c^2+8)^2}]$

63. $\sqrt{1.21h^4k^4}$

53. $\sqrt{25x^{12}(y-1)^4}$

54. $\sqrt{64a^{10}(a+4)^{14}}$

64. $\sqrt{2.25m^6p^6}$

55. $\sqrt{9m^6n^4(m+n)^{18}}$

65. $-\sqrt{\dfrac{169a^2b^4c^6}{196x^4y^6z^8}}$

56. $\sqrt{25m^{26}n^{42}r^{66}s^{84}}$

66. $-\left[-\sqrt{\dfrac{81y^4(z-1)^2}{225x^8z^4w^6}}\right]$

EXERCISES FOR REVIEW

(7.3)   **67.** Find the quotient. $\dfrac{x^2 - 1}{4x^2 - 1} \div \dfrac{x - 1}{2x + 1}$.

(7.5)   **68.** Find the sum. $\dfrac{1}{x + 1} + \dfrac{3}{x + 1} + \dfrac{2}{x^2 - 1}$.

(7.6)   **69.** Solve the equation, if possible: $\dfrac{1}{x - 2} = \dfrac{3}{x^2 - x - 2} - \dfrac{3}{x + 1}$.

(7.9)   **70.** Perform the division: $\dfrac{15x^3 - 5x^2 + 10x}{5x}$.

(7.9)   **71.** Perform the division: $\dfrac{x^3 - 5x^2 + 13x - 21}{x - 3}$.

★ Answers to Sample Sets (8.1)

A.   **1.** 6 and -6 **2.** 5 and -5 **3.** 10 and -10 **4.** 8 and -8 **5.** 1 and -1 **6.** $\dfrac{1}{2}$ and $-\dfrac{1}{2}$

  7. $\dfrac{3}{4}$ and $-\dfrac{3}{4}$ **8.** 0.1 and -0.1 **9.** 0.3 and -0.3

B.   **1.** $\sqrt{100} = 10$ and $-\sqrt{100} = -10$ **2.** $\sqrt{121} = 11$ and $-\sqrt{121} = -11$ **3.** $\sqrt{35}$ and $-\sqrt{35}$
  4. 5.92 and -5.92

C.   **1.** $x \geq -5$ **2.** $y \geq 8$ **3.** $a \geq -\dfrac{2}{3}$ **4.** $m \geq \dfrac{6}{5}$

D.   **1.** y^4 **2.** $4a^2$ **3.** $7x^2y^3$ **4.** $-10x^4y^6z$ **5.** $-6(a + 5)^2$ **6.** $15w^2(z^2 - 1)$ **7.** $0.5y^3z^7$
  8. x^n **9.** x^{2n}

8.2 Simplifying Square Root Expressions

Section Overview

- ☐ **PERFECT SQUARES**
- ☐ **THE PRODUCT PROPERTY OF SQUARE ROOTS**
- ☐ **THE QUOTIENT PROPERTY OF SQUARE ROOTS**
- ☐ **SQUARE ROOTS NOT INVOLVING FRACTIONS**
- ☐ **SQUARE ROOTS INVOLVING FRACTIONS**

To begin our study of the process of simplifying a square root expression, we must note three facts: one fact concerning perfect squares and two concerning properties of square roots.

☐ PERFECT SQUARES

Perfect Squares

Real numbers that are squares of rational numbers are called **perfect squares.** The numbers 25 and $\dfrac{1}{4}$ are examples of perfect squares since $25 = 5^2$ and $\dfrac{1}{4} = \left(\dfrac{1}{2}\right)^2$, and 5 and $\dfrac{1}{2}$ are rational numbers. The number 2 is *not* a perfect square since $2 = (\sqrt{2})^2$ and $\sqrt{2}$ is not a rational number.

Although we will not make a detailed study of irrational numbers, we will make the following observation:

Any indicated square root whose radicand is not a perfect square is an irrational number.

The numbers $\sqrt{6}$, $\sqrt{15}$, and $\sqrt{\dfrac{3}{4}}$ are each irrational since each radicand $\left(6, 15, \dfrac{3}{4}\right)$ is not a perfect square.

☐ THE PRODUCT PROPERTY OF SQUARE ROOTS

Notice that

$$\sqrt{9 \cdot 4} = \sqrt{36} = 6 \qquad \text{and}$$
$$\sqrt{9} \, \sqrt{4} = 3 \cdot 2 = 6$$

Since both $\sqrt{9 \cdot 4}$ and $\sqrt{9} \, \sqrt{4}$ equal 6, it must be that

$$\sqrt{9 \cdot 4} = \sqrt{9} \, \sqrt{4}$$

The Product Property
$\sqrt{xy} = \sqrt{x} \, \sqrt{y}$

This suggests that in general, if x and y are positive real numbers,
$$\sqrt{xy} = \sqrt{x} \, \sqrt{y}$$
The square root of the product is the product of the square roots.

☐ THE QUOTIENT PROPERTY OF SQUARE ROOTS

We can suggest a similar rule for quotients. Notice that

$$\sqrt{\dfrac{36}{4}} = \sqrt{9} = 3 \qquad \text{and}$$

$$\dfrac{\sqrt{36}}{\sqrt{4}} = \dfrac{6}{2} = 3$$

Since both $\dfrac{36}{4}$ and $\dfrac{\sqrt{36}}{\sqrt{4}}$ equal 3, it must be that

$$\sqrt{\dfrac{36}{4}} = \dfrac{\sqrt{36}}{\sqrt{4}}$$

The Quotient Property
$\sqrt{\dfrac{x}{y}} = \dfrac{\sqrt{x}}{\sqrt{y}}$

This suggests that in general, if x and y are positive real numbers,
$$\sqrt{\dfrac{x}{y}} = \dfrac{\sqrt{x}}{\sqrt{y}}, \qquad y \neq 0$$
The square root of the quotient is the quotient of the square roots.

CAUTION

It is extremely important to remember that

$$\sqrt{x + y} \neq \sqrt{x} + \sqrt{y} \qquad \text{or} \qquad \sqrt{x - y} \neq \sqrt{x} - \sqrt{y}$$

For example, notice that $\sqrt{16 + 9} = \sqrt{25} = 5$, but $\sqrt{16} + \sqrt{9} = 4 + 3 = 7$.

We shall study the process of simplifying a square root expression by distinguishing between two types of square roots: square roots not involving a fraction and square roots involving a fraction.

☐ SQUARE ROOTS NOT INVOLVING FRACTIONS

A square root that does not involve fractions is in **simplified form** if there are no perfect squares in the radicand.

The square roots $\sqrt{x}$, $\sqrt{ab}$, $\sqrt{5mn}$, $\sqrt{2(a + 5)}$ are in simplified form since none of the radicands contains a perfect square.

The square roots $\sqrt{x^2}$, $\sqrt{a^3} = \sqrt{a^2 a}$ are *not* in simplified form since each radicand contains a perfect square.

To simplify a square root expression that does not involve a fraction, we can use the following two rules:

SIMPLIFYING SQUARE ROOTS WITHOUT FRACTIONS

1. If a factor of the radicand contains a variable with an *even* exponent, the square root is obtained by dividing the exponent by 2.
2. If a factor of the radicand contains a variable with an *odd* exponent, the square root is obtained by first factoring the variable factor into two factors so that one has an even exponent and the other has an exponent of 1, then using the product property of square roots.

☆ SAMPLE SET A

Simplify each square root.

1. $\sqrt{a^4}$.   The exponent is even: $\frac{4}{2} = 2$. The exponent on the square root is 2.

$\sqrt{a^4} = a^2$

2. $\sqrt{a^6 b^{10}}$.   Both exponents are even: $\frac{6}{2} = 3$ and $\frac{10}{2} = 5$. The exponent on the square root of a^6 is 3. The exponent on the square root if b^{10} is 5.

$\sqrt{a^6 b^{10}} = a^3 b^5$

3. $\sqrt{y^5}$.   The exponent is odd: $y^5 = y^4 y$. Then

$\sqrt{y^5} = \sqrt{y^4 y} = \sqrt{y^4} \sqrt{y} = y^2 \sqrt{y}$

4. $\sqrt{36 a^7 b^{11} c^{20}} = \sqrt{6^2 a^6 a b^{10} b c^{20}}$ $a^7 = a^6 a$, $b^{11} = b^{10} b$

$= \sqrt{6^2 a^6 b^{10} c^{20} \cdot ab}$ by the commutative property of multiplication.

$= \sqrt{6^2 a^6 b^{10} c^{20}} \sqrt{ab}$ by the product property of square roots.

$= 6 a^3 b^5 c^{10} \sqrt{ab}$

5. $\sqrt{49 x^8 y^3 (a - 1)^6} = \sqrt{7^2 x^8 y^2 y (a - 1)^6}$

$= \sqrt{7^2 x^8 y^2 (a - 1)^6} \sqrt{y}$

$= 7 x^4 y (a - 1)^3 \sqrt{y}$

6. $\sqrt{75} = \sqrt{25 \cdot 3} = \sqrt{5^2 \cdot 3} = \sqrt{5^2} \sqrt{3} = 5\sqrt{3}$

★ **PRACTICE SET A**

Simplify each square root.

1. $\sqrt{m^8}$ **2.** $\sqrt{h^{14}k^{22}}$ **3.** $\sqrt{81a^{12}b^6c^{38}}$ **4.** $\sqrt{144x^4y^{80}(b+5)^{16}}$

5. $\sqrt{w^5}$ **6.** $\sqrt{w^7z^3k^{13}}$ **7.** $\sqrt{27a^3b^4c^5d^6}$ **8.** $\sqrt{180m^4n^{15}(a-12)^{15}}$

☐ SQUARE ROOTS INVOLVING FRACTIONS

A square root expression is in simplified form if there are

1. no perfect squares in the radicand,
2. no fractions in the radicand, or
3. no square root expressions in the denominator.

The square root expressions $\sqrt{5a}$, $\dfrac{4\sqrt{3xy}}{5}$, and $\dfrac{11m^2n\sqrt{a-4}}{2x^2}$ are in simplified form.

The square root expressions $\sqrt{\dfrac{3x}{8}}$, $\sqrt{\dfrac{4a^4b^3}{5}}$, and $\dfrac{2y}{\sqrt{3x}}$ are *not* in simplified form.

SIMPLIFYING SQUARE ROOTS WITH FRACTIONS

To simplify the square root expression $\sqrt{\dfrac{x}{y}}$,

1. Write the expression as $\dfrac{\sqrt{x}}{\sqrt{y}}$ using the rule $\sqrt{\dfrac{x}{y}} = \dfrac{\sqrt{x}}{\sqrt{y}}$.

2. Multiply the fraction by 1 in the form of $\dfrac{\sqrt{y}}{\sqrt{y}}$.

3. Simplify the remaining fraction, $\dfrac{\sqrt{xy}}{y}$.

Rationalizing the Denominator

The process involved in step 2 is called **rationalizing the denominator.** This process removes square root expressions from the denominator using the fact that $(\sqrt{y})(\sqrt{y}) = y$.

☆ **SAMPLE SET B**

Simplify each square root.

1. $\sqrt{\dfrac{9}{25}} = \dfrac{\sqrt{9}}{\sqrt{25}} = \dfrac{3}{5}$

2. $\sqrt{\dfrac{3}{5}} = \dfrac{\sqrt{3}}{\sqrt{5}} = \dfrac{\sqrt{3}}{\sqrt{5}} \cdot \dfrac{\sqrt{5}}{\sqrt{5}} = \dfrac{\sqrt{15}}{5}$

3. $\sqrt{\dfrac{9}{8}} = \dfrac{\sqrt{9}}{\sqrt{8}} = \dfrac{\sqrt{9}}{\sqrt{8}} \cdot \dfrac{\sqrt{8}}{\sqrt{8}} = \dfrac{3\sqrt{8}}{8} = \dfrac{3\sqrt{4 \cdot 2}}{8} = \dfrac{3\sqrt{4}\,\sqrt{2}}{8} = \dfrac{3 \cdot 2\sqrt{2}}{8} = \dfrac{3\sqrt{2}}{4}$

Continued

4. $\sqrt{\dfrac{k^2}{m^3}} = \dfrac{\sqrt{k^2}}{\sqrt{m^3}} = \dfrac{k}{\sqrt{m^3}} = \dfrac{k}{\sqrt{m^2 m}} = \dfrac{k}{\sqrt{m^2}\sqrt{m}} = \dfrac{k}{m\sqrt{m}} = \dfrac{k}{m\sqrt{m}} \cdot \dfrac{\sqrt{m}}{\sqrt{m}} = \dfrac{k\sqrt{m}}{m\sqrt{m}\sqrt{m}} = \dfrac{k\sqrt{m}}{m \cdot m} = \dfrac{k\sqrt{m}}{m^2}$

5. $\sqrt{x^2 - 8x + 16} = \sqrt{(x-4)^2}$
$\qquad\qquad\qquad\quad = x - 4$

★ **PRACTICE SET B**

Simplify each square root.

1. $\sqrt{\dfrac{81}{25}}$
2. $\sqrt{\dfrac{2}{7}}$
3. $\sqrt{\dfrac{4}{5}}$
4. $\sqrt{\dfrac{10}{4}}$
5. $\sqrt{\dfrac{9}{4}}$
6. $\sqrt{\dfrac{a^3}{6}}$

7. $\sqrt{\dfrac{y^4}{x^3}}$
8. $\sqrt{\dfrac{32a^5}{b^7}}$
9. $\sqrt{(x+9)^2}$
10. $\sqrt{x^2 + 14x + 49}$

Answers to Practice Sets are on p. 422.

Section 8.2 EXERCISES

For problems 1–66, simplify each of the radical expressions.

1. $\sqrt{8b^2}$
2. $\sqrt{20a^2}$

3. $\sqrt{24x^4}$
4. $\sqrt{27y^6}$

5. $\sqrt{a^5}$
6. $\sqrt{m^7}$

7. $\sqrt{x^{11}}$
8. $\sqrt{y^{17}}$

9. $\sqrt{36n^9}$
10. $\sqrt{49x^{13}}$

11. $\sqrt{100x^5 y^{11}}$
12. $\sqrt{64a^7 b^3}$

13. $5\sqrt{16m^6 n^7}$
14. $8\sqrt{9a^4 b^{11}}$

15. $3\sqrt{16x^3}$
16. $8\sqrt{25y^3}$

17. $\sqrt{12a^4}$
18. $\sqrt{32m^8}$

19. $\sqrt{32x^7}$
20. $\sqrt{12y^{13}}$

21. $\sqrt{50a^3 b^9}$
22. $\sqrt{48p^{11} q^5}$

23. $4\sqrt{18a^5 b^{17}}$
24. $8\sqrt{108x^{21} y^3}$

25. $-4\sqrt{75a^4 b^6}$
26. $-6\sqrt{72x^2 y^4 z^{10}}$

27. $-\sqrt{b^{12}}$

28. $-\sqrt{c^{18}}$

45. $\sqrt{\dfrac{4}{3}}$

46. $-\sqrt{\dfrac{2}{5}}$

29. $\sqrt{a^2b^2c^2}$

30. $\sqrt{4x^2y^2z^2}$

47. $-\sqrt{\dfrac{3}{10}}$

48. $\sqrt{\dfrac{16a^2}{5}}$

31. $-\sqrt{9a^2b^3}$

32. $-\sqrt{16x^4y^5}$

49. $\sqrt{\dfrac{24a^5}{7}}$

50. $\sqrt{\dfrac{72x^2y^3}{5}}$

33. $\sqrt{m^6n^8p^{12}q^{20}}$

34. $\sqrt{r^2}$

51. $\sqrt{\dfrac{2}{a}}$

52. $\sqrt{\dfrac{5}{b}}$

35. $\sqrt{p^2}$

36. $\sqrt{\dfrac{1}{4}}$

53. $\sqrt{\dfrac{6}{x^3}}$

54. $\sqrt{\dfrac{12}{y^5}}$

37. $\sqrt{\dfrac{1}{16}}$

38. $\sqrt{\dfrac{4}{25}}$

55. $\sqrt{\dfrac{49x^2y^5z^9}{25a^3b^{11}}}$

56. $\sqrt{\dfrac{27x^6y^{15}}{3^3x^3y^5}}$

39. $\sqrt{\dfrac{9}{49}}$

40. $\dfrac{5\sqrt{8}}{\sqrt{3}}$

57. $\sqrt{(b+2)^4}$

58. $\sqrt{(a-7)^8}$

41. $\dfrac{2\sqrt{32}}{\sqrt{3}}$

42. $\sqrt{\dfrac{5}{6}}$

43. $\sqrt{\dfrac{2}{7}}$

44. $\sqrt{\dfrac{3}{10}}$

59. $\sqrt{(x+2)^6}$

60. $\sqrt{(x+2)^2(x+1)^2}$

61. $\sqrt{(a-3)^4(a-1)^2}$ **62.** $\sqrt{(b+7)^8(b-7)^6}$ **65.** $\sqrt{(a^2-2a+1)^4}$ **66.** $\sqrt{(x^2+2x+1)^{12}}$

63. $\sqrt{a^2-10a+25}$ **64.** $\sqrt{b^2+6b+9}$

EXERCISES FOR REVIEW

(4.6) **67.** Solve the inequality $3(a+2) \le 2(3a+4)$

(6.1) **68.** Graph the inequality $6x \le 5(x+1) - 6$.

$\longleftarrow\hspace{6cm}\longrightarrow$

(6.4) **69.** Supply the missing words. When looking at a graph from left-to-right, lines with _____ slope rise, while lines with _____ slope fall.

(7.8) **70.** Simplify the complex fraction $\dfrac{5+\dfrac{1}{x}}{5-\dfrac{1}{x}}$.

(8.1) **71.** Simplify $\sqrt{121x^4w^6z^8}$ by removing the radical sign.

★ **Answers to Practice Sets (8.2)**

A. **1.** m^4 **2.** h^7k^{11} **3.** $9a^6b^3c^{19}$ **4.** $12x^2y^{40}(b+5)^8$ **5.** $w^2\sqrt{w}$ **6.** $w^3zk^6\sqrt{wzk}$

  7. $3ab^2c^2d^3\sqrt{3ac}$ **8.** $6m^2n^7(a-12)^7\sqrt{5n(a-12)}$

B. **1.** $\dfrac{9}{5}$ **2.** $\dfrac{\sqrt{14}}{7}$ **3.** $\dfrac{2\sqrt{5}}{5}$ **4.** $\dfrac{\sqrt{10}}{2}$ **5.** $\dfrac{3}{2}$ **6.** $\dfrac{a\sqrt{6a}}{6}$ **7.** $\dfrac{y^2\sqrt{x}}{x^2}$ **8.** $\dfrac{4a^2\sqrt{2ab}}{b^4}$ **9.** $x+9$

  10. $x+7$

8.3 Multiplication of Square Root Expressions

Section Overview

☐ **THE PRODUCT PROPERTY OF SQUARE ROOTS**
☐ **MULTIPLICATION RULE FOR SQUARE ROOT EXPRESSIONS**

☐ THE PRODUCT PROPERTY OF SQUARE ROOTS

In our work with simplifying square root expressions, we noted that

$$\sqrt{xy} = \sqrt{x}\,\sqrt{y}$$

Since this is an equation, we may write it as

$$\sqrt{x}\,\sqrt{y} = \sqrt{xy}$$

To multiply two square root expressions, we use the product property of square roots.

The Product Property
$\sqrt{x}\,\sqrt{y} = \sqrt{xy}$

$$\sqrt{x}\,\sqrt{y} = \sqrt{xy}$$

The product of the square roots is the square root of the product.

In practice, it is usually easier to simplify the square root expressions before actually performing the multiplication. To see this, consider the following product:

$$\sqrt{8}\,\sqrt{48}$$

We can multiply these square roots in *either* of two ways:

1. Simplify then multiply.

$$\sqrt{4 \cdot 2}\,\sqrt{16 \cdot 3} = (2\sqrt{2})(4\sqrt{3}) = 2 \cdot 4\sqrt{2 \cdot 3} = 8\sqrt{6}$$

2. Multiply then simplify.

$$\sqrt{8}\,\sqrt{48} = \sqrt{8 \cdot 48} = \sqrt{384} = \sqrt{64 \cdot 6} = 8\sqrt{6}$$

Notice that in the second method, the expanded term (the third expression, $\sqrt{384}$) may be difficult to factor into a perfect square and some other number.

☐ MULTIPLICATION RULE FOR SQUARE ROOT EXPRESSIONS

The preceding example suggests that the following rule for multiplying two square root expressions.

Rule for Multiplying Square Root Expressions

1. Simplify each square root expression, if necessary.
2. Perform the multiplication.
3. Simplify, if necessary.

☆ SAMPLE SET A

Find each of the following products.

1. $\sqrt{3}\,\sqrt{6} = \sqrt{3 \cdot 6} = \sqrt{18} = \sqrt{9 \cdot 2} = 3\sqrt{2}$

2. $\sqrt{8}\,\sqrt{2} = 2\sqrt{2}\,\sqrt{2} = 2\sqrt{2 \cdot 2} = 2\sqrt{4} = 2 \cdot 2 = 4$

This product might be easier if we were to multiply first and then simplify.

$$\sqrt{8}\,\sqrt{2} = \sqrt{8 \cdot 2} = \sqrt{16} = 4$$

3. $\sqrt{20}\,\sqrt{7} = \sqrt{4}\,\sqrt{5}\,\sqrt{7} = 2\sqrt{5 \cdot 7} = 2\sqrt{35}$

4. $\sqrt{5a^3}\,\sqrt{27a^5} = (a\sqrt{5a})(3a^2\sqrt{3a}) = 3a^3\sqrt{15a^2}$
$$= 3a^3 \cdot a\sqrt{15}$$
$$= 3a^4\sqrt{15}$$

5. $\sqrt{(x+2)^7}\,\sqrt{(x-1)} = \sqrt{(x+2)^6(x+2)}\,\sqrt{x-1} = (x+2)^3\sqrt{(x+2)}\,\sqrt{(x-1)}$
$$= (x+2)^3\sqrt{(x+2)(x-1)}$$
$$\text{or} \qquad = (x+2)^3\sqrt{x^2+x-2}$$

Continued

6. $\sqrt{3}(7 + \sqrt{6}) = 7\sqrt{3} + \sqrt{3}\,\sqrt{6} = 7\sqrt{3} + \sqrt{18}$
$$= 7\sqrt{3} + \sqrt{9 \cdot 2}$$
$$= 7\sqrt{3} + 3\sqrt{2}$$

7. $\sqrt{6}(\sqrt{2} - \sqrt{10}) = \sqrt{6}\,\sqrt{2} - \sqrt{6}\,\sqrt{10}$
$$= \sqrt{12} - \sqrt{60}$$
$$= \sqrt{4 \cdot 3} - \sqrt{4 \cdot 15}$$
$$= 2\sqrt{3} - 2\sqrt{15}$$

8. $\sqrt{45a^6b^3}[\sqrt{9ab} - \sqrt{5(b-3)^3}] = 3a^3b\sqrt{5b}[3\sqrt{ab} - (b-3)\sqrt{5(b-3)}]$
$$= 9a^3b\sqrt{5ab^2} - 3a^3b(b-3)\sqrt{25b(b-3)}$$
$$= 9a^3b^2\sqrt{5a} - 3a^3b(b-3) \cdot 5\sqrt{b(b-3)}$$
$$= 9a^3b^2\sqrt{5a} - 15a^3b(b-3)\sqrt{b(b-3)}$$

★ **PRACTICE SET A**

Find each of the following products.

1. $\sqrt{5}\,\sqrt{6}$

2. $\sqrt{32}\,\sqrt{2}$

3. $\sqrt{x+4}\,\sqrt{x+3}$

4. $\sqrt{8m^5n}\,\sqrt{20m^2n}$

5. $\sqrt{9(k-6)^3}\,\sqrt{k^2 - 12k + 36}$

6. $\sqrt{3}(\sqrt{2} + \sqrt{5})$

7. $\sqrt{2a}(\sqrt{5a} - \sqrt{8a^3})$

8. $\sqrt{32m^5n^3}(\sqrt{2mn^2} - \sqrt{10n^7})$

Answers to the Practice Set are on p. 427.

Section 8.3 EXERCISES

For problems 1–70, find each of the products.

1. $\sqrt{2}\,\sqrt{10}$

2. $\sqrt{3}\,\sqrt{15}$

7. $\sqrt{5}\,\sqrt{5}$

8. $\sqrt{7}\,\sqrt{7}$

3. $\sqrt{7}\,\sqrt{8}$

4. $\sqrt{20}\,\sqrt{3}$

9. $\sqrt{8}\,\sqrt{8}$

10. $\sqrt{15}\,\sqrt{15}$

5. $\sqrt{32}\,\sqrt{27}$

6. $\sqrt{45}\,\sqrt{50}$

11. $\sqrt{48}\,\sqrt{27}$

12. $\sqrt{80}\,\sqrt{20}$

13. $\sqrt{5}\,\sqrt{m}$ **14.** $\sqrt{7}\,\sqrt{a}$ **35.** $\sqrt{a^3}\,\sqrt{a^5}$ **36.** $\sqrt{x^3}\,\sqrt{x^7}$

15. $\sqrt{6}\,\sqrt{m}$ **16.** $\sqrt{10}\,\sqrt{h}$ **37.** $\sqrt{x^9}\,\sqrt{x^3}$ **38.** $\sqrt{y^7}\,\sqrt{y^9}$

17. $\sqrt{20}\,\sqrt{a}$ **18.** $\sqrt{48}\,\sqrt{x}$ **39.** $\sqrt{y^3}\,\sqrt{y^4}$ **40.** $\sqrt{x^8}\,\sqrt{x^5}$

19. $\sqrt{75}\,\sqrt{y}$ **20.** $\sqrt{200}\,\sqrt{m}$ **41.** $\sqrt{x+2}\,\sqrt{x-3}$ **42.** $\sqrt{a-6}\,\sqrt{a+1}$

21. $\sqrt{a}\,\sqrt{a}$ **22.** $\sqrt{x}\,\sqrt{x}$ **43.** $\sqrt{y+3}\,\sqrt{y-2}$ **44.** $\sqrt{h+1}\,\sqrt{h-1}$

23. $\sqrt{y}\,\sqrt{y}$ **24.** $\sqrt{h}\,\sqrt{h}$ **45.** $\sqrt{x+9}\,\sqrt{(x+9)^2}$ **46.** $\sqrt{y-3}\,\sqrt{(y-3)^5}$

25. $\sqrt{3}\,\sqrt{3}$ **26.** $\sqrt{6}\,\sqrt{6}$ **47.** $\sqrt{3a^2}\,\sqrt{15a^3}$ **48.** $\sqrt{2m^4n^3}\,\sqrt{14m^5n}$

27. $\sqrt{k}\,\sqrt{k}$ **28.** $\sqrt{m}\,\sqrt{m}$ **49.** $\sqrt{12(p-q)^3}\,\sqrt{3(p-q)^5}$

29. $\sqrt{m^2}\,\sqrt{m}$ **30.** $\sqrt{a^2}\,\sqrt{a}$ **50.** $\sqrt{15a^2(b+4)^4}\,\sqrt{21a^3(b+4)^5}$

31. $\sqrt{x^3}\,\sqrt{x}$ **32.** $\sqrt{y^3}\,\sqrt{y}$ **51.** $\sqrt{125m^5n^4r^8}\,\sqrt{8m^6r}$

33. $\sqrt{y}\,\sqrt{y^4}$ **34.** $\sqrt{k}\,\sqrt{k^6}$ **52.** $\sqrt{7(2k-1)^{11}(k+1)^3}\,\sqrt{14(2k-1)^{10}}$

53. $\sqrt{y^3}\,\sqrt{y^5}\,\sqrt{y^2}$

54. $\sqrt{x^6}\,\sqrt{x^2}\,\sqrt{x^9}$

55. $\sqrt{2a^4}\,\sqrt{5a^3}\,\sqrt{2a^7}$

56. $\sqrt{x^n}\,\sqrt{x^n}$

57. $\sqrt{y^{2n}}\,\sqrt{y^{4n}}$

58. $\sqrt{a^{2n+5}}\,\sqrt{a^3}$

59. $\sqrt{2m^{3n+1}}\,\sqrt{10m^{n+3}}$

60. $\sqrt{75(a-2)^7}\,\sqrt{48a-96}$

61. $\sqrt{2}(\sqrt{8}+\sqrt{6})$

62. $\sqrt{5}(\sqrt{3}+\sqrt{7})$

63. $\sqrt{3}(\sqrt{x}+\sqrt{2})$

64. $\sqrt{11}(\sqrt{y}+\sqrt{3})$

65. $\sqrt{8}(\sqrt{a}-\sqrt{3a})$

66. $\sqrt{x}(\sqrt{x^3}-\sqrt{2x^4})$

67. $\sqrt{y}(\sqrt{y^5}+\sqrt{3y^3})$

68. $\sqrt{8a^5}(\sqrt{2a}-\sqrt{6a^{11}})$

69. $\sqrt{12m^3}(\sqrt{6m^7}-\sqrt{3m})$

70. $\sqrt{5x^4y^3}(\sqrt{8xy}-5\sqrt{7x})$

EXERCISES FOR REVIEW

(5.5) **71.** Factor $a^4y^4-25w^2$.

(6.4) **72.** Find the slope of the line that passes through the points $(-5, 4)$ and $(-3, 4)$.

(7.3) **73.** Perform the indicated operations:

$$\frac{15x^2-20x}{6x^2+x-12}\cdot\frac{8x+12}{x^2-2x-15}\div\frac{5x^2+15x}{x^2-25}$$

(8.1) **74.** Simplify $\sqrt{x^4y^2z^6}$ by removing the radical sign.

(8.2) **75.** Simplify $\sqrt{12x^3y^5z^8}$.

★ **Answers to Practice Set (8.3)**

A. **1.** $\sqrt{30}$ **2.** 8   **3.** $\sqrt{(x+4)(x+3)}$ **4.** $4m^3n\sqrt{10m}$ **5.** $3(k-6)^2\sqrt{k-6}$ **6.** $\sqrt{6}+\sqrt{15}$

   7. $a\sqrt{10}-4a^2$ **8.** $8m^3n^2\sqrt{n}-8m^2n^5\sqrt{5m}$

8.4 Division of Square Root Expressions

Section Overview

- ☐ THE DIVISION PROPERTY OF SQUARE ROOTS
- ☐ RATIONALIZING THE DENOMINATOR
- ☐ CONJUGATES AND RATIONALIZING THE DENOMINATOR

☐ THE DIVISION PROPERTY OF SQUARE ROOTS

In our work with simplifying square root expressions, we noted that

$$\sqrt{\frac{x}{y}} = \frac{\sqrt{x}}{\sqrt{y}}$$

Since this is an equation, we may write it as

$$\frac{\sqrt{x}}{\sqrt{y}} = \sqrt{\frac{x}{y}}$$

To divide two square root expressions, we use the division property of square roots.

The Division Property

$$\frac{\sqrt{x}}{\sqrt{y}} = \sqrt{\frac{x}{y}}$$

> $$\frac{\sqrt{x}}{\sqrt{y}} = \sqrt{\frac{x}{y}}$$
>
> The quotient of the square roots is the square root of the quotient.

☐ RATIONALIZING THE DENOMINATOR

As we can see by observing the right side of the equation governing the division of square roots, the process may produce a fraction in the radicand. This means, of course, that the square root expression is not in simplified form. It is sometimes more useful to rationalize the denominator of a square root expression before actually performing the division.

✰ **SAMPLE SET A**

Simplify the square root expressions.

1. $\sqrt{\dfrac{3}{7}}$.   This radical expression is not in simplified form since there is a fraction under the radical sign. We can eliminate this problem using the division property of square roots.

$$\sqrt{\frac{3}{7}} = \frac{\sqrt{3}}{\sqrt{7}} = \frac{\sqrt{3}}{\sqrt{7}} \cdot \frac{\sqrt{7}}{\sqrt{7}} = \frac{\sqrt{3}\sqrt{7}}{7} = \frac{\sqrt{21}}{7}$$

2. $\dfrac{\sqrt{5}}{\sqrt{3}}$.   A direct application of the rule produces $\sqrt{\dfrac{5}{3}}$, which must be simplified. Let us rationalize the denominator before we perform the division.

$$\frac{\sqrt{5}}{\sqrt{3}} = \frac{\sqrt{5}}{\sqrt{3}} \cdot \frac{\sqrt{3}}{\sqrt{3}} = \frac{\sqrt{5}\,\sqrt{3}}{3} = \frac{\sqrt{15}}{3}$$

3. $\dfrac{\sqrt{21}}{\sqrt{7}} = \sqrt{\dfrac{21}{7}} = \sqrt{3}.$ The rule produces the quotient quickly. We could also rationalize the denominator first and produce the same result.

$$\frac{\sqrt{21}}{\sqrt{7}} = \frac{\sqrt{21}}{\sqrt{7}} \cdot \frac{\sqrt{7}}{\sqrt{7}} = \frac{\sqrt{21 \cdot 7}}{7} = \frac{\sqrt{3 \cdot 7 \cdot 7}}{7} = \frac{\sqrt{3 \cdot 7^2}}{7} = \frac{7\sqrt{3}}{7} = \sqrt{3}$$

4. $\dfrac{\sqrt{80x^9}}{\sqrt{5x^4}} = \sqrt{\dfrac{80x^9}{5x^4}} = \sqrt{16x^5} = \sqrt{16}\,\sqrt{x^4 x} = 4x^2\sqrt{x}$

5. $\dfrac{\sqrt{50a^3b^7}}{\sqrt{5ab^5}} = \sqrt{\dfrac{50a^3b^7}{5ab^5}} = \sqrt{10a^2b^2} = ab\sqrt{10}$

6. $\dfrac{\sqrt{5a}}{\sqrt{b}}$.   Some observation shows that a direct division of the radicands will produce a fraction. This suggests that we rationalize the denominator first.

$$\frac{\sqrt{5a}}{\sqrt{b}} = \frac{\sqrt{5a}}{\sqrt{b}} \cdot \frac{\sqrt{b}}{\sqrt{b}} = \frac{\sqrt{5a}\,\sqrt{b}}{b} = \frac{\sqrt{5ab}}{b}$$

7. $\dfrac{\sqrt{m-6}}{\sqrt{m+2}} = \dfrac{\sqrt{m-6}}{\sqrt{m+2}} \cdot \dfrac{\sqrt{m+2}}{\sqrt{m+2}} = \dfrac{\sqrt{m^2-4m-12}}{m+2}$

8. $\dfrac{\sqrt{y^2-y-12}}{\sqrt{y+3}} = \sqrt{\dfrac{y^2-y-12}{y+3}} = \sqrt{\dfrac{(y+3)(y-4)}{(y+3)}} = \sqrt{\dfrac{\cancel{(y+3)}(y-4)}{\cancel{(y+3)}}} = \sqrt{y-4}$

★ **PRACTICE SET A**

Simplify the square root expressions.

1. $\dfrac{\sqrt{26}}{\sqrt{13}}$

2. $\dfrac{\sqrt{7}}{\sqrt{3}}$

3. $\dfrac{\sqrt{80m^5n^8}}{\sqrt{5m^2n}}$

4. $\dfrac{\sqrt{196(x+7)^8}}{\sqrt{2(x+7)^3}}$

5. $\dfrac{\sqrt{n+4}}{\sqrt{n-5}}$

6. $\dfrac{\sqrt{a^2-6a+8}}{\sqrt{a-2}}$

7. $\dfrac{\sqrt{x^{3n}}}{\sqrt{x^n}}$

8. $\dfrac{\sqrt{a^{3m-5}}}{\sqrt{a^{m-1}}}$

☐ CONJUGATES AND RATIONALIZING THE DENOMINATOR

To perform a division that contains a binomial in the denominator, such as $\dfrac{3}{4 + \sqrt{6}}$, we multiply the numerator and denominator by a *conjugate* of the denominator.

Conjugate

> A conjugate of the binomial $a + b$ is $a - b$. Similarly, a conjugate of $a - b$ is $a + b$.

Notice that when the conjugates $a + b$ and $a - b$ are multiplied together, they produce a difference of two squares.

$$(a + b)(a - b) = a^2 - ab + ab - b^2 = a^2 - b^2$$

This principle helps us eliminate square root radicals, as shown in these examples that illustrate finding the product of conjugates.

1. $(5 + \sqrt{2})(5 - \sqrt{2}) = 5^2 - (\sqrt{2})^2$
    $= 25 - 2$
    $= 23$
2. $(\sqrt{6} - \sqrt{7})(\sqrt{6} + \sqrt{7}) = (\sqrt{6})^2 - (\sqrt{7})^2$
    $= 6 - 7$
    $= -1$

☆ SAMPLE SET B

Simplify the following expressions.

1. $\dfrac{3}{4 + \sqrt{6}}$.   The conjugate of the denominator is $4 - \sqrt{6}$. Multiply the fraction by 1 in the form of $\dfrac{4 - \sqrt{6}}{4 - \sqrt{6}}$.

$$\frac{3}{4 + \sqrt{6}} \cdot \frac{4 - \sqrt{6}}{4 - \sqrt{6}} = \frac{3(4 - \sqrt{6})}{4^2 - (\sqrt{6})^2}$$

$$= \frac{12 - 3\sqrt{6}}{16 - 6}$$

$$= \frac{12 - 3\sqrt{6}}{10}$$

2. $\dfrac{\sqrt{2x}}{\sqrt{3} - \sqrt{5x}}$.   The conjugate of the denominator is $\sqrt{3} + \sqrt{5x}$. Multiply the fraction by 1 in the form of $\dfrac{\sqrt{3} + \sqrt{5x}}{\sqrt{3} + \sqrt{5x}}$.

$$\frac{\sqrt{2x}}{\sqrt{3} - \sqrt{5x}} \cdot \frac{\sqrt{3} + \sqrt{5x}}{\sqrt{3} + \sqrt{5x}} = \frac{\sqrt{2x}(\sqrt{3} + \sqrt{5x})}{(\sqrt{3})^2 - (\sqrt{5x})^2}$$

$$= \frac{\sqrt{2x}\,\sqrt{3} + \sqrt{2x}\,\sqrt{5x}}{3 - 5x}$$

$$= \frac{\sqrt{6x} + \sqrt{10x^2}}{3 - 5x}$$

$$= \frac{\sqrt{6x} + x\sqrt{10}}{3 - 5x}$$

★ PRACTICE SET B

Simplify the following expressions.

1. $\dfrac{5}{9 + \sqrt{7}}$ **2.** $\dfrac{-2}{1 - \sqrt{3x}}$ **3.** $\dfrac{\sqrt{8}}{\sqrt{3x} + \sqrt{2x}}$ **4.** $\dfrac{\sqrt{2m}}{m - \sqrt{3m}}$

Answers to Practice Sets are on p. 433.

Section 8.4 EXERCISES

For problems 1–62, simplify each expression.

1. $\dfrac{\sqrt{28}}{\sqrt{2}}$ **2.** $\dfrac{\sqrt{200}}{\sqrt{10}}$ **11.** $\sqrt{\dfrac{49}{225}}$ **12.** $\sqrt{\dfrac{3}{5}}$

3. $\dfrac{\sqrt{28}}{\sqrt{7}}$ **4.** $\dfrac{\sqrt{96}}{\sqrt{24}}$ **13.** $\sqrt{\dfrac{3}{7}}$ **14.** $\sqrt{\dfrac{1}{2}}$

5. $\dfrac{\sqrt{180}}{\sqrt{5}}$ **6.** $\dfrac{\sqrt{336}}{\sqrt{21}}$ **15.** $\sqrt{\dfrac{5}{2}}$ **16.** $\sqrt{\dfrac{11}{25}}$

7. $\dfrac{\sqrt{162}}{\sqrt{18}}$ **8.** $\sqrt{\dfrac{25}{9}}$ **17.** $\sqrt{\dfrac{15}{36}}$ **18.** $\sqrt{\dfrac{5}{16}}$

9. $\sqrt{\dfrac{36}{35}}$ **10.** $\sqrt{\dfrac{225}{16}}$ **19.** $\sqrt{\dfrac{7}{25}}$ **20.** $\sqrt{\dfrac{32}{49}}$

21. $\sqrt{\dfrac{50}{81}}$

22. $\dfrac{\sqrt{125x^5}}{\sqrt{5x^3}}$

35. $\dfrac{\sqrt{5x}}{\sqrt{2}}$

36. $\dfrac{\sqrt{2a^3b}}{\sqrt{14a}}$

23. $\dfrac{\sqrt{72m^7}}{\sqrt{2m^3}}$

24. $\dfrac{\sqrt{162a^{11}}}{\sqrt{2a^5}}$

37. $\dfrac{\sqrt{3m^4n^3}}{\sqrt{6mn^5}}$

38. $\dfrac{\sqrt{5(p-q)^6(r+s)^4}}{\sqrt{25(r+s)^3}}$

25. $\dfrac{\sqrt{75y^{10}}}{\sqrt{3y^4}}$

26. $\dfrac{\sqrt{48x^9}}{\sqrt{3x^2}}$

39. $\dfrac{\sqrt{m(m-6)-m^2+6m}}{\sqrt{3m-7}}$

27. $\dfrac{\sqrt{125a^{14}}}{\sqrt{5a^5}}$

28. $\dfrac{\sqrt{27a^{10}}}{\sqrt{3a^5}}$

40. $\dfrac{\sqrt{r+1}}{\sqrt{r-1}}$

41. $\dfrac{\sqrt{s+3}}{\sqrt{s-3}}$

29. $\dfrac{\sqrt{108x^{21}}}{\sqrt{3x^4}}$

30. $\dfrac{\sqrt{48x^6y^7}}{\sqrt{3xy}}$

42. $\dfrac{\sqrt{a^2+3a+2}}{\sqrt{a+1}}$

43. $\dfrac{\sqrt{x^2-10x+24}}{\sqrt{x-4}}$

31. $\dfrac{\sqrt{45a^3b^8c^2}}{\sqrt{5ab^2c}}$

32. $\dfrac{\sqrt{66m^{12}n^{15}}}{\sqrt{11mn^8}}$

44. $\dfrac{\sqrt{x^2-2x-8}}{\sqrt{x+2}}$

45. $\dfrac{\sqrt{x^2-4x+3}}{\sqrt{x-3}}$

33. $\dfrac{\sqrt{30p^5q^{14}}}{\sqrt{5q^7}}$

34. $\dfrac{\sqrt{b}}{\sqrt{5}}$

46. $\dfrac{\sqrt{2x^2-x-1}}{\sqrt{x-1}}$

47. $\dfrac{-5}{4+\sqrt{5}}$

48. $\dfrac{1}{1 + \sqrt{x}}$

49. $\dfrac{2}{1 - \sqrt{a}}$

56. $\dfrac{\sqrt{7x}}{2 - \sqrt{5x}}$

57. $\dfrac{\sqrt{6y}}{1 + \sqrt{3y}}$

50. $\dfrac{-6}{\sqrt{5} - 1}$

51. $\dfrac{-6}{\sqrt{7} + 2}$

58. $\dfrac{\sqrt{2}}{\sqrt{3} - \sqrt{2}}$

59. $\dfrac{\sqrt{a}}{\sqrt{a} + \sqrt{b}}$

52. $\dfrac{3}{\sqrt{3} - \sqrt{2}}$

53. $\dfrac{4}{\sqrt{6} + \sqrt{2}}$

60. $\dfrac{\sqrt{8a^3 b^5}}{4 - \sqrt{2ab}}$

61. $\dfrac{\sqrt{7x}}{\sqrt{5x} + \sqrt{x}}$

54. $\dfrac{\sqrt{5}}{\sqrt{8} - \sqrt{6}}$

55. $\dfrac{\sqrt{12}}{\sqrt{12} - \sqrt{8}}$

62. $\dfrac{\sqrt{3y}}{\sqrt{2y} - \sqrt{y}}$

EXERCISES FOR REVIEW

(1.5) **63.** Simplify $x^8 y^7 \left(\dfrac{x^4 y^8}{x^3 y^4} \right)$.

(4.6) **64.** Solve the compound inequality $-8 \le 7 - 5x \le -23$.

(6.5) **65.** Construct the graph of $y = \dfrac{2}{3}x - 4$.

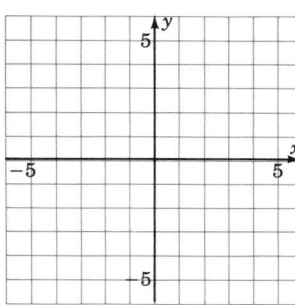

(8.1) **66.** The symbol $\sqrt{x}$ represents which square root of the number x, $x \geq 0$?

(8.3) **67.** Simplify $\sqrt{a^2 + 8a + 16}$.

★ **Answers to Practice Sets (8.4)**

A. 1. $\sqrt{2}$ **2.** $\dfrac{\sqrt{21}}{3}$ **3.** $4mn^3\sqrt{mn}$ **4.** $7(x+7)^2\sqrt{2(x+7)}$ **5.** $\dfrac{\sqrt{n^2 - n - 20}}{n - 5}$ **6.** $\sqrt{a - 4}$ **7.** x^n

8. a^{m-2}

B. 1. $\dfrac{45 - 5\sqrt{7}}{74}$ **2.** $\dfrac{-2 - 2\sqrt{3x}}{1 - 3x}$ **3.** $\dfrac{2\sqrt{6x} - 4\sqrt{x}}{x}$ **4.** $\dfrac{\sqrt{2m} + \sqrt{6}}{m - 3}$

8.5 Addition and Subtraction of Square Root Expressions

Section Overview	☐ THE LOGIC BEHIND THE PROCESS
	☐ THE PROCESS

☐ THE LOGIC BEHIND THE PROCESS

Now we will study methods of simplifying radical expressions such as

$$4\sqrt{3} + 8\sqrt{3} \qquad \text{or} \qquad 5\sqrt{2x} - 11\sqrt{2x} + 4(\sqrt{2x} + 1)$$

The procedure for adding and subtracting square root expressions will become apparent if we think back to the procedure we used for simplifying polynomial expressions such as

$$4x + 8x \qquad \text{or} \qquad 5a - 11a + 4(a + 1)$$

The variables x and a are letters representing some unknown quantities (perhaps x represents $\sqrt{3}$ and a represents $\sqrt{2x}$). Combining like terms gives us

$$4x + 8x = 12x \qquad \text{or} \qquad 4\sqrt{3} + 8\sqrt{3} = 12\sqrt{3}$$

and

$$5a - 11a + 4(a + 1) \qquad \text{or} \qquad 5\sqrt{2x} - 11\sqrt{2x} + 4(\sqrt{2x} + 1)$$
$$5a - 11a + 4a + 4 \qquad\qquad\qquad 5\sqrt{2x} - 11\sqrt{2x} + 4\sqrt{2x} + 4$$
$$-2a + 4 \qquad\qquad\qquad\qquad\qquad -2\sqrt{2x} + 4$$

❏ THE PROCESS

Let's consider the expression $4\sqrt{3} + 8\sqrt{3}$. There are two ways to look at the simplification process:

1. We are asking, "How many square roots of 3 do we have?"

   $4\sqrt{3}$ means we have 4 "square roots of 3"
   $8\sqrt{3}$ means we have 8 "square roots of 3"

   Thus, altogether we have 12 "square roots of 3."

2. We can also use the idea of combining like terms. If we recall, the process of combining like terms is based on the distributive property

   $$4x + 8x = 12x \qquad \text{because} \qquad 4x + 8x = (4 + 8)x = 12x$$

   We could simplify $4\sqrt{3} + 8\sqrt{3}$ using the distributive property.

   $$4\sqrt{3} + 8\sqrt{3} = (4 + 8)\sqrt{3} = 12\sqrt{3}$$

Both methods will give us the same result. The first method is probably a bit quicker, but keep in mind, however, that the process works because it is based on one of the basic rules of algebra, the distributive property of real numbers.

☆ SAMPLE SET A

Simplify the following radical expressions.

1. $-6\sqrt{10} + 11\sqrt{10} = 5\sqrt{10}$

2. $4\sqrt{32} + 5\sqrt{2}$.   Simplify $\sqrt{32}$.

$$\begin{aligned}
4\sqrt{16 \cdot 2} + 5\sqrt{2} &= 4\sqrt{16}\,\sqrt{2} + 5\sqrt{2} \\
&= 4 \cdot 4\sqrt{2} + 5\sqrt{2} \\
&= 16\sqrt{2} + 5\sqrt{2} \\
&= 21\sqrt{2}
\end{aligned}$$

3. $-3x\sqrt{75} + 2x\sqrt{48} - x\sqrt{27}$.   Simplify each of the three radicals.

$$\begin{aligned}
&= -3x\sqrt{25 \cdot 3} + 2x\sqrt{16 \cdot 3} - x\sqrt{9 \cdot 3} \\
&= -15x\sqrt{3} + 8x\sqrt{3} - 3x\sqrt{3} \\
&= (-15x + 8x - 3x)\sqrt{3} \\
&= -10x\sqrt{3}
\end{aligned}$$

4. $5a\sqrt{24a^3} - 7\sqrt{54a^5} + a^2\sqrt{6a} + 6a$.   Simplify each radical.

$$\begin{aligned}
&= 5a\sqrt{4 \cdot 6 \cdot a^2 \cdot a} - 7\sqrt{9 \cdot 6 \cdot a^4 \cdot a} + a^2\sqrt{6a} + 6a \\
&= 10a^2\sqrt{6a} - 21a^2\sqrt{6a} + a^2\sqrt{6a} + 6a \\
&= (10a^2 - 21a^2 + a^2)\sqrt{6a} + 6a \\
&= -10a^2\sqrt{6a} + 6a
\end{aligned}$$

Factor out $-2a$. (This step is optional.)

$$= -2a(5a\sqrt{6a} - 3)$$

★ **PRACTICE SET A**

Find each sum or difference.

1. $4\sqrt{18} - 5\sqrt{8}$ **2.** $6x\sqrt{48} + 8x\sqrt{75}$ **3.** $-7\sqrt{84x} - 12\sqrt{189x} + 2\sqrt{21x}$

4. $9\sqrt{6} - 8\sqrt{6} + 3$ **5.** $\sqrt{a^3} + 4a\sqrt{a}$ **6.** $4x\sqrt{54x^3} + \sqrt{36x^2} + 3\sqrt{24x^5} - 3x$

☆ **SAMPLE SET B**

1. $\sqrt{7}(\sqrt{8} - 3)$.   **Use the distributive property to distribute the $\sqrt{7}$.**

$$\sqrt{7}(\sqrt{8} - 3) = \sqrt{7}\,\sqrt{8} - \sqrt{7} \cdot 3$$
$$= \sqrt{56} - 3\sqrt{7} \qquad \text{Simply } \sqrt{56}.$$
$$= \sqrt{4 \cdot 14} - 3\sqrt{7}$$
$$= 2\sqrt{14} - 3\sqrt{7}$$

2. $(\sqrt{2} + \sqrt{3})(\sqrt{5} + \sqrt{12})$.   **Use the distributive property.**

$$(\sqrt{2} + \sqrt{3})(\sqrt{5} + \sqrt{12}) = \sqrt{2}\,\sqrt{5} + \sqrt{2}\,\sqrt{12} + \sqrt{3}\,\sqrt{5} + \sqrt{3}\,\sqrt{12}$$
$$= \sqrt{10} + \sqrt{24} + \sqrt{15} + \sqrt{36}$$
$$= \sqrt{10} + 2\sqrt{6} + \sqrt{15} + 6$$

3. $(4\sqrt{2} - 3\sqrt{6})(5\sqrt{2} + \sqrt{6})$

$$(4\sqrt{2} - 3\sqrt{6})(5\sqrt{2} + \sqrt{6})$$

$$= 4 \cdot 5\sqrt{2}\,\sqrt{2} + 4\sqrt{2}\,\sqrt{6} - 3 \cdot 5\sqrt{6}\,\sqrt{2} - 3\sqrt{6}\,\sqrt{6}$$
$$= 20\sqrt{4} + 4\sqrt{12} - 15\sqrt{12} - 3\sqrt{36}$$
$$= 20 \cdot 2 + 4\sqrt{4 \cdot 3} - 15\sqrt{4 \cdot 3} - 3 \cdot 6$$
$$= 40 + 8\sqrt{3} - 30\sqrt{3} - 18$$
$$= 22 - 22\sqrt{3} \qquad \text{or} \qquad 22(1 - \sqrt{3})$$

Continued

4. $\dfrac{3 + \sqrt{8}}{3 - \sqrt{8}}$.   We'll rationalize the denominator by multiplying this fraction by 1 in the form $\dfrac{3 + \sqrt{8}}{3 + \sqrt{8}}$.

$$
\begin{aligned}
\frac{3 + \sqrt{8}}{3 - \sqrt{8}} \cdot \frac{3 + \sqrt{8}}{3 + \sqrt{8}} &= \frac{(3 + \sqrt{8})(3 + \sqrt{8})}{3^2 - (\sqrt{8})^2} \\
&= \frac{9 + 3\sqrt{8} + 3\sqrt{8} + \sqrt{8}\,\sqrt{8}}{9 - 8} \\
&= \frac{9 + 6\sqrt{8} + 8}{1} \\
&= 17 + 6\sqrt{8} \\
&= 17 + 6\sqrt{4 \cdot 2} \\
&= 17 + 12\sqrt{2}
\end{aligned}
$$

5. $\dfrac{2 + \sqrt{7}}{4 - \sqrt{3}}$.   Rationalize the denominator by multiplying this fraction by 1 in the form $\dfrac{4 + \sqrt{3}}{4 + \sqrt{3}}$.

$$
\begin{aligned}
\frac{2 + \sqrt{7}}{4 - \sqrt{3}} \cdot \frac{4 + \sqrt{3}}{4 + \sqrt{3}} &= \frac{(2 + \sqrt{7})(4 + \sqrt{3})}{4^2 - (\sqrt{3})^2} \\
&= \frac{8 + 2\sqrt{3} + 4\sqrt{7} + \sqrt{21}}{16 - 3} \\
&= \frac{8 + 2\sqrt{3} + 4\sqrt{7} + \sqrt{21}}{13}
\end{aligned}
$$

★ **PRACTICE SET B**

Simplify each by performing the indicated operation.

1. $\sqrt{5}(\sqrt{6} - 4)$ 2. $(\sqrt{5} + \sqrt{7})(\sqrt{2} + \sqrt{8})$ 3. $(3\sqrt{2} - 2\sqrt{3})(4\sqrt{3} + \sqrt{8})$ 4. $\dfrac{4 + \sqrt{5}}{3 - \sqrt{8}}$

Answers to Practice Sets are on p. 441.

Section 8.5 EXERCISES

For problems 1–75, simplify each expression by performing the indicated operation.

1. $4\sqrt{5} - 2\sqrt{5}$ 2. $10\sqrt{2} + 8\sqrt{2}$ 3. $-3\sqrt{6} - 12\sqrt{6}$ 4. $-\sqrt{10} - 2\sqrt{10}$

5. $3\sqrt{7x} + 2\sqrt{7x}$ **6.** $6\sqrt{3a} + \sqrt{3a}$ **19.** $\sqrt{24} - 2\sqrt{54} - 4\sqrt{12}$

7. $2\sqrt{18} + 5\sqrt{32}$ **8.** $4\sqrt{27} - 3\sqrt{48}$ **20.** $6\sqrt{18} + 5\sqrt{32} + 4\sqrt{50}$

9. $\sqrt{200} - \sqrt{128}$ **10.** $4\sqrt{300} + 2\sqrt{500}$ **21.** $-8\sqrt{20} - 9\sqrt{125} + 10\sqrt{180}$

11. $6\sqrt{40} + 8\sqrt{80}$ **12.** $2\sqrt{120} - 5\sqrt{30}$ **22.** $2\sqrt{27} + 4\sqrt{3} - 6\sqrt{12}$

13. $8\sqrt{60} - 3\sqrt{15}$ **14.** $\sqrt{a^3} - 3a\sqrt{a}$ **23.** $\sqrt{14} + 2\sqrt{56} - 3\sqrt{136}$

15. $\sqrt{4x^3} + x\sqrt{x}$ **16.** $2b\sqrt{a^3b^5} + 6a\sqrt{ab^7}$ **24.** $3\sqrt{2} + 2\sqrt{63} + 5\sqrt{7}$

17. $5xy\sqrt{2xy^3} - 3y^2\sqrt{2x^3y}$ **25.** $4ax\sqrt{3x} + 2\sqrt{3a^2x^3} + 7\sqrt{3a^2x^3}$

18. $5\sqrt{20} + 3\sqrt{45} - 3\sqrt{40}$ **26.** $3by\sqrt{5y} + 4\sqrt{5b^2y^3} - 2\sqrt{5b^2y^3}$

27. $\sqrt{2}(\sqrt{3} + 1)$

35. $(3 - \sqrt{2})(4 - \sqrt{2})$

28. $\sqrt{3}(\sqrt{5} - 3)$

36. $(5 + \sqrt{7})(4 - \sqrt{7})$

29. $\sqrt{5}(\sqrt{3} - \sqrt{2})$

37. $(\sqrt{2} + \sqrt{5})(\sqrt{2} + 3\sqrt{5})$

30. $\sqrt{7}(\sqrt{6} - \sqrt{3})$

38. $(2\sqrt{6} - \sqrt{3})(3\sqrt{6} + 2\sqrt{3})$

31. $\sqrt{8}(\sqrt{3} + \sqrt{2})$

39. $(4\sqrt{5} - 2\sqrt{3})(3\sqrt{5} + \sqrt{3})$

32. $\sqrt{10}(\sqrt{10} - \sqrt{5})$

40. $(3\sqrt{8} - 2\sqrt{2})(4\sqrt{2} - 5\sqrt{8})$

33. $(1 + \sqrt{3})(2 - \sqrt{3})$

41. $(\sqrt{12} + 5\sqrt{3})(2\sqrt{3} - 2\sqrt{12})$

34. $(5 + \sqrt{6})(4 - \sqrt{6})$

42. $(1 + \sqrt{3})^2$

43. $(3 + \sqrt{5})^2$

44. $(2 - \sqrt{6})^2$

45. $(2 - \sqrt{7})^2$

46. $(1 + \sqrt{3x})^2$

47. $(2 + \sqrt{5x})^2$

48. $(3 - \sqrt{3x})^2$

49. $(8 - \sqrt{6b})^2$

50. $(2a + \sqrt{5a})^2$

51. $(3y - \sqrt{7y})^2$

52. $(3 + \sqrt{3})(3 - \sqrt{3})$

53. $(2 + \sqrt{5})(2 - \sqrt{5})$

54. $(8 + \sqrt{10})(8 - \sqrt{10})$

55. $(6 + \sqrt{7})(6 - \sqrt{7})$

56. $(\sqrt{2} + \sqrt{3})(\sqrt{2} - \sqrt{3})$

57. $(\sqrt{5} + \sqrt{2})(\sqrt{5} - \sqrt{2})$

58. $(\sqrt{a} + \sqrt{b})(\sqrt{a} - \sqrt{b})$

59. $(\sqrt{x} + \sqrt{y})(\sqrt{x} - \sqrt{y})$

68. $\dfrac{2 - \sqrt{8}}{2 + \sqrt{8}}$

69. $\dfrac{4 + \sqrt{5}}{4 - \sqrt{5}}$

60. $\dfrac{2}{5 + \sqrt{3}}$

61. $\dfrac{4}{6 + \sqrt{2}}$

70. $\dfrac{1 + \sqrt{6}}{1 - \sqrt{6}}$

71. $\dfrac{8 - \sqrt{3}}{2 + \sqrt{18}}$

62. $\dfrac{1}{3 - \sqrt{2}}$

63. $\dfrac{1}{4 - \sqrt{3}}$

72. $\dfrac{6 - \sqrt{2}}{4 + \sqrt{12}}$

73. $\dfrac{\sqrt{3} - \sqrt{2}}{\sqrt{3} + \sqrt{2}}$

64. $\dfrac{8}{2 - \sqrt{6}}$

65. $\dfrac{2}{3 - \sqrt{7}}$

66. $\dfrac{\sqrt{5}}{3 + \sqrt{3}}$

67. $\dfrac{\sqrt{3}}{6 + \sqrt{6}}$

74. $\dfrac{\sqrt{6a} - \sqrt{8a}}{\sqrt{8a} + \sqrt{6a}}$

75. $\dfrac{\sqrt{2b} - \sqrt{3b}}{\sqrt{3b} + \sqrt{2b}}$

EXERCISES FOR REVIEW

(1.6) **76.** Simplify $\left(\dfrac{x^5y^3}{x^2y}\right)^5$.

(1.6) **77.** Simplify $(8x^3y)^2(x^2y^3)^4$.

(2.6) **78.** Write $(x-1)^4(x-1)^{-7}$ so that only positive exponents appear.

(8.2) **79.** Simplify $\sqrt{27x^5y^{10}z^3}$.

(8.4) **80.** Simplify $\dfrac{1}{2+\sqrt{x}}$ by rationalizing the denominator.

★ Answers to Practice Sets (8.5)

A. **1.** $2\sqrt{2}$　　**2.** $64x\sqrt{3}$　　**3.** $-48\sqrt{21x}$　　**4.** $\sqrt{6}+3$　　**5.** $5a\sqrt{a}$　　**6.** $18x^2\sqrt{6x}+3x$

B. **1.** $\sqrt{30}-4\sqrt{5}$　　**2.** $3\sqrt{10}+3\sqrt{14}$　　**3.** $8\sqrt{6}-12$　　**4.** $12+8\sqrt{2}+3\sqrt{5}+2\sqrt{10}$

8.6 Square Root Equations with Applications

Section Overview

☐ **SQUARE ROOT EQUATIONS AND EXTRANEOUS SOLUTIONS**
☐ **METHOD FOR SOLVING SQUARE ROOT EQUATIONS**

☐ SQUARE ROOT EQUATIONS AND EXTRANEOUS SOLUTIONS

Square Root Equation

A **square root equation** is an equation that contains a variable under a square root sign. The fact that

$$\sqrt{x} \cdot \sqrt{x} = (\sqrt{x})^2 = x$$

Extraneous Solutions

suggests that we can solve a square root equation by squaring both sides of the equation. Squaring both sides of an equation can, however, introduce extraneous solutions. Consider the equation

$$x = -6$$

The solution is -6. Square both sides.

$$x^2 = (-6)^2$$
$$x^2 = 36$$

This equation has two solutions, -6 and $+6$. The $+6$ is an extraneous solution since it does not check in the original equation: $+6 \neq -6$.

☐ METHOD FOR SOLVING SQUARE ROOT EQUATIONS

Solving Square Root Equations

1. Isolate a radical. This means get a square root expression by itself on one side of the equal sign.
2. Square both sides of the equation.
3. Simplify the equation by combining like terms.
4. Repeat step 1 if radicals are still present.
5. Obtain potential solutions by solving the resulting non-square root equation.
6. Check each potential solution by substitution into the original equation.

☆ SAMPLE SET A

Solve each square root equation.

1. $\sqrt{x} = 8$.   The radical is isolated. Square both sides.

$(\sqrt{x})^2 = 8^2$

$x = 64$ Check this potential solution.

Check: $\sqrt{64} \stackrel{?}{=} 8$

$8 \stackrel{\checkmark}{=} 8$

64 is the solution.

2. $\sqrt{y-3} = 4$.   The radical is isolated. Square both sides.

$y - 3 = 16$ Solve this nonradical equation.
$y = 19$ Check this potential solution.

Check: $\sqrt{19-3} \stackrel{?}{=} \sqrt{16}$
$\sqrt{16} \stackrel{?}{=} 4$
$4 \stackrel{\checkmark}{=} 4$

19 is the solution.

3. $\sqrt{2m+3} - \sqrt{m-8} = 0$.   Isolate either radical.

$\sqrt{2m+3} = \sqrt{m-8}$ Square both sides.

$2m + 3 = m - 8$ Solve this nonradical equation.

$m = -11$ Check this potential solution.

Check: $\sqrt{2(-11)+3} - \sqrt{(-11)-8} \stackrel{?}{=} 0$
$\sqrt{-22+3} - \sqrt{-19} \stackrel{?}{=} 0$

Since $\sqrt{-19}$ is not a real number, the potential solution of $m = -11$ does not check. This equation has no real solution.

4. $\sqrt{4x-5} = -6$.   By inspection, this equation has no real solution.

The symbol, $\sqrt{}$, signifies the *positive* square root and not the negative square root.

★ PRACTICE SET A

Solve each square root equation.

1. $\sqrt{y} = 14$ **2.** $\sqrt{a-7} = 5$ **3.** $\sqrt{3a+8} - \sqrt{2a+5} = 0$ **4.** $\sqrt{m-4} = -11$

Answers to the Practice Set are on p. 445.

Section 8.6 EXERCISES

For problems 1–29, solve the square root equations.

1. $\sqrt{x} = 5$

2. $\sqrt{y} = 7$

3. $\sqrt{a} = 10$

4. $\sqrt{c} = 12$

5. $\sqrt{x} = -3$

6. $\sqrt{y} = -6$

7. $\sqrt{x} = 0$

8. $\sqrt{x} = 1$

9. $\sqrt{x + 3} = 3$

10. $\sqrt{y - 5} = 5$

11. $\sqrt{a + 2} = 6$

12. $\sqrt{y + 7} = 9$

13. $\sqrt{y - 4} - 4 = 0$

14. $\sqrt{x - 10} - 10 = 0$

15. $\sqrt{x - 16} = 0$

16. $\sqrt{y - 25} = 0$

17. $\sqrt{6m - 4} = \sqrt{5m - 1}$

18. $\sqrt{5x + 6} = \sqrt{3x + 7}$

19. $\sqrt{7a + 6} = \sqrt{3a - 18}$

20. $\sqrt{4x + 3} = \sqrt{x - 9}$

21. $\sqrt{10a - 7} - \sqrt{2a + 9} = 0$

22. $\sqrt{12k - 5} - \sqrt{9k + 10} = 0$

23. $\sqrt{x - 6} - \sqrt{3x - 8} = 0$

24. $\sqrt{4a - 5} - \sqrt{7a - 20} = 0$

25. $\sqrt{2m - 6} = \sqrt{m - 2}$

26. $\sqrt{6r - 11} = \sqrt{5r + 3}$

27. $\sqrt{3x + 1} = \sqrt{2x - 6}$

28. $\sqrt{x-7} - \sqrt{5x+1} = 0$

29. $\sqrt{2a+9} - \sqrt{a-4} = 0$

30. At a certain electronics company, the daily output Q is related to the number of people A on the assembly line by $Q = 400 + 10\sqrt{A + 125}$.
   (a) Determine the daily output if there are 44 people on the assembly line.

   (b) Determine how many people are needed on the assembly line if the daily output is to be 520.

31. At a store, the daily number of sales S is approximately related to the number of employees E by $S = 100 + 15\sqrt{E + 6}$
   (a) Determine the approximate number of sales if there are 19 employees.

   (b) Determine the number of employees the store would need to produce 310 sales.

32. ▦ The resonance frequency f in an electronic circuit containing inductance L and capacitance C in series is given by

$$f = \frac{1}{2\pi\sqrt{LC}}$$

   (a) Determine the resonance frequency in an electronic circuit if the inductance is 4 and the capacitance is 0.0001. Use $\pi = 3.14$.

   (b) Determine the inductance in an electric circuit if the resonance frequency is 7.12 and the capacitance is 0.0001. Use $\pi = 3.14$.

33. If two magnetic poles of strength m and m' units are at a distance r centimeters (cm) apart, the force F of repulsion in air between them is given by

$$F = \frac{mm'}{r^2}$$

   (a) Determine the force of repulsion if two magnetic poles of strengths 20 and 40 units are 5 cm apart in air.

   (b) Determine how far apart are two magnetic poles of strengths 30 and 40 units if the force of repulsion in air between them is 18.75.

34. The velocity V in feet per second of outflow of a liquid from an orifice is given by $V = 8\sqrt{h}$, where h is the height in feet of the liquid above the opening.
   (a) Determine the velocity of outflow of a liquid from an orifice that is 9 feet below the top surface of a liquid (V is in feet/sec).

(b) Determine how high a liquid is above an orifice if the velocity of outflow is 81 feet/second.

35. ▦ The period T in seconds of a simple pendulum of length L in feet is given by $T = 2\pi\sqrt{\dfrac{L}{32}}$.

(a) Determine the period of a simple pendulum that is 2 feet long. Use $\pi = 3.14$.

(b) Determine the length in feet of a simple pendulum whose period is 10.8772 seconds. Use $\pi = 3.14$.

36. The kinetic energy KE in foot pounds of a body of mass m in slugs moving with a velocity v in feet/sec is given by

$$KE = \frac{1}{2}mv^2$$

(a) Determine the kinetic energy of a 2-slug body moving with a velocity of 4 ft/sec.

(b) Determine the velocity in feet/sec of a 4-slug body if its kinetic energy is 50 foot pounds.

EXERCISES FOR REVIEW

(2.6) **37.** Write $\dfrac{x^{10}y^3(x+7)^4}{x^{-2}y^3(x+7)^{-1}}$ so that only positive exponents appear.

(4.1) **38.** Classify $x + 4 = x + 7$ as an identity, a contradiction, or a conditional equation.

(6.4) **39.** Supply the missing words. In the coordinate plane, lines with _____ slope rise and lines with _____ slope fall.

(8.2) **40.** Simplify $\sqrt{(x+3)^4(x-2)^6}$.

(8.5) **41.** Simplify $(3 + \sqrt{5})(4 - \sqrt{5})$.

★ **Answers to Practice Set (8.6)**

A. **1.** $y = 196$ **2.** $a = 32$ **3.** $a = -3$ is extraneous, no real solution **4.** no real solution

Square Root (8.1)

The square root of a positive number x is a number such that when it is squared, the number x results.

Every positive number has two square roots, one positive and one negative. They are opposites of each other.

Principal Square Root $\sqrt{x}$ (8.1)

If x is a positive real number, then

$\sqrt{x}$ represents the positive square root of x. The positive square root of a number is called the *principal square root* of the number.

Secondary Square Root $-\sqrt{x}$ (8.1)

$-\sqrt{x}$ represents the negative square root of x. The negative square root of a number is called the *secondary square root* of the number.

Radical Sign, Radicand, and Radical (8.1)

In the expression $\sqrt{x}$,

$\sqrt{}$ is called the *radical sign*.
x is called the *radicand*.
$\sqrt{x}$ is called a *radical*.
The horizontal bar that appears attached to the radical sign, $\sqrt{}$, is a grouping symbol that specifies the radicand.

Meaningful Expressions (8.1)

A radical expression will only be meaningful if the radicand (the expression under the radical sign) is *not* negative:

$\sqrt{-25}$ is not meaningful   and $\sqrt{-25}$ is not a real number

Simplifying Square Root Expressions (8.1)

If a is a nonnegative number, then

$\sqrt{a^2} = a$

Perfect Squares (8.2)

Real numbers that are squares of rational numbers are called *perfect squares*.

Irrational Numbers (8.2)

Any indicated square root whose radicand is not a perfect square is an irrational number.

$\sqrt{2}$, $\sqrt{5}$, and $\sqrt{10}$ are irrational numbers

The Product Property (8.2)

$\sqrt{xy} = \sqrt{x}\,\sqrt{y}$

The Quotient Property (8.2)

$\sqrt{\dfrac{x}{y}} = \dfrac{\sqrt{x}}{\sqrt{y}}, \qquad y \neq 0$

Be Careful (8.2)

$\sqrt{x+y} \neq \sqrt{x} + \sqrt{y}$ $(\sqrt{16+9} \neq \sqrt{16} + \sqrt{9})$
$\sqrt{x-y} \neq \sqrt{x} - \sqrt{y}$ $(\sqrt{25-16} \neq \sqrt{25} - \sqrt{16})$

Simplified Form (8.2)

A square root that does not involve fractions is in simplified form if there are no perfect squares in the radicand.

A square root involving a fraction is in simplified form if there are no

1. perfect squares in the radicand,
2. fractions in the radicand, or
3. square root expressions in the denominator.

Rationalizing the Denominator (8.2)

The process of eliminating radicals from the denominator is called *rationalizing the denominator*.

Multiplying Square Root Expressions (8.3)

The product of the square roots is the square root of the product.

$\sqrt{x}\,\sqrt{y} = \sqrt{xy}$

1. Simplify each square root, if necessary.
2. Perform the multiplication.
3. Simplify, if necessary.

Dividing Square Root Expressions **(8.4)**

The quotient of the square roots is the square root of the quotient.

$$\frac{\sqrt{x}}{\sqrt{y}} = \sqrt{\frac{x}{y}}$$

Addition and Subtraction of Square Root Expressions **(8.5)**

$$a\sqrt{x} + b\sqrt{x} = (a + b)\sqrt{x}$$
$$a\sqrt{x} - b\sqrt{x} = (a - b)\sqrt{x}$$

Square Root Equation **(8.6)**

A *square root equation* is an equation that contains a variable under a square root radical sign.

Solving Square Root Equations **(8.6)**

1. Isolate a radical.
2. Square both sides of the equation.
3. Simplify by combining like terms.
4. Repeat step 1 if radical are still present.
5. Obtain potential solution by solving the resulting non-square root equation.
6. Check potential solutions by substitution.

EXERCISE SUPPLEMENT

Sections 8.1 – 8.5

For problems 1–65, simplify the expressions.

1. $\sqrt{10}\,\sqrt{2}$
2. $\sqrt{6}\,\sqrt{8}$
3. $\sqrt{18}\,\sqrt{40}$
4. $\sqrt{11}\,\sqrt{11}$
5. $\sqrt{y}\,\sqrt{y}$
6. $\sqrt{r^3}\,\sqrt{r^3}$
7. $\sqrt{m+3}\,\sqrt{m+3}$
8. $\sqrt{a-7}\,\sqrt{a-7}$
9. $\sqrt{x^2+4x+4}$
10. $\sqrt{y^2-12y+36}$
11. $\dfrac{\sqrt{x+5}}{\sqrt{x+2}}$
12. $\dfrac{\sqrt{n-3}}{\sqrt{n-1}}$
13. $\dfrac{\sqrt{50}}{\sqrt{2}}$
14. $\dfrac{\sqrt{75}}{5\sqrt{3}}$
15. $\dfrac{\sqrt{a^2+6a+9}}{\sqrt{a+3}}$
16. $\dfrac{\sqrt{4x^2+4x+1}}{\sqrt{2x+1}}$
17. $\dfrac{\sqrt{x^2-11x+24}}{\sqrt{x-8}}$
18. $\dfrac{\sqrt{y^2+11y+28}}{\sqrt{y+4}}$
19. $\sqrt{3}(\sqrt{5}+\sqrt{3})$
20. $\sqrt{5}(\sqrt{6}-\sqrt{10})$
21. $\sqrt{a}(\sqrt{a}-\sqrt{bc})$
22. $\sqrt{x}(\sqrt{x^5}-\sqrt{3x})$
23. $\sqrt{7a^3}(\sqrt{2a}-\sqrt{4a^3})$
24. $\dfrac{3}{\sqrt{7}}$
25. $\dfrac{2}{\sqrt{5}}$
26. $\dfrac{6}{\sqrt{2}}$
27. $\dfrac{8y}{\sqrt{y}}$
28. $\dfrac{16a^2}{\sqrt{5a}}$
29. $(2+\sqrt{3})(2-\sqrt{3})$
30. $(x+\sqrt{8})(3x+\sqrt{8})$
31. $(4y-\sqrt{3x})(4y+\sqrt{3x})$
32. $(6r+\sqrt{2s})(4r+\sqrt{2s})$
33. $\dfrac{2}{2+\sqrt{7}}$
34. $\dfrac{4}{1-\sqrt{6}}$
35. $\dfrac{6}{x+\sqrt{y}}$
36. $\dfrac{10}{a-\sqrt{2b}}$
37. $\dfrac{\sqrt{5}}{a+\sqrt{3}}$
38. $\dfrac{\sqrt{2}}{1+\sqrt{10}}$
39. $\dfrac{8+\sqrt{3}}{2+\sqrt{6}}$
40. $\dfrac{4+\sqrt{11}}{4-\sqrt{11}}$
41. $\sqrt{\dfrac{36a^4b^5c^{11}}{x^2y^5}}$
42. $\sqrt{x^{12}y^{10}z^8w^7}$
43. $\sqrt{32x^5y(x-2)^3}$
44. $-2\sqrt{60r^4s^3}$
45. $\sqrt{\dfrac{3}{16}}$
46. $\sqrt{\dfrac{4}{25}}$
47. $\sqrt{\dfrac{9}{16}}$
48. $\sqrt{\dfrac{5}{36}}$
49. $\sqrt{\dfrac{1}{6}}$
50. $\sqrt{\dfrac{3}{10}}$
51. $\sqrt{(x+4)^4(x-1)^5}$
52. $\sqrt{(3x+5)^3(2x-7)^3}$
53. $\sqrt{(y-3z)^{12}(y+3z)^{10}(y-5z)^3}$
54. $\sqrt{(8a-5b)^{26}(2a-9b)^{40}(a-b)^{15}}$
55. $4\sqrt{11}+8\sqrt{11}$
56. $-\sqrt{6}+5\sqrt{6}$
57. $5\sqrt{60}-7\sqrt{15}$
58. $4ax^2\sqrt{75x^4}+6a\sqrt{3x^8}$
59. $-3\sqrt{54}-16\sqrt{96}$
60. $\sqrt{18x^2y}\,\sqrt{2x^2y}$
61. $\sqrt{4x^2+32x+64}+\sqrt{10x^2+80x+160}$
62. $-2\sqrt{9x^2-42x+49}+5\sqrt{18x^2-84x+98}$
63. $-10\sqrt{56a^3b^7}+2a^2b\sqrt{126ab^5}$
64. $\dfrac{\sqrt{3x}-\sqrt{5x}}{\sqrt{7x}+\sqrt{2x}}$
65. $\dfrac{\sqrt{6a}+\sqrt{2a}}{\sqrt{3a}-\sqrt{5a}}$

Section 8.6

For problems 66–85, solve the equations.

66. $\sqrt{3x}=9$
67. $\sqrt{4a}=16$
68. $\sqrt{x+7}=4$
69. $\sqrt{a+6}=-5$
70. $\sqrt{4a+5}=21$
71. $\sqrt{3m+7}=10$
72. $\sqrt{y+10}=5$
73. $\sqrt{a-7}=6$
74. $\sqrt{4x-8}=x-2$
75. $\sqrt{2x+3}+8=11$
76. $\sqrt{a^2+5}+5=a$
77. $\sqrt{5b+4}-5=-2$
78. $\sqrt{2a+1}-10=-3$

79. $\sqrt{2x+5} = \sqrt{x+3}$

80. $\sqrt{5a-11} = \sqrt{2a-16}$

81. At a small business, the monthly number of sales S is approximately related to the number of employees E by $S = 140 + 8\sqrt{E-2}$.
  (a) Determine the approximate number of sales if the number of employees is 27.
  (b) Determine the approximate number of employees if the monthly sales are 268.

82. The resonance frequency f in an electronic circuit containing inductance L and capacitance C in series is given by

$$f = \frac{1}{2\pi\sqrt{LC}}$$

  (a) Determine the resonance frequency in an electronic circuit if the inductance is 9 and the capacitance is 0.0004. Use $\pi = 3.14$.
  (b) Determine the inductance in an electric circuit if the resonance frequency is 5.308 and the capacitance is 0.0001. Use $\pi = 3.14$.

83. If two magnetic poles of strengths m and m' units are at a distance r centimeters (cm) apart, the force F of repulsion in air between them is given by

$$F = \frac{mm'}{r^2}$$

  (a) Determine the force of repulsion if two magnetic poles of strengths 22 and 46 units are 8 cm apart.
  (b) Determine how far apart are two magnetic poles of strengths 14 and 16 units if the force of repulsion in air between them is 42 units.

For problems 1–17, simplify each of the square root expressions.

1. _____

1. **(8.3)** $\sqrt{8} \cdot \sqrt{5}$

2. _____

2. **(8.4)** $\dfrac{\sqrt{80}}{\sqrt{12}}$

3. _____

3. **(8.2, 8.4)** $\dfrac{\sqrt{n^2 + n - 12}}{\sqrt{n - 3}}$

4. _____

4. **(8.1, 8.2)** $\sqrt{24a^3b^5c^8}$

5. _____

5. **(8.1, 8.2)** $\sqrt{\dfrac{64x^4y^5z^6}{49a^3b^2c^9}}$

6. _____

6. **(8.2)** $\sqrt{(x - 2)^2(x + 1)^4}$

7. _____

7. **(8.2)** $\sqrt{a^2 - 8a + 16}$

8. _____

8. **(8.4)** $\dfrac{4}{2 + \sqrt{x}}$

9. _____

9. **(8.4)** $\dfrac{\sqrt{3a}}{\sqrt{2a} + \sqrt{5a}}$

10. _____

10. **(8.5)** $2x\sqrt{27} + x\sqrt{12}$

11. _____

11. **(8.5)** $-3a\sqrt{a^5b^3} + 2a^3b\sqrt{ab}$

12. _____

12. **(8.4, 8.5)** $\sqrt{10}(\sqrt{8} - \sqrt{2})$

13. _____

13. **(8.4, 8.5)** $(3 + \sqrt{6})(2 + \sqrt{5})$

14. _____

14. **(8.4, 8.5)** $(\sqrt{10} - \sqrt{3})(\sqrt{5} + \sqrt{2})$

15. _____

15. **(8.4, 8.5)** $(4 - \sqrt{5y})^2$

16. _____

16. (8.4, 8.5) $\dfrac{6 - \sqrt{3}}{4 + \sqrt{2}}$

17. _____

17. (8.4, 8.5) $\dfrac{\sqrt{2} + \sqrt{3}}{\sqrt{3} - \sqrt{5}}$

18. _____

For problems 18–24, solve the equations.

18. (8.6) $\sqrt{x + 8} = 4$

19. _____

19. (8.6) $\sqrt{3a + 1} = 4$

20. _____

20. (8.6) $\sqrt{2x} = -3$

21. _____

21. (8.6) $\sqrt{3x + 18} + 7 = 0$

22. _____

22. (8.6) $\sqrt{3m - 5} = \sqrt{2m + 1}$

23. _____

23. (8.6) $2\sqrt{a + 2} - 2 = 0$

24. _____

24. (8.6) $\sqrt{b - 7} - \sqrt{5b + 1} = 0$

25. _____

25. (8.6) At a small business, the number of monthly sales S is approximately related to the number of employees E by $S = 175 + 7\sqrt{E - 3}$.
   (a) Determine the approximate number of sales if the number of employees is 39.
   (b) Determine the approximate number of employees if the number of sales is 224.

9 Quadratic Equations

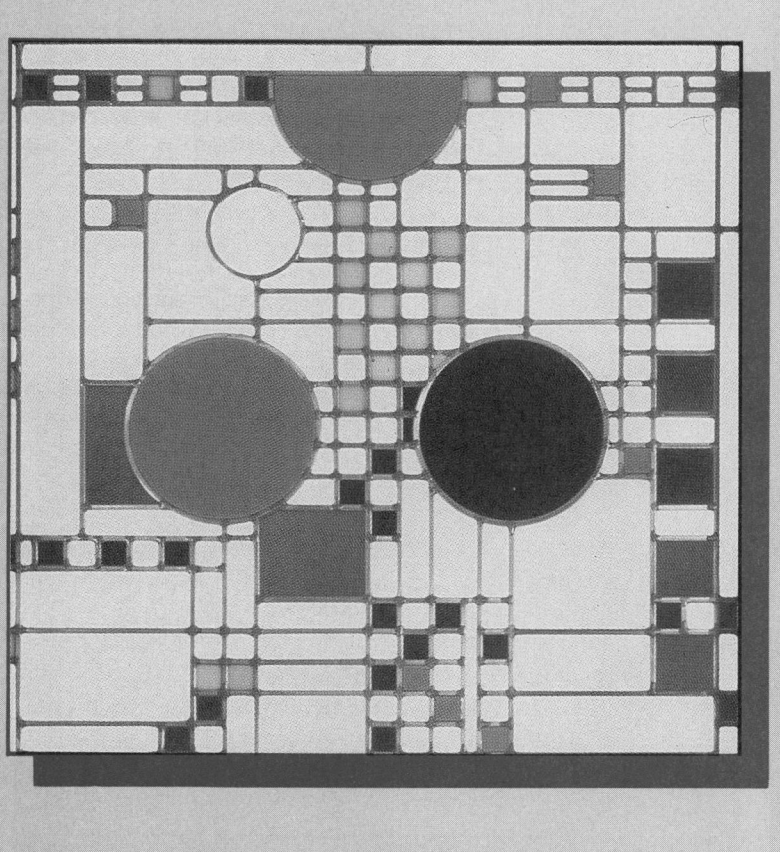

After completing this chapter, you should

Section 9.1 Solving Quadratic Equations
- be able to place a quadratic equation into standard form
- be familiar with the zero-factor property of real numbers

Section 9.2 Solving Quadratic Equations by Factoring
- be able to solve quadratic equations by factoring

Section 9.3 Solving Quadratic Equations Using the Method of Extraction of Roots
- be able to solve quadratic equations using the method of extraction of roots
- be able to determine the nature of the solutions to a quadratic equation

Section 9.4 Solving Quadratic Equations Using the Method of Completing the Square
- understand the logic underlying the method of completing the square
- be able to solve a quadratic equation using the method of completing the square

Section 9.5 Solving Quadratic Equations Using the Quadratic Formula
- recognize the standard form of a quadratic equation
- understand the derivation of the quadratic formula
- solve quadratic equations using the quadratic formula

Section 9.6 Applications
- become more proficient at using the five-step method for solving applied problems

Section 9.7 Graphing Quadratic Equations
- be able to construct the graph of a parabola

9.1 Solving Quadratic Equations

Section Overview
- ☐ STANDARD FORM OF A QUADRATIC EQUATION
- ☐ ZERO-FACTOR PROPERTY OF REAL NUMBERS

☐ STANDARD FORM OF A QUADRATIC EQUATION

In Chapter 4 we studied linear equations in one and two variables and methods for solving them. We observed that a linear equation in one variable was any equation that could be written in the form $ax + b = 0$, $a \neq 0$, and a linear equation in two variables was any equation that could be written in the form $ax + by = c$, where a and b are not both 0. We now wish to study quadratic equations in one variable.

Quadratic Equation

> A **quadratic equation** is an equation of the form $ax^2 + bx + c = 0$, $a \neq 0$.
>
> The **standard form** of the quadratic equation is $ax^2 + bx + c = 0$, $a \neq 0$.

For a quadratic equation in standard form $ax^2 + bx + c = 0$,

a is the coefficient of x^2.
b is the coefficient of x.
c is the constant term.

☆ SAMPLE SET A

The following *are* quadratic equations.

1. $3x^2 + 2x - 1 = 0$. $a = 3$, $b = 2$, $c = -1$.

2. $5x^2 + 8x = 0$. $a = 5$, $b = 8$, $c = 0$.

Notice that this equation could be written $5x^2 + 8x + 0 = 0$. Now it is clear that $c = 0$.

3. $x^2 + 7 = 0$. $a = 1$, $b = 0$, $c = 7$.

Notice that this equation could be written $x^2 + 0x + 7 = 0$. Now it is clear that $b = 0$.

The following are *not* quadratic equations.

4. $3x + 2 = 0$. $a = 0$.   This equation is linear.

5. $8x^2 + \dfrac{3}{x} - 5 = 0$.   The expression on the left side of the equal sign has a variable in the denominator and, therefore, is not a quadratic.

★ PRACTICE SET A

Which of the following equations are quadratic equations? Answer "yes" or "no" to each equation.

1. $6x^2 - 4x + 9 = 0$ **2.** $5x + 8 = 0$ **3.** $4x^3 - 5x^2 + x + 6 = 8$

4. $4x^2 - 2x + 4 = 1$ **5.** $\dfrac{2}{x} - 5x^2 = 6x + 4$ **6.** $9x^2 - 2x + 6 = 4x^2 + 8$

❑ ZERO-FACTOR PROPERTY

Our goal is to solve quadratic equations. The method for solving quadratic equations is based on the *zero-factor* property of real numbers. We were introduced to the zero-factor property in Section 7.1. We state it again.

Zero-Factor Property

> If two numbers a and b are multiplied together and the resulting product is 0, then at least one of the numbers must be 0. Algebraically, if $a \cdot b = 0$, then $a = 0$ or $b = 0$, or both $a = 0$ and $b = 0$.

☆ SAMPLE SET B

Use the zero-factor property to solve each equation.

1. If $9x = 0$, then x must be 0.

2. If $-2x^2 = 0$, then $x^2 = 0$, $x = 0$.

3. If $5(x - 1) = 0$, then $x - 1$ must be 0, since 5 is not zero.

$$x - 1 = 0$$
$$x = 1$$

4. If $x(x + 6) = 0$, then

$$x = 0 \quad \text{or} \quad x + 6 = 0$$
$$x = -6$$

$$x = 0, -6.$$

5. If $(x + 2)(x + 3) = 0$, then

$$x + 2 = 0 \quad \text{or} \quad x + 3 = 0$$
$$x = -2 \qquad\qquad x = -3$$

$$x = -2, -3.$$

6. If $(x + 10)(4x - 5) = 0$, then

$$x + 10 = 0 \quad \text{or} \quad 4x - 5 = 0$$
$$x = -10 \qquad\qquad 4x = 5$$
$$x = \frac{5}{4}$$

$$x = -10, \frac{5}{4}.$$

★ PRACTICE SET B

Use the zero-factor property to solve each equation.

1. $6(a - 4) = 0$ **2.** $(y + 6)(y - 7) = 0$ **3.** $(x + 5)(3x - 4) = 0$

Answers to Practice Sets are on p. 458.

Section 9.1 EXERCISES

For problems 1–20, write the values of a, b, and c in the quadratic equations.

1. $3x^2 + 4x - 7 = 0$

2. $7x^2 + 2x + 8 = 0$

3. $2y^2 - 5y + 5 = 0$

4. $7a^2 + a - 8 = 0$

5. $-3a^2 + 4a - 1 = 0$

6. $7b^2 + 3b = 0$

7. $2x^2 + 5x = 0$

8. $4y^2 + 9 = 0$

9. $8a^2 - 2a = 0$

10. $6x^2 = 0$

11. $4y^2 = 0$

12. $5x^2 - 3x + 9 = 4x^2$

13. $7x^2 + 2x + 1 = 6x^2 + x - 9$

14. $-3x^2 + 4x - 1 = -4x^2 - 4x + 12$

15. $5x - 7 = -3x^2$

16. $3x - 7 = -2x^2 + 5x$

17. $0 = x^2 + 6x - 1$

18. $9 = x^2$

19. $x^2 = 9$

20. $0 = -x^2$

For problems 21–53, use the zero-factor property to solve the equations.

21. $4x = 0$

22. $16y = 0$

23. $9a = 0$

24. $4m = 0$

25. $3(k + 7) = 0$

26. $8(y - 6) = 0$

27. $-5(x + 4) = 0$

28. $-6(n + 15) = 0$

29. $y(y - 1) = 0$

30. $a(a - 6) = 0$

31. $n(n + 4) = 0$

32. $x(x + 8) = 0$

33. $9(a - 4) = 0$

34. $-2(m + 11) = 0$

35. $x(x + 7) = 0$

36. $n(n - 10) = 0$

37. $(y - 4)(y - 8) = 0$

38. $(k - 1)(k - 6) = 0$

39. $(x + 5)(x + 4) = 0$

40. $(y + 6)(2y + 1) = 0$

41. $(x - 3)(5x - 6) = 0$

42. $(5a + 1)(2a - 3) = 0$

43. $(6m + 5)(11m - 6) = 0$

44. $(2m - 1)(3m + 8) = 0$

45. $(4x + 5)(2x - 7) = 0$

46. $(3y + 1)(2y + 1) = 0$

47. $(7a + 6)(7a - 6) = 0$

48. $(8x + 11)(2x - 7) = 0$

49. $(5x - 14)(3x + 10) = 0$

50. $(3x - 1)(3x - 1) = 0$

51. $(2y + 5)(2y + 5) = 0$

52. $(7a - 2)^2 = 0$

53. $(5m - 6)^2 = 0$

EXERCISES FOR REVIEW

(5.5)   **54.** Factor $12ax - 3x + 8a - 2$ by grouping.

(6.5)   **55.** Construct the graph of $6x + 10y - 60 = 0$.

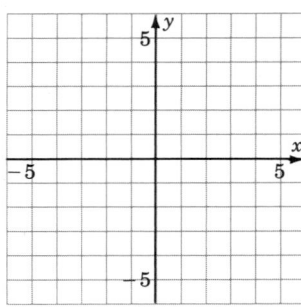

(7.5)   **56.** Find the difference: $\dfrac{1}{x^2 + 2x + 1} - \dfrac{1}{x^2 - 1}$.

(8.5)   **57.** Simplify $\sqrt{7}(\sqrt{2} + 2)$.

(8.6)   **58.** Solve the radical equation $\sqrt{3x + 10} = x + 4$.

★ **Answers to Practice Sets (9.1)**

A.   **1.** yes   **2.** no   **3.** no   **4.** yes   **5.** no   **6.** yes

B.   **1.** $a = 4$ **2.** $y = -6, 7$ **3.** $x = -5, \dfrac{4}{3}$

9.2 Solving Quadratic Equations by Factoring

Section Overview

  ❑ **FACTORING METHOD**
  ❑ **SOLVING MENTALLY AFTER FACTORING**

❑ FACTORING METHOD

To solve quadratic equations by factoring, we must make use of the zero-factor property.

1. Set the equation equal to zero, that is, get all the nonzero terms on one side of the equal sign and 0 on the other.

    $ax^2 + bx + c = 0$

2. Factor the quadratic expression.

    $(\quad)(\quad) = 0$

3. By the zero-factor property, at least one of the factors must be zero, so, set each of the factors equal to 0 and solve for the variable.

☆ **SAMPLE SET A**

Solve the following quadratic equations. (We will show the check for problem 1.)

1. $x^2 - 7x + 12 = 0$.   The equation is already set equal to 0. Factor.

$(x - 3)(x - 4) = 0$ Set each factor equal to 0.

$x - 3 = 0$ or $x - 4 = 0$
$x = 3$ or $x = 4$

Check: If $x = 3$, $x^2 - 7x + 12 = 0$ If $x = 4$, $x^2 - 7x + 12 = 0$
$3^2 - 7 \cdot 3 + 12 \overset{?}{=} 0$ $4^2 - 7 \cdot 4 + 12 \overset{?}{=} 0$
$9 - 21 + 12 \overset{?}{=} 0$ $16 - 28 + 12 \overset{?}{=} 0$
$0 \overset{\checkmark}{=} 0$ $0 = 0$

Thus, the solutions to this equation are $x = 3, 4$.

2. $x^2 = 25$.   Set the equation equal to 0.

$x^2 - 25 = 0$ Factor.

$(x + 5)(x - 5) = 0$ Set each factor equal to 0.

$x + 5 = 0$ or $x - 5 = 0$
$x = -5$ or $x = 5$

Thus, the solutions to this equation are $x = 5, -5$.

3. $x^2 = 2x$.   Set the equation equal to 0.

$x^2 - 2x = 0$ Factor.

$x(x - 2)$ Set each factor equal to 0.

$x = 0$ or $x - 2 = 0$
             $x = 2$

Thus, the solutions to this equation are $x = 0, 2$.

4. $2x^2 + 7x - 15 = 0$.   Factor.

$(2x - 3)(x + 5) = 0$ Set each factor equal to 0.

$2x - 3 = 0$ or $x + 5 = 0$
$2x = 3$ or $x = -5$

$x = \dfrac{3}{2}$

Thus, the solutions to this equation are $x = \dfrac{3}{2}, -5$.

5. $63x^2 = 13x + 6$

$63x^2 - 13x - 6 = 0$
$(9x + 2)(7x - 3) = 0$
$9x + 2 = 0$ or $7x - 3 = 0$
$9x = -2$ or $7x = 3$

$x = \dfrac{-2}{9}$ or $x = \dfrac{3}{7}$

Thus, the solutions to this equation are $x = \dfrac{-2}{9}, \dfrac{3}{7}$.

★ PRACTICE SET A

Solve the following equations, if possible.

1. $(x - 7)(x + 4) = 0$ **2.** $(2x + 5)(5x - 7) = 0$ **3.** $x^2 + 2x - 24 = 0$

4. $6x^2 + 13x - 5 = 0$ **5.** $5y^2 + 2y = 3$ **6.** $m(2m - 11) = 0$

7. $6p^2 = -(5p + 1)$ **8.** $r^2 - 49 = 0$

☐ SOLVING MENTALLY AFTER FACTORING

Let's consider problems 4 and 5 of Sample Set A in more detail. Let's look particularly at the factorizations $(2x - 3)(x + 5) = 0$ and $(9x + 2)(7x - 3) = 0$. The next step is to set each factor equal to zero and solve. We can solve mentally if we understand how to solve linear equations: we transpose the constant from the variable term and then divide by the coefficient of the variable.

☆ SAMPLE SET B

Solve the following equation mentally.

$(2x - 3)(x + 5) = 0$

$2x - 3 = 0$ **Mentally add 3 to both sides. The constant changes sign.**

$2x = 3$ **Divide by 2, the coefficient of x. The 2 divides the constant 3 into $\dfrac{3}{2}$. The coefficient becomes the denominator.**

$x = \dfrac{3}{2}$

$x + 5 = 0$ **Mentally subtract 5 from both sides. The constant changes sign.**

$x = -5$ **Divide by the coefficient of x, 1. The coefficient becomes the denominator.**

$x = \dfrac{-5}{1} = -5$

$x = -5$

Now, we can immediately write the solution to the equation after factoring by looking at each factor, changing the sign of the constant, then dividing by the coefficient.

★ **PRACTICE SET B**

Solve $(9x + 2)(7x - 3) = 0$ using this mental method.

Answers to Practice Sets are on p. 465.

Section 9.2 EXERCISES

For problems 1–76, solve the equations, if possible.

1. $(x + 1)(x + 3) = 0$

2. $(x + 4)(x + 9) = 0$

3. $(x - 5)(x - 1) = 0$

4. $(x - 6)(x - 3) = 0$

5. $(x - 4)(x + 2) = 0$

6. $(x + 6)(x - 1) = 0$

7. $(2x + 1)(x - 7) = 0$

8. $(3x + 2)(x - 1) = 0$

9. $(4x + 3)(3x - 2) = 0$

10. $(5x - 1)(4x + 7) = 0$

11. $(6x + 5)(9x - 4) = 0$

12. $(3a + 1)(3a - 1) = 0$

13. $x(x + 4) = 0$

14. $y(y - 5) = 0$

15. $y(3y - 4) = 0$

16. $b(4b + 5) = 0$

17. $x(2x + 1)(2x + 8) = 0$

28. $x^2 + 9 = 0$

18. $y(5y + 2)(2y - 1) = 0$

29. $x^2 + 36 = 0$

19. $(x - 8)^2 = 0$

30. $x^2 - 25 = 0$

20. $(x - 2)^2 = 0$

31. $a^2 - 100 = 0$

21. $(b + 7)^2 = 0$

32. $a^2 - 81 = 0$

22. $(a + 1)^2 = 0$

33. $b^2 - 49 = 0$

23. $x(x - 4)^2 = 0$

34. $y^2 - 1 = 0$

24. $y(y + 9)^2 = 0$

35. $3a^2 - 75 = 0$

25. $y(y - 7)^2 = 0$

36. $5b^2 - 20 = 0$

26. $y(y + 5)^2 = 0$

37. $y^3 - y = 0$

27. $x^2 - 4 = 0$

38. $a^2 = 9$

39. $b^2 = 4$

40. $b^2 = 1$

41. $a^2 = 36$

42. $3a^2 = 12$

43. $-2x^2 = -4$

44. $-2a^2 = -50$

45. $-7b^2 = -63$

46. $-2x^2 = -32$

47. $3b^2 = 48$

48. $a^2 - 8a + 16 = 0$

49. $y^2 + 10y + 25 = 0$

50. $y^2 + 9y + 16 = 0$

51. $x^2 - 2x - 1 = 0$

52. $a^2 + 6a + 9 = 0$

53. $a^2 + 4a + 4 = 0$

54. $x^2 + 12x = -36$

55. $b^2 - 14b = -49$

56. $3a^2 + 18a + 27 = 0$

57. $2m^3 + 4m^2 + 2m = 0$

58. $3mn^2 - 36mn + 36m = 0$

59. $a^2 + 2a - 3 = 0$

60. $a^2 + 3a - 10 = 0$

61. $x^2 + 9x + 14 = 0$

62. $x^2 - 7x + 12 = 3$

63. $b^2 + 12b + 27 = 0$

64. $b^2 - 3b + 2 = 0$

65. $x^2 - 13x = -42$

66. $a^3 = -8a^2 - 15a$

67. $6a^2 + 13a + 5 = 0$

68. $6x^2 - 4x - 2 = 0$

69. $12a^2 + 15a + 3 = 0$

70. $18b^2 + 24b + 6 = 0$

71. $12a^2 + 24a + 12 = 0$

72. $4x^2 - 4x = -1$

73. $2x^2 = x + 15$

74. $4a^2 = 4a + 3$

75. $4y^2 = -4y - 2$ **76.** $9y^2 = 9y + 18$

EXERCISES FOR REVIEW

(1.6)   **77.** Simplify $(x^4y^3)^2(xy^2)^4$.

(2.6)   **78.** Write $(x^{-2}y^3w^4)^{-2}$ so that only positive exponents appear.

(7.5)   **79.** Find the sum: $\dfrac{x}{x^2 - x - 2} + \dfrac{1}{x^2 - 3x + 2}$.

(7.8)   **80.** Simplify $\dfrac{\dfrac{1}{a} + \dfrac{1}{b}}{\dfrac{1}{a} - \dfrac{1}{b}}$.

(9.1)   **81.** Solve $(x + 4)(3x + 1) = 0$.

★ **Answers to Practice Sets (9.2)**

A. 1. $x = 7, -4$ **2.** $x = \dfrac{-5}{2}, \dfrac{7}{5}$ **3.** $x = 4, -6$ **4.** $x = \dfrac{1}{3}, \dfrac{-5}{2}$ **5.** $y = \dfrac{3}{5}, -1$ **6.** $m = 0, \dfrac{11}{2}$

7. $p = \dfrac{-1}{3}, \dfrac{-1}{2}$ **8.** $r = 7, -7$

B. $x = -\dfrac{2}{9}, \dfrac{3}{7}$

9.3 Solving Quadratic Equations Using the Method of Extraction of Roots

Section Overview

☐ **THE METHOD OF EXTRACTION OF ROOTS**
☐ **THE NATURE OF SOLUTIONS**

☐ **THE METHOD OF EXTRACTION OF ROOTS**

Extraction of Roots

Quadratic equations of the form $x^2 - K = 0$ can be solved by *the method of extraction of roots* by rewriting it in the form $x^2 = K$.

To solve $x^2 = K$, we are required to find some number, x, that when squared produces K. This number, x, must be a square root of K. If K is greater than zero, we know that it possesses two square roots, $\sqrt{K}$ and $-\sqrt{K}$. We also know that

$$(\sqrt{K})^2 = (\sqrt{K})(\sqrt{K}) = K \qquad \text{and} \qquad (-\sqrt{K})^2 = (-\sqrt{K})(-\sqrt{K}) = K$$

We now have two replacements for x that produce true statements when substituted into the equation. Thus, $x = \sqrt{K}$ and $x = -\sqrt{K}$ are both solutions to $x^2 = K$. We use the notation $x = \pm\sqrt{K}$ to denote both the principal and the secondary square roots.

❑ THE NATURE OF SOLUTIONS

Solutions of

$x^2 = K$

For quadratic equations of the form $x^2 = K$,

1. If K is greater than or equal to zero, the solutions are $\pm\sqrt{K}$.
2. If K is negative, no real number solutions exist.
3. If K is zero, the only solution is 0.

☆ SAMPLE SET A

Solve each of the following quadratic equations using the method of extraction of roots.

1. $x^2 - 49 = 0$.   **Rewrite.**

$x^2 = 49$

$x = \pm\sqrt{49}$
$x = \pm 7$

Check: $(7)^2 \overset{?}{=} 49$ $(-7)^2 \overset{?}{=} 49$
          $49 \overset{\checkmark}{=} 49$ $49 \overset{\checkmark}{=} 49$

2. $25a^2 = 36$

$a^2 = \dfrac{36}{25}$

$a = \pm\sqrt{\dfrac{36}{25}}$

$a = \pm\dfrac{6}{5}$

Check: $25\left(\dfrac{6}{5}\right)^2 \overset{?}{=} 36$ $25\left(\dfrac{-6}{5}\right)^2 \overset{?}{=} 36$

          $25\left(\dfrac{36}{25}\right) \overset{?}{=} 36$ $25\left(\dfrac{36}{25}\right) \overset{?}{=} 36$

          $36 \overset{\checkmark}{=} 36$ $36 \overset{\checkmark}{=} 36$

3. $4m^2 - 32 = 0$

$4m^2 = 32$

$m^2 = \dfrac{32}{4}$

$m^2 = 8$
$m = \pm\sqrt{8}$
$m = \pm 2\sqrt{2}$

Check: $4(2\sqrt{2})^2 \overset{?}{=} 32$ $4(-2\sqrt{2})^2 \overset{?}{=} 32$
          $4[2^2(\sqrt{2})^2] \overset{?}{=} 32$ $4[(-2)^2(\sqrt{2})^2] \overset{?}{=} 32$
          $4[4 \cdot 2] \overset{?}{=} 32$ $4[4 \cdot 2] \overset{?}{=} 32$
          $4 \cdot 8 \overset{?}{=} 32$ $4 \cdot 8 \overset{?}{=} 32$
          $32 \overset{\checkmark}{=} 32$ $32 \overset{\checkmark}{=} 32$

4. Solve $5x^2 - 15y^2z^7 = 0$ for x.

$5x^2 = 15y^2z^7$ **Divide both sides by 5.**

$x^2 = 3y^2z^7$
$x = \pm\sqrt{3y^2z^7}$
$x = \pm yz^3\sqrt{3z}$

5. 🖩 Calculator problem.   Solve $14a^2 - 235 = 0$. Round to the nearest hundredth.

$14a^2 - 235 = 0.$ **Rewrite.**

$14a^2 = 235$ **Divide both sides by 14.**

$a^2 = \dfrac{235}{14}$

On the Calculator

Type   235

Press $\boxed{\div}$

Type   14

Press $\boxed{=}$

Press $\boxed{\sqrt{}}$

Display reads: 4.0970373

Rounding to the nearest hundredth produces 4.10. We must be sure to insert the $\pm$ symbol.

$a \approx \pm 4.10$

6. $k^2 = -64$

$k = \pm\sqrt{-64}$

The radicand is negative so no real number solutions exist.

★ **PRACTICE SET A**

Solve each of the following quadratic equations using the method of extraction of roots.

1. $x^2 - 144 = 0$ **2.** $9y^2 - 121 = 0$ **3.** $6a^2 = 108$

4. Solve $4n^2 = 24m^2p^8$ for n.   **5.** Solve $5p^2q^2 = 45p^2$ for q.

6. 🖩 Solve $16m^2 - 2206 = 0$. Round to the nearest hundredth.   **7.** $h^2 = -100$

☆ SAMPLE SET B

Solve each of the following quadratic equations using the method of extraction of roots.

1. $(x + 2)^2 = 81$

$x + 2 = \pm\sqrt{81}$

$x + 2 = \pm 9$ **Subtract 2 from both sides.**

$x = -2 \pm 9$
$x = -2 + 9$ and $x = -2 - 9$
$x = 7$ $x = -11$

2. $(a + 3)^2 = 5$

$a + 3 = \pm\sqrt{5}$ **Subtract 3 from both sides.**

$a = -3 \pm\sqrt{5}$

★ PRACTICE SET B

Solve each of the following quadratic equations using the method of extraction of roots.

1. $(a + 6)^2 = 64$ **2.** $(m - 4)^2 = 15$ **3.** $(y - 7)^2 = 49$ **4.** $(k - 1)^2 = 12$

5. $(x - 11)^2 = 0$

Answers to Practice Sets are on p. 470.

Section 9.3 EXERCISES

For problems 1–25, solve each of the quadratic equations using the method of extraction of roots.

5. $b^2 = 1$ **6.** $a^2 = 1$

1. $x^2 = 36$ **2.** $x^2 = 49$

3. $a^2 = 9$ **4.** $a^2 = 4$ **7.** $x^2 = 25$ **8.** $x^2 = 81$

9. $a^2 = 5$ **10.** $a^2 = 10$ **28.** $a^2 = 25c^2$,   for a

11. $b^2 = 12$ **12.** $b^2 = 6$ **29.** $k^2 = m^2n^2$,   for k

13. $y^2 = 3$ **14.** $y^2 = 7$ **30.** $k^2 = p^2q^2r^2$,   for k

15. $a^2 - 8 = 0$ **16.** $a^2 - 3 = 0$ **31.** $2y^2 = 2a^2n^2$,   for y

17. $a^2 - 5 = 0$ **18.** $y^2 - 1 = 0$ **32.** $9y^2 = 27x^2z^4$,   for y

19. $x^2 - 10 = 0$ **20.** $x^2 - 11 = 0$ **33.** $x^2 - z^2 = 0$,   for x

21. $3x^2 - 27 = 0$ **22.** $5b^2 - 5 = 0$ **34.** $x^2 - z^2 = 0$,   for z

23. $2x^2 = 50$ **24.** $4a^2 = 40$ **35.** $5a^2 - 10b^2 = 0$,   for a

25. $2x^2 = 24$

For problems 36–50, solve each of the quadratic equations using the method of extraction of roots.

36. $(x - 1)^2 = 4$ **37.** $(x - 2)^2 = 9$

For problems 26–35, solve for the indicated variable.

26. $x^2 = 4a^2$,   for x

27. $x^2 = 9b^2$,   for x

38. $(x - 3)^2 = 25$ **39.** $(a - 5)^2 = 36$

40. $(a + 3)^2 = 49$ **41.** $(a + 9)^2 = 1$ **49.** $(x - a)^2 = b^2$, for x

42. $(a - 6)^2 = 3$ **43.** $(x + 4)^2 = 5$ **50.** $(x + c)^2 = a^2$, for x

44. $(b + 6)^2 = 7$ **45.** $(x + 1)^2 = a$, for x

🖩 **Calculator Problems**

For problems 51–55, round each result to the nearest hundredth.

46. $(y + 5)^2 = b$, for y

51. $8a^2 - 168 = 0$ **52.** $6m^2 - 5 = 0$

47. $(y + 2)^2 = a^2$, for y

53. $0.03y^2 = 1.6$ **54.** $0.048x^2 = 2.01$

48. $(x + 10)^2 = c^2$, for x

55. $1.001x^2 - 0.999 = 0$

EXERCISES FOR REVIEW

(6.1)   **56.** Graph the linear inequality $3(x + 2) < 2(3x + 4)$.

⟵————————————————⟶

(7.6)   **57.** Solve the fractional equation $\dfrac{x - 1}{x + 4} = \dfrac{x + 3}{x - 1}$.

(8.3)   **58.** Find the product: $\sqrt{32x^3y^5}\ \sqrt{2x^3y^3}$.

(9.2)   **59.** Solve $x^2 - 4x = 0$.

(9.2)   **60.** Solve $y^2 - 8y = -12$.

★ **Answers to Practice Sets (9.3)**

A. **1.** $x = \pm 12$ **2.** $y = \pm\dfrac{11}{3}$ **3.** $a = \pm 3\sqrt{2}$ **4.** $n = \pm mp^4\sqrt{6}$ **5.** $q = \pm 3$ **6.** $m = \pm 11.74$

B. **1.** $a = 2, -14$ **2.** $m = 4 \pm \sqrt{15}$ **3.** $y = 0, 14$ **4.** $k = 1 \pm 2\sqrt{3}$ **5.** $x = 11$

9.4 Solving Quadratic Equations Using the Method of Completing the Square

Section Overview

☐ **THE LOGIC BEHIND THE METHOD**
☐ **THE METHOD OF COMPLETING THE SQUARE**

☐ THE LOGIC BEHIND THE METHOD

Suppose we wish to solve the quadratic equation $x^2 - 3x - 1 = 0$. Since the equation is not of the form $x^2 = K$, we cannot use extraction of roots. Next, we try factoring, but after a few trials we see that $x^2 - 3x - 1$ is not factorable. We need another method for solving quadratic equations.

The method we shall study is based on perfect square trinomials and extraction of roots. The method is called solving quadratic equations by *completing the square*. Consider the equation $x^2 + 6x + 5 = 0$.

This quadratic equation could be solved by factoring, but we'll use the method of completing the square. We will explain the method in detail after we look at this example. First we'll rewrite the equation as

$$x^2 + 6x = -5$$

Then, we'll *add* 9 to each side. We get

$$x^2 + 6x + 9 = -5 + 9$$

The left side factors as a perfect square trinomial.

$$(x + 3)^2 = 4$$

We can solve this by extraction of roots.

$$x + 3 = +\sqrt{4}$$
$$x + 3 = \pm 2$$
$$x = \pm 2 - 3$$
$$x = +2 - 3 \quad \text{and} \quad x = -2 - 3$$
$$x = -1 \quad \text{and} \quad -5$$

Notice that when the roots are rational numbers, the equation is factorable.

The big question is, "How did we know to *add* 9 to each side of the equation?" We can convert *any* quadratic trinomial appearing in an equation into a perfect square trinomial if we know what number to add to both sides. We can determine that particular number by observing the following situation:

Consider the square of the binomial and the resulting perfect square trinomial

$$(x + p)^2 = x^2 + 2px + p^2$$

Notice that the constant term (the number we are looking for) can be obtained from the linear term $2px$. If we take one half the coefficient of x, $\dfrac{2p}{2} = p$, and square it, we get the constant term p^2. This is true for every perfect square trinomial with leading coefficient 1.

> In a perfect square trinomial with leading coefficient 1, the constant term is the square of one half the coefficient of the linear term.

Study these examples to see what constant term will make the given binomial into a perfect square trinomial.

1. $x^2 + 6x$.   The constant must be the square of one half the coefficient of x. Since the coefficient of x is 6, we have

   $$\frac{6}{2} = 3 \quad \text{and} \quad 3^2 = 9$$

   The constant is 9.

   $$x^2 + 6x + 9 = (x + 3)^2$$

   This is a perfect square trinomial.

2. $a^2 + 10a$.   The constant must be the square of one half the coefficient of a. Since the coefficient of a is 10, we have

   $$\frac{10}{2} = 5 \quad \text{and} \quad 5^2 = 25$$

   The constant is 25.

   $$a^2 + 10a + 25 = (a + 5)^2$$

3. $y^2 + 3y$.   The constant must be the square of one half the coefficient of y. Since the coefficient of y is 3, we have

   $$\frac{3}{2} \quad \text{and} \quad \left(\frac{3}{2}\right)^2 = \frac{9}{4}$$

   The constant is $\frac{9}{4}$.

   $$y^2 + 3y + \frac{9}{4} = \left(y + \frac{3}{2}\right)^2$$

❏ THE METHOD OF COMPLETING THE SQUARE

Now, with these observations, we can describe the method of completing the square.

The Method of Completing the Square

1. Write the equation so that the constant term appears on the right side of the equation.
2. If the leading coefficient is different from 1, divide each term of the equation by that coefficient.
3. Take one half of the coefficient of the linear term, square it, then *add* it to *both* sides of the equation.
4. The trinomial on the left is now a perfect square trinomial and can be factored as (   )². The first term in the parentheses is the square root of the quadratic term. The last term in the parentheses is one-half the coefficient of the linear term.
5. Solve this equation by extraction of roots.

☆ **SAMPLE SET A**

Solve the following equations.

1. $x^2 + 8x - 9 = 0.$ Add 9 to both sides.

$x^2 + 8x = 9$ One half the coefficient of x is 4, and 4^2 is 16. Add 16 to both sides.

$x^2 + 8x + 16 = 9 + 16$

$x^2 + 8x + 16 = 25$ Factor.

$(x + 4)^2 = 25$ Take square roots.

$x + 4 = \pm 5$
$x = \pm 5 - 4$ $+5 - 4 = 1, -5 - 4 = -9$
$x = 1, \quad -9$

2. $x^2 - 3x - 1 = 0.$ Add 1 to both sides.

$x^2 - 3x = 1$ One half the coefficient of x is $\dfrac{-3}{2}$. Square it: $\left(\dfrac{-3}{2}\right)^2 = \dfrac{9}{4}$. Add $\dfrac{9}{4}$ to each side.

$x^2 - 3x + \dfrac{9}{4} = 1 + \dfrac{9}{4}$

$x^2 - 3x + \dfrac{9}{4} = \dfrac{13}{4}$ Factor. Notice that since the sign of the middle term of the trinomial is "$-$", its factored form has a "$-$" sign.

$\left(x - \dfrac{3}{2}\right)^2 = \dfrac{13}{4}$ Now take square roots.

$x - \dfrac{3}{2} = \pm \sqrt{\dfrac{13}{4}}$

$x - \dfrac{3}{2} = \pm \dfrac{\sqrt{13}}{2}$

$x = \pm \dfrac{\sqrt{13}}{2} + \dfrac{3}{2}$

$x = \dfrac{\pm \sqrt{13} + 3}{2}$

$x = \dfrac{3 \pm \sqrt{13}}{2}$

3. $3a^2 - 36a - 39 = 0.$ Add 39 to both sides.

$3a^2 - 36a = 39$ The leading coefficient is 3 and we need it to be 1. Divide each term by 3.

$a^2 - 12a = 13$ One half the coefficient of a is -6. Square it: $(-6)^2 = 36$. Add 36 to each side.

$a^2 - 12a + 36 = 13 + 36$
$a^2 - 12a + 36 = 49$
$(a - 6)^2 = 49$ Factor.
$a - 6 = \pm 7$

$a = \pm 7 + 6$ $+7 + 6 = 13, -7 + 6 = -1$

$a = 13, \quad -1$

Continued

4. $2x^2 + x + 4 = 0$

$2x^2 + x = -4$

$x^2 + \dfrac{1}{2}x = -2$

$x^2 + \dfrac{1}{2}x + \left(\dfrac{1}{4}\right)^2 = -2 + \left(\dfrac{1}{4}\right)^2$

$\left(x + \dfrac{1}{4}\right)^2 = -2 + \dfrac{1}{16} = \dfrac{-32}{16} + \dfrac{1}{16} = \dfrac{-31}{16}$

Since we know that the square of any number is positive, this equation has no real number solution.

5. ▦ Calculator problem.　　Solve $7a^2 - 5a - 1 = 0$. Round each solution to the nearest tenth.

$7a^2 - 5a - 1 = 0$

$7a^2 - 5a = 1$

$a^2 - \dfrac{5}{7}a = \dfrac{1}{7}$

$a^2 - \dfrac{5}{7}a + \left(\dfrac{5}{14}\right)^2 = \dfrac{1}{7} + \left(\dfrac{5}{14}\right)^2$

$\left(a - \dfrac{5}{14}\right)^2 = \dfrac{1}{7} + \dfrac{25}{196} = \dfrac{28}{196} + \dfrac{25}{196} = \dfrac{53}{196}$

$a - \dfrac{5}{14} = \pm\sqrt{\dfrac{53}{196}} = \pm\dfrac{\sqrt{53}}{14}$

$a = \dfrac{5}{14} \pm \dfrac{\sqrt{53}}{14} = \dfrac{5 \pm \sqrt{53}}{14}$

1. We will first compute the value of the square root.

Type　　　　　　53

Press　　　　　 $\boxed{\sqrt{x}}$

Display reads:　　7.2801099

Press the key that places this value into memory.

2. For $a = \dfrac{5 + \sqrt{53}}{14}$,

Type　　　　　　5

Press　　　　　 $\boxed{+}$

Press the key that recalls the value in memory.

Press　　　　　 $\boxed{=}$

Press　　　　　 $\boxed{\div}$

Type　　　　　　14

Press　　　　　 $\boxed{=}$

Display reads:　　.87715071

Rounding to tenths, we get $a \approx 0.9$.

3. For $a = \dfrac{5 - \sqrt{53}}{14}$

Type   5

Press $\boxed{-}$

Press the key that recalls the value in memory.

Press $\boxed{=}$

Press $\boxed{\div}$

Type   14

Display reads: $-.16286499$

Rounding to tenths, we get $a \approx -0.2$. Thus, $a \approx 0.9$ and -0.2 to the nearest tenth.

★ **PRACTICE SET A**

Solve each of the following quadratic equations using the method of completing the square.

1. $x^2 - 2x - 48 = 0$ **2.** $x^2 + 3x - 5 = 0$ **3.** $4m^2 + 5m = -1$ **4.** $5y^2 - 2y - 4 = 0$

5. Calculator problem.   Solve $3x^2 - x - 1 = 0$. Round each solution to the nearest tenth.

Answers to the Practice Set are on p. 478.

Section 9.4 EXERCISES

For problems 1–29, solve the equations by completing the square.

1. $x^2 + 2x - 8 = 0$ **3.** $a^2 + 7a + 12 = 0$

2. $y^2 - 5y - 6 = 0$ **4.** $x^2 - 10x + 16 = 0$

5. $y^2 - 2y - 24 = 0$

6. $a^2 + 2a - 35 = 0$

7. $x^2 + 2x + 5 = 0$

8. $x^2 - 6x + 1 = 0$

9. $x^2 + 4x + 4 = 0$

10. $a^2 + 4a + 7 = 0$

11. $b^2 + 5b - 3 = 0$

12. $b^2 - 6b = 72$

13. $a^2 + 10a - 9 = 0$

14. $a^2 - 2a - 3 = 0$

15. $x^2 - 10x = 0$

16. $y^2 - 8y = 0$

17. $a^2 - 6a = 0$

18. $b^2 + 6b = 0$

19. $-x^2 - 14x = 13$

20. $-x^2 + 8x = -84$

21. $2a^2 + 2a - 1 = 0$

22. $4b^2 - 8b = 16$

23. $9x^2 + 12x - 5 = 0$

24. $16y^2 - 8y - 3 = 0$

25. $2x^2 + 5x - 4 = 0$

26. $3a^2 + 2a - 24 = 0$

27. $x^2 + 2x + 8 = 0$

28. $y^2 - 3y + 10 = 0$

29. $7a^2 + 3a - 1 = 0$

🖩 **Calculator Problems**

For problems 30–33, round each solution to the nearest hundredth.

30. $5m^2 - 2m - 6 = 0$

31. $3y^2 + 5y = 7$

32. $1.8x^2 + 2.3x - 4.1 = 0$

33. $0.04a^2 - 0.03a + 0.02 = 0$

EXERCISES FOR REVIEW

(5.4) **34.** Factor $12ax - 6bx + 20ay - 10by$ by grouping.

(6.1) **35.** Graph the compound inequality $-6 \leq -2x + 2 < -4$.

$$\longleftarrow\joinrel\longrightarrow$$

(6.6) **36.** Find the equation of the line that passes through the points $(1, -2)$ and $(0, 4)$.

(7.3) **37.** Find the product: $\dfrac{x^2 - 4x - 12}{x^2 - 2x - 8} \cdot \dfrac{x^2 - 3x - 4}{x^2 - 3x - 18}$.

(9.3) **38.** Use the method of extraction of roots to solve $(x - 2)^2 = 25$.

★ **Answers to Practice Set (9.4)**

A. **1.** $x = -6, 8$ **2.** $x = \dfrac{-3 \pm \sqrt{29}}{2}$ **3.** $m = \dfrac{-1}{4}, -1$ **4.** $y = \dfrac{1 \pm \sqrt{21}}{5}$ **5.** $x = 0.8, -0.4$

9.5 Solving Quadratic Equations Using the Quadratic Formula

Section Overview

- ❑ **STANDARD FORM OF A QUADRATIC EQUATION**
- ❑ **THE QUADRATIC FORMULA**
- ❑ **DERIVATION OF THE QUADRATIC FORMULA**

❑ STANDARD FORM OF A QUADRATIC EQUATION

We have observed that a quadratic equation is an equation of the form

$$ax^2 + bx + c = 0, \qquad a \neq 0$$

where

a is the coefficient of the quadratic term,
b is the coefficient of the linear term, and
c is the constant term.

Standard Form   The equation $ax^2 + bx + c = 0$ is the *standard form* of a quadratic equation.

☆ SAMPLE SET A

Determine the values of a, b, and c.

1. In the equation $3x^2 + 5x + 2 = 0$,

$a = 3$
$b = 5$
$c = 2$

2. In the equation $12x^2 - 2x - 1 = 0$,

$a = 12$
$b = -2$
$c = -1$

3. In the equation $2y^2 + 3 = 0$,

$a = 2$

$b = 0$ **Because the equation could be written**

$2y^2 + 0y + 3 = 0$

$c = 3$

4. In the equation $-8y^2 + 11y = 0$,

$a = -8$
$b = 11$

$c = 0$ **Since $-8y^2 + 11y + 0 = 0$.**

5. In the equation $z^2 = z + 8$,

$a = 1$
$b = -1$

$c = -8$ **When we write the equation in standard form, we get**

$z^2 - z - 8 = 0$

★ **PRACTICE SET A**

Determine the values of a, b, and c in the following quadratic equations.

1. $4x^2 - 3x + 5 = 0$ **2.** $3y^2 - 2y + 9 = 0$ **3.** $x^2 - 5x - 1 = 0$

4. $z^2 - 4 = 0$ **5.** $x^2 - 2x = 0$ **6.** $y^2 = 5y - 6$ **7.** $2x^2 - 4x = -1$

8. $5x - 3 = -3x^2$ **9.** $2x - 11 - 3x^2 = 0$ **10.** $y^2 = 0$

The solutions to *all* quadratic equations depend only and completely on the values a, b, and c.

❑ **THE QUADRATIC FORMULA**

When a quadratic equation is written in standard form so that the values a, b, and c are readily determined, the equation can be solved using the *quadratic formula*. The values that satisfy the equation are found by substituting the values a, b, and c into the formula

Quadratic Formula

$$x = \frac{-b \pm \sqrt{b^2 - 4ac}}{2a}$$

Keep in mind that the plus or minus symbol, $\pm$, is just a shorthand way of denoting the two possibilities:

$$x = \frac{-b + \sqrt{b^2 - 4ac}}{2a} \quad \text{and} \quad x = \frac{-b - \sqrt{b^2 - 4ac}}{2a}$$

The quadratic formula can be derived by using the method of completing the square.

☐ DERIVATION OF THE QUADRATIC FORMULA

Solve $ax^2 + bx + c = 0$ for x by completing the square.

1. Subtract c from both sides.

   $$ax^2 + bx = -c$$

2. Divide both sides by a, the coefficient of x^2.

   $$x^2 + \frac{b}{a}x = \frac{-c}{a}$$

3. Now we have the proper form to complete the square. Take one half the coefficient of x, square it, and add the result to both sides of the equation found in step 2.

   (a) $\dfrac{1}{2} \cdot \dfrac{b}{a} = \dfrac{b}{2a}$ is one half the coefficient of x.

   (b) $\left(\dfrac{b}{2a}\right)^2$ is the square of one half the coefficient of x.

   $$x^2 + \frac{b}{a}x + \left(\frac{b}{2a}\right)^2 = \frac{-c}{a} + \left(\frac{b}{2a}\right)^2$$

4. The left side of the equation is now a perfect square trinomial and can be factored. This gives us

   $$\left(x + \frac{b}{2a}\right)^2 = \frac{-c}{a} + \frac{b^2}{4a^2}$$

5. Add the two fractions on the right side of the equation. The LCD $= 4a^2$.

   $$\left(x + \frac{b}{2a}\right)^2 = \frac{-4ac}{4a^2} + \frac{b^2}{4a^2}$$

   $$\left(x + \frac{b}{2a}\right)^2 = \frac{-4ac + b^2}{4a^2}$$

   $$\left(x + \frac{b}{2a}\right)^2 = \frac{b^2 - 4ac}{4a^2}$$

6. Solve for x using the method of extraction of roots.

   $$x + \frac{b}{2a} = \pm\sqrt{\frac{b^2 - 4ac}{4a^2}}$$

   $$x + \frac{b}{2a} = \pm\frac{\sqrt{b^2 - 4ac}}{\sqrt{4a^2}} \qquad \sqrt{4a^2} = |2a| = 2|a| = \pm 2a$$

   $$x + \frac{b}{2a} = \pm\frac{\sqrt{b^2 - 4ac}}{2a}$$

   $$x = -\frac{b}{2a} \pm \frac{\sqrt{b^2 - 4ac}}{2a} \qquad \textbf{Add these two fractions.}$$

   $$x = \frac{-b \pm \sqrt{b^2 - 4ac}}{2a}$$

⭐ **SAMPLE SET B**

Solve each of the following quadratic equations using the quadratic formula.

1. $3x^2 + 5x + 2 = 0$.

   1. Identify a, b, and c.

   $a = 3$, $b = 5$,   and $c = 2$

   2. Write the quadratic formula.

$$x = \frac{-b \pm \sqrt{b^2 - 4ac}}{2a}$$

   3. Substitute.

$$x = \frac{-5 \pm \sqrt{(5)^2 - 4(3)(2)}}{2(3)}$$

$$= \frac{-5 \pm \sqrt{25 - 24}}{6}$$

$$= \frac{-5 \pm \sqrt{1}}{6}$$

$$= \frac{-5 \pm 1}{6} \qquad \mathbf{-5 + 1 = -4 \text{ and } -5 - 1 = -6}$$

$$= \frac{-4}{6}, \ \frac{-6}{6}$$

$$x = \frac{-2}{3}, \ -1$$

Note: Since these roots are rational numbers, this equation could have been solved by factoring.

2. $12x^2 - 2x - 1 = 0$.

   1. Identify a, b, and c.

   $a = 12$, $b = -2$,   and $c = -1$

   2. Write the quadratic formula.

$$x = \frac{-b \pm \sqrt{b^2 - 4ac}}{2a}$$

   3. Substitute.

$$x = \frac{-(-2) \pm \sqrt{(-2)^2 - 4(12)(-1)}}{2(12)}$$

$$= \frac{2 \pm \sqrt{4 + 48}}{24} \qquad \mathbf{Simplify.}$$

$$= \frac{2 \pm \sqrt{52}}{24} \qquad \mathbf{Simplify.}$$

$$= \frac{2 \pm \sqrt{4 \cdot 13}}{24} \qquad \mathbf{Simplify.}$$

Continued

$$= \frac{2 \pm 2\sqrt{13}}{24} \qquad \text{Reduce. Factor 2 from the terms of the numerator.}$$

$$= \frac{2(1 \pm \sqrt{13})}{24}$$

$$x = \frac{1 \pm \sqrt{13}}{12}$$

3. $2y^2 + 3 = 0$

  1. Identify a, b, and c.

    $a = 2,\quad b = 0,\quad \text{and}\quad c = 3$

  2. Write the quadratic formula.

$$x = \frac{-b \pm \sqrt{b^2 - 4ac}}{2a}$$

  3. Substitute.

$$x = \frac{-0 \pm \sqrt{0^2 - 4(2)(3)}}{2(2)}$$

$$x = \frac{0 \pm \sqrt{-24}}{4}$$

This equation has no real number solution since we have obtained a negative number under the radical sign.

4. $-8x^2 + 11x = 0$

  1. Identify a, b, and c.

    $a = -8,\quad b = 11,\quad \text{and}\quad c = 0$

  2. Write the quadratic formula.

$$x = \frac{-b \pm \sqrt{b^2 - 4ac}}{2a}$$

  3. Substitute.

$$x = \frac{-11 \pm \sqrt{11^2 - 4(-8)(0)}}{2(-8)}$$

$$= \frac{-11 \pm \sqrt{121 - 0}}{-16} \qquad \text{Simplify.}$$

$$= \frac{-11 \pm \sqrt{121}}{-16} \qquad \text{Simplify.}$$

$$= \frac{-11 \pm 11}{-16}$$

$$x = 0, \quad \frac{11}{8}$$

5. $(3x + 1)(x - 4) = x^2 + x - 2$

1. Write the equation in standard form.

$3x^2 - 11x - 4 = x^2 + x - 2$
$2x^2 - 12x - 2 = 0$
$x^2 - 6x - 1 = 0$

2. Identify a, b, and c.

$a = 1$, $b = -6$, and $c = -1$

3. Write the quadratic formula.

$x = \dfrac{-b \pm \sqrt{b^2 - 4ac}}{2a}$

4. Substitute.

$x = \dfrac{-(-6) \pm \sqrt{(-6)^2 - 4(1)(-1)}}{2(1)}$

$= \dfrac{6 \pm \sqrt{36 + 4}}{2}$

$= \dfrac{6 \pm \sqrt{40}}{2}$

$= \dfrac{6 \pm \sqrt{4 \cdot 10}}{2}$

$= \dfrac{6 \pm 2\sqrt{10}}{2}$

$= \dfrac{2(3 \pm \sqrt{10})}{2}$

$x = 3 \pm \sqrt{10}$

★ **PRACTICE SET B**

Solve each of the following quadratic equations using the quadratic formula.

1. $2x^2 + 3x - 7 = 0$ **2.** $5a^2 - 2a - 1 = 0$ **3.** $6y^2 + 5 = 0$ **4.** $-3m^2 + 2m = 0$

Answers to Practice Sets are on p. 486.

Section 9.5 EXERCISES

For problems 1–40, solve the equations using the quadratic formula.

1. $x^2 - 2x - 3 = 0$

2. $x^2 + 5x + 6 = 0$

3. $y^2 - 5y + 4 = 0$

4. $a^2 + 4a - 21 = 0$

5. $a^2 + 12a + 20 = 0$

6. $b^2 - 4b + 4 = 0$

7. $b^2 + 4b + 4 = 0$

8. $x^2 + 10x + 25 = 0$

9. $2x^2 - 5x - 3 = 0$

10. $6y^2 + y - 2 = 0$

11. $4x^2 - 2x - 1 = 0$

12. $3y^2 + 2y - 1 = 0$

13. $5a^2 - 2a - 3 = 0$

14. $x^2 - 3x + 1 = 0$

15. $x^2 - 5x - 4 = 0$

16. $(x + 2)(x - 1) = 1$

17. $(a + 4)(a - 5) = 2$

18. $(x - 3)(x + 3) = 7$

19. $(b - 4)(b + 4) = 9$

20. $x^2 + 8x = 2$

21. $y^2 = -5y + 4$

22. $x^2 = -3x + 7$

23. $x^2 = -2x - 1$

24. $x^2 + x + 1 = 0$

31. $4a^2 + 7a - 2 = -2a + a$

25. $a^2 + 3a - 4 = 0$

26. $y^2 + y = -4$

32. $(2x + 5)(x - 4) = x^2 - x + 2$

27. $b^2 + 3b = -2$

33. $(x - 4)^2 = 3$

28. $x^2 + 6x + 8 = -x - 2$

34. $(x + 2)^2 = 4$

29. $x^2 + 4x = 2x - 5$

35. $(b - 6)^2 = 8$

30. $6b^2 + 5b - 4 = b^2 + b + 1$

36. $(3 - x)^2 = 6$

37. $3(x^2 + 1) = 2(x + 7)$

39. $-4(a^2 + 2) + 3 = 5$

38. $2(y^2 - 3) = -3(y - 1)$

40. $-(x^2 + 3x - 1) = 2$

EXERCISES FOR REVIEW

(1.6) **41.** Simplify $\left(\dfrac{x^8 y^7 z^5}{x^4 y^6 z^2}\right)^2$.

(2.6) **42.** Write $4a^{-6}b^2 c^3 a^5 b^{-3}$ so that only positive exponents appear.

(3.5) **43.** Find the product: $(2y + 7)(3y - 1)$.

(8.5) **44.** Simplify $\sqrt{80} - \sqrt{45}$.

(9.4) **45.** Solve $x^2 - 4x - 12 = 0$ by completing the square.

★ Answers to Practice Sets (9.5)

A. **1.** $a = 4$ **2.** $a = 3$ **3.** $a = 1$ **4.** $a = 1$ **5.** $a = 1$
$\phantom{\textbf{1.}}\ b = -3$ $\phantom{\textbf{2.}}\ b = -2$ $\phantom{\textbf{3.}}\ b = -5$ $\phantom{\textbf{4.}}\ b = 0$ $\phantom{\textbf{5.}}\ b = -2$
$\phantom{\textbf{1.}}\ c = 5$ $\phantom{\textbf{2.}}\ c = 9$ $\phantom{\textbf{3.}}\ c = -1$ $\phantom{\textbf{4.}}\ c = -4$ $\phantom{\textbf{5.}}\ c = 0$

    6. $a = 1$ **7.** $a = 2$ **8.** $a = 3$ **9.** $a = -3$ **10.** $a = 1$
$\phantom{\textbf{6.}}\ b = -5$ $\phantom{\textbf{7.}}\ b = -4$ $\phantom{\textbf{8.}}\ b = 5$ $\phantom{\textbf{9.}}\ b = 2$ $\phantom{\textbf{10.}}\ b = 0$
$\phantom{\textbf{6.}}\ c = 6$ $\phantom{\textbf{7.}}\ c = 1$ $\phantom{\textbf{8.}}\ c = -3$ $\phantom{\textbf{9.}}\ c = -11$ $\phantom{\textbf{10.}}\ c = 0$

B. **1.** $x = \dfrac{-3 \pm \sqrt{65}}{4}$ **2.** $a = \dfrac{1 \pm \sqrt{6}}{5}$ **3.** no real number solution   **4.** $m = 0, \dfrac{2}{3}$

9.6 Applications

Section Overview	☐ **THE FIVE-STEP METHOD**
	☐ **EXAMPLES**

☐ THE FIVE-STEP METHOD

We are now in a position to study some applications of quadratic equations. Quadratic equations can arise from a variety of physical (applied) and mathematical (logical) problems.

We will, again, apply the *five-step method* for solving word problems.

Five-Step Method of Solving Word Problems

Step 1: Let x (or some other letter) represent the unknown quantity.
Step 2: Translate the verbal expression to mathematical symbols and form an equation.
Step 3: Solve this equation.
Step 4: Check the solution by substituting the result into the equation found in step 2.
Step 5: Write a conclusion.

Remember, step 1 is very important.

ALWAYS START BY INTRODUCING A VARIABLE.

Once the quadratic equation is developed (step 2), try to solve it by factoring. If factoring doesn't work, use the quadratic formula. A calculator may help to make some of the calculations a little less tedious.

☐ EXAMPLES

☆ SAMPLE SET A

A producer of personal computer mouse covers determines that the number N of covers sold is related to the price x of a cover by $N = 35x - x^2$. At what price should the producer price a mouse cover in order to sell 216 of them?

Step 1: Let x = the price of a mouse cover.
Step 2: Since N is to be 216, the equation is

$$216 = 35x - x^2$$

Step 3: $216 = 35x - x^2$ **Rewrite in standard form.**

$x^2 - 35x + 216 = 0$ **Try factoring.**

$(x - 8)(x - 27) = 0$
$x - 8 = 0$ or $x - 27 = 0$
$x = 8$ or $x = 27$

Check these potential solutions.

Step 4:   If $x = 8$,   If $x = 27$,

$35 \cdot 8 - 8^2 \overset{?}{=} 216$ $35 \cdot 27 - 27^2 \overset{?}{=} 216$
$280 - 64 \overset{?}{=} 216$ $945 - 729 \overset{?}{=} 216$
$216 \overset{✓}{=} 216$ $216 \overset{✓}{=} 216$

These solutions check.
Step 5:   The computer mouse covers can be priced at either \$8 or \$27 in order to sell 216 of them.

★ PRACTICE SET A

1. A manufacturer of cloth personal computer dust covers notices that the number N of covers sold is related to the price of covers by $N = 30x - x^2$. At what price should the manufacturer price the covers in order to sell 216 of them?

Step 1:

Step 2:

Step 3:

Step 4:

Step 5:   In order to sell 216 covers, the manufacturer should price them at either _____ or _____ .

2. It is estimated that t years from now the population of a particular city will be

$P = t^2 - 24t + 96{,}000.$

How many years from now will the population be 95,865?

Step 1:

Step 2:

Step 3:

Step 4:

Step 5:

☆ SAMPLE SET B

The length of a rectangle is 4 inches more than twice its width. The area is 30 square inches. Find the dimensions (length and width).

Step 1:   Let x = the width. Then, $2x + 4$ = the length.

Step 2:   The area of a rectangle is defined to be the length of the rectangle times the width of the rectangle. Thus,

$$x(2x + 4) = 30$$

Step 3: $x(2x + 4) = 30$

$2x^2 + 4x = 30$

$2x^2 + 4x - 30 = 0$ **Divide each side by 2.**

$x^2 + 2x - 15 = 0$ **Factor.**

$(x + 5)(x - 3) = 0$
$x = -5, \quad 3$

$x = -5$ has no physical meaning so we disregard it. Check $x = 3$.
$x = 3$
$2x + 4 = 2 \cdot 3 + 4 = 10$

Step 4: $x(2x + 4) = 30$
$3(2 \cdot 3 + 4) \overset{?}{=} 30$
$3(6 + 4) \overset{?}{=} 30$
$3(10) \overset{?}{=} 30$
$30 \overset{\checkmark}{=} 30$

Step 5:   Width = 3 inches and length = 10 inches.

★ PRACTICE SET B

1. The length of a rectangle is 3 feet more than twice its width. The area is 14 square feet. Find the dimensions.

2. The area of a triangle is 24 square meters. The base is 2 meters longer than the height. Find the base and height.

The formula for the area of a triangle is $A = \dfrac{1}{2} b \cdot h$.

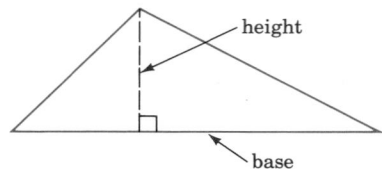

☆ **SAMPLE SET C**

The product of two consecutive integers is 156. Find them.

Step 1:   Let x = the smaller integer.
                $x + 1$ = the next integer.

Step 2: $x(x + 1) = 156$

Step 3: $x(x + 1) = 156$

$x^2 + x = 156$
$x^2 + x - 156 = 0$
$(x - 12)(x + 13) = 0$
$x = 12, \quad -13$

This factorization may be hard to guess. We could also use the quadratic formula.

$x^2 + x - 156 = 0$
$a = 1, \qquad b = 1, \qquad c = -156$

$$x = \frac{-1 \pm \sqrt{1^2 - 4(1)(-156)}}{2(1)}$$

$$= \frac{-1 \pm \sqrt{1 + 624}}{2}$$

$$= \frac{-1 \pm 25}{2} \qquad \frac{-1 + 25}{2} = \frac{24}{2} = 12 \text{ and } \frac{-1 - 25}{2} = \frac{-26}{2} = -13$$

$x = 12, \quad -13 \qquad$ Check 12, 13 and $-13, -12$.

$x + 1 = 13, -12$

Step 4:   If $x = 12$: $12(12 + 1) \overset{?}{=} 156$
                             $12(13) \overset{?}{=} 156$
                                $156 \overset{\checkmark}{=} 156$

        If $x = -13$ $-13(-13 + 1) \overset{?}{=} 156$
                            $-13(-12) \overset{?}{=} 156$
                                $156 \overset{\checkmark}{=} 156$

Step 5:   There are *two* solutions: 12, 13 and $-13, -12$.

★ **PRACTICE SET C**

1. The product of two consecutive integers is 210. Find them.

2. Four is added to an integer and that sum is tripled. When this result is multiplied by the original integer, the product is −12. Find the integer.

☆ **SAMPLE SET D**

A box with no top and a square base is to be made by cutting out 2-inch squares from each corner and folding up the sides of a piece of a square cardboard. The volume of the box is to be 8 cubic inches. What size should the piece of cardboard be?

Step 1:   Let $x = $ the length (and width) of the piece of cardboard.

Step 2:   The volume of a rectangular box is

$V = $ (length) (width) (height)
$8 = (x-4)(x-4)2$

Step 3: $8 = (x-4)(x-4)2$

$8 = (x^2 - 8x + 16)2$
$8 = 2x^2 - 16x + 32$

$2x^2 - 16x + 24 = 0$ **Divide each side by 2.**

$x^2 - 8x + 12 = 0$ **Factor.**

$(x-6)(x-2) = 0$
$x = 6, \quad 2$

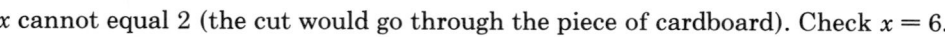

x cannot equal 2 (the cut would go through the piece of cardboard). Check $x = 6$.

Step 4: $(6-4)(6-4)2 \overset{?}{=} 8$
$\qquad (2)(2)2 \overset{?}{=} 8$
$\qquad\qquad 8 \overset{\checkmark}{=} 8$

Step 5:   The piece of cardboard should be 6 inches by 6 inches.

★ PRACTICE SET D

A box with no top and a square base is to be made by cutting 3-inch squares from each corner and folding up the sides of a piece of cardboard. The volume of the box is to be 48 cubic inches. What size should the piece of cardboard be?

☆ SAMPLE SET E

A study of the air quality in a particular city by an environmental group suggests that t years from now the level of carbon monoxide, in parts per million, in the air will be

$A = 0.3t^2 + 0.1t + 4.2$

(a) What is the level, in parts per million, of carbon monoxide in the air now?

Since the equation $A = 0.3t^2 + 0.1t + 4.2$ specifies the level t years from now, we have $t = 0$.

$A = 0.3(0)^2 + 0.1(0) + 4.2$
$A = 4.2$

(b) How many years from now will the level of carbon monoxide be at 8 parts per million?

Step 1: $t =$ the number of years when the level is 8.
Step 2: $8 = 0.3t^2 + 0.1t + 4.2$
Step 3: $8 = 0.3t^2 + 0.1t + 4.2$

$0 = 0.3t^2 + 0.1t - 3.8$ This does not readily factor, so we'll use the quadratic formula.

$a = 0.3, \quad b = 0.1, \quad c = -3.8$

$t = \dfrac{-0.1 \pm \sqrt{(0.1)^2 - 4(0.3)(-3.8)}}{2(0.3)}$

$= \dfrac{-0.1 \pm \sqrt{0.01 + 4.56}}{0.6} = \dfrac{-0.1 \pm \sqrt{4.57}}{0.6}$

$= \dfrac{-0.1 \pm 2.14}{0.6}$

$t = 3.4 \quad$ and $\quad -3.73$

$t = -3.73$ has no physical meaning. Check $t = 3.4$

Step 4: This value of t has been rounded to the nearest tenth. It does check (pretty closely).
Step 5: About 3.4 years from now the carbon monoxide level will be 8.

★ **PRACTICE SET E**

A study of the air quality in a particular city by an environmental group suggests that t years from now the level of carbon monoxide, in parts per million, in the air will be

$A = 0.2t^2 + 0.1t + 5.1$

(a) What is the level, in parts per million, now?

(b) How many years from now will the level of carbon monoxide be at 8 parts per million? Round to the nearest tenth.

⭐ **SAMPLE SET F**

A contractor is to pour a concrete walkway around a swimming pool that is 20 feet wide and 40 feet long. The area of the walkway is to be 544 square feet. If the walkway is to be of uniform width, how wide should the contractor make it?

Step 1:   Let $x =$ the width of the walkway.
Step 2:   A diagram will help us to get the equation.

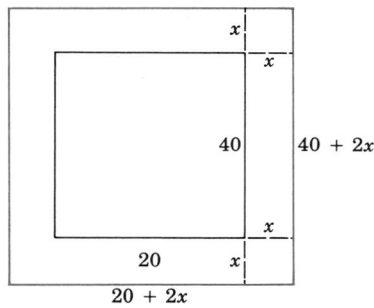

(Area of pool and walkway) − (area of pool) = (area of walkway)
$(20 + 2x)(40 + 2x) - 20 \cdot 40 = 544$

Step 3: $(20 + 2x)(40 + 2x) - 20 \cdot 40 = 544$

$800 + 120x + 4x^2 - 800 = 544$
$120x + 4x^2 = 544$

$4x^2 + 120x - 544 = 0$ **Divide each term by 4.**

$x^2 + 30x - 136 = 0$ **Solve by factoring.**

$(x - 4)(x + 34) = 0$ **(This is difficult to factor so we may wish to use the quadratic formula.)**

$x - 4 = 0$ or $x + 34 = 0$
$x = 4$ or $x = -34$ has no physical meaning.

Check a width of 4 feet as a solution.

Continued

Step 4:   Area of pool and walkway $= (20 + 2 \cdot 4)(40 + 2 \cdot 4)$
$= (28)(48)$
$= 1344$

Area of pool $= (20)(40) = 800$
Area of walkway $= 1344 - 800 \underset{\checkmark}{=} 544$
This solution checks.

Step 5:   The contractor should make the walkway 4 feet wide.

★ PRACTICE SET F

A contractor is to pour a concrete walkway around a swimming pool that is 15 feet wide and 25 feet long. The area of the walkway is to be 276 square feet. If the walkway is to be of uniform width, how wide should the contractor make it?

Answers to Practice Sets are on p. 498.

Section 9.6 EXERCISES

Some of the following problems have actual applications and some are intended only as logic developers. A calculator may be helpful. The problems appear in groups and correspond to the noted Sample Set problem.

Sample Set A – Type Problems

1. The manufacturer of electronic fuel injectors determines that the number N of injectors sold is related to the price x per injector by $N = 22x - x^2$. At what price should the manufacturer price the injectors so that 112 of them are sold?

2. The owner of a stained-glass shop determines that the number N of pieces of a particular type of glass sold in a month is related to the price x per piece by $N = 21x - x^2$. At what price should the shop buyer price the glass so that 162 pieces sell?

3. It is estimated that t years from now the population of a certain city will be

   $$P = t^2 - 15t + 12{,}036$$

   (a) What is the population now?

   (b) How many years from now will the population be 12,000?

4. It is estimated that t years from now the population of a certain city will be

$P = t^2 - 16t + 24{,}060$

(a) What is the population now?

(b) How many years from now will the population be 24,000?

5. If an object is thrown vertically upward, its height h, above the ground, in feet, after t seconds is given by $h = h_0 + v_0 t - 16t^2$, where h_0 is the initial height from which the object is thrown and v_0 is the initial velocity of the object. Using this formula and an approach like that of Sample Set A, solve this problem.

A ball thrown vertically into the air has the equation of motion $h = 48 + 32t - 16t^2$.
(a) How high is the ball at $t = 0$ (the initial height of the ball)?

(b) How high is the ball at $t = 1$ (after 1 second in the air)?

(c) When does the ball hit the ground? (*Hint:* Determine the appropriate value for h then solve for t.)

6. A woman's glasses accidently fall off her face while she is looking out of a window in a tall building. The equation relating h, the height above the ground in feet, and t, the time in seconds her glasses have been falling, is $h = 64 - 16t^2$.

(a) How high was the woman's face when her glasses fell off?

(b) How many seconds after the glasses fell did they hit the ground?

Sample Set B – Type Problems

7. The length of a rectangle is 6 feet more than twice its width. The area is 8 square feet. Find the dimensions.

8. The length of a rectangle is 18 inches more than three times its width. The area is 81 square inches. Find the dimensions.

9. The length of a rectangle is two thirds its width. The area is 14 square meters. Find the dimensions.

10. The length of a rectangle is four ninths its width. The area is 144 square feet. Find the dimensions.

11. The area of a triangle is 14 square inches. The base is 3 inches longer than the height. Find both the length of the base and height.

12. The area of a triangle is 34 square centimeters. The base is 1 cm longer than twice the height. Find both the length of the base and the height.

Sample Set C – Type Problems

13. The product of two consecutive integers is 72. Find them.

14. The product of two consecutive negative integers is 42. Find them.

15. The product of two consecutive odd integers is 143. Find them. (*Hint:* The quadratic equation is factorable, but the quadratic formula may be quicker.)

16. The product of two consecutive even integers is 168. Find them.

17. Three is added to an integer and that sum is doubled. When this result is multiplied by the original integer the product is 20. Find the integer.

18. Four is added to three times an integer. When this sum and the original integer are multiplied, the product is −1. Find the integer.

Sample Set D – Type Problems

19. A box with no top and a square base is to be made by cutting out 2-inch squares from each corner and folding up the sides of a piece of cardboard.

The volume of the box is to be 25 cubic inches. What size should the piece of cardboard be?

(b) How many years from now will the level of carbon monoxide be at 3.1 parts per million?

20. A box with no top and a square base is to made by cutting out 8-inch squares from each corner and folding up the sides of a piece of cardboard. The volume of the box is to be 124 cubic inches. What size should the piece of cardboard be?

Sample Set F – Type Problems

23. A contractor is to pour a concrete walkway around a wading pool that is 4 feet wide and 8 feet long. The area of the walkway and pool is to be 96 square feet. If the walkway is to be of uniform width, how wide should it be?

Sample Set E – Type Problems

21. A study of the air quality in a particular city by an environmental group suggests that t years from now the level of carbon monoxide, in parts per million, will be $A = 0.1t^2 + 0.1t + 2.2$.
(a) What is the level, in parts per million, of carbon monoxide in the air now?

(b) How many years from now will the level of carbon monoxide be at 3 parts per million?

Astrophysical Problem

24. A very interesting application of quadratic equations is determining the length of a solar eclipse (the moon passing between the earth and sun). The length of a solar eclipse is found by solving the quadratic equation

$$(a + bt)^2 + (c + dt)^2 = (e + ft)^2$$

for t. The letters a, b, c, d, e, and f are constants that pertain to a particular eclipse. The equation is a quadratic equation in t and can be solved by the quadratic formula (and definitely a calculator). Two values of t will result. The length of the eclipse is just the difference of these t-values.

The following constants are from a solar eclipse that occurred on August 3, 431 B.C.

$$a = -619 \qquad b = 1438$$
$$c = 912 \qquad d = -833$$
$$e = 1890.5 \qquad f = -2$$

Determine the length of this particular solar eclipse.

22. A similar study to that of problem 21 suggests $A = 0.3t^2 + 0.25t + 3.0$.
(a) What is the level, in parts per million, of carbon monoxide in the air now?

EXERCISES FOR REVIEW

(7.5) **25.** Find the sum: $\dfrac{2x + 10}{x^2 + x - 2} + \dfrac{x + 3}{x^2 - 3x + 2}$.

(7.6) **26.** Solve the fractional equation $\dfrac{4}{x + 12} + \dfrac{3}{x + 3} = \dfrac{4}{x^2 + 5x + 6}$.

(*Hint:* Check for extraneous solutions.)

(7.7) **27.** One pipe can fill a tank in 120 seconds and another pipe can fill the same tank in 90 seconds. How long will it take both pipes working together to fill the tank?

(9.5) **28.** Use the quadratic formula to solve $10x^2 - 3x - 1 = 0$.

(9.5) **29.** Use the quadratic formula to solve $4x^2 - 3x = 0$.

★ Answers to Practice Sets (9.6)

A. **1.** 12 or 18   **2.** In 9 and 15 years, the population of the city will be 95,865.

B. **1.** width $=$ 2 feet, length $=$ 7 feet   **2.** height $=$ 6 meters, base $=$ 8 meters

C. **1.** 14 and 15, and -14 and -15 **2.** -2

D. **1.** 10 in. by 10 in.; 2 by 2 is not physically possible.

E. **(a).** 5.1 parts per million   **(b).** 3.6 years

F. 3 ft wide

9.7 Graphing Quadratic Equations

Section Overview

☐ **PARABOLAS**
☐ **CONSTRUCTING GRAPHS OF PARABOLAS**

☐ PARABOLAS

We will now study the graphs of quadratic equations in two variables with general form

$$y = ax^2 + bx + c, \qquad a \neq 0, \quad a, b, c \text{ are real numbers}$$

Parabola

All such graphs have a similar shape. The graph of a quadratic equation of this type is called a *parabola* and it will assume one of the following shapes.

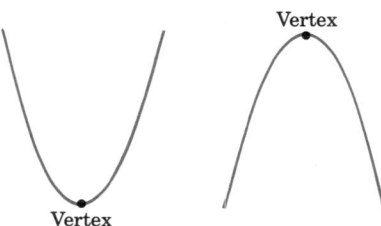

Vertex

The high point or low point of a parabola is called the *vertex* of the parabola.

☐ CONSTRUCTING GRAPHS OF PARABOLAS

We will construct the graph of a parabola by choosing several x-values, computing to find the corresponding y-values, plotting these ordered pairs, then drawing a smooth curve through them.

☆ **SAMPLE SET A**

1. Graph $y = x^2$.   Construct a table to exhibit several ordered pairs.

x	$y = x^2$
0	0
1	1
2	4
3	9
-1	1
-2	4
-3	9

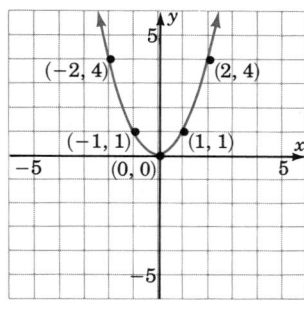

This is the most basic parabola. Although other parabolas may be wider, narrower, moved up or down, moved to the left or right, or inverted, they will all have this same basic shape. We will need to plot as many ordered pairs as necessary to ensure this basic shape.

2. Graph $y = x^2 - 2$.   Construct a table of ordered pairs.

x	$y = x^2 - 2$
0	-2
1	-1
2	2
3	7
-1	-1
-2	2
-3	7

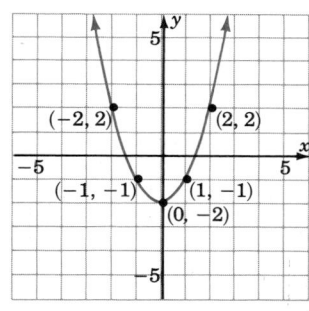

Notice that the graph of $y = x^2 - 2$ is precisely the graph of $y = x^2$ but translated 2 units down. Compare the equations $y = x^2$ and $y = x^2 - 2$. Do you see what causes the 2 unit downward translation?

★ **PRACTICE SET A**

Use the idea suggested in Sample Set A to sketch (quickly and perhaps not perfectly accurately) the graphs of

$y = x^2 + 1$ and $y = x^2 - 3$

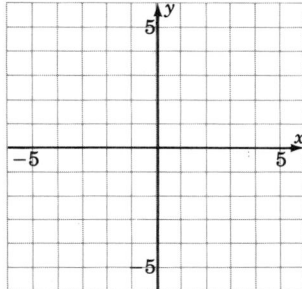
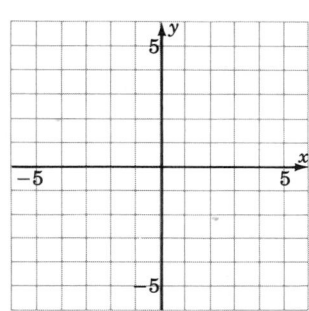

Graph $y = (x + 2)^2$.

Do we expect the graph to be similar to the graph of $y = x^2$? Make a table of ordered pairs.

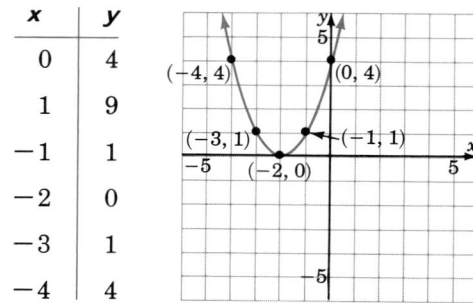

x	y
0	4
1	9
−1	1
−2	0
−3	1
−4	4

Notice that the graph of $y = (x + 2)^2$ is precisely the graph of $y = x^2$ but translated 2 units to the left. The $+2$ inside the parentheses moves $y = x^2$ two units to the left. A negative value inside the parentheses makes a move to the right.

★ **PRACTICE SET B**

1. Use the idea suggested in Sample Set B to sketch the graphs of

$$y = (x - 3)^2 \qquad \text{and} \qquad y = (x + 1)^2$$

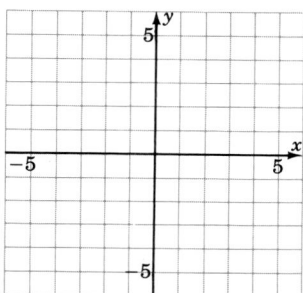

2. Graph $y = (x - 2)^2 + 1$

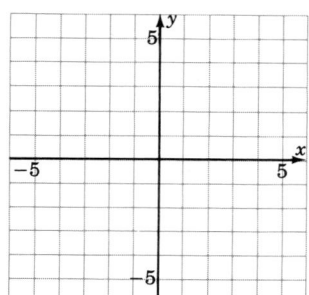

Answers to Practice Sets are on p. 503.

Section 9.7 EXERCISES

For problems 1–21, graph the quadratic equations.

1. $y = x^2$

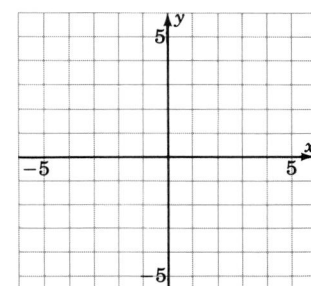

5. $y = (x + 3)^2$

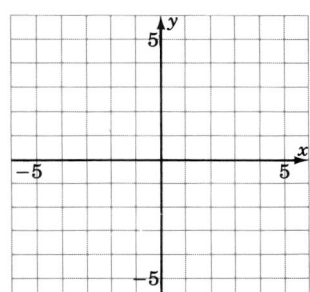

9. $y = x^2 + 2$

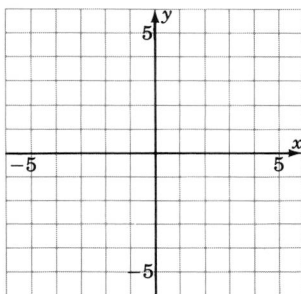

2. $y = -x^2$

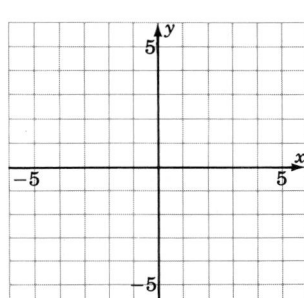

6. $y = (x + 1)^2$

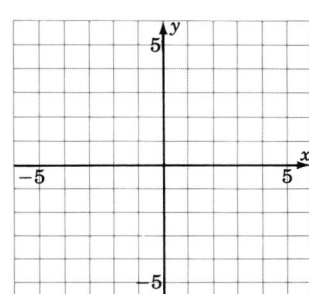

10. $y = x^2 + \dfrac{1}{2}$

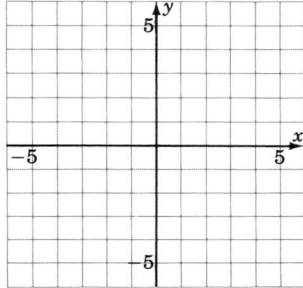

3. $y = (x - 1)^2$

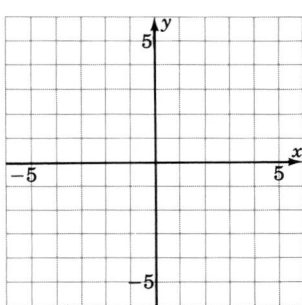

7. $y = x^2 - 3$

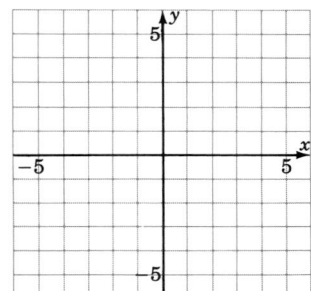

11. $y = x^2 - \dfrac{1}{2}$

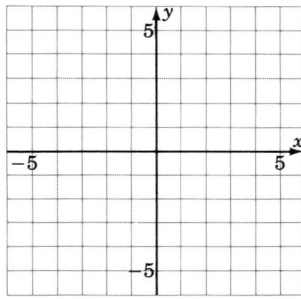

4. $y = (x - 2)^2$

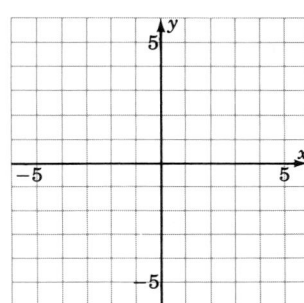

8. $y = x^2 - 1$

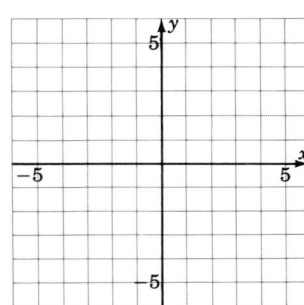

12. $y = -x^2 + 1$ (Compare with problem 2.)

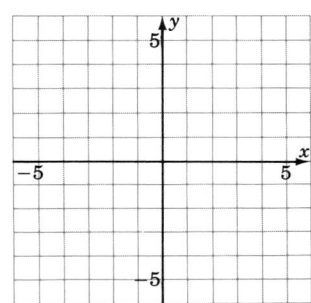

13. $y = -x^2 - 1$ (Compare with problem 1.)

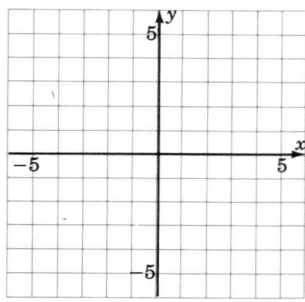

17. $y = -(x + 3)^2$

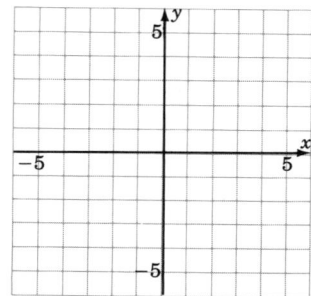

21. $y = \frac{1}{3}x^2$

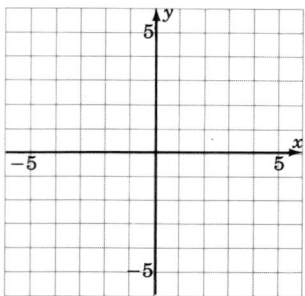

14. $y = (x - 1)^2 - 1$

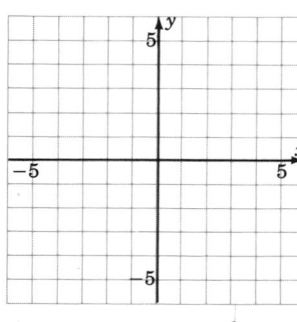

18. $y = 2x^2$

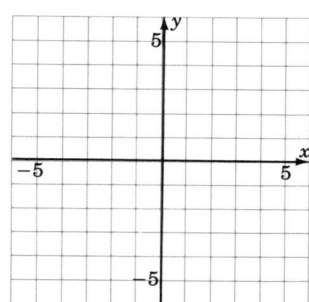

For problems 22–25, try to guess the quadratic equation that corresponds to the given graph.

22.

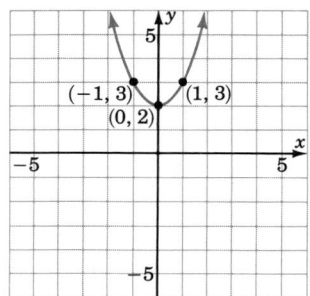

15. $y = (x + 3)^2 + 2$

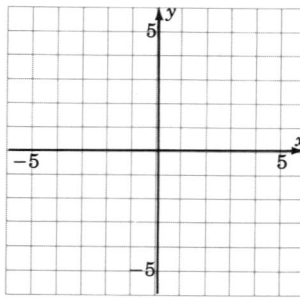

19. $y = 3x^2$

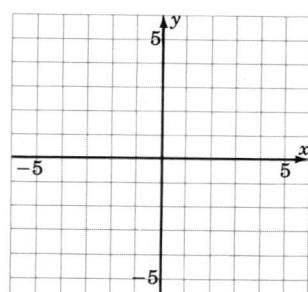

23.

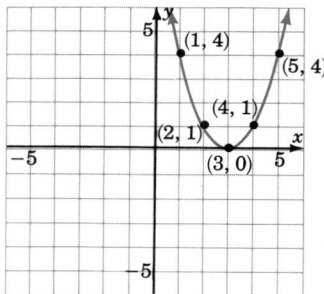

16. $y = -(x + 1)^2$

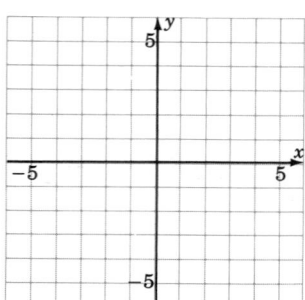

20. $y = \frac{1}{2}x^2$

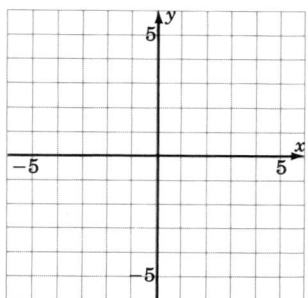

24.

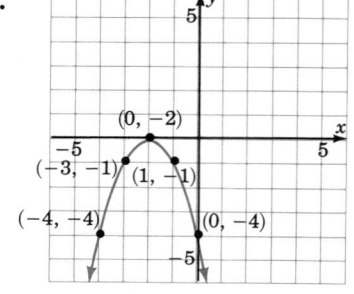

25.

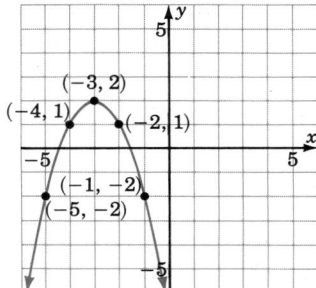

EXERCISES FOR REVIEW

(2.5)   **26.** Simplify and write $(x^{-4}y^5)^{-3}(x^{-6}y^4)^2$ so that only positive exponents appear.

(5.6)   **27.** Factor $y^2 - y - 42$.

(7.5)   **28.** Find the sum: $\dfrac{2}{a-3} + \dfrac{3}{a+3} + \dfrac{18}{a^2-9}$.

(8.4)   **29.** Simplify $\dfrac{2}{4+\sqrt{5}}$.

(9.6)   **30.** Four is added to an integer and that sum is doubled. When this result is multiplied by the original integer, the product is -6. Find the integer.

★ Answers to Practice Sets (9.7)

A.

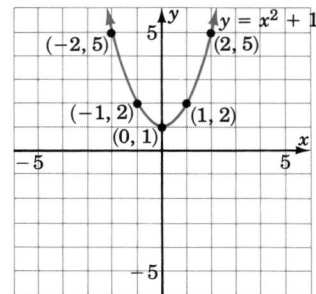

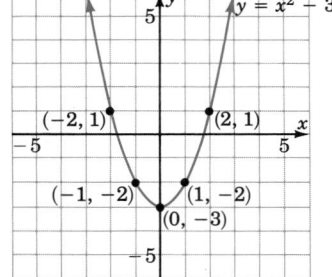

B. 1.

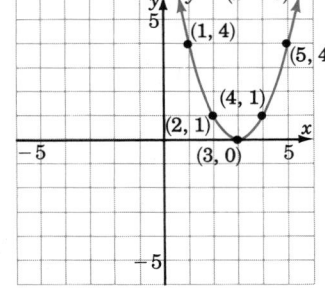

2.

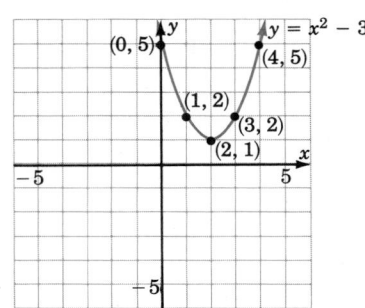

Quadratic Equation (9.1)

A *quadratic* equation is an equation of the form $ax^2 + bx + c = 0$, where $a \neq 0$. This form is the *standard form* of a quadratic equation.

a is the coefficient of x^2.
b is the coefficient of x.
c is the constant term.

Zero-Factor Property (9.1)

If two numbers a and b are multiplied together and the resulting product is 0, then at least one of the numbers must be 0.

Solving Quadratic Equations by Factoring (9.2)

1. Set the equation equal to 0.
2. Factor the quadratic expression.
3. By the zero-factor property, at least one of the factors must be zero, so, set each factor equal to zero and solve for the variable.

Extraction of Roots (9.3)

Quadratic equations of the form $x^2 - K = 0$ or $x^2 = K$ can be solved by the *method of extraction of roots*. We do so by taking both the positive and negative square roots of each side. If K is a positive real number then $x = \sqrt{K}, -\sqrt{K}$. If K is a negative real number, no real number solution exists.

Completing the Square (9.4)

The quadratic equation $ax^2 + bx + c = 0$ can be solved by *completing the square*.

1. Write the equation so that the constant term appears on the right side of the equal sign.
2. If the leading coefficient is different from 1, divide each term of the equation by that coefficient.
3. Find one half of the coefficient of the linear term, square it, then *add* it to *both* sides of the equation.
4. The trinomial on the left side of the equation is now a perfect square trinomial and can be factored as $(\qquad)^2$.
5. Solve the equation by extraction of roots.

Quadratic Formula (9.5)

The quadratic equation $ax^2 + bx + c = 0$ can be solved using the quadratic formula.

a is the coefficient of x^2.
b is the coefficient of x.
c is the constant term.

$$x = \frac{-b \pm \sqrt{b^2 - 4ac}}{2a}$$

Parabola (9.6)

The graph of a quadratic equation of the form $y = ax^2 + bx + c$ is a parabola.

Vertex of a Parabola (9.7)

The high point or low point of a parabola is the *vertex* of the parabola.

EXERCISE SUPPLEMENT

Sections 9.1 and 9.2

For problems 1–35, solve the equations.

1. $(x - 2)(x - 5) = 0$
2. $(b + 1)(b - 6) = 0$
3. $(a + 10)(a - 5) = 0$
4. $(y - 3)(y - 4) = 0$
5. $(m - 8)(m + 1) = 0$
6. $(4y + 1)(2y + 3) = 0$
7. $(x + 2)(3x - 1) = 0$
8. $(5a - 2)(3a - 10) = 0$
9. $x(2x + 3) = 0$
10. $(a - 5)^2 = 0$
11. $(y + 3)^2 = 0$
12. $c^2 = 36$
13. $16y^2 - 49 = 0$
14. $6r^2 - 36 = 0$
15. $a^2 + 6a + 8 = 0$
16. $r^2 + 7r + 10 = 0$
17. $s^2 - 9s + 8 = 0$
18. $y^2 = -10y - 9$
19. $11y - 2 = -6y^2$
20. $16x^2 - 3 = -2x$
21. $m^2 = 4m - 4$
22. $3(y^2 - 8) = -7y$
23. $a(4b + 7) = 0$
24. $x^2 - 64 = 0$
25. $m^2 - 81 = 0$
26. $9x^2 - 25 = 0$
27. $5a^2 - 125 = 0$
28. $8r^3 - 6r = 0$
29. $m^2 - 6m + 5 = 0$
30. $x^2 + 2x - 24 = 0$
31. $x^2 + 3x = 28$
32. $20a^2 - 3 = 7a$
33. $2y^2 - 6y = 8$
34. $a^2 + 2a = -1$
35. $2r^2 = 5 - 3r$

Section 9.3

For problems 36–59, solve the equations using extraction of roots.

36. $y^2 = 81$
37. $a^2 = 121$
38. $x^2 = 35$
39. $m^2 = 2$
40. $r^2 = 1$
41. $s^2 - 10 = 0$
42. $4x^2 - 64 = 0$
43. $-3y^2 = -75$
44. Solve $y^2 = 4a^2$ for y.
45. Solve $m^2 = 16n^2p^4$ for m.
46. Solve $x^2 = 25y^4z^{10}w^8$ for x.
47. Solve $x^2 - y^2 = 0$ for y.
48. Solve $a^4b^8 - x^6y^{12}z^2 = 0$ for a^2.
49. $(x - 2)^2 = 9$
50. $(y + 3)^2 = 25$
51. $(a + 10)^2 = 1$
52. $(m + 12)^2 = 6$
53. $(r - 8)^2 = 10$
54. $(x - 1)^2 = 5$
55. $(a - 2)^2 = -2$
56. Solve $(x - 2b)^2 = b^2$ for x.
57. Solve $(y + 6)^2 = a$ for y.
58. Solve $(2a - 5)^2 = c$ for a.
59. Solve $(3m - 11)^2 = 2a^2$ for m.

Sections 9.4 and 9.5

For problems 60–84, solve the equations by completing the square or by using the quadratic formula.

60. $y^2 - 8y - 12 = 0$
61. $s^2 + 2s - 24 = 0$
62. $a^2 + 3a - 9 = 0$
63. $b^2 + b - 8 = 0$
64. $3x^2 - 2x - 1 = 0$
65. $5a^2 + 2a - 6 = 0$
66. $a^2 = a + 4$

67. $y^2 = 2y + 1$

68. $m^2 - 6 = 0$

69. $r^2 + 2r = 9$

70. $3p^2 + 2p = 7$

71. $10x^3 + 2x^2 - 22x = 0$

72. $6r^3 + 6r^2 - 3r = 0$

73. $15x^2 + 2x^3 = 12x^4$

74. $6x^3 - 6x = -6x^2$

75. $(x + 3)(x - 4) = 3$

76. $(y - 1)(y - 2) = 6$

77. $(a + 3)(a + 4) = -10$

78. $(2m + 1)(3m - 1) = -2$

79. $(5r + 6)(r - 1) = 2$

80. $4x^2 + 2x - 3 = 3x^2 + x + 1$

81. $5a^2 + 5a + 4 = 3a^2 + 2a + 5$

82. $(m + 3)^2 = 11$

83. $(r - 8)^2 = 70$

84. $(2x + 7)^2 = 51$

Section 9.6

For problems 85–93, find the solution.

85. The revenue R, in dollars, collected by a certain manufacturer of inner tubes is related to the number x of inner tubes sold by $R = 1400 - 16x + 3x^2$. How many inner tubes must be sold to produce a profit of $1361?

86. A study of the air quality in a particular city by an environmental group suggests that t years from now the level of carbon monoxide, in parts per million, in the air will be $A = 0.8t^2 + 0.5t + 3.3$.
  (a) What is the level, in parts per million, of carbon monoxide in the air now?
  (b) How many years from now will the carbon monoxide level be at 6 parts per million?

87. A contractor is to pour a concrete walkway around a community garden that is 15 feet wide and 50 feet long. The area of the walkway and garden is to be 924 square feet and of uniform width. How wide should the contractor make it?

88. A ball thrown vertically into the air has the equation of motion $h = 144 + 48t - 16t^2$
  (a) How high is the ball at $t = 0$?
  (b) How high is the ball at $t = 1$?
  (c) When does the ball hit the ground?

89. The length of a rectangle is 5 feet longer than three times its width. Find the dimensions if the area is to be 138 square feet.

90. The area of a triangle is 28 square centimeters. The base is 3 cm longer than the height. Find both the length of the base and the height.

91. The product of two consecutive integers is 210. Find them.

92. The product of two consecutive negative integers is 272. Find them.

93. A box with no top and a square base is to be made by cutting out 3-inch squares from each corner and folding up the sides of a piece of cardboard. The volume of the box is to be 25 cubic inches. What size should the piece of cardboard be?

For the quadratic equations in problems 1 and 2, specify the values of a, b, and c.

1.   **1. (9.1)** $2y^2 - 3y + 10 = 0$

2.   **2. (9.1)** $10b^2 = 3b$

3.   For problems 3–5, use the zero-factor property to solve each quadratic equation.

      3. (9.1) $(3x + 5)(x - 1) = 0$

4.   **4. (9.1)** $3b(2b - 1) = 0$

5.   **5. (9.1)** $(a - 8)^2 = 0$

6.   For problems 6–11, solve each quadratic equation by factoring.

      6. (9.2) $4x^2 - 16 = 0$

7.   **7. (9.2)** $y^2 - 12y + 32 = 0$

8.   **8. (9.2)** $a^2 - 5a = 14$

9.   **9. (9.2)** $6a^2 = 10 - 11a$

10.   **10. (9.2)** $2x^2 = -2 - 5x$

11.   **11. (9.2)** $x^3 - 25x = 0$

12.   For problems 12–18, solve each quadratic equation by extraction of roots.

      12. (9.3) $c^2 = 81$

13.   **13. (9.3)** $x^2 = 15$

14.   **14. (9.3)** $3a^2 - 18 = 0$

15.   **15. (9.3)** $(x - 5)^2 = 1$

16.   **16. (9.3)** $(y + 11)^2 - 9 = 0$

17. _____

18. _____

19. _____

20. _____

21. _____

22. _____

23. _____

24. _____

25. _____

26. _____

27. _____

28. _____

29. _____

30. _____

17. **(9.3)** $y^2 - 25z^2 = 0$ for y

18. **(9.3)** $6a^2 - 18b^2c^2$ for a

For problems 19–25, solve each quadratic equation using the quadratic formula.

19. **(9.5)** $x^2 - 6x - 16 = 0$

20. **(9.5)** $y^2 - 2y - 7 = 0$

21. **(9.5)** $(m + 2)^2 - 5 = 0$

22. **(9.5)** $(x + b)^2 = c^2$

23. **(9.5)** $(x + 1)(x + 4) = 6$

24. **(9.5)** $5z^2 - 5z - 5 = 2z^2 - z$

25. **(9.5)** $2m^2 = 5m$

For problems 26 and 27, solve each quadratic equation by completing the square.

26. **(9.4)** $x^2 + 6x - 8 = 0$

27. **(9.4)** $2x^2 + 7x - 12 = 0$

28. **(9.6)** The product of two consecutive odd integers is 143. What are they?

29. **(9.6)** A study of the air quality by an environmental group suggests that t years from now the level of carbon monoxide in the air, in parts per million, will be given by the quadratic equation

$$A = 0.4t^2 + 0.1t + 3.1$$

where A represents the amount of carbon monoxide in the air.
(a) What is the level, in parts per million, of carbon monoxide in the air now?
(b) How many years from now will the level of carbon monoxide be at 18.1 parts per million?

30. **(9.6)** The length of a rectangle is 6 inches longer than the width of the rectangle. Find the dimensions of the rectangle if the area is 112 square feet.

For problems 31–33, construct the graphs of the following equations.

31. _____

31. (9.7) $y = x^2 - 3$

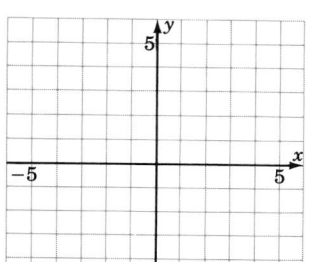

32. (9.7) $y = (x + 1)^2$

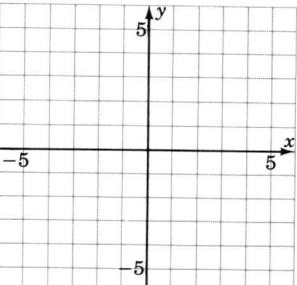

32. _____

33. (9.7) $y = (x - 2)^2 + 3$

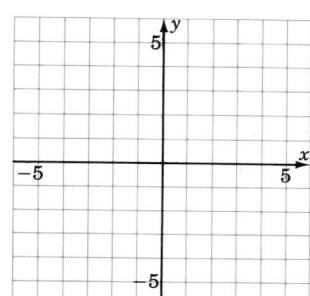

33. _____

For problems 34 and 35, write the equation that corresponds to each graph.

34. _____

34. (9.7)

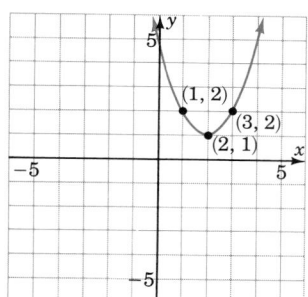

35. (9.7)

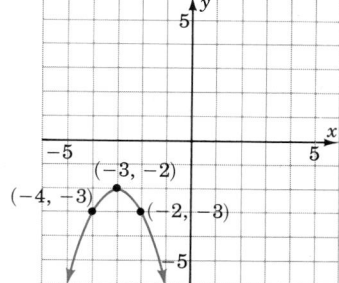

35. _____

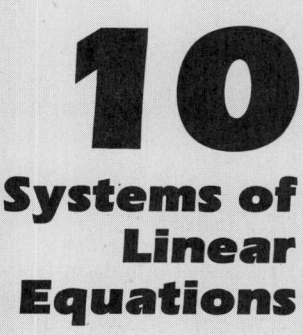

10
Systems of Linear Equations

After completing this chapter, you should

Section 10.1 Solutions by Graphing
- be able to recognize a system of equations and a solution to it
- be able to graphically interpret independent, inconsistent, and dependent systems
- be able to solve a system of linear equations graphically

Section 10.2 Elimination by Substitution
- know when the substitution method works best
- be able to use the substitution method to solve a system of linear equations
- know what to expect when using substitution with a system that consists of parallel lines or coincident lines

Section 10.3 Elimination by Addition
- know the properties used in the addition method
- be able to use the addition method to solve a system of linear equations
- know what to expect when using the addition method with a system that consists of parallel or coincident lines

Section 10.4 Applications
- become more familiar with the five-step method for solving applied problems
- be able to solve number problems
- be able to solve value and rate problems

10.1 Solutions by Graphing

Section Overview

☐ SYSTEMS OF EQUATIONS
☐ SOLUTION TO A SYSTEM OF EQUATIONS
☐ GRAPHS OF SYSTEMS OF EQUATIONS
☐ INDEPENDENT, INCONSISTENT, AND DEPENDENT SYSTEMS
☐ THE METHOD OF SOLVING A SYSTEM GRAPHICALLY

☐ SYSTEMS OF EQUATIONS

System of Equations

A collection of two linear equations in two variables is called a *system of linear equations in two variables,* or more briefly, **a system of equations.** The *pair* of equations

$$\begin{cases} 5x - 2y = 5 \\ x + y = 8 \end{cases}$$

is a system of equations. The brace { is used to denote that the two equations occur together (simultaneously).

☐ SOLUTION TO A SYSTEM OF EQUATIONS

Solution to a System

We know that one of the infinitely many solutions to one linear equation in two variables is an ordered pair. An ordered pair that is a solution to both of the equations in a system is called a *solution to the system of equations.* For example, the ordered pair (3, 5) is a solution to the system

$$\begin{cases} 5x - 2y = 5 \\ x + y = 8 \end{cases}$$

since (3, 5) is a solution to both equations.

$5x - 2y = 5$ $x + y = 8$
$5(3) - 2(5) \overset{?}{=} 5$ $3 + 5 \overset{?}{=} 8$
$15 - 10 \overset{?}{=} 5$ $8 \overset{\checkmark}{=} 8$
$5 \overset{\checkmark}{=} 5$

☐ GRAPHS OF SYSTEMS OF EQUATIONS

One method of solving a system of equations is by graphing. We know that the graph of a linear equation in two variables is a straight line. The graph of a system will consist of two straight lines. When two straight lines are graphed, one of three possibilities may result.

1. The lines intersect at the point (a, b). The point (a, b) is the solution to the corresponding system.

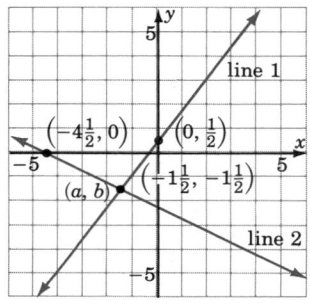

2. The lines are parallel. They do not intersect. The system has no solution.

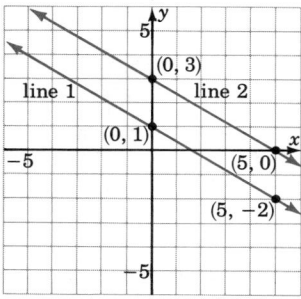

3. The lines are coincident (one on the other). They intersect at infinitely many points. The system has infinitely many solutions.

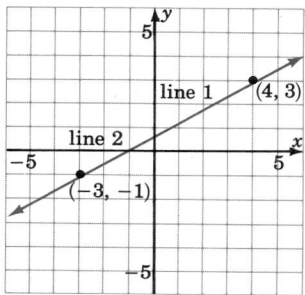

☐ INDEPENDENT, INCONSISTENT, AND DEPENDENT SYSTEMS

Independent Systems

Systems in which the lines intersect at precisely one point are called **independent systems.** In applications, independent systems can arise when the collected data are accurate and complete. For example,

The sum of two numbers is 10 and the product of the two numbers is 21. Find the numbers.

In this application, the data are accurate and complete. The solution is 7 and 3.

Inconsistent Systems

Systems in which the lines are parallel are called **inconsistent systems.** In applications, inconsistent systems can arise when the collected data are contradictory. For example,

The sum of two even numbers is 30 and the difference of the same two numbers is 0. Find the numbers.

The data are contradictory. There is no solution to this application.

Dependent Systems

Systems in which the lines are coincident are called **dependent systems.** In applications, dependent systems can arise when the collected data are incomplete. For example.

The difference of two numbers is 9 and twice one number is 18 more than twice the other.

The data are incomplete. There are infinitely many solutions.

☐ **THE METHOD OF SOLVING A SYSTEM GRAPHICALLY**

The Method of Solving a System Graphically

To solve a system of equations graphically: Graph both equations.

1. If the lines intersect, the solution is the ordered pair that corresponds to the point of intersection. The system is independent.
2. If the lines are parallel, there is no solution. The system is inconsistent.
3. If the lines are coincident, there are infinitely many solutions. The system is dependent.

☆ **SAMPLE SET A**

Solve each of the following systems by graphing.

1. $\begin{cases} 2x + y = 5 & (1) \\ x + y = 2 & (2) \end{cases}$

Write each equation in slope-intercept form.

$(1) \quad -2x + y = 5 \qquad (2) \quad x + y = 2$
$\qquad\quad y = 2x + 5 \qquad\qquad\quad\; y = -x + 2$

Graph each of these equations.

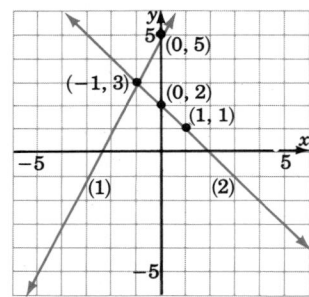

The lines appear to intersect at the point $(-1, 3)$. The solution to this system is $(-1, 3)$, or

$x = -1, \quad y = 3$

Check: Substitute $x = -1$, $y = 3$ into each equation.

$(1) \quad -2x + y = 5 \qquad\qquad (2) \quad x + y = 2$
$\qquad\quad -2(-1) + 3 \overset{?}{=} 5 \qquad\qquad\qquad -1 + 3 \overset{?}{=} 2$
$\qquad\quad 2 + 3 \overset{?}{=} 5 \qquad\qquad\qquad\qquad\quad 2 \overset{\checkmark}{=} 2$
$\qquad\quad 5 \overset{\checkmark}{=} 5$

2. $\begin{cases} -x + y = -1 & (1) \\ -x + y = 2 & (2) \end{cases}$

Write each equation in slope-intercept form.

$(1) \quad -x + y = -1 \qquad (2) \quad -x + y = 2$
$\qquad\qquad\; y = x - 1 \qquad\qquad\qquad\; y = x + 2$

Graph each of these equations.

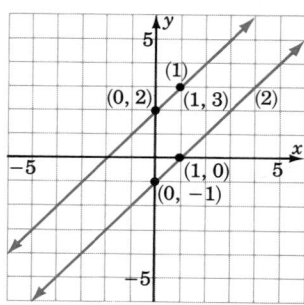

These lines are parallel. This system has no solution. We denote this fact by writing *inconsistent*.

We are sure that these lines are parallel because we notice that they have the same slope, $m = 1$ for both lines. The lines are not coincident because the y-intercepts are different.

3. $\begin{cases} -2x + 3y = -2 & (1) \\ -6x + 9y = -6 & (2) \end{cases}$

Write each equation in slope-intercept form.

(1) $-2x + 3y = -2$
  $3y = 2x - 2$

  $y = \dfrac{2}{3}x - \dfrac{2}{3}$

(2) $-6x + 9y = -6$
  $9y = 6x - 6$

  $y = \dfrac{2}{3}x - \dfrac{2}{3}$

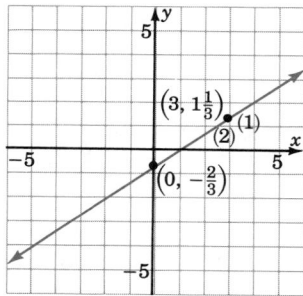

Both equations are the same. This system has infinitely many solutions. We write *dependent*.

★ **PRACTICE SET A**

Solve each of the following systems by graphing. Write the ordered pair solution or state that the system is inconsistent, or dependent.

1. $\begin{cases} 2x + y = 1 \\ -x + y = -5 \end{cases}$

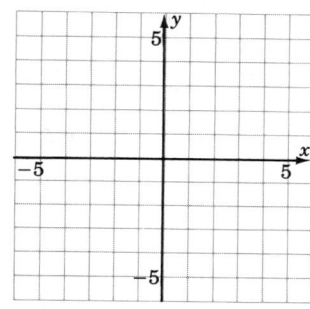

2. $\begin{cases} -2x + 3y = 6 \\ 6x - 9y = -18 \end{cases}$

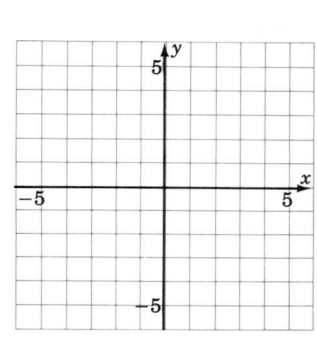

3. $\begin{cases} 3x + 5y = 15 \\ 9x + 15y = 15 \end{cases}$

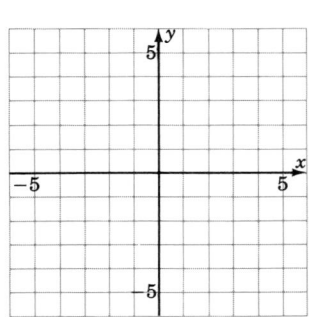

4. $\begin{cases} y = -3 \\ x + 2y = -4 \end{cases}$

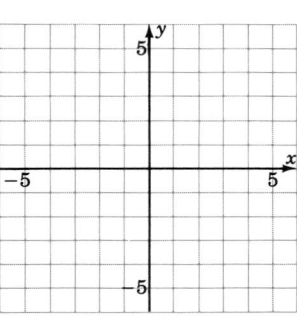

Answers to the Practice Set are on p. 518.

Section 10.1 EXERCISES

For problems 1–12, solve the systems by graphing. Write the ordered pair solution, or state that the system is inconsistent or dependent.

1. $\begin{cases} x + y = -5 \\ -x + y = 1 \end{cases}$

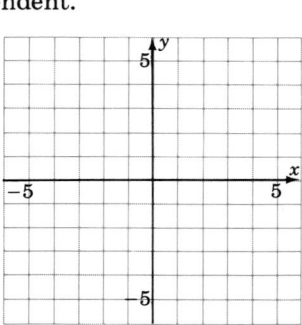

3. $\begin{cases} -3x + y = 5 \\ -x + y = 3 \end{cases}$

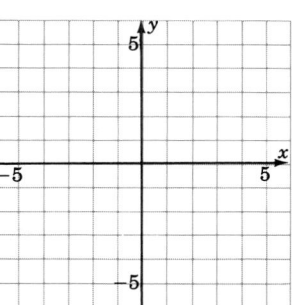

2. $\begin{cases} x + y = 4 \\ x + y = 0 \end{cases}$

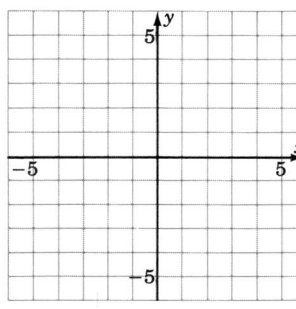

4. $\begin{cases} x - y = -6 \\ x + 2y = 0 \end{cases}$

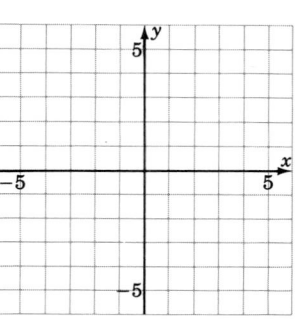

5. $\begin{cases} 3x + y = 0 \\ 4x - 3y = 12 \end{cases}$

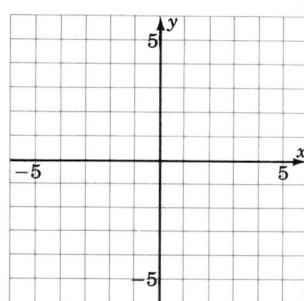

6. $\begin{cases} -4x + y = 7 \\ -3x + y = 2 \end{cases}$

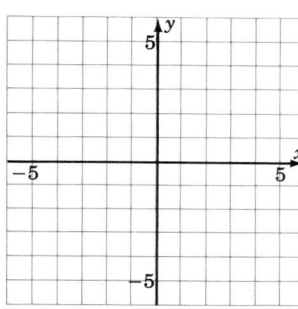

7. $\begin{cases} 2x + 3y = 6 \\ 3x + 4y = 6 \end{cases}$

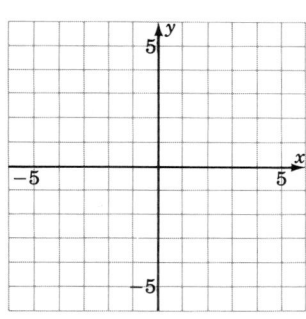

8. $\begin{cases} x + y = -3 \\ 4x + 4y = -12 \end{cases}$

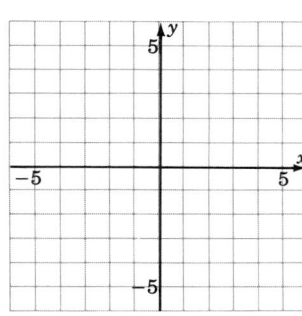

9. $\begin{cases} 2x - 3y = 1 \\ 4x - 6y = 4 \end{cases}$

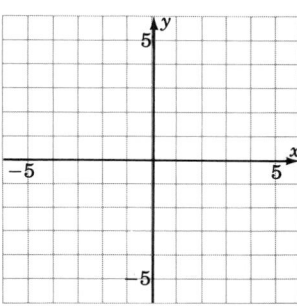

10. $\begin{cases} x + 2y = 3 \\ -3x - 6y = -9 \end{cases}$

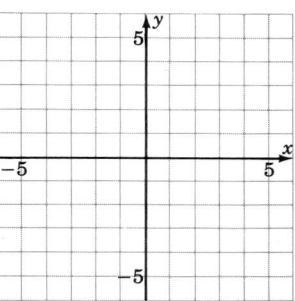

11. $\begin{cases} x - 2y = 6 \\ 3x - 6y = 18 \end{cases}$

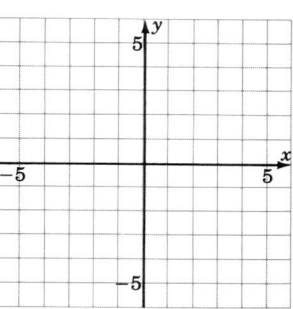

12. $\begin{cases} 2x + 3y = 6 \\ -10x - 15y = 30 \end{cases}$

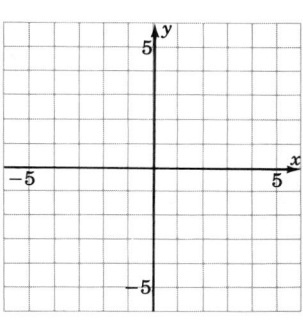

EXERCISES FOR REVIEW

(2.7)   **13.** Express 0.000426 in scientific notation.

(3.6)   **14.** Find the product: $(7x - 3)^2$.

(6.4)   **15.** Supply the missing word. The _____ of a line is a measure of the steepness of the line.

(9.1)   **16.** Supply the missing word. An equation of the form $ax^2 + bx + c = 0$, $a \neq 0$, is called a _____ equation.

(9.7)   **17.** Construct the graph of the quadratic equation $y = x^2 - 3$.

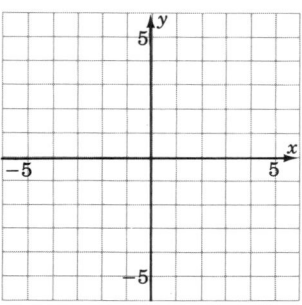

★ Answers to Practice Set (10.1)

A. **1.** $x = 2, y = -3$

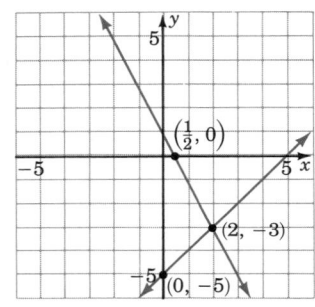

2. dependent

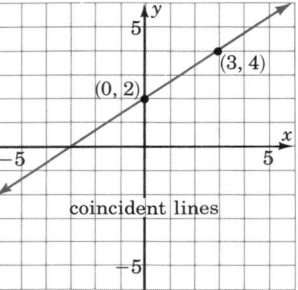

3. inconsistent

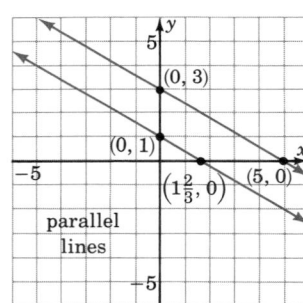

4. $x = 2, y = -3$

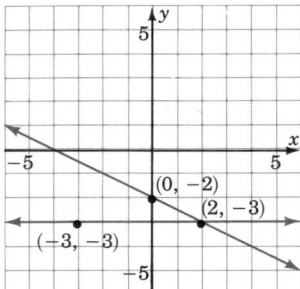

10.2 Elimination by Substitution

Section Overview

- ❑ **WHEN SUBSTITUTION WORKS BEST**
- ❑ **THE SUBSTITUTION METHOD**
- ❑ **SUBSTITUTION AND PARALLEL LINES**
- ❑ **SUBSTITUTION AND COINCIDENT LINES**

❑ WHEN SUBSTITUTION WORKS BEST

We know how to solve a linear equation in one variable. We shall now study a method for solving a system of two linear equations in two variables by transforming the two equations in two variables into one equation in one variable.

To make this transformation, we need to eliminate one equation and one variable. We can make this *elimination by substitution.*

When Substitution Works Best

The substitution method works best when *either* of these conditions exists:

1. One of the variables has a coefficient of 1, *or*
2. One of the variables can be made to have a coefficient of 1 without introducing fractions.

❑ THE SUBSTITUTION METHOD

The Substitution Method

To solve a system of two linear equations in two variables,

1. Solve one of the equations for one of the variables.
2. Substitute the expression for the variable chosen in step 1 into the other equation.
3. Solve the resulting equation in one variable.
4. Substitute the value obtained in step 3 into the equation obtained in step 1 and solve to obtain the value of the other variable.
5. Check the solution in both equations.
6. Write the solution as an ordered pair.

☆ SAMPLE SET A

Solve the system $\begin{cases} 2x + 3y = 14 & (1) \\ 3x + y = 7 & (2) \end{cases}$

Step 1: Since the coefficient of y in equation 2 is 1, we will solve equation 2 for y.

$$y = -3x + 7$$

Step 2: Substitute the expression $-3x + 7$ for y in equation 1.

$$2x + 3(-3x + 7) = 14$$

Step 3: Solve the equation obtained in step 2.

$$2x + 3(-3x + 7) = 14$$
$$2x - 9x + 21 = 14$$
$$-7x + 21 = 14$$
$$-7x = -7$$
$$x = 1$$

Continued

Step 4:   Substitute $x = 1$ into the equation obtained in step 1, $y = -3x + 7$.

$y = -3(1) + 7$
$y = -3 + 7$
$y = 4$

We now have $x = 1$ and $y = 4$.

Step 5:   Substitute $x = 1$, $y = 4$ into each of the original equations for a check.

(1) $2x + 3y = 14$ (2) $3x + y = 7$
$2(1) + 3(4) \stackrel{?}{=} 14$ $3(1) + (4) \stackrel{?}{=} 7$
$2 + 12 \stackrel{?}{=} 14$ $3 + 4 \stackrel{?}{=} 7$
$14 \stackrel{\checkmark}{=} 14$ $7 \stackrel{\checkmark}{=} 7$

Step 6:   The solution is $(1, 4)$. The point $(1, 4)$ is the point of intersection of the two lines of the system.

★ PRACTICE SET A

Solve the system $\begin{cases} 5x - 8y = 18 \\ 4x + y = 7 \end{cases}$

☐ SUBSTITUTION AND PARALLEL LINES

The following rule alerts us to the fact that the two lines of a system are parallel.

Substitution and Parallel Lines | **If computations eliminate all the variables and produce a contradiction, the two lines of a system are parallel, and the system is called inconsistent.**

☆ SAMPLE SET B

Solve the system $\begin{cases} 2x - y = 1 \\ 4x - 2y = 4 \end{cases}$ (1)
(2)

Step 1:   Solve equation 1 for y.

$2x - y = 1$
$-y = -2x + 1$
$y = 2x - 1$

Step 2:   Substitute the expression $2x - 1$ for y into equation 2.

$4x - 2(2x - 1) = 4$

Step 3:   Solve the equation obtained in step 2.

$4x - 2(2x - 1) = 4$
$4x - 4x + 2 = 4$
$2 \neq 4$

Computations have eliminated all the variables and produce a contradiction. These lines are parallel.

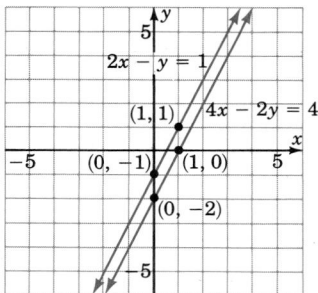

This system is inconsistent.

★ **PRACTICE SET B**

Solve the system $\begin{cases} 7x - 3y = 2 \\ 14x - 6y = 1 \end{cases}$

☐ SUBSTITUTION AND COINCIDENT LINES

The following rule alerts us to the fact that the two lines of a system are coincident.

Substitution and Coincident Lines

> If computations eliminate all the variables and produce an identity, the two lines of a system are coincident and the system is called dependent.

☆ **SAMPLE SET C**

Solve the system $\begin{cases} 4x + 8y = 8 & (1) \\ 3x + 6y = 6 & (2) \end{cases}$

Step 1:   Divide equation 1 by 4 and solve for x.

$4x + 8y = 8$
$x + 2y = 2$
$x = -2y + 2$

Step 2:   Substitute the expression $-2y + 2$ for x in equation 2.

$3(-2y + 2) + 6y = 6$

Continued

522   Chapter 10 Systems of Linear Equations

Step 3:   Solve the equation obtained in step 2.

$$3(-2y + 2) + 6y = 6$$
$$-6y + 6 + 6y = 6$$
$$6 = 6$$

Computations have eliminated all the variables and produced an identity. These lines are coincident.

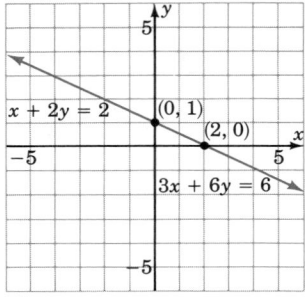

This system is dependent.

★ **PRACTICE SET C**

Solve the system $\begin{cases} 4x + 3y = 1 \\ -8x - 6y = -2 \end{cases}$

Systems in which a coefficient of one of the variables is not 1 or cannot be made to be 1 without introducing fractions are not well suited for the substitution method. The problem in Sample Set D illustrates this "messy" situation.

☆ **SAMPLE SET D**

Solve the system $\begin{cases} 3x + 2y = 1 & (1) \\ 4x - 3y = 3 & (2) \end{cases}$

Step 1:   We will solve equation (1) for y.

$$3x + 2y = 1$$
$$2y = -3x + 1$$
$$y = \frac{-3}{2}x + \frac{1}{2}$$

Step 2: Substitute the expression $\dfrac{-3}{2}x + \dfrac{1}{2}$ for y in equation (2).

$$4x - 3\left(\dfrac{-3}{2}x + \dfrac{1}{2}\right) = 3$$

Step 3: Solve the equation obtained in step 2.

$$4x - 3\left(\dfrac{-3}{2}x + \dfrac{1}{2}\right) = 3 \qquad \textbf{Multiply both sides by the LCD, 2.}$$

$$4x + \dfrac{9}{2}x - \dfrac{3}{2} = 3$$

$$8x + 9x - 3 = 6$$

$$17x - 3 = 6$$

$$17x = 9$$

$$x = \dfrac{9}{17}$$

Step 4: Substitute $x = \dfrac{9}{17}$ into the equation obtained in step 1, $y = \dfrac{-3}{2}x + \dfrac{1}{2}$.

$$y = \dfrac{-3}{2}\left(\dfrac{9}{17}\right) + \dfrac{1}{2}$$

$$y = \dfrac{-27}{34} + \dfrac{17}{34} = \dfrac{-10}{34} = \dfrac{-5}{17}$$

We now have $x = \dfrac{9}{17}$ and $y = \dfrac{-5}{17}$.

Step 5: Substitution will show that these values of x and y check.

Step 6: The solution is $\left(\dfrac{9}{17}, \dfrac{-5}{17}\right)$.

★ **PRACTICE SET D**

Solve the system $\begin{cases} 9x - 5y = -4 \\ 2x + 7y = -9 \end{cases}$

Answers to Practice Sets are on p. 525.

Section 10.2 EXERCISES

For problems 1–26, solve the systems by substitution.

1. $\begin{cases} 3x + 2y = 9 \\ \qquad y = -3x + 6 \end{cases}$ **2.** $\begin{cases} 5x - 3y = -6 \\ \qquad y = -4x + 2 \end{cases}$ **11.** $\begin{cases} 4x - 2y = 8 \\ 6x + 3y = 0 \end{cases}$ **12.** $\begin{cases} 2x + 3y = 12 \\ 2x + 4y = 18 \end{cases}$

3. $\begin{cases} 2x + 2y = 0 \\ x \qquad = 3y - 4 \end{cases}$ **4.** $\begin{cases} 3x + 5y = 9 \\ x \qquad = 4y - 14 \end{cases}$ **13.** $\begin{cases} 3x - \ 9y = 6 \\ 6x - 18y = 5 \end{cases}$ **14.** $\begin{cases} -x + \ 4y = \ 8 \\ 3x - 12y = 10 \end{cases}$

5. $\begin{cases} -3x + \ y = -4 \\ 2x + 3y = \ 10 \end{cases}$ **6.** $\begin{cases} -4x + \ y = -7 \\ 2x + 5y = \ 9 \end{cases}$ **15.** $\begin{cases} x + y = -6 \\ x - y = \ 4 \end{cases}$ **16.** $\begin{cases} 2x + \ y = 0 \\ x - 3y = 0 \end{cases}$

7. $\begin{cases} 6x - 6 \ = 18 \\ x + 3y = \ 3 \end{cases}$ **8.** $\begin{cases} -x - y = 5 \\ 2x + y = 5 \end{cases}$ **17.** $\begin{cases} 4x - 2y = 7 \\ \qquad y = 4 \end{cases}$ **18.** $\begin{cases} x + 6y = \ 11 \\ x \qquad = -1 \end{cases}$

9. $\begin{cases} -5x + \ y = \ 4 \\ 10x - 2y = -8 \end{cases}$ **10.** $\begin{cases} x + \ 4y = \ 1 \\ -3x - 12y = -1 \end{cases}$ **19.** $\begin{cases} 2x - 4y = 10 \\ 3x = 5y + 12 \end{cases}$ **20.** $\begin{cases} y + 7x + 4 = 0 \\ x \qquad = -7y + 28 \end{cases}$

21. $\begin{cases} x + 4y = 0 \\ x + \dfrac{2}{3}y = \dfrac{10}{3} \end{cases}$ **22.** $\begin{cases} x = 24 - 5y \\ x - \dfrac{5}{4}y = \dfrac{3}{2} \end{cases}$ **25.** $\begin{cases} \dfrac{4}{5}x + \dfrac{1}{2}y = \dfrac{3}{10} \\ \dfrac{1}{3}x + \dfrac{1}{2}y = \dfrac{-1}{6} \end{cases}$ **26.** $\begin{cases} x - \dfrac{1}{3}y = \dfrac{-8}{3} \\ -3x + y = 1 \end{cases}$

23. $\begin{cases} x = 11 - 6y \\ 3x + 18y = -33 \end{cases}$ **24.** $\begin{cases} 2x + \dfrac{1}{3}y = 4 \\ 3x + 6y = 39 \end{cases}$

EXERCISES FOR REVIEW

(7.3) **27.** Find the quotient: $\dfrac{x^2 - x - 12}{x^2 - 2x - 15} \div \dfrac{x^2 - 3x - 10}{x^2 - 2x - 8}$.

(7.5) **28.** Find the difference: $\dfrac{x + 2}{x^2 + 5x + 6} - \dfrac{x + 1}{x^2 + 4x + 3}$.

(8.2) **29.** Simplify $-\sqrt{81x^8 y^5 z^4}$.

(9.5) **30.** Use the quadratic formula to solve $2x^2 + 2x - 3 = 0$.

(10.1) **31.** Solve by graphing $\begin{cases} x - y = 1 \\ 2x + y = 5 \end{cases}$

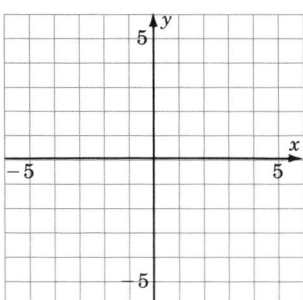

★ Answers to Practice Sets (10.2)

A. The point $(2, -1)$ is the point of intersection of the two lines.

B. Substitution produces $4 \neq 1$, or $\dfrac{1}{2} \neq 2$, a contradiction. These lines are parallel and the system is inconsistent.

C. Computations produce $-2 = -2$, an identity. These lines are coincident and the system is dependent.

D. These lines intersect at the point $(-1, -1)$.

10.3 Elimination by Addition

☐ **THE PROPERTIES USED IN THE ADDITION METHOD**
☐ **THE ADDITION METHOD**
☐ **ADDITION AND PARALLEL OR COINCIDENT LINES**

☐ THE PROPERTIES USED IN THE ADDITION METHOD

Another method of solving a system of two linear equations in two variables is called the *method of elimination by addition*. It is similar to the method of elimination by substitution in that the process eliminates one equation and one variable. The method of elimination by addition makes use of the following two properties.

> 1. If A, B, and C are algebraic expressions such that
> $$\begin{aligned} A &= B \quad &\text{and} \\ C &= D \quad &\text{then} \end{aligned}$$
> $$\overline{A + C = B + D}$$
> 2. $ax + (-ax) = 0$

Property 1 states that if we add the left sides of two equations together and the right sides of the same two equations together, the resulting sums will be equal. We call this *adding equations*. Property 2 states that the sum of two opposites is zero.

☐ THE ADDITION METHOD

The Addition Method

> To solve a system of two linear equations in two variables by addition,
> 1. Write, if necessary, both equations in general form, $ax + by = c$.
> 2. If necessary, multiply one or both equations by factors that will produce opposite coefficients for one of the variables.
> 3. Add the equations to eliminate one equation and one variable.
> 4. Solve the equation obtained in step 3.
> 5. Do one of the following:
>   (a) Substitute the value obtained in step 4 into either of the original equations and solve to obtain the value of the other variable,
>   or
>   (b) Repeat steps 1–5 for the other variable.
> 6. Check the solutions in both equations.
> 7. Write the solution as an ordered pair.

The addition method works well when the coefficient of one of the variables is 1 or a number other than 1.

☆ SAMPLE SET A

Solve $\begin{cases} x - y = 2 & (1) \\ 3x + y = 14 & (2) \end{cases}$

Step 1: Both equations appear in the proper form.
Step 2: The coefficients of y are already opposites, 1 and -1, so there is no need for a multiplication.

Step 3:   Add the equations.

$$x - y = \ \ 2$$
$$3x + y = 14$$
$$\overline{4x + 0 = 16}$$

Step 4:   Solve the equation $4x = 16$.

$$4x = 16$$
$$x = 4$$

The problem is not solved yet; we still need the value of y.

Step 5:   Substitute $x = 4$ into either of the original equations. We will use equation 1.

$$4 - y = 2 \quad \textbf{Solve for } y.$$

$$-y = -2$$
$$y = 2$$

We now have $x = 4, \quad y = 2$.

Step 6:   Substitute $x = 4$ and $y = 2$ into both the original equations for a check.

(1) $x - y = 2$ (2) $3x + y = 14$
        $4 - 2 \overset{?}{=} 2$ $3(4) + 2 \overset{?}{=} 14$
        $2 \overset{\checkmark}{=} 2$ $12 + 2 \overset{?}{=} 14$
                                    $14 \overset{\checkmark}{=} 14$

Step 7:   The solution is (4, 2).

The two lines of this system intersect at (4, 2).

★ **PRACTICE SET A**

Solve each system by addition.

1. $\begin{cases} x + y = 6 \\ 2x - y = 0 \end{cases}$ **2.** $\begin{cases} x + 6y = 8 \\ -x - 2y = 0 \end{cases}$

☆ **SAMPLE SET B**

Solve the following systems using the addition method.

1. Solve $\begin{cases} 6a - 5b = \ \ 14 & (1) \\ 2a + 2b = -10 & (2) \end{cases}$

Step 1:   The equations are already in the proper form, $ax + by = c$.

Continued

Step 2:   If we multiply equation (2) by -3, the coefficients of a will be opposites and become 0 upon addition, thus eliminating a.

$$\begin{cases} 6a - 5b = 14 \\ -3(2a + 2b) = -3(-10) \end{cases} \rightarrow \begin{cases} 6a - 5b = 14 \\ -6a - 6b = 30 \end{cases}$$

Step 3:   Add the equations.

$$\begin{aligned} 6a - 5b &= 14 \\ -6a - 6b &= 30 \\ \hline 0 - 11b &= 44 \end{aligned}$$

Step 4:   Solve the equation $-11b = 44$.

$$-11b = 44$$
$$b = -4$$

Step 5:   Substitute $b = -4$ into either of the original equations. We will use equation 2.

$$2a + 2b = -10$$
$$2a + 2(-4) = -10 \quad \textbf{Solve for } a.$$
$$2a - 8 = -10$$
$$2a = -2$$
$$a = -1$$

We now have $a = -1$ and $b = -4$.

Step 6:   Substitute $a = -1$ and $b = -4$ into both the original equations for a check.

(1) $6a - 5b = 14$ (2) $2a + 2b = -10$
     $6(-1) - 5(-4) \overset{?}{=} 14$ $2(-1) + 2(-4) \overset{?}{=} -10$
     $-6 + 20 \overset{?}{=} 14$ $-2 - 8 \overset{?}{=} -10$
     $14 \overset{\checkmark}{=} 14$ $-10 \overset{\checkmark}{=} -10$

Step 7:   The solution is $(-1, -4)$.

2. Solve $\begin{cases} 3x + 2y = -4 & (1) \\ 4x = 5y + 10 & (2) \end{cases}$

Step 1:   Rewrite the system in the proper form.

$$\begin{cases} 3x + 2y = -4 & (1) \\ 4x - 5y = 10 & (2) \end{cases}$$

Step 2:   Since the coefficients of y already have opposite signs, we will eliminate y.
Multiply equation (1) by 5, the coefficient of y in equation 2.
Multiply equation (2) by 2, the coefficient of y in equation 1.

$$\begin{cases} 5(3x + 2y) = 5(-4) \\ 2(4x - 5y) = 2(10) \end{cases} \rightarrow \begin{cases} 15x + 10y = -20 \\ 8x - 10y = 20 \end{cases}$$

Step 3:   Add the equations.

$$\begin{aligned} 15x + 10y &= -20 \\ 8x - 10y &= 20 \\ \hline 23x + 0 &= 0 \end{aligned}$$

Step 4:   Solve the equation $23x = 0$

$$23x = 0$$
$$x = 0$$

Step 5: Substitute $x = 0$ into either of the original equations. We will use equation 1.

$3x + 2y = -4$

$3(0) + 2y = -4$ **Solve for y.**

$0 + 2y = -4$
$y = -2$

We now have $x = 0$ and $y = -2$.

Step 6: Substitution will show that these values check.

Step 7: The solution is $(0, -2)$.

★ PRACTICE SET B

Solve each of the following systems using the addition method.

1. $\begin{cases} 3x + y = 1 \\ 5x + y = 3 \end{cases}$ 2. $\begin{cases} x + 4y = 1 \\ x - 2y = -5 \end{cases}$ 3. $\begin{cases} 2x + 3y = -10 \\ -x + 2y = -2 \end{cases}$

4. $\begin{cases} 5x - 3y = 1 \\ 8x - 6y = 4 \end{cases}$ 5. $\begin{cases} 3x - 5y = 9 \\ 4x + 8y = 12 \end{cases}$

☐ ADDITION AND PARALLEL OR COINCIDENT LINES

When the lines of a system are parallel or coincident, the method of elimination produces results identical to that of the method of elimination by substitution.

Addition and Parallel Lines

Addition and Coincident Lines

If computations eliminate all variables and produce a contradiction, the two lines of the system are parallel and the system is called inconsistent.

If computations eliminate all variables and produce an identity, the two lines of the system are coincident and the system is called dependent.

☆ **SAMPLE SET C**

1. Solve $\begin{cases} 2x - y = 1 & (1) \\ 4x - 2y = 4 & (2) \end{cases}$

Step 1:   The equations are in the proper form.
Step 2:   We can eliminate x by multiplying equation (1) by -2.

$$\begin{cases} -2(2x - y) = -2(1) \\ 4x - 2y \quad = 4 \end{cases} \rightarrow \begin{cases} -4x + 2y = -2 \\ \ \ 4x - 2y = \ \ 4 \end{cases}$$

Step 3:   Add the equations.

$$\begin{array}{r} -4x + 2y = -2 \\ \underline{4x - 2y = \ \ 4} \\ 0 + 0 \ = \ \ 2 \\ 0 \quad\ \ = \ \ 2 \end{array}$$

This is false and is therefore a contradiction. The lines of this system are parallel.

This system is inconsistent.

2. Solve $\begin{cases} 4x + 8y = 8 & (1) \\ 3x + 6y = 6 & (2) \end{cases}$

Step 1:   The equations are in the proper form.
Step 2:   We can eliminate x by multiplying equation (1) by -3 and equation (2) by 4.

$$\begin{cases} -3(4x + 8y) = -3(8) \\ \ \ 4(3x + 6y) = 4(6) \end{cases} \rightarrow \begin{cases} -12x - 24y = -24 \\ \ \ 12x + 24y = 24 \end{cases}$$

Step 3:   Add the equations.

$$\begin{array}{r} -12x - 24y = -24 \\ \underline{12x + 24y = \ \ 24} \\ 0 \ + \ 0 \ = \ \ 0 \\ 0 \ = \ \ 0 \end{array}$$

This is true and is an identity. The lines of this system are coincident.

This system is dependent.

★ **PRACTICE SET C**

Solve each of the following systems using the addition method.

1. $\begin{cases} -x + \ \ 2y = 6 \\ -6x + 12y = 1 \end{cases}$ **2.** $\begin{cases} 4x - 28y = -4 \\ \ \ x - \ \ 7y = -1 \end{cases}$

Answers to Practice Sets are on p. 533.

Section 10.3 EXERCISES

For problems 1–33, solve the systems using elimination by addition.

1. $\begin{cases} x + y = 11 \\ x - y = -1 \end{cases}$

2. $\begin{cases} x + 3y = 13 \\ x - 3y = -11 \end{cases}$

11. $\begin{cases} 3x + 4y = 7 \\ x + 5y = 6 \end{cases}$

12. $\begin{cases} 4x - 2y = 2 \\ 7x + 4y = 26 \end{cases}$

3. $\begin{cases} 3x - 5y = -4 \\ -4x + 5y = 2 \end{cases}$

4. $\begin{cases} 2x - 7y = 1 \\ 5x + 7y = -22 \end{cases}$

13. $\begin{cases} 3x + y = -4 \\ 5x - 2y = -14 \end{cases}$

14. $\begin{cases} 5x - 3y = 20 \\ -x + 6y = -4 \end{cases}$

5. $\begin{cases} -3x + 4y = -24 \\ 3x - 7y = 42 \end{cases}$

6. $\begin{cases} 8x + 5y = 3 \\ 9x - 5y = -71 \end{cases}$

15. $\begin{cases} 6x + 2y = -18 \\ -x + 5y = 19 \end{cases}$

16. $\begin{cases} x - 11y = 17 \\ 2x - 22y = 4 \end{cases}$

7. $\begin{cases} -x + 2y = -6 \\ x + 3y = -4 \end{cases}$

8. $\begin{cases} 4x + y = 0 \\ 3x + y = 0 \end{cases}$

17. $\begin{cases} -2x + 3y = 20 \\ -3x + 2y = 15 \end{cases}$

18. $\begin{cases} -5x + 2y = -4 \\ -3x - 5y = 10 \end{cases}$

9. $\begin{cases} x + y = -4 \\ -x - y = 4 \end{cases}$

10. $\begin{cases} -2x - 3y = -6 \\ 2x + 3y = 6 \end{cases}$

19. $\begin{cases} -3x - 4y = 2 \\ -9x - 12y = 6 \end{cases}$

20. $\begin{cases} 3x - 5y = 28 \\ -4x - 2y = -20 \end{cases}$

21. $\begin{cases} 6x - 3y = 3 \\ 10x - 7y = 3 \end{cases}$ **22.** $\begin{cases} -4x + 12y = 0 \\ -8x + 16y = 0 \end{cases}$

29. $\begin{cases} x + \dfrac{1}{3}\,y = \dfrac{4}{3} \\ -x + \dfrac{1}{6}\,y = \dfrac{2}{3} \end{cases}$ **30.** $\begin{cases} 8x - 3y = 25 \\ 4x - 5y = -5 \end{cases}$

23. $\begin{cases} 3x + y = -1 \\ 12x + 4y = 6 \end{cases}$ **24.** $\begin{cases} 8x + 5y = -23 \\ -3x - 3y = 12 \end{cases}$

31. $\begin{cases} -10x - 4y = 72 \\ 9x + 5y = 39 \end{cases}$ **32.** $\begin{cases} 12x + 16y = -36 \\ -10x + 12y = 30 \end{cases}$

25. $\begin{cases} 2x + 8y = 10 \\ 3x + 12y = 15 \end{cases}$ **26.** $\begin{cases} 4x + 6y = 8 \\ 6x + 8y = 12 \end{cases}$

33. $\begin{cases} 25x - 32y = 14 \\ -50x + 64y = -28 \end{cases}$

27. $\begin{cases} 10x + 2y = 2 \\ -15x - 3y = 3 \end{cases}$ **28.** $\begin{cases} x + \dfrac{3}{4}\,y = -\dfrac{1}{2} \\ \dfrac{3}{5}\,x + y = -\dfrac{7}{5} \end{cases}$

EXERCISES FOR REVIEW

(2.5) **34.** Simplify and write $(2x^{-3}y^4)^5(2xy^{-6})^{-5}$ so that only positive exponents appear.

(8.5) **35.** Simplify $\sqrt{8} + 3\sqrt{50}$.

(8.6) **36.** Solve the radical equation $\sqrt{2x + 3} + 5 = 8$.

(10.1)   37. Solve by graphing $\begin{cases} x + y = 4 \\ 3x - y = 0 \end{cases}$

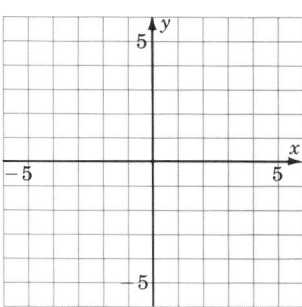

(10.2)   38. Solve using the substitution method: $\begin{cases} 3x - 4y = -11 \\ 5x + \ y = \ -3 \end{cases}$

★ **Answers to Practice Sets (10.3)**

A. 1. $(2, 4)$ **2.** $(-4, 2)$

B. 1. $(1, -2)$ **2.** $(-3, 1)$ **3.** $(-2, -2)$ **4.** $(-1, -2)$ **5.** $(3, 0)$

C. 1. inconsistent   **2.** dependent

10.4 Applications

Section Overview

- ☐ **THE FIVE-STEP METHOD**
- ☐ **NUMBER PROBLEMS**
- ☐ **VALUE AND RATE PROBLEMS: COIN PROBLEMS AND MIXTURE PROBLEMS**

☐ THE FIVE-STEP METHOD

When solving practical problems, it is often more convenient to introduce two variables rather than only one. Two variables should be introduced only when two relationships can be found within the problem. Each relationship will produce an equation, and a system of two equations in two variables will result.

We will use the five-step method to solve these problems.

1. Introduce two variables, one for each unknown quantity.
2. Look for two relationships within the problem. Translate the verbal phrases into mathematical expressions to form two equations.
3. Solve the resulting system of equations.
4. Check the solution.
5. Write a conclusion.

❑ NUMBER PROBLEMS

The sum of two numbers is 37. One number is 5 larger than the other. What are the numbers?

Step 1:   Let x = smaller number.
y = larger number.

Step 2:   There are two relationships.
(a)   The sum is 37.

$$x + y = 37$$

(b)   One is 5 larger than the other.

$$y = x + 5$$

$$\begin{cases} x + y = 37 & (1) \\ y = x + 5 & (2) \end{cases}$$

Step 3: $\begin{cases} x + y = 37 & (1) \\ y = x + 5 & (2) \end{cases}$

We can easily solve this system by substitution. Substitute $x + 5$ for y in equation 1.

$$x + (x + 5) = 37$$
$$x + x + 5 = 37$$
$$2x + 5 = 37$$
$$2x = 32$$
$$x = 16 \qquad \text{Then,} \quad y = 16 + 5 = 21.$$
$$x = 16, \quad y = 21$$

Step 4:   The sum is 37.

$$x + y = 37$$
$$16 + 21 \stackrel{?}{=} 37$$
$$37 \stackrel{\checkmark}{=} 37$$

One is 5 larger than the other.

$$y = x + 5$$
$$21 \stackrel{?}{=} 16 + 5$$
$$21 \stackrel{\checkmark}{=} 21$$

Step 5:   The two numbers are 16 and 21.

★ **PRACTICE SET A**

The difference of two numbers is 9, and the sum of the same two numbers is 19. What are the two numbers?

Step 1:

Step 2:

Step 3:

Step 4:

Step 5:

☐ VALUE AND RATE PROBLEMS: COIN PROBLEMS AND MIXTURE PROBLEMS

Value and Rate Problems

The problems in Sample Sets B and C are *value problems*. They are referred to as value problems because one of the equations of the system used in solving them is generated by considering a **value, or rate, or amount times a quantity.**

COIN PROBLEMS

☆ SAMPLE SET B

1. A parking meter contains 27 coins consisting only of dimes and quarters. If the meter contains $4.35, how many of each type of coin is there?

Step 1:   Let D = number of dimes.
               Q = number of quarters.

Step 2:   There are two relationships.

(a)   There are 27 coins. $D + Q = 27$.

(b) Contribution due to dimes $= 10D$
                              (Value) · (Quantity)

Contribution due to quarters $= 25Q$.
                              (Value) · (Quantity)

$10D + 25Q = 435$

$$\begin{cases} D + Q = 27 & (1) \\ 10D + 25Q = 435 & (2) \end{cases}$$

Continued

Step 3: $\begin{cases} D + Q = 27 & (1) \\ 10D + 25Q = 435 & (2) \end{cases}$

We can solve this system using elimination by addition. Multiply both sides of equation (1) by -10 and add.

$$-10D - 10Q = -270$$
$$\underline{10D + 25Q = 435}$$
$$15Q = 165$$
$$Q = 11 \qquad \text{Then, } D + 11 = 27$$
$$D = 16$$

$$D = 16, \quad Q = 11$$

Step 4:   16 dimes and 11 quarters is 27 coins.

$$10(16) + 11(25) \overset{?}{=} 435$$
$$160 + 275 \overset{?}{=} 435$$
$$435 \overset{\checkmark}{=} 435$$

The solution checks.

Step 5:   There are 11 quarters and 16 dimes.

★ **PRACTICE SET B**

A bag contains only nickels and dimes. The value of the collection is \$2. If there are 26 coins in all, how many of each coin are there?

MIXTURE PROBLEMS

☆ **SAMPLE SET C**

A chemistry student needs 40 milliliters (ml) of a 14% acid solution. She had two acid solutions, A and B, to mix together to form the 40 ml acid solution. Acid solution A is 10% acid and acid solution B is 20% acid. How much of each solution should be used?

Step 1:   Let $x =$ number of ml of solution A.
$ y =$ number of ml of solution B.

Step 2:   There are two relationships.

(a)   The sum of the number of ml of the two solutions is 40.

$$x + y = 40$$

(b)   To determine the second equation, draw a picture of the situation.

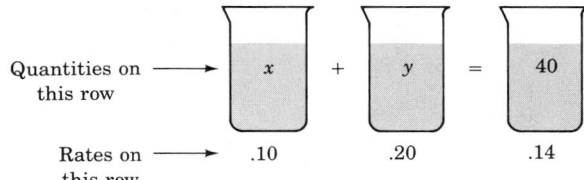

Quantities on → x + y = 40
this row

Rates on → .10 .20 .14
this row

The equation follows directly from the drawing if we use the idea of *amount times quantity*.

$$.10x + .20y = .14(40)$$

$$\begin{cases} x + y = 40 & (1) \\ .10x + .20y = .14(40) & (2) \end{cases}$$

Step 3: $\begin{cases} x + y = 40 & (1) \\ .10x + .20y = .14(40) & (2) \end{cases}$

Solve this system by addition. First, eliminate decimals in equation 2 by multiplying both sides by 100.

$$\begin{cases} x + y = 40 & (1) \\ 10x + 20y = 14(40) & (2) \end{cases}$$

Eliminate x by multiplying equation 1 by -10 and then adding.

$$-10x - 10y = -400$$
$$\underline{10x + 20y = 560}$$
$$10y = 160$$
$$y = 16 \qquad \text{Then,} \quad x + 16 = 40$$
$$x = 24$$

$$x = 24, \quad y = 16$$

Step 4: 24 ml and 16 ml to add to 40 ml.

$$10(24) + 20(16) \overset{?}{=} 560$$
$$240 + 320 \overset{?}{=} 560$$
$$560 \overset{\checkmark}{=} 560$$

The solution checks.

Step 5: The student should use 24 ml of acid solution A and 16 ml of acid solution B.

★ **PRACTICE SET C**

A chemistry student needs 60 ml of a 26% salt solution. He has two salt solutions, A and B, to mix together to form the 60 ml solution. Salt solution A is 30% salt and salt solution B is 20% salt. How much of each solution should be used?

Answers to Practice Sets are on p. 540.

Section 10.4 EXERCISES

1. The sum of two numbers is 22. One number is 6 more than the other. What are the numbers?

2. The sum of two numbers is 32. One number is 8 more than the other. What are the numbers?

3. The difference of two numbers is 12 and one number is three times as large as the other. What are the numbers?

4. The difference of two numbers is 9 and one number is 10 times larger than the other. What are the numbers?

5. Half the sum of two numbers is 14 and half the difference is 2. What are the numbers?

6. One third of the sum of two numbers is 6 and one fifth of the difference is 2. What are the numbers?

7. A 14 pound mixture of grapes sells for $3.10. Type 1 grape sells for 25¢ a pound and type 2 grape sells for 20¢ a pound. How many pounds of each type of grape were used?

8. The cost of 80 liters of a blended cleaning solution is $28. Type 1 solution costs 20¢ a liter and type 2 solution costs 40¢ a liter. How many liters of each solution were used to form the blended solution?

9. The cost of 42 grams of a certain chemical compound is $14.40. Type 1 chemical costs 45¢ a gram and type 2 chemical costs 30¢ a gram. How many grams of each chemical were used to form the compound?

10. A play was attended by 342 people, some adults and some children. Admission for adults was $1.50 and for children 75¢. How many adults and how many children attended the play?

11. 200 tickets were sold to a college's annual musical performance. Tickets for students were $2.50 and for nonstudents $3.50. The total amount collected was $537. How many nonstudents purchased tickets for the performance?

12. A chemistry student needs 22 ml of a 38% acid solution. She has two acid solutions, A and B, to mix together to form the solution. Acid solution A is 40% acid and acid solution B is 30% acid. How much of each solution should be used?

13. A chemistry student needs 50 ml of a 72% salt solution. He has two salt solutions, A and B, to mix together to form the solution. Salt solution A is 60% salt and salt solution B is 80% salt. How much of each solution should be used?

14. A chemist needs 2 liters of an 18% acid solution. He has two solutions, A and B, to mix together to form the solution. Acid solution A is 10% acid and acid solution B is 15% acid. Can the chemist form the needed 18% acid solution? (Verify by calculation.) If the chemist locates a 20% acid solution, how much would have to be mixed with the 10% solution to obtain the needed 2-liter 18% solution?

15. A chemist needs 3 liters of a 12% acid solution. She has two acid solutions, A and B, to mix together to form the solution. Acid solution A is 14% acid and acid solution B is 20% acid. Can the chemist form the needed 12% solution? (Verify by calculation.) If the chemist locates a 4% acid solution, how much would have to be mixed with the 14% acid solution to obtain the needed 3-liter 12% solution?

16. A chemistry student needs 100 ml of a 16% acid solution. He has a bottle of 20% acid solution. How much pure water and how much of the 20% solution should be mixed to dilute the 20% acid solution to a 16% acid solution?

17. A chemistry student needs 1 liter of a 78% salt solution. She has a bottle of 80% salt solution. How much pure water and how much of the 80% salt solution should be mixed to dilute the 80% salt solution to a 78% salt solution?

18. A parking meter contains 42 coins. The total value of the coins is $8.40. If the meter contains only dimes and quarters, how many of each type are there?

19. A child's bank contains 78 coins. The coins are only pennies and nickels. If the value of the coins is $1.50, how many of each coin are there?

EXERCISES FOR REVIEW

(1.6) **20.** Simplify $\dfrac{x^{n+6}}{x^2}$.

(3.6) **21.** Find the product: $(3x - 5)^2$.

(7.5) **22.** Find the difference: $\dfrac{2x - 1}{x^2 + x - 6} - \dfrac{x + 2}{x^2 + 5x + 6}$.

(10.2) **23.** Use the substitution method to solve $\begin{cases} 4x - y = 3 \\ 2x - 5y = 15 \end{cases}$

(10.3) **24.** Use the addition method to solve $\begin{cases} 3x + 4y = -7 \\ 4x - 5y = 1 \end{cases}$

★ **Answers to Practice Sets (10.4)**

A. The two numbers are 14 and 5.

B. There are 14 dimes and 12 nickels.

C. The student should use 36 ml of salt solution A and 24 ml of salt solution B.

Chapter 10   SUMMARY OF KEY CONCEPTS

System of Equations (10.1)

A collection of two linear equations in two variables is called a *system of equations.*

Solution to a System (10.1)

An ordered pair that is a solution to both equations in a system is called a *solution to the system of equations.* The values $x = 3$, $y = 1$ are a solution to the system

$$\begin{cases} x - y = 2 \\ x + y = 4 \end{cases}$$

Independent Systems (10.1)

Systems in which the lines intersect at precisely one point are *independent systems.* In applications, independent systems can arise when the collected data are accurate and complete.

Inconsistent Systems (10.1)

Systems in which the lines are parallel are *inconsistent systems.* In applications, inconsistent systems can arise when the collected data are contradictory.

Dependent Systems (10.1)

Systems in which the lines are coincident (one on the other) are *dependent systems.* In applications, dependent systems can arise when the collected data are incomplete.

Solving a System by Graphing (10.1)

To solve a system by graphing:

1. Graph each equation of the same set of axes.
2. If the lines intersect, the solution is the point of intersection.

Solving a System by Substitution (10.2)

To solve a system using substitution,

1. Solve one of the equations for one of the variables.
2. Substitute the expression for the variable chosen in step 1 into the other equation.
3. Solve the resulting equation in one variable.
4. Substitute the value obtained in step 3 into the equation obtained in step 1 and solve to obtain the value of the other variable.
5. Check the solution in both equations.
6. Write the solution as an ordered pair.

Solving a System by Addition (10.3)

To solve a system using addition,

1. Write, if necessary, both equations in general form
   $ax + by = c$
2. If necessary, multiply one or both equations by factors that will produce opposite coefficients for one of the variables.
3. Add the equations to eliminate one equation and one variable.
4. Solve the equation obtained in step 3.
5. Substitute the value obtained in step 4 into either of the original equations and solve to obtain the value of the other variable.
6. Check the solution in both equations.
7. Write the solution as an ordered pair.

Substitution and Addition and Parallel Lines (10.2, 10.3)

If computations eliminate all variables and produce a contradiction, the two lines of the system are parallel and no solution exists. The system is inconsistent.

Substitution and Addition and Coincident Lines (10.2, 10.3)

If computations eliminate all variables and produce an identity, the two lines of the system are coincident and the system has infinitely many solutions. The system is dependent.

Applications (10.4)

The five-step method can be used to solve applied problems that involve linear systems that consist of two equations in two variables. The solutions of number problems, mixture problems, and value and rate problems are examined in this section. The rate problems have particular use in chemistry.

EXERCISE SUPPLEMENT

For problems 1–20, solve the systems of equations.

1. $\begin{cases} 4x + y = 5 \\ -2x + 3y = -13 \end{cases}$

2. $\begin{cases} -5x + 2y = 5 \\ x + 7y = -1 \end{cases}$

3. $\begin{cases} x - 3y = 17 \\ 8x + 2y = 46 \end{cases}$

4. $\begin{cases} 6m + 5n = -9 \\ 2m - 4n = 14 \end{cases}$

5. $\begin{cases} 3x - 9y = 5 \\ -x + 3y = 0 \end{cases}$

6. $\begin{cases} y = 2x - 5 \\ 8x - 75 = 5 \end{cases}$

7. $\begin{cases} x = 8 \\ 9y = 5x - 76 \end{cases}$

8. $\begin{cases} 7x - 2y = 4 \\ -14x + 4y = -8 \end{cases}$

9. $\begin{cases} y = -x - 7 \\ x = y - 5 \end{cases}$

10. $\begin{cases} 20x + 15y = -13 \\ 5x - 20y = 13 \end{cases}$

11. $\begin{cases} x - 6y = 12 \\ 4x + 6y = 18 \end{cases}$

12. $\begin{cases} 8x + 9y = 0 \\ 4x + 3y = 0 \end{cases}$

13. $\begin{cases} -5x + 2y = 1 \\ 10x - 4y = -2 \end{cases}$

14. $\begin{cases} 2x - 5y = 3 \\ 5x + 2y = -7 \end{cases}$

15. $\begin{cases} 6x + 5y = 14 \\ 4x - 8y = 32 \end{cases}$

16. $\begin{cases} 5x - 7y = 4 \\ 10x - 14y = 1 \end{cases}$

17. $\begin{cases} 2m + 10n = 0 \\ -4m - 20n = -6 \end{cases}$

18. $\begin{cases} 7r - 2s = 6 \\ -3r + 5s = -15 \end{cases}$

19. $\begin{cases} 28a - 21b = -19 \\ 21a + 7b = 15 \end{cases}$

20. $\begin{cases} 72x - 108y = 21 \\ 18x + 36y = 25 \end{cases}$

21. The sum of two numbers is 35. One number is 7 larger than the other. What are the numbers?

22. The difference of two numbers is 48. One number is three times larger than the other. What are the numbers?

23. A 35 pound mixture of two types of cardboard sells for $30.15. Type I cardboard sells for 90¢ a pound and type II cardboard sells for 75¢ a pound. How many pounds of each type of cardboard were used?

24. The cost of 34 calculators of two different types is $1139. Type I calculator sells for $35 each and type II sells for $32 each. How many of each type of calculators were used?

25. A chemistry student needs 46 ml of a 15% salt solution. She has two salt solutions, A and B, to mix together to form the needed 46 ml solution. Salt solution A is 12% salt and salt solution B is 20% salt. How much of each solution should be used?

26. A chemist needs 100 ml of a 78% acid solution. He has two acid solutions to mix together to form the needed 100-ml solution. One solution is 50% acid and the other solution is 90% acid. How much of each solution should be used?

27. One third the sum of two numbers is 12 and half the difference is 14. What are the numbers?

28. Two angles are said to be complementary if their measures add to 90°. If one angle measures 8 more than four times the measure of its complement, find the measure of each of the angles.

29. A chemist needs 4 liters of a 20% acid solution. She has two solutions to mix together to form the 20% solution. One solution is 30% acid and the other solution is 24% acid. Can the chemist form the needed 20% acid solution? If the chemist locates a 14% acid solution, how much would have to be mixed with the 24% acid solution to obtain the needed 20% solution?

30. A chemist needs 80 ml of a 56% salt solution. She has a bottle of 60% salt solution. How much pure water and how much of the 60% salt solution should be mixed to dilute the 60% salt solution to a 56% salt solution?

1. _____

1. (10.1) Solve using graphing: $\begin{cases} 3x + 2y = 4 \\ 15x + 10y = -10 \end{cases}$

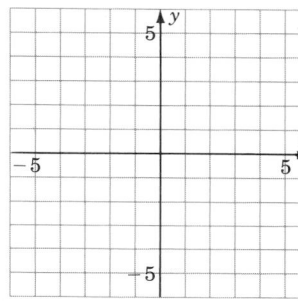

2. (10.1) Solve using graphing: $\begin{cases} 2x - 3y = 2 \\ x + 2y = 8 \end{cases}$

2. _____

3. _____

4. _____

3. (10.2) Solve using substitution: $\begin{cases} 2x + 6y = 16 \\ x - 4y = -13 \end{cases}$

4. (10.3) Solve using addition: $\begin{cases} 3x + 8y = -5 \\ x - 2y = 3 \end{cases}$

5. _____

5. (10.2, 10.3) Solve using either substitution or addition: $\begin{cases} 4x - 4y = 8 \\ x - y = 5 \end{cases}$

6. (10.2, 10.3) Solve using either substitution or addition: $\begin{cases} 9x + 3y = 12 \\ 3x - y = 4 \end{cases}$

6. _____

7. _____

7. (10.4) The sum of two numbers is 43 and the difference of the same two numbers is 7. What are the numbers?

8. (10.4) A chemist needs 80 ml of an 18% acid solution. She has two acid solutions, A and B, to mix together to form the 80-ml solution. Acid solution A is 15% acid and acid solution B is 20% acid. How much of each solution should be used?

8. _____

9. _____

9. **(10.4)** A parking meter contains 32 coins. If the meter contains only nickels and quarters, and the total value of the coins is $4.60, how many of each type of coin are there?

10. **(10.4)** A person has $15,000 to invest. If he invests part at 8% and the rest at 12%, how much should he invest at each rate to produce the same return as if he had invested it all at 9%?

10. _____

Answers To Selected Exercises

CHAPTER R

Section R.1   Exercises

1. 8^3 **3.** 5^7 **5.** $3^5 \cdot 4^2$ **7.** $2^3 \cdot 9^8$
9. $x^3 \cdot y^2$ **11.** $3 \cdot 3 \cdot 3 \cdot 3$ **13.** $2 \cdot 2 \cdot 2 \cdot 2 \cdot 2$
15. $5 \cdot 5 \cdot 5 \cdot 6 \cdot 6$ **17.** $x \cdot x \cdot x \cdot x \cdot y \cdot y \cdot y \cdot y$
19. 1, 2, 4, 5, 10, 20   **21.** 1, 2, 3, 4, 6, 12   **23.** 1, 3, 7, 21
25. 1, 11   **27.** 1, 19

Section R.2   Exercises

1. prime   **3.** composite   **5.** prime   **7.** prime
9. prime   **11.** composite   **13.** composite
15. composite   **17.** $2 \cdot 19$ **19.** $2 \cdot 31$ **21.** $2^4 \cdot 11$
23. $3^2 \cdot 7 \cdot 13$ **25.** $5^2 \cdot 7^2 \cdot 11^2$

Section R.3   Exercises

1. $2^3 \cdot 3$ **3.** $2^2 \cdot 3$ **5.** $2 \cdot 3 \cdot 5$ **7.** $2^2 \cdot 3^2 \cdot 7$
9. $2^2 \cdot 3 \cdot 7$ **11.** $2 \cdot 3 \cdot 5^2$ **13.** $2^4 \cdot 3$
15. $3 \cdot 5 \cdot 7$ **17.** $2^3 \cdot 3 \cdot 5$ **19.** $2 \cdot 3^2 \cdot 5 \cdot 7^2$
21. $2^4 \cdot 3 \cdot 5$ **23.** $2^4 \cdot 3^2$ **25.** $2^5 \cdot 7$

Section R.4   Exercises

1. $\dfrac{3}{4}$ **3.** $\dfrac{3}{7}$ **5.** $\dfrac{3}{2}$ **7.** $\dfrac{5}{3}$ **9.** $\dfrac{5}{6}$ **11.** $\dfrac{18}{5}$
13. $\dfrac{2}{3}$ **15.** $\dfrac{3}{4}$ **17.** $\dfrac{1}{2}$ **19.** $\dfrac{1}{3}$ **21.** 3
23. $\dfrac{11}{12}$ **25.** $\dfrac{27}{19}$ **27.** 6   **29.** 12   **31.** 20
33. 75   **35.** 48

Section R.5   Exercises

1. $\dfrac{4}{9}$ **3.** $\dfrac{1}{3}$ **5.** $\dfrac{5}{12}$ **7.** $\dfrac{9}{10}$ **9.** 1   **11.** $\dfrac{2}{3}$
13. $\dfrac{10}{9}$ **15.** $\dfrac{10}{9}$ **17.** $\dfrac{57}{7}$ **19.** $\dfrac{5}{8}$ **21.** 1
23. $\dfrac{13}{23}$ **25.** $\dfrac{19}{20}$ **27.** $\dfrac{15}{16}$ **29.** $\dfrac{5}{8}$ **31.** $\dfrac{31}{24}$
33. $\dfrac{5}{6}$ **35.** $\dfrac{-1}{180}$ **37.** $\dfrac{7}{30}$ **39.** $\dfrac{47}{18}$

Section R.6   Exercises

1. 8.95   **3.** 2.234   **5.** 6.6312   **7.** 3.68406
9. 13.632   **11.** 25.355   **13.** 0.324   **15.** 12,008
17. 4.2   **19.** 0.0032   **21.** $\dfrac{3}{50}$ **23.** $3\dfrac{7}{10}$
25. $712\dfrac{1}{25000}$ **27.** 0.45   **29.** 0.636

Section R.7   Exercises

1. 40%   **3.** 12.5%   **5.** 68.18%   **7.** 22.22%
9. 49.09%   **11.** 1500%   **13.** 36%   **15.** 44.6%
17. 425%   **19.** 8698%   **21.** 1400%   **23.** 0.35
25. 0.186   **27.** 0.090145   **29.** 0.0000005

CHAPTER 1

Section 1.1   Exercises

1. 20   **3.** 7   **5.** 8   **7.** 78   **9.** 203   **11.** 29
13. 1   **15.** $91\dfrac{2}{3}$ **17.** 508   **19.** 24.4   **21.** 55
23. 1   **25.** 0   **27.** different   **29.** same
31. $a + b, b + a$ **33.** $x + 16$ **35.** $81x$
37. $(x + b)(x + 7)$ **39.** $\dfrac{x}{7b}$ **41.** $x - 8 = 17$
43. $\dfrac{x}{6} \geq 44$ **45.** true   **47.** true   **49.** false
51. 120   **53.** 0.00024, or $\dfrac{1}{4165}$

Section 1.2   Exercises

1. Q, R **3.** W, Z, Q, R **5.** Q, R **7.** Q, R
9.
11. neither
13.
15.
17.
no
19. $<$ **21.** $>$ **23.** no   **25.** 99   **27.** yes, 0
29. $-6, -5, -4, -3, -2$

31. There are no natural numbers between -15 and -1.

33. $\left(\dfrac{95}{1}\right)^{\circ}$

35. Yes, every integer is a rational number.

37. Yes. $\dfrac{1}{2} + \dfrac{1}{2} = 1$ or $1 + 1 = 2$ **39.** 5 units

41. 8 units   **43.** $m - n$ units   **45.** 23

47. different   **49.** true

Section 1.3   Exercises

1. $3 + x$ **3.** $x10$ **5.** $6r$ **7.** cx **9.** $(s + 1)6$

11. $(a + 7)(x + 16)$ **13.** $m(0.06)$ **15.** $(6h + 1)5$

17. $(10a - b)k$ **19.** $(4)(-16)$ **21.** $\bigcirc \cdot \square$

23. $18xy$ **25.** $24abc$ **27.** $30mnruz$

29. $\dfrac{1}{16}ade$ **31.** $9(x + 2y)(6 + z)(3x + 5y)$

33. $br + 5b$ **35.** $jk + k$ **37.** $xz + 9wz$

39. $8g + 2fg$ **41.** $30xy + 45xz$ **43.** $xz + yz + mz$

45. $ax + bx + cx + 10a + 10b + 10c$ **47.** $a + 16$

49. $348.3a + 81.7b + 15.05c$ **51.** $2L_m z_t + 16kz_t$

53. false   **55.** $<$

Section 1.4   Exercises

1. b^4 **3.** x^8 **5.** $5s^2$ **7.** $a^3 - (b + 7)^2$ **9.** x^5

11. $2(3^4)x^2y^5$ **13.** $7x^2(a + 8)^2$ **15.** $(4x)^5$ or 4^5x^5

17. $(-7)^4a^5b^5$ **19.** $(z + w)^3(z - w)^2$

21. $3x^3y^2 - (x + 1)^3$ **23.** $6 \cdot 6$

25. $8 \cdot x \cdot x \cdot x \cdot y \cdot y$

27. $(9aaabb)(9aaabb)(9aaabb)$ or $9 \cdot 9 \cdot 9aaaaaaaaaabbbbbb$

29. $10aaabb(3c)(3c)$ or $10 \cdot 3 \cdot 3aaaabbcc$

31. $(xx - yy)(xx + yy)$

33. Select $x = 2$. Then, $196 \neq 28$.   **35.** zero

37. 16   **39.** 105   **41.** 59   **43.** 4   **45.** 1

47. 4   **49.** 71   **51.** $\dfrac{51}{19}$ **53.** 5

55. $\dfrac{1070}{11}$ or $97.\overline{27}$

57.

59. xy

Section 1.5   Exercises

1. $3^5 = 243$ **3.** $9^2 = 81$ **5.** $2^9 = 512$ **7.** x^5

9. y^{12} **11.** k^{11} **13.** $6x^7$ **15.** $20y^{10}$

17. $144x^4y^5z^3$ **19.** $32x^3y^3$ **21.** $\dfrac{1}{8}a^2b^8$

23. $8^2 = 64$ **25.** $2^5 = 32$ **27.** x^2

29. y^5 **31.** x^2 **33.** m^7 **35.** y^2w^5 **37.** x^2y^3

39. $e^0 = 1$ **41.** $x^0 = 1$ **43.** $2a$ **45.** t^2y^4

47. a^9b^{15} **49.** $(x + 3y)^8(2x - 1)^3$ **51.** x^{n+r}

53. x^{2n+3} **55.** x **57.** $m^{\diamond + \star + \triangle}$

59. $a^{\triangle + \triangledown}b^{\square + \diamond}$ **61.** $8ax + 12bx$ **63.** 8

Section 1.6   Exercises

1. a^5c^5 **3.** $8a^3$ **5.** $81x^4y^4$ **7.** $81a^4b^4$

9. $49y^6$ **11.** $125x^{18}$ **13.** $100a^4b^2$ **15.** $x^8y^{12}z^{20}$

17. $x^{15}y^{10}z^{20}$ **19.** $a^{32}b^{56}c^{48}d^{64}$ **21.** 1

23. $\dfrac{1}{64}c^{20}d^{16}e^8f^{18}$ **25.** x^6y^8 **27.** $a^{18}b^{21}$

29. $x^{26}y^{14}z^8$ **31.** $4a^2b^6$ **33.** x^8 **35.** $m^4n^4p^{12}$

37. $1000x^8y^7z^{33}$ **39.** $25x^{14}y^{22}$ **41.** $\dfrac{27a^3b^3}{64x^3y^3}$

43. $\dfrac{27a^6b^9}{c^{12}}$ **45.** $\dfrac{x^8(y - 1)^{12}}{(x + 6)^4}$ **47.** x^{n+6}

49. $2\star a \star b \star$ **51.** $2m^{\triangle - \star}$ **53.** $\dfrac{2^{\square}x^{\triangle\square}}{y^{\triangledown\square}}$ **55.** no

57. 147   **59.** $4x^5y^6z^2$

Chapter 1   Exercise Supplement

1. 61   **3.** 438   **5.** 2   **7.** 79   **9.** $\dfrac{37}{48}$

11. 8   **13.** 43   **15.** $252 > 246$ **17.** $2 > 0$

19. different   **21.** $c + d$; $d + c$ **23.** $\dfrac{62}{f}$ or $62 \div f$

25. $6x - 2$

27. $(y + 11) \div (y + 10) - 12$ or $\dfrac{y + 11}{y + 10} - 12$

29. yes

31.

33.

35.

37.

39.

41. $-3 < 0$ **43.** $-8 < -5$ **45.** yes; -99

47. 4, 5, 6, or 7   **49.** $-2, -1, 0, 1,$ or 2 **51.** $-6°$

53. 4   **55.** commutative, multiplication

57. $4b + a$ **59.** $(a-1)2$

61. $(-9)(6)(-2)$ or $(-9)(-2)(6)$ or $(6)(-2)(-9)$ or $(-2)(-9)(6)$

63. $\diamond \cdot \triangle$ **65.** $32abc$ **67.** 0   **69.** $3a + 12$

71. $8gh + 4gk$ **73.** $6xy + 12yz + 15wy$

75. $4ax + 3bx + 4ay + 3by$ **77.** x^5

79. $(a + 2b)^2 - (a + 3b)^4$ **81.** a^7 **83.** $(-8)^4 x^3 y^5$

85. $2y^4z^3 + 7yz^3(a-6)^3$ **87.** $3xxx$ **89.** $4b \cdot 4b$

91. $(xxx + 7)(xxx + 7)(yy - 3)(yy - 3)(yy - 3)(z + 10)$

93. (a) any value except zero   (b) only zero

95. 366   **97.** 180.3   **99.** 10   **101.** $4b^8$

103. $6x^5y^{13}$ **105.** $81a^4$ **107.** $x^{12}y^{24}$

109. $\dfrac{9}{16} x^{16}y^{12}a^{20}b^{30}$ **111.** $7a^3b^3c^5$ **113.** x^{11}

115. $x^{13}y^9z^{19}$ **117.** $\dfrac{9x^4}{16y^6}$ **119.** x^{n+m}

121. $48b^{7n+9}$ **123.** $x^{35t}y^{28r}$ **125.** u^{w-k}

Chapter 1   Proficiency Exam

1. 40   **2.** 1   **3.** $\dfrac{137}{68}$ **4.** 75   **5.** $>$

6. $(x-1)(3x+2)$ **7.** $\dfrac{x}{12} \le (x+4)$

8.
```
        -1.6
 -5 -4 -3 -2 -1  0  1  2  3  4  5
```

9. Zero is neither positive nor negative.

10.
```
    14 15 16 17 18 19 20
```

11.
```
 -4 -3 -2 -1  0  1  2  3  4  5
```

12. 0, 1, 2, 3, 4, 5   **13.** yes; 10   **14.** $(a + 3)m$

15. $96abcd$ **16.** $24y^2(x-9)^2$ **17.** $16x^3y^5$

18. $81a^5b^6$ **19.** $72a^{11}b^7$ **20.** x^8y^7

21. $13x^7y^4(y-x^4)^2(y+x)^4$ **22.** $x^{4n}y^{12m}z^{8p}$

23. 1   **24.** $x^{\nabla+\square-\triangle}y^{\triangle-\nabla}$ **25.** a variable

CHAPTER 2

Section 2.1   Exercises

1. a plus sign or no sign at all   **3.** a negative five

5. twelve   **7.** negative negative four

9. five plus seven   **11.** eleven plus negative two

13. six minus negative eight   **15.** $-(-8) = 8$

17. 2   **19.** 1   **21.** -3 **23.** -6 **25.** 26

27. 31   **29.** 12   **31.** 17

33. $5 - (-2) = 5 + 2 = 7$ **35.** 16   **37.** 32

39. 13   **41.** 26   **43.** 0   **45.** x^{n+8} **47.** $\dfrac{16a^4b^2}{9x^2y^6}$

Section 2.2   Exercises

1. 5   **3.** 6   **5.** 8   **7.** 16   **9.** -12 **11.** -9

13. 1   **15.** -3 **17.** -14 **19.** -28

21. -68 **23.** 26   **25.** -8 **27.** -1 **29.** 4

31. 6   **33.** 3   **35.** 12   **37.** 4   **39.** 4

41. 5   **43.** 3   **45.** 6   **47.** 100   **49.** 92

51. -1 **53.** $|\$-2,400,000|$ **55.** $4(a+b)$

57. commutative property of addition   **59.** 4

Section 2.3   Exercises

1. 16   **3.** 8   **5.** -15 **7.** -12 **9.** -24

11. 11   **13.** 8   **15.** 2   **17.** -15 **19.** 8

21. -8 **23.** -15 **25.** -25 **27.** -12

29. 24   **31.** -21 **33.** 0   **35.** 0   **37.** 23

39. 328   **41.** 876   **43.** -1265 **45.** -6.084

47. -20 **49.** -17 **51.** -18 **53.** 14

55. 16   **57.** 4   **59.** $-\$28.50$ **61.** $\$3.00$

63. $5a^6c$ **65.** 8

Section 2.4   Exercises

1. 5   **3.** -1 **5.** -13 **7.** -6 **9.** -11

11. -13 **13.** -14 **15.** -2 **17.** 5   **19.** 11

21. 5   **23.** -5 **25.** -8 **27.** -6 **29.** 7

31. 29   **33.** -324 **35.** -429 **37.** -71

39. 164   **41.** 1   **43.** 1   **45.** 8   **47.** 4

49. 6   **51.** -27 **53.** -12 **55.** -11

57. $20xy + 44x$ **59.** 11   **61.** -3

Section 2.5   Exercises

1. 16   **3.** 32   **5.** 54   **7.** 32   **9.** -36

11. -32 **13.** -18 **15.** -24 **17.** -90

19. -60 **21.** -12 **23.** 3   **25.** -13 **27.** 9

29. -5 **31.** 11   **33.** 28   **35.** -4 **37.** -12

39. -4 **41.** -6 **43.** 15   **45.** 49   **47.** -140

49. -7 **51.** -3 **53.** -5 **55.** 13   **57.** -3

59. 4   **61.** -4 **63.** 15   **65.** 2   **67.** -2

69. -3 **71.** 1458   **73.** -120 **75.** 40

77. $(x + 2y)^2(3x - 1)$ **79.** -11

Section 2.6   Exercises

1. $\dfrac{1}{x^2}$ **3.** $\dfrac{1}{x^7}$ **5.** $\dfrac{1}{a^{10}}$ **7.** $\dfrac{1}{b^{14}}$ **9.** $\dfrac{1}{y^5}$

11. $\dfrac{1}{(x-5)^3}$ **13.** $\dfrac{1}{(a+9)^{10}}$ **15.** $\dfrac{1}{(a-1)^{12}}$

17. $\dfrac{x^7}{y^5}$ **19.** $\dfrac{a^7}{b^8}$ **21.** $\dfrac{x^3y^2}{z^6}$ **23.** $\dfrac{a^7zw^3}{b^9}$

25. $\dfrac{x^5}{y^5z^2}$ **27.** $\dfrac{d^4}{a^4b^6c}$ **29.** $\dfrac{4y^2}{x^6}$ **31.** $\dfrac{7b^2c^2}{a^2}$

33. $\dfrac{7a^2(a-4)^3}{b^6c^7}$ **35.** $\dfrac{7(w+1)^3}{(w+2)^2}$ **37.** $\dfrac{(x^2+3)^3}{(x^2-1)^4}$

39. $\dfrac{1}{(3x^2-4x-8)^9(2x+11)^3}$ **41.** $\dfrac{7a}{(a^2-4)^2(b^2-1)^2}$

43. $\dfrac{5y^3}{(y^3+1)z^4w^2(y^3-1)^2}$ **45.** $\dfrac{27x}{y^3}$ **47.** $\dfrac{4}{a^3}$

49. 1 **51.** $\dfrac{1}{(x+5)^4}$ **53.** $\dfrac{8}{(b+2)^9}$ **55.** $\dfrac{-8a^5b^2}{c^2}$

57. -5 **59.** 1 **61.** a^4 **63.** $4x^6$ **65.** $23y$

67. $\dfrac{3b^3c^5}{a^3}$ **69.** $\dfrac{4bc^9y^2}{a^2d^3z^8}$ **71.** $\dfrac{4a^3}{b^2c}$ **73.** $\dfrac{3x^3}{y^2z^4}$

75. $\dfrac{3}{a^7b^5}$ **77.** $\dfrac{128}{a^7bx}$ **79.** $\dfrac{4x^3}{y^7}$ **81.** $\dfrac{23a^4b^5x^6}{c^2y^5}$

83. $\dfrac{10}{3x^2y^7z^2}$ **85.** $\dfrac{224b^3}{c^{12}(a^2+21)^4(a+6)^3}$ **87.** $\dfrac{1}{7}$

89. $\dfrac{1}{32}$ **91.** $\dfrac{2}{9}$ **93.** 2 **95.** $\dfrac{1}{24}$ **97.** $\dfrac{1}{9}$

99. 24 **101.** 36 **103.** 63 **105.** $\dfrac{1}{a^{15}}$

107. $\dfrac{1}{x^{32}}$ **109.** b^4 **111.** y^{27} **113.** b

115. 1 **117.** $\dfrac{x^{12}y^{12}}{z^2}$ **119.** $x^{20}y^{15}$ **121.** $\dfrac{16a^4}{b^{12}}$

123. $\dfrac{a^6x^4}{25b^{12}y^{18}}$ **125.** $\dfrac{n^{28}s^{16}}{m^{32}r^{20}}$ **127.** $64x^{15}y^9$

129. 20 **131.** 1

Section 2.7   Exercises

1. 5.89×10^3 **3.** 2.5×10^{23} **5.** 2×10^{-4}

7. 5.7×10^4 **9.** $2.8\times10^{12}, 2.463\times10^{25}$

11. 3.36×10^3 **13.** 8×10^6 **15.** 1.5×10^{62}

17. 8×10^{-9} **19.** 1.03×10^{-18} **21.** 3.1×10^{-26}

23. 3.16×10^2 **25.** 7.4×10^5 **27.** 3.16×10^{-5}

29. 1×10^{-16} **31.** $100,000,000$

33. $5,866,000,000,000$

35. $43,000,000,000,000,000,000$ **37.** $60,000,000$

39. $0.00000000000000000000047$ **41.** 0.0125

43. 3.2×10^9 **45.** 2.4×10^3 **47.** 9×10^{-16}

49. 1.512×10^{-31} **51.** 3.481×10^{29}

53. 7.744×10^{-99} **55.** 6.687×10^{16}

57. 1.47×10^{-5} **59.** 4.5×10^{37} **61.** $-5,-4,-3$

63. -5 **65.** $\dfrac{1}{(2z+1)^2}$

Chapter 2   Exercise Supplement

1. -27 **3.** $\dfrac{8}{9}$ **5.** $-k$ **7.** 3 **9.** 12

11. -2 **13.** 17 **15.** 18 **17.** -12 **19.** 10

21. $-\dfrac{18}{5}$ **23.** -7 **25.** -48 **27.** -1

29. $\dfrac{1}{a}$ **31.** $\dfrac{a^3}{b^2c^5}$ **33.** $\dfrac{x^3y^2}{(x-3)^7}$ **35.** $\dfrac{1}{2x}$

37. $\dfrac{1}{-2}$ **39.** $7xy^3z^2$ **41.** $\dfrac{c^2y^4z}{9a^5b^9x^2}$ **43.** $-8a^2b^3$

45. $\dfrac{1}{64}$ **47.** $\dfrac{1}{32}$ **49.** $\dfrac{1}{a^6b^3}$ **51.** c^4

53. $\dfrac{y^{24}z^{12}}{x^{18}}$ **55.** $\dfrac{b^{28}c^{32}y^{12}z^4}{16d^{16}x^8}$ **57.** 7.3567×10^4

59. 7.46×10^5 **61.** 3.87×10^{-2} **63.** 3.76×10^{-1}

65. 5.37×10^{-13} **67.** $387,000$ **69.** $60,090,000$

71. 0.0000388 **73.** 0.000000000008002

75. $2,101,000,000,000,000$ **77.** $1,000,000$

79. $1,000,000,000$ **81.** 2.1×10^{14} **83.** 2.7×10^{78}

85. 2.4×10^{-8} **87.** 2.1×10^1 **89.** 2.9839×10^{-28}

Chapter 2   Proficiency Exam

1. -6 **2.** -15 **3.** -4 **4.** -7 **5.** -7

6. -7 **7.** 3 **8.** 5 **9.** -3 **10.** 125

11. $\dfrac{5x^3z^4}{y^2}$ **12.** $\dfrac{m^6}{4(n-4)^3}$ **13.** $\dfrac{8}{a^{11}}$ **14.** $\dfrac{1}{6}$

15. $\dfrac{1}{(k-6)^5}$ **16.** $\dfrac{(y-3)^{12}}{(y+1)^2}$ **17.** 1 **18.** $\dfrac{1}{a^{12}}$

19. $\dfrac{n^{16}s^8}{m^{20}r^{24}}$ **20.** 1 **21.** 2.71×10^{-4}

22. $890,000$ **23.** 6000 **24.** 1.6×10^{-31}

25. a positive integer

CHAPTER 3

Section 3.1 Exercises

1. An algebraic expression is a number, a letter, or a collection of numbers and letters along with meaningful signs of operation.

3. x is an expression because it is a letter (see the definition).

5. two: $6x$, -10 **7.** three: $5x^2$, $6x$, -2

9. one: $5cz$ **11.** one: 61 **13.** one: $4y^3$

15. two: a, 1 **17.** three: $2x$, x, 7

19. two: $(a+1)$, $(a-1)$ **21.** x^2 **23.** $9b^2$

25. $(a+b)$ **27.** $2ab^2c^2$ **29.** $(a-3)$

31. no common factors **33.** $(a+7)$ **35.** $\dfrac{9}{32}$

37. 12 **39.** 6 **41.** 10 **43.** 8 **45.** xy^5

47. $5x$ **49.** $x^2(x-7)$ **51.** 7 **53.** a

55. $6x^2b^2$ **57.** $3ab^2$ **59.** $(-4)a^5b$

61. $16x^{16}(x-1)^{12}$ **63.** -50 **65.** 1.52×10^{-5}

Section 3.2 Exercises

1. The value of x is equal to six times the value of y.

3. e is equal to 9 less than the value of g.

5. The value of three times t is equal to six times s.

7. The value of r is equal to two ninths times the value of s.

9. The value of f is equal to 55 more than $\dfrac{97}{100}$ times the value of k.

11. The value of q^2 is equal to nine times the value of x^8 plus two times the value of y.

13. 31.4 **15.** 3 **17.** 360 **19.** 48 **21.** 8

23. $42\dfrac{5}{18}$ **25.** 448 **27.** 205.44 **29.** 396

31. 150 **33.** 588.49 **35.** 238,328 **37.** 18.4191

39. 650 **41.** 379.94 **43.** 2,043

45. 1.7298×10^{11} **47.** 6 units **49.** 99.33 tons

51. 195.46474 feet **53.** -8 **55.** $2; 5(a+b), 2x^2$

Section 3.3 Exercises

1. binomial; first (linear); 5, 7

3. binomial; second (quadratic); 4, 9

5. binomial; fourth; 1, 1

7. monomial; first (linear); 5

9. trinomial; third (cubic); 5, 2, 3

11. trinomial; third (cubic); 41, 22, 1

13. monomial; sixth; 2

15. monomial; first (linear); 9

17. trinomial; second (quadratic); 3, -6, 11

19. binomial; ninth; 1, 9

21. binomial; eighth; 6, 3

23. monomial; zero; 5

25. monomial; ninth; 4 **27.** linear

29. quadratic **31.** linear **33.** cubic

35. quadratic **37.** linear **39.** linear

41. cubic **43.** fifth degree **45.** 19th degree

47. . . . there is a variable in the denominator

49. yes **51.** $11-3x=5$ **53.** $z=2$

55. The value of y is 5 more than three times the value of x.

Section 3.4 Exercises

1. $4x$ **3.** $21a$ **5.** $3x$ **7.** $-5k$ **9.** $21y$

11. $-7h$ **13.** $11ab$ **15.** $11a^2$ **17.** $-5y$

19. $622ab^4 + 100a^4b$ **21.** $-7w^2$

23. $4x^3 - 7x^2 - 9x - 11$ **25.** $x-y$

27. $2z^3 - 5z^2 + 16z$ **29.** $-18w^5 + w^4$

31. $16d^3r - 24d^2r - 2$ **33.** 8 **35.** $5z$

37. $6x^3 + 12x + 5$ **39.** $7a + 18$ **41.** $-28b + 18$

43. $31x + 8$ **45.** $10a$ **47.** $80x + 24y^2$

49. $5x + 3$ **51.** $3a^2b + 8b^2 - 4b + 7x$ **53.** yes

55. $3x^2 + 8x - 8$ **57.** $2b^4 + 5b^3 - 5b^2 + 2b + 2$

59. $-3x^3 + 21x^2 + 4x$ **61.** $x^2y^2 - 3x^2y - xy^2$

63. $52k^2 + 6k$ **65.** $128n^2 - 90n - 3m$

67. $144c^2 - 112a + 77b + 1728c$

69. $18x^2y^3 + 5xy^3 + 4y^3$ **71.** $-24a - 16$

73. $-24x^2y^4 - 16x^2y^3 + 104xy^2$ **75.** 4 **77.** $3x$

Section 3.5 Exercises

1. $7x + 42$ **3.** $6y + 24$ **5.** $5a - 30$

7. $12x + 6$ **9.** $36y - 27$ **11.** $-9a - 63$

13. $-4x - 8$ **15.** $-3a + 18$ **17.** $-10a - 5$

19. $-30y + 18$ **21.** $x^2 + 6x$ **23.** $m^2 - 4m$

25. $3x^2 + 6x$ **27.** $6a^2 - 30a$ **29.** $15x^2 + 12x$

31. $2b^2 - 2b$ **33.** $15x^4 + 12x^2$

35. $20a^7 + 12a^6 + 8a^5$ **37.** $-5x^3 - 10x^2$

39. $6x^4y^3 - 12x^3y$ **41.** $2b^6x^3 - 11b^5x^2$

43. $18y^7 - 27y^6 + 72y^5 + 9y^4 - 54y^3$

45. $a^2 + 6a + 8$ **47.** $y^2 + 3y - 18$

49. $i^2 + 2i - 15$ **51.** $6a^2 - 20a + 6$

53. $18y^2 + 93y + 110$ **55.** $-x^2 - x + 12$

57. $x^3 + x^2 + 2x + 2$ **59.** $6x^4 - 7x^2 - 5$

61. $12x^5y^7 + 30x^3y^5 + 12x^3y^3 + 30xy$

63. $4a^2 - 28a - 32$ **65.** $x^3 + 5x^2 + 4x$

67. $y^5 - 5y^4 + 6y^3$ **69.** $5y^8 + 40y^7 + 35y^6$

71. $10x^6y^5 - 5x^5y^4 - 6x^4y^3 + 3x^3y^2$

73. $8c^3 + 40c + 88$

75. $24a^5b^9 + 42a^4b^{11} + 12a^3b^{13} + 84a^3b^3$

77. $x^3 - 6x^2 - 10x + 21$

79. $21a^7 - 22a^5 - 15a^3 - 7a^2 - 2a - 2$

81. $10a^3 + 8a^3b + 4a^2b^2 + 5a^2b - b^2 - 8a - 4b - 2ab$

83. $x^2 + 2x + 1$ **85.** $a^2 + 4a + 4$

87. $-9x^2 + 30x - 25$ **89.** $7x^2 - 5x - 7$

91. $2y^2 - 5xy - 12x$ **93.** $6x^2 - 33x - 1$

95. $9a^2b + 13ab^2 - ab$ **97.** $17a^3 - 11a^2 - 7a - 6$

99. $12x - 9$ **101.** $7x - 10$ **103.** $26x + 5$

105. $-20x - 18$ **107.** 0 **109.** 1.98×10^5

111. degree is 6;   4, 3, -5

Section 3.6   Exercises

1. $x^2 + 6x + 9$ **3.** $x^2 + 16x + 64$

5. $y^2 + 18y + 81$ **7.** $a^2 - 8a + 16$

9. $a^2 - 14a + 49$ **11.** $b^2 + 30b + 225$

13. $x^2 - 24x + 144$ **15.** $y^2 - 40y + 400$

17. $16x^2 + 16x + 4$ **19.** $49x^2 - 28x + 4$

21. $9a^2 - 54a + 81$ **23.** $25a^2 - 30ab + 9b^2$

25. $4h^2 - 32hk + 64k^2$ **27.** $a^2 + \dfrac{2}{3}a + \dfrac{1}{9}$

29. $x^2 + \dfrac{4}{5}x + \dfrac{4}{25}$ **31.** $y^2 - \dfrac{5}{3}y + \dfrac{25}{36}$

33. $x^2 + 2.6x + 1.69$ **35.** $a^2 + a + 0.25$

37. $x^2 - 6.2x + 9.61$ **39.** $b^2 - 0.08b + 0.0016$

41. $x^2 - 25$ **43.** $x^2 - 1$ **45.** $f^2 - 81$

47. $4y^2 - 9$ **49.** $4a^2 - 49b^2$ **51.** $25h^2 - 4k^2$

53. $a^2 - \dfrac{4}{81}$ **55.** $4b^2 - \dfrac{36}{49}$

57. $(a - b)(a - b) = a^2 - ab - ab + b^2 = a^2 - 2ab + b^2$

59. first term squared

61. (a) Square the first term.
    (b) Square the second term and subtract it from the first term.

63. $\dfrac{1}{80}$ **65.** $10m^2 + 9m - 9$

Section 3.7   Exercises

1. $x =$ all real numbers   **3.** $x =$ all real numbers

5. $x =$ all real numbers

7. $x =$ all real numbers except zero

9. $h =$ all real numbers except zero

11. $s =$ all real numbers except 6

13. commutative property of multiplication

15. "how far"   **17.** $16x^2 - 9$

Chapter 3   Exercise Supplement

1. three: $4x^2, 7x, 12$ **3.** two: $c, 8$ **5.** a^2

7. $12y^3$ **9.** $2(a + 2b)$ **11.** no common factors

13. 12   **15.** $2a^2b$ **17.** 8   **19.** $(y + 3)$

21. The value of a is equal to three times the value of b.

23. The value of f is equal to six times g more than one half times the value of m squared.

25. The value of P squared is equal to the value of a cubed times k.   **27.** 10

29. 4.1515×10^{12} **31.** $\dfrac{7}{50}$ or 0.14

33. trinomial;   cubic; 4, 3, 1

35. monomial;   zero;   147

37. binomial;   tenth;   9, 10

39. . . . because there is a variable in the denominator

41. linear   **43.** linear   **45.** quadratic

47. $22x^2y^3 + 3xy + 6$ **49.** $11y^2 + 38y + 14$

51. $120y^3 + 86y^2 + 24y + 45$ **53.** $5x^2 + 2x + 3$

55. $60c^2 + 120a + 60b$

57. $36m^4 + 7m^3 + 4m^2 + 16m$

59. $3a^2b^2c^2 + 7abc^2 + ab^2c + 12abc$ **61.** $6a^6 + 30a^4$

63. $10m^{13} + 15m^{10} + 5m^8 + 5m^7 + 5m^6$

65. $x^2 + 5x + 6$ **67.** $a^2 + 4a + 3$ **69.** $-6xy$

71. $-8x^4$ **73.** $4x - 32$ **75.** $-15a^2 + 18a$

77. $-11y^3 + 15y^2 + 16y + 10$ **79.** -5

81. $-6a^2b - 8q^2b - 9b^2$ **83.** $-a - 27$

85. $2x - 4$ **87.** $2x - 29$ **89.** $x^2 - 11x + 24$

91. $16b^2 - 4bc - 2c^2$ **93.** $a^2 - 6a + 9$

95. $36x^2 - 48x + 16$ **97.** $x^2 + 2xy + y^2$

99. $m^2 - 1$ **101.** $9c^2 - 100$ **103.** $25 - 4b^2$

105. $2y^2 + 7ay + 3a^2$ **107.** $x^4 - x^2 - 6$

109. $48y^2 + 32y - 256$

111. $m^4n + 3m^3n^2 + 2m^2n^3$

113. $3p^5 + 21p^4 + 63p^3 + 129p^2 + 84p$

115. $x^2 - 4x + 4$ **117.** $x^4 + 2x^2y + y^2$

119. $9x^4y^6 - 24x^6y^4 + 16x^8y^2$

121. all real numbers

123. all real numbers except 2   **125.** x can equal any real number; y can equal any real number except -10

Chapter 3   Proficiency Exam

1. two: $3a(a+1), -(a+2)(a-3)$ **2.** $5x^3$

3. $8x(b-6)$ **4.** $8y^3z$ **5.** 108

6. trinomial; 5th degree; numerical coefficients: 3, 4, 8

7. $15x^2 + 5x - 3$ **8.** $6a^2$ **9.** $x^2 + 6x + 8$

10. $6a^2 + 16a - 70$ **11.** $y^2 + 6y + 9$

12. $36a^2 + 84ay + 49y^2$ **13.** $16x^2 - 72xy + 81y^2$

14. $18x^4 + 51x^3 + 15x^2$ **15.** $12a^2 - 13ab + 3b^2$

16. $-18y^4 - 12y^3 + 24y^2$ **17.** $-4b^7 + 8b^5 - 4b^3$

18. $4a^6 + 12a^3b^2 + 9b^4$ **19.** $4a^2 - 13a + 11$

20. $25h^2 - 4k^2$ **21.** $-2a^2 + 6a + 11$ **22.** $10x + 7$

23. -3 **24.** 392   **25.** all real numbers except (-3)

CHAPTER 4

Section 4.1   Exercises

1. conditional   **3.** identity   **5.** identity

7. solved   **9.** not solved   **11.** not solved

13. not solved   **15.** $k = -9$ **17.** $y = -17$

19. $x = -14$ **21.** $g = -287$ **23.** $x = -443$

25. $y = -18.059$ **27.** $n = 4 - m$

29. $b = -a + 3c + d - 2f$ **31.** $c = 2a - 3b - 11$

33. $\dfrac{4y^4}{x^2}$ **35.** $9x^2 - 6x + 1$

Section 4.2   Exercises

1. $x = 14$ **3.** $x = 8$ **5.** $x = 14$ **7.** $a = -16$

9. $p = -18$ **11.** $a = -4$ **13.** $x = 7$

15. $k = -42$ **17.** $x = 6$ **19.** $k = 42$

21. $x = 768$ **23.** $m = -56$ **25.** $f = -6386$

27. $k = 0.06$ **29.** $y = 9.453$ **31.** $m = \dfrac{-10}{3}$

33. $h = \dfrac{21}{8}$ **35.** $p = \dfrac{7r}{q}$ **37.** $b = \dfrac{2d}{a}$

39. $b = \dfrac{15c}{8}$ **41.** $t = -\dfrac{3p^2}{4}$ **43.** $\square = \dfrac{2\triangledown}{3}$

45. binomial; 3rd degree; $10, -7$

47. all real numbers except -7

Section 4.3   Exercises

1. $x = 5$ **3.** $a = 7$ **5.** $y = -5$ **7.** $x = -3$

9. $y = -2$ **11.** $x = 36$ **13.** $y = 24$

15. $m = -44$ **17.** $k = -5$ **19.** $x = -8$

21. $k = -14$ **23.** $y = -5$ **25.** $y = 20$

27. $k = -1$ **29.** contradiction   **31.** $x = -3$

33. $x = \dfrac{19}{14}$ **35.** $k = 3$ **37.** $m = 2$

39. $R = 112$ **41.** $S_x{}^2 = F \cdot S_y{}^2$; $S_x{}^2 = 7.1604$

43. $y = \dfrac{x - 7}{4}$ **45.** $y = \dfrac{-2x + 12}{5}$

47. $n = \dfrac{5m + h}{2}$ **49.** $j = \dfrac{\star\triangle - \square}{9}$

51. $(x + 3)^3(x - 2)^7$ **53.** $4x^2 - 4x + 1$ **55.** $x = \dfrac{-15}{4}$

Section 4.4   Exercises

1. $a - 4$ **3.** $b + 7$ **5.** $-5 + c$ **7.** $2d + 14$

9. $\dfrac{1}{3} - e$ **11.** $\dfrac{4}{9}f = 21$ **13.** $3g = 2g + 9$

15. $2h + 6 = 30$ **17.** $k - 25 = 3.019$

19. $\dfrac{m}{4} = 68$ **21.** $\dfrac{n}{10} = n - 1$ **23.** $x = x + 4x$

25. $\dfrac{Q + 7}{2} = 22$ **27.** $\dfrac{\frac{r + r}{3}}{3} = 15$ **29.** $\dfrac{5}{s + 6} = 14$

31. $\dfrac{20}{8x} + 1 = 9$ **33.** $\dfrac{v}{10} + 4 = 24$

35. $\dfrac{w + 6}{2} + 5 = 43$ **37.** $7y + 2y = 90$

39. $\dfrac{(z + 6)13}{6z} = 59$ **41.** $\dfrac{x - 11}{15} - 1 = 5$

43. $n - 5$ **45.** $m - 1$ **47.** $p - 8 = 3$

49. $\dfrac{23}{2n - 2} = 34$ **51.** graph

53. $|a| = \begin{cases} a, \text{ if } a \geq 0 \\ -a, \text{ if } a < 0 \end{cases}$ **55.** $x = 3$

Section 4.5   Exercises

1. Step 1:   Let $x =$ the unknown quantity.
Step 2:   The equation is $x - 18 = 52$.
Step 3:   (Solve the equation.) Add 18 to each side.
$x - 18 + 18 = 52 + 18$
$x = 70$
Step 4:   (Check) $70 - 18 = 52$;   True.
Step 5:   The number is 70.

3. Step 5:   The number is 30.

5. Step 5:   The original quantity is 29.

7. Step 5:   Last year's output was 50 items.

9. Step 5:   A proton weighs 4923.16 units.

11. Step 5:   The radius of the earth is 6378 km.

13. Step 5:   The length of the shorter piece is 3 feet, and the length of the longer piece is 9 feet.

15. Step 5:   The statistician has collected 260 pieces of data.

17. Step 5:   Each beaker will hold $263\frac{1}{3}$ ml of chloride solution.

19. Step 5:   The unknown number is 26.

21. Step 5:   The unknown number is 21.

23. Step 5:   The unknown number is $\frac{41}{3}$.

25. Step 5:   One unknown number is 6; the other is 21.

27. Step 5:   The unknown number is $\frac{245}{41}$.

29. Step 5:   The age of D is 3 months; C is 6 months; B is 36 months; A is 144 months.

31. Step 5:   The first integer is 11; second is 12; third is 13.

33. **(a)** Step 5:   The time passed in space is 0.62 days.
    (b) Step 5:   9.3 years have passed on the spacecraft.
    (c) Step 5:   96.77 years have passed on earth.
    (d) Step 5:   Earth year when she returns will be 2387.

35. conditional   **37.** $x = -2$

Section 4.6   Exercises

1. $x < 5$ **3.** $y \geq -17$ **5.** $x \leq 3$ **7.** $z < -\frac{1}{2}$

9. $x \geq -4$ **11.** $z > -11$ **13.** $x \geq 48$

15. $x \geq 18$ **17.** $x \leq \frac{6}{5}$ **19.** $b > -10$

21. $x < -\frac{15}{4}$ **23.** $y > \frac{16}{21}$ **25.** $y \geq 2$

27. $x \leq 15$ **29.** $x \geq 4$ **31.** $x \geq 5$ **33.** $y < -3$

35. $x < 6$ **37.** $x \leq -7$ **39.** $x \leq -4$ **41.** $x > 8$

43. $y \geq -9$ **45.** $x < -2$ **47.** $x \geq -3$

49. $x < -2$ **51.** $x > -2$

53. First number: any number strictly smaller than 6.
Second number: any number strictly smaller than 30.
No smallest possible value for either number.
No largest possible value for either number.

55. $x^{10}y^{15}z^{10}$ **57.** $2x^2 + x - 28$ **59.** 16 inches

Section 4.7   Exercises

1. $(1, 22)$ **3.** $(4, 26)$ **5.** $\left(-3, \frac{9}{4}\right)$

7. $\left(-6, -\frac{29}{3}\right)$ **9.** $\left(\frac{1}{2}, 0\right)$ **11.** $(1, 1)$

13. $\left(\frac{9}{4}, \frac{3}{2}\right)$ **15.** $\left(8, \frac{111}{20}\right)$ **17.** $(-40, -1)$

19. $\left(a, \frac{3 - 2a}{31}\right)$ **21.** $(2, -30)$ **23.** $\left(2, -\frac{9}{5}\right)$

25. $\left(6, -\frac{21}{2}\right)$ **27.** $(-7, -40)$ **29.** $\left(0, \frac{7}{4}\right)$

31. $\left(1, -\frac{38}{3}\right)$ **33.** $(1, 5)$ **35.** $(3, -10)$ **37.** $(1, 6)$

39. **(a)** Approximately 147 mph using $(5, 147.25)$
    (b) Approximately 157 mph using $(10, 156.9)$
    (c) Approximately 188 mph using $(26, 187.78)$
    (d) Approximately 196 mph using $(30, 195.5)$

41. **(a)** Approximately 5 sinkings using $(4, 4.92)$
    (b) Approximately 10 sinkings using $(9, 10.12)$
    (c) Approximately 11 sinkings using $(10, 11.16)$

43. **(a)** Approximately 6.4 using $(40, 6.4)$
    (b) Approximately 4.752 using $(42, 7.752)$

45. **(a)** Approximately 29 items using $(80, 29.42)$
    (b) Approximately 43 items using $(95, 42.77)$

47. **(a)** Approximately 87 units using $(12, 87.31)$
    (b) Approximately 45 units using $(15, 44.86)$

49. $12x^2 + 17x - 5$ **51.** $x = 0$ **53.** $-3 < y < 4$

Chapter 4   Exercise Supplement

1. $y = 8$ **3.** $r = 17$ **5.** $x = -10$ **7.** $x = 0$

9. $x = 16$ **11.** $r = -4$ **13.** $x = 6$ **15.** $x = -6$

17. $x = 4$ **19.** $x = -1$ **21.** $x = -2$

23. $m = 24$ **25.** $y = 486$ **27.** $c = 90$

29. $y = -15$ **31.** $a = -\frac{55}{6}$ **33.** $c = 16$

35. $x = 28$ **37.** $r = -\frac{7}{3}$ **39.** $x = -\frac{45}{4}$

41. $x = 3$ **43.** $a = -4$ **45.** $r = -\frac{17}{4}$

47. $W = 13$ **49.** $R = 690$ **51.** $x = \frac{y - 8}{5}$

53. $y = -\frac{1}{2}x - 2$ **55.** $b = -\frac{2ct - 10a}{3}$

57. $x(4 + 7)$ **59.** $\frac{2}{5}x - 5$ **61.** $3x = 40$

63. $\dfrac{(4x-5)}{7}=x+10$ **65.** $x(x+11)=6$

67. $\dfrac{x}{2x}+8x=-1$ **69.** $\dfrac{x-1}{5x}=x^3$

71. last year's price $=\$96$ **73.** $x=860$

75. $x=11$
$x+2=13$
$x+4=15$
$x+6=17$

77. $x=116.13139$ **79.** 82 pieces of data

81. $y<12$ **83.** $x>5$ **85.** $a\le-\dfrac{7}{2}$

87. $a\ge11$ **89.** $b\ge-12$ **91.** $c\ge-9$

93. $y<-1$ **95.** $x>-\dfrac{9}{14}$ **97.** $x>\dfrac{3}{2}$

99. $x<\dfrac{7}{3}$ **101.** $(-3,19)$ **103.** $(2,4)$

105. $(4,0)$ **107.** $\left(\dfrac{7}{2},-1\right)$ **109.** $(-5,0)$

Chapter 4   Proficiency Exam

1. $x=6$ **2.** $a=\dfrac{-13}{6}$ **3.** $a=-16$ **4.** $x=10$

5. $y=-3$ **6.** $b=-10$ **7.** $a=-\dfrac{3}{2}$ **8.** $y=-7$

9. $x=9$ **10.** $p=\dfrac{6q-3}{2}$ **11.** $T=\dfrac{Vp}{nR}$

12. $\triangle=\star\nabla-\square$ **13.** $a\ge12$ **14.** $a>2$

15. $a\ge-23$ **16.** $x<6$ **17.** $3+2a$

18. $\dfrac{2}{3}x-8$ **19.** $2+4x$ **20.** $2x(x^3)=12$

21. $\dfrac{x-5}{x+10}=6x$ **22.** $x=15$

23. There are no three consecutive odd integers that add to 38.

24. $x<4$ **25.** $(3,13)$

CHAPTER 5

Section 5.1   Exercises

1. 5 **3.** $2a$ **5.** 3 **7.** $5x^3$ **9.** $2x^3$

11. $2xy$ **13.** $3b^3c$ **15.** $-9xb$ **17.** $4x^3bf^7$

19. $7a^{17}b^5c^{18}d$ **21.** $\dfrac{1}{4}x^3$ **23.** d **25.** 1

27. $4(x+y)^3$ **29.** $-13(x-5y)^3(x-3y)^5$

31. $(x+y)$ **33.** $6x^ny^{n-6}$ **35.** $x^{12}z^6$
37. $4x^2-16x+16$

Section 5.2   Exercises

1. $2x+5$ **3.** $x+5$ **5.** $a+3$ **7.** $3x+4$

9. $9a-2$ **11.** $x-2$ **13.** $2a+1$

15. $\dfrac{3}{2}x^2+x+3$ **17.** $3x^2+x-2$

19. y^2-2y+1 **21.** $2y^2-5y-6$ **23.** $9x^2+10$

25. $4x-3$ **27.** $5a^3+6a^2+8a+4$

29. $2a^3-a^2-3a+4$ **31.** x^5+4x^3-4

33. $2x-7$ **35.** $5a+2b$ **37.** $4x^4y^4+x+1$

39. $8a^2bc^6+7ab^2c^5-6b^3c^4-1$ **41.** $-a-2$

43. $-2x-3$ **45.** $-a-b$ **47.** $-a+b-c$

49. $a+b+c$ **51.** $6x^2y$ **53.** $a=3$

Section 5.3   Exercises

1. $9(a+2)$ **3.** $4(2b+3)$ **5.** $2(2x-3)$

7. $7(3y-4)$ **9.** $6x(2x+3)$ **11.** $2(4y^2+9)$

13. $3(y^2-2)$ **15.** $6y(y-1)$ **17.** $b(y^2+1)$

19. $5x(a^2x+2)$ **21.** $5(2x^2+x-3)$

23. $3(5y^3-8y+3)$ **25.** $b(y^3+y^2+y+1)$

27. $x(9x+6y+4)$ **29.** $13x^2y^5(-c-3)$

31. $-2(3y^3+4y^2+7y-5)$ **33.** $N(x+y)$

35. $A(x-y)$ **37.** $(x-9)(a+b)$

39. $(9a-b)(w-x)$

41. $3x^2y^3(x^3y-4xy+9x^3-2y^3)$

43. $-x^2y(11xy-16x^2y^2-2)$ **45.** $t=3$

Section 5.4   Exercises

1. $(2b+3)(a+9)$ **3.** $(y+1)(x+3)$

5. $(a+5b)(r+4s)$

7. $3(4mx-2bx+7ay-3by)$ Not factorable by grouping

9. $(a^2+3)(b^2+2)$ **11.** Not factorable by grouping

13. $(x+y)(x-3)$ **15.** Not factorable by grouping

17. $(4s-9)(3s-2t)$ **19.** $a^2b^2(a^2b^2+2)(1+3ab)$

21. $y(5y^2z+3xw)(x^2-2z)$

23. $(m^6n^7p^3-8qt^2)(5m^4n^{10}-p)$ **25.** 6×10^{-3}

27.

Section 5.5   Exercises

1. $(a + 3)(a - 3)$ **3.** $(x + 4)(x - 4)$

5. $(a + 10)(a - 10)$ **7.** $4(a + 4)(a - 4)$

9. $3(x + 3)(x - 3)$ **11.** $(2a + 5)(2a - 5)$

13. $(6y + 5)(6y - 5)$ **15.** $3(2a + 5)(2a - 5)$

17. $2(2y + 5)(2y - 5)$ **19.** $(xy + 5)(xy - 5)$

21. $(x^2y^2 + 3a)(x^2y^2 - 3a)$ **23.** $b^2(2a + 3)(2a - 3)$

25. $(a + b)(a - b)$ **27.** $(x^2 + y^2)(x + y)(x - y)$

29. $(a^4 + y)(a^4 - y)$ **31.** $(b^3 + x^2)(b^3 - x^2)$

33. $(5 + a)(5 - a)$ **35.** $4(5 + 3b^2)(5 - 3b^2)$

37. $(x^2 + 4)(x + 2)(x - 2)$

39. $(a^2 + b^2)(a + b)(a - b)$

41. $(x^6 + x^6)(x^3 + y^3)(x^3 - y^3)$

43. $ac^2(a + 5)(a - 5)$

45. $(7xy^2z^3 + 8a^2bc^4d^5)(7xy^2z^3 - 8a^2bc^4d^5)$

47. $(x + 5)^2$ **49.** $(a + 6)^2$ **51.** $(y + 10)^2$

53. $(a - 2)^2$ **55.** $(x - 5)^2$ **57.** $(a - 12)^2$

59. $(x + 1)^2$ **61.** $(b - 1)^2$ **63.** $(3x + 1)^2$

65. $(4a - 3)^2$ **67.** $(3x + y)^2$ **69.** $(6a + 5b)^2$

71. $3(2a - 5)^2$ **73.** $2(4x + 1)^2$

75. not factorable   **77.** $x^3(x + 4)^2$

79. $(m - 3)(x - y)$

Section 5.6   Exercises

1. $(x + 3)(x + 1)$ **3.** $(x + 3)(x + 4)$

5. $(y + 6)(y + 2)$ **7.** $(y - 4)(y - 1)$

9. $(a + 4)(a - 1)$ **11.** $(x - 7)(x + 3)$

13. $(y + 8)(y + 2)$ **15.** $(y - 7)(y - 1)$

17. $(a + 6)(a - 5)$ **19.** $(a - 10)(a - 2)$

21. $(x + 6)(x + 7)$ **23.** $(x + 5)(x + 8)$

25. $(b + 8)(b + 7)$ **27.** $4(x + 2)(x + 1)$

29. $5(y^2 - 14y + 88)$ **31.** $x(x + 4)(x + 2)$

33. $x^2(x + 7)(x + 2)$ **35.** $4a(a - 7)(a - 3)$

37. $2y^2(n - 8)(n + 3)$ **39.** $y^3(y + 6)(y + 7)$

41. $(2x - a)(3y + a)$ **43.** $(4x - 3)(x + 5)$

Section 5.7   Exercises

1. $(x + 2)(x + 1)$ **3.** $(2x + 5)(x + 1)$

5. $(2x + 3)(x + 4)$ **7.** $(3x - 4)(x + 1)$

9. $(2x - 3)(2x + 7)$ **11.** not factorable

13. $(8y - 1)(2y - 3)$ **15.** $(5x + 2)(2x + 5)$

17. not factorable   **19.** $(6x - 1)(4x - 5)$

21. $(6x - 1)(4x + 5)$ **23.** $(3x + y)(2x + y)$

25. $(5r + 1)(r - 5)$ **27.** $(x + 29)(x - 5)$

29. $(4x + y)(4x - 3y)$ **31.** not factorable

33. $2(4a^2 + 5ab - 3b^2)$ **35.** $6(2b^2 + a)(b^2 + 2a)$

37. $3(a^3 + 2b^2)(a^3 - 3b^2)$

39. $2a^2z^2(7 - 20a - 23a^2)$ or $-2a^2z^2(23a^2 + 20a - 7)$

41. $x^4 + x^3 - 12x^2$ **43.** $x = \dfrac{11}{2}$

Chapter 5   Exercise Supplement

1. $16a^4$ **3.** $4a^2c$ **5.** $17(a + 1)(b + 3)^4$

7. $4x^4y^4(x + y)^3(x + 3y)^2$ **9.** $-4x^5y^{n-8}$

11. $(x - 2)$ **13.** $4a - 1$ **15.** $-3y^2 + 9y - 12$

17. $x^2y - 2x + 3$ **19.** $x + 2y + c^2$ **21.** $4(2a + 1)$

23. $3y(y + 9)$ **25.** $3(7x + 38)$ **27.** $m(a^3 - 1)$

29. $(r + 1)[r^2(r + 1)^2 - 3r(r + 1) + 1]$

31. $(10 - 3x)(23 + 4x)$ **33.** $(x - 3)(y + 4)$

35. $(a + b)(a - 7)$ **37.** $(r - 1)(r + s)$

39. $2(a + 1)(a - 1)$ **41.** $(m + 6)(m - 6)$

43. $(a + 4)^2$ **45.** not factorable

47. $(a + 5)(a + 4)$ **49.** $(x + 10)(x + 4)$

51. $(n - 7)^2$ **53.** $(a - 5)(a - 4)$

55. $(4a + 3)(a - 3)$ **57.** $(6a - 1)(7a + 2)$

59. $2(28m^2 + 13m + 3)$ **61.** $(2x + 3y)(2x - y)$

63. $2(x - 2)(x + 5)$ **65.** $x(x + 4)(x - 1)$

67. $b(15a^2b - a - 2)$ **69.** $(6a - 1)\left(3a - \dfrac{1}{2}\right)$

71. $(2x - 3y)^2$ **73.** $(r^3s^4 + 3p^2q^6)^2$

75. $(9a^4b^6c^5 + 5x^{10}y^9)(9a^4b^6c^5 - 5x^{10}y^9)$

Chapter 5   Proficiency Exam

1. $9a^2 + 3a + 3$ **2.** $5x^ny^{2n-3}$

3. $-14x^2y^2(y^2b + 2yb + 3)$ **4.** $(a + c)(y + 2)$

5. $xy(xy - 2)(6z + 5y)$ **6.** $4(a + 2c)(a - 2c)$

7. $(m^2 + n^2)(m + n)(m - n)$ **8.** $(b + 4)^2$

9. $(3y - 5)^2$ **10.** not factorable

11. $(x - 6)(x + 5)$ **12.** $4x^2(x^2 - 5)(x + 2)(x - 2)$

13. $(9x - 2)(x + 3)$

CHAPTER 6

Section 6.1   Exercises

1. $x = 3$

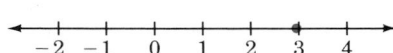

3. $x = \dfrac{1}{2}$

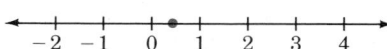

5. $y = 1$

7. $z = \dfrac{1}{3}$

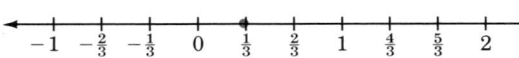

9. $r = \dfrac{1}{28}$

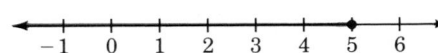

11. $x \le 5$

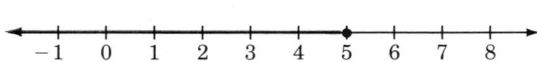

13. $x > -17$

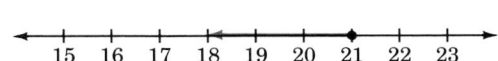

15. $m \le 5$

17. $x \le -5$

19. $y \le 21$

21. $y \ge -\dfrac{32}{5}$

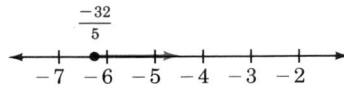

23. $2 \le x < 3$

25. $3 \le x < 5$

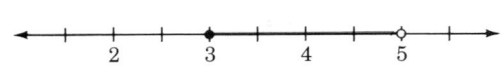

27. $5x^2$ **29.** $(-2, 18)$

Section 6.2   Exercises

1.
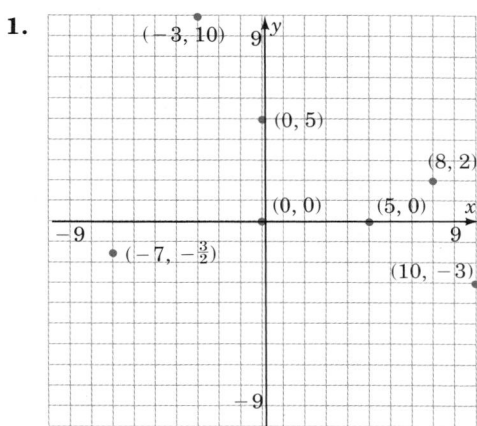

3. Coordinates of the origin are $(0, 0)$.

5.

I	II	III	IV
$x > 0$	$x < 0$	$x < 0$	$x > 0$
$y > 0$	$y > 0$	$y < 0$	$y < 0$

7. Yes, there does appear to be a relation.

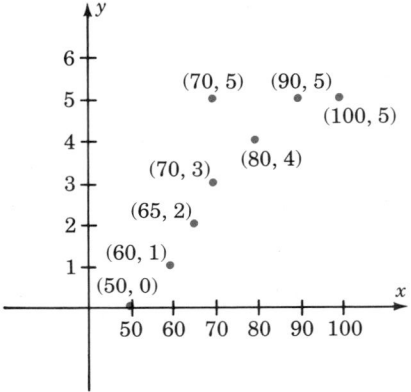

9. Yes, there does appear to be a relation.

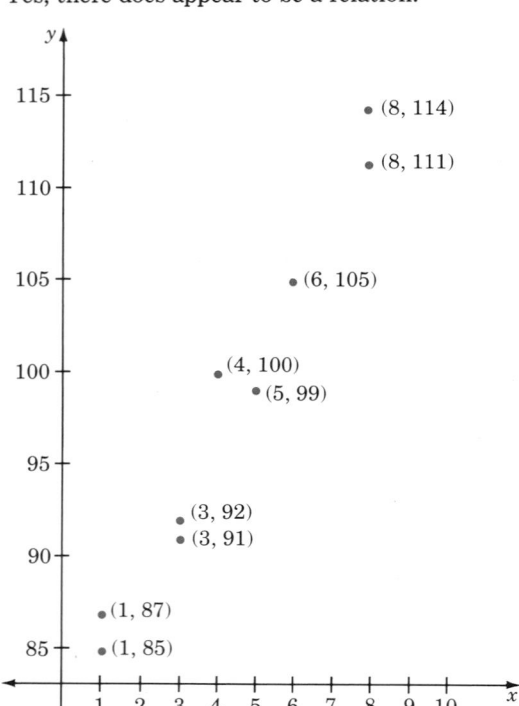

11. equation **13.** contradiction

Section 6.3 Exercises

1.

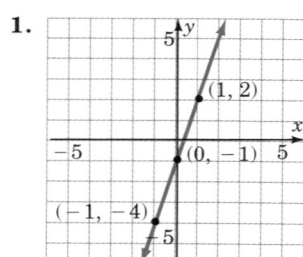

3.

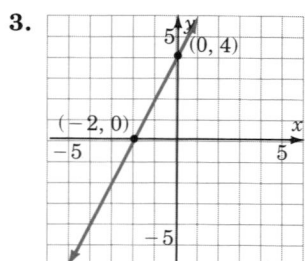

5.

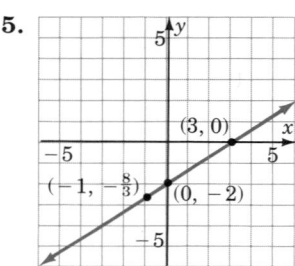

7.

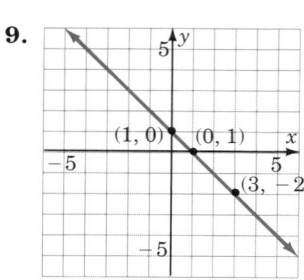

9.

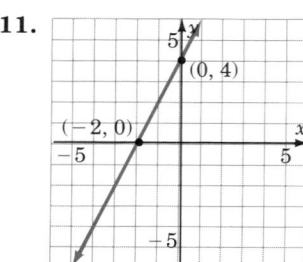

11.

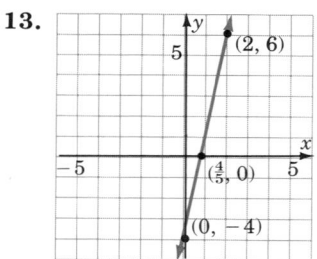

13.

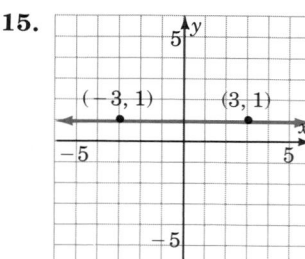

15.

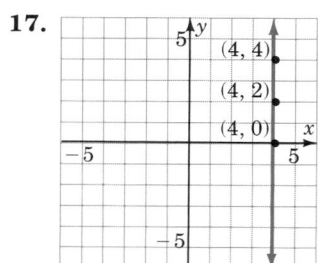

17.

19. $x = \dfrac{3}{2}$

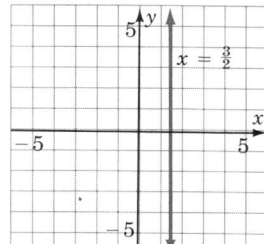

21. $y = -2$

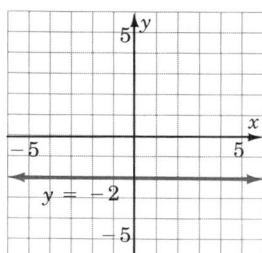

23.

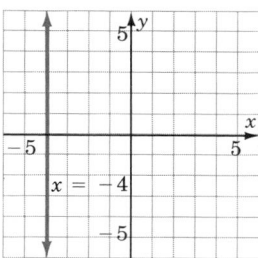

25.

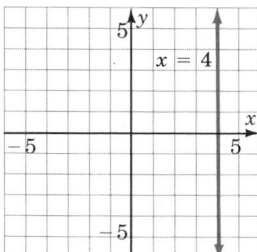

27.

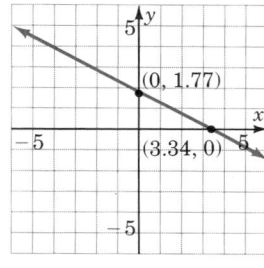

29.

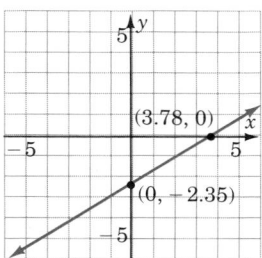

31. commutative property of addition

33. $3x^2 - 19x - 14$ **35.** quadrants

Section 6.4   Exercises

1. slope $= 3$; y-intercept $= (0, 4)$

3. slope $= 9$; y-intercept $= (0, 1)$

5. slope $= -4$; y-intercept $= (0, 5)$

7. slope $= -6$; y-intercept $= (0, -1)$

9. slope $= -1$; y-intercept $= (0, 2)$

11. slope $= 4$; y-intercept $= (0, 5)$

13. slope $= -4$; y-intercept $= (0, 9)$

15. slope $= \dfrac{2}{7}$; y-intercept $= (0, -12)$

17. slope $= -\dfrac{4}{5}$; y-intercept $= \left(0, -\dfrac{4}{7}\right)$

19. slope $= \dfrac{6}{5}$; y-intercept $= \left(0, -\dfrac{1}{10}\right)$

21. slope $= 1$; y-intercept $= (0, -3)$

23. slope $= -\dfrac{5}{3}$; y-intercept $= (0, 2)$

25. slope $= \dfrac{1}{4}$; y-intercept $= \left(0, -\dfrac{1}{4}\right)$

27. $m = \dfrac{4}{3}$ **29.** $m = -\dfrac{7}{4}$ **31.** $m = 2$

33. $m = -\dfrac{7}{6}$ **35.** $m = -4$ **37.** $m = 1$

39. $m = 2$ **41.** $m = 2$

43. $m = 0$ (horizontal line $y = 3$)

45. No slope (vertical line at $x = 8$)

47. $m = 0$ (horizontal line at $y = -6$)

49. decline

51. slope $= -0.31$
  y-intercept $= (0, 0.35)$

53. slope $= 0.64$
  y-intercept $= (0, 2.71)$

55. $m = 1.31$

57. $m = 0$ (horizontal line at $y = 227.61$)

59. No slope (vertical line $x = 88.81$)

61. 1 if $xyw \neq 0$ **63.** 10

65.

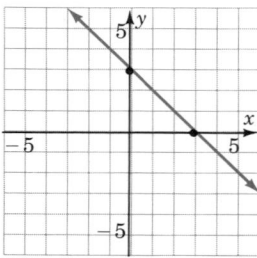

Section 6.5   Exercises

1.

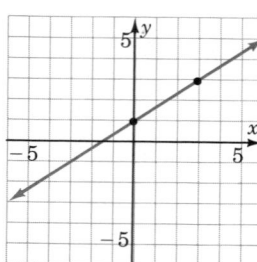

3.

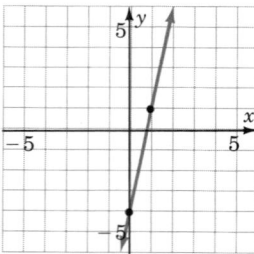

5.

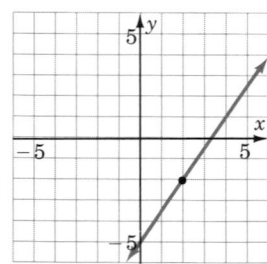

7.

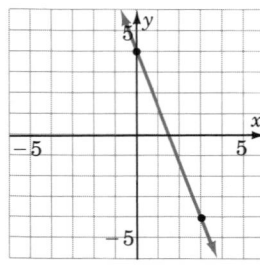

9.

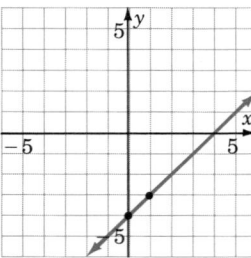

11.

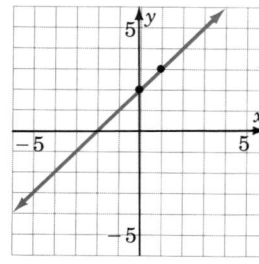

13.

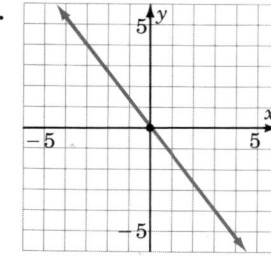

15.

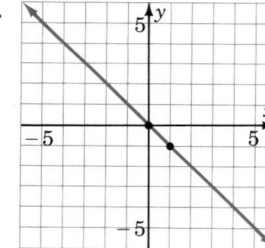

17.

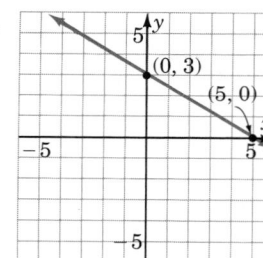

19. $x \leq 1$

21.

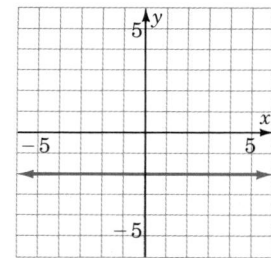

23. $m = \dfrac{-2}{3}$

Section 6.6   Exercises

1. $y = 3x + 4$ **3.** $y = 8x + 1$ **5.** $y = -6x - 1$

7. $y = -\dfrac{3}{2}x$ **9.** $y = x + 5$ **11.** $y = 8x - 32$

13. $y = -x + 6$ **15.** $y = -2x + 1$ **17.** $y = \dfrac{8}{5}x$

19. $y = x + 3$ **21.** $y = 3$ (horizontal line)

23. $x = 4$ (vertical line)   **25.** $y = x$

27. $y = -\dfrac{9}{5}x + \dfrac{2}{5}$ **29.** $y = x - 2$

31. $y = \dfrac{2}{5}x + 1$ **33.** $y = \dfrac{1}{4}x + 1$ **35.** $x = -4$

37. $y = -3x - 1$ **39.** y-intercept   **41.** $m = 1$

Section 6.7   Exercises

1.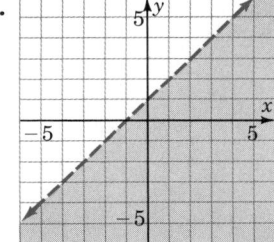

3.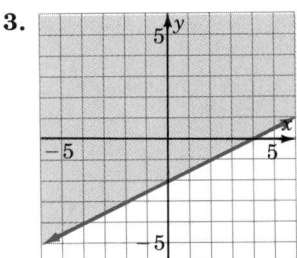

5.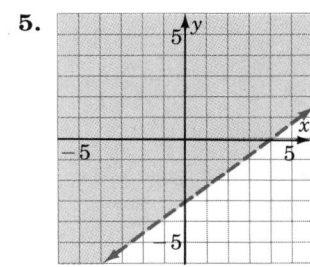

7.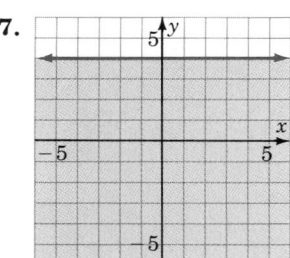

9.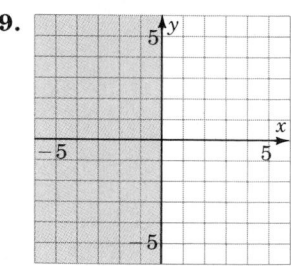

11.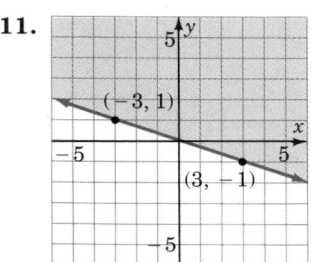

13.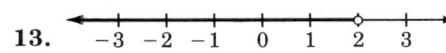

15. $m = \dfrac{y_2 - y_1}{x_2 - x_1}$

17. $y = 4x + 6$

Chapter 6   Exercise Supplement

1. $x = 4$

3. $x = \dfrac{3}{5}$

5. $y \ge -2$

7. $x \le \dfrac{8}{3}$

9. $0 \le y < 3$

11.

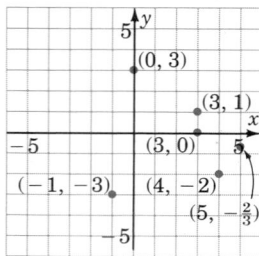

13. a straight line

15.

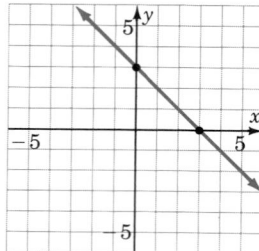

17.

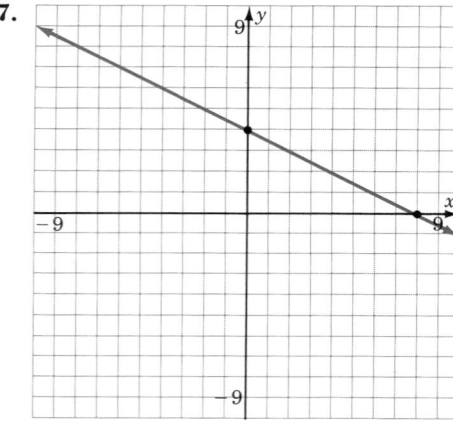

19.

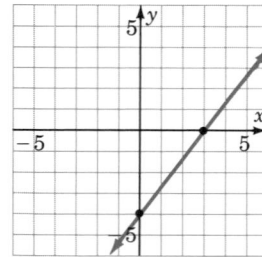

21.

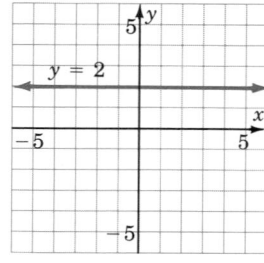

23.

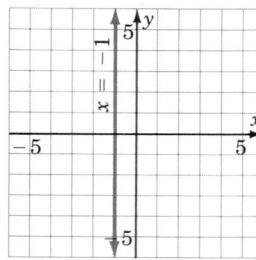

25.

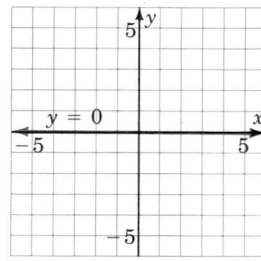

27. measure

29. slope: 4
y-intercept: $(0, 10)$

31. slope: 9
y-intercept: $(0, -1)$

33. slope: -5
y-intercept: $(0, -4)$

35. slope: -6
y-intercept: $(0, 0)$

37. slope: $\dfrac{5}{4}$

y-intercept: $\left(0, \dfrac{1}{4}\right)$

39. slope: $-\dfrac{4}{5}$

y-intercept: $\left(0, \dfrac{6}{5}\right)$

41. slope: 2
y-intercept: $(0, 4)$

43. slope: -1

y-intercept: $\left(0, \dfrac{1}{3}\right)$

45. slope: 0
y-intercept: $(0, 4)$

47. slope: -2 **49.** slope: $\dfrac{7}{2}$ **51.** slope: $\dfrac{3}{2}$

53. No slope **55.** slope: $\dfrac{57}{4}$

57. The slopes of parallel lines are equal.

59. $y = 3x - 6$ **61.** $y = x - 2$ **63.** $y = -11x - 4$

65. $y = -x$ **67.** $y = 2x + 3$ **69.** $y = -5x + 7$

71. $y = -2x + 2$ **73.** $y = 2x - 1$

75. $y = -2x + 13$ **77.** $y = \dfrac{2}{5}x + \dfrac{11}{5}$

79. $y = \dfrac{2}{3}x - 5$ **81.** $y = 7$ (zero slope)

83. $x = -1$ (no slope)   **85.** $y = -2x + 2$

87. $y = \dfrac{2}{3}x - 2$ **89.** $y = -2$ **91.** $y = 1$

93.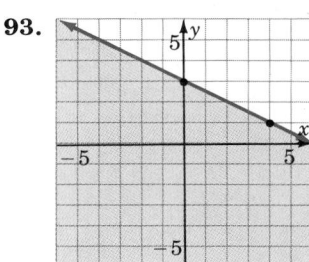

95.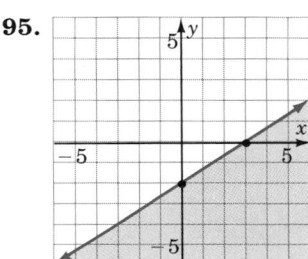

97.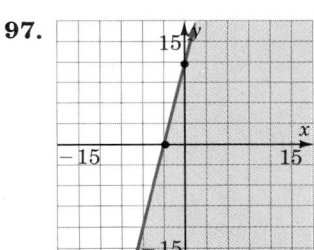

99.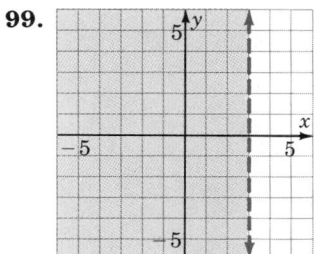

Chapter 6   Proficiency Exam

1. $x < 3$

2. $-14 < x \le -10$

3.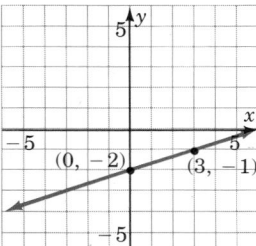

4. $(0, 1), (3, 3), (-3, 0), (2, -3)$

5. straight line   **6.** general form

7. slope-intercept   **8.** It lies on the line.

9. 2 units up   **10.** $\dfrac{3}{2}$

11. no slope; vertical line at $x = -2$

12. slope $= -\dfrac{2}{3}$, y-intercept is $\left(0, -\dfrac{1}{3}\right)$ **13.** rise

14. $y = 4x - 3$ **15.** $y = -\dfrac{3}{2}x + \dfrac{4}{3}$

16. $y = \dfrac{2}{3}x + \dfrac{8}{3}$ **17.** $y = 7x$ **18.** $y = \dfrac{1}{3}x + \dfrac{1}{3}$

19. $y = \dfrac{1}{3}x - 2$

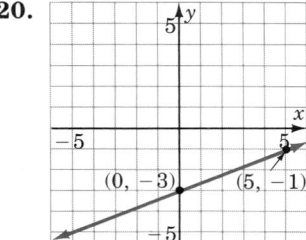

20.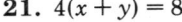

21. $4(x + y) = 8$

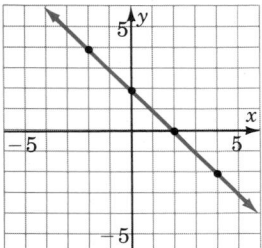

22.

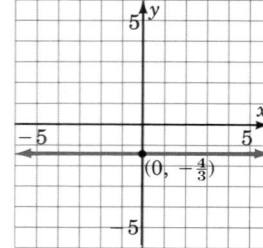

$(0, -\frac{4}{3})$

23. $x = -2$

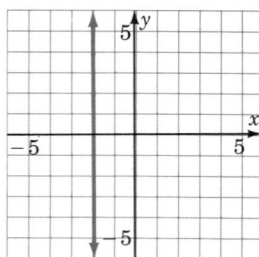

24.

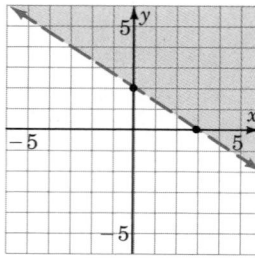

25. $y = -\frac{1}{3}x + 3$

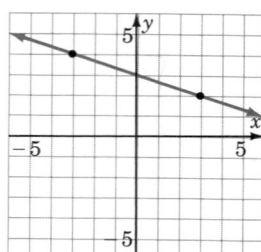

CHAPTER 7

Section 7.1 Exercises

1. $x \neq 4$ **3.** $x \neq -1$ **5.** $x \neq -2, 2$

7. $x \neq -6, 6$ **9.** $b \neq 0, -6$ **11.** $x \neq 0, 10, -1$

13. $x \neq 0, 5, 7$ **15.** $b \neq 1, 3$ **17.** $y \neq 5, -4$

19. $x \neq 0, \frac{1}{2}, -\frac{2}{3}$

21. $(-3)5 = -15, -(3 \cdot 5) = -15$

23. $-(1 \cdot 4) = -4, 4(-1) = -4$

25. $(-9)(-10) = 90$ and $(9)(10) = 90$

27. -2 **29.** $-2x - 7$ **31.** $-x + 4$

33. $a + 1$ **35.** $-y - 10$ **37.** $1 \leq x < 3$

39. $(x - 6)^2$

Section 7.2 Exercises

1. $\dfrac{2}{(x - 4)}$ **3.** $\dfrac{3}{(y - 7)}$ **5.** $\dfrac{1}{(x - 2)}$ **7.** $\dfrac{1}{4}y$

9. $8ab$ **11.** $\dfrac{x - 2}{x + 5}$ **13.** $\dfrac{a + 6}{a + 2}$ **15.** 1

17. $-3x(x + 4)$ **19.** $-3xy^4(x - 1)$ **21.** $(x + 10)^2$

23. $(x - 8)(x + 6)^3$ **25.** $(y - 2)^3(y - 1)^2$

27. $\dfrac{(a - 7)^4}{(a + 6)^3}$ **29.** $\dfrac{(a + 2)(a - 1)^2}{(a + 1)}$

31. $4(x + 2)^2(x - 5)^4$ **33.** $\dfrac{(x + 4)}{(x - 1)}$ **35.** $\dfrac{(x - 3)}{(x + 1)}$

37. $\dfrac{(x + 2)}{x}$ **39.** $\dfrac{b + 3}{b + 2}$

41. $\dfrac{(4a - 3)}{(a - 1)}$ **43.** $\dfrac{(x + 3)}{(x + 4)}$ **45.** $a + 2$

47. $-(a - 1)$ or $-a + 1$ **49.** $-2(x - 2)$

51. $\dfrac{-3x + 10}{10}$ **53.** -1 **55.** $y^3 - 1$

57. $a(a + 1)(a - 1)$ **59.** $2a^2 + 5$ **61.** $\dfrac{x^3}{x^4 - 3}$

63. $\dfrac{1}{16a^2b^8}$ **65.** $(5x - 1)(2x - 3)$ **67.** $x \neq -3, 6$

Section 7.3 Exercises

1. $\dfrac{6a^2}{5}$ **3.** 1 **5.** $\dfrac{16a^2}{5}$ **7.** $\dfrac{9x^4}{14}$ **9.** $\dfrac{20x^5y^2}{3}$

11. $\dfrac{14}{3x^4y}$ **13.** $-15a^2bpq$ **15.** $3a$ **17.** $\dfrac{9a^3b}{2}$

19. $\dfrac{9p^4y}{11x}$ **21.** $\dfrac{-6b^3x}{y^4}$ **23.** $\dfrac{x + 2}{x + 1}$ **25.** $\dfrac{2x + 5}{x - 2}$

27. $\dfrac{4}{x - 1}$ **29.** $\dfrac{3}{4}$ **31.** $4abx^2$ **33.** $\dfrac{4a^3b}{3}$

35. $49m^3n$ **37.** $x - 1$ **39.** $(b + 1)^3(b - 7)^3$

41. $(x^3 - 7)^2(x + 1)(x - 1)$ **43.** $(y - 1)$

45. $(a - 2b)^2(a + b)$ **47.** $\dfrac{6(x - 1)}{(x - 3)}$

49. $\dfrac{(a + 2)(a + 3)}{(a - 3)(a + 1)}$ **51.** 1 **53.** $\dfrac{(2a + 1)(a - 6)(a + 1)}{(3a + 1)(a - 1)(a - 2)}$

55. $\dfrac{x(x - y)}{1 - y}$ **57.** $\dfrac{x - 2}{x + 1}$ **59.** $\dfrac{a + 3b}{a + b}$

61. $\dfrac{2ab^2}{5}$ **63.** $-6x^2y^4$ **65.** $\dfrac{-3(a - 2)(a - 1)}{2(a - 6)(a + 1)}$

67. $\dfrac{-5(x^2+4x+1)}{(x-3)(x-1)^2}$ **69.** $\dfrac{-(b-8)}{3(b+2)}$

71. $\dfrac{-2(x+3)}{(x+1)}$ **73.** $\dfrac{3(x-3)(x+1)(2x-3)}{2(x^2-3x-2)(x-2)}$

75. $\dfrac{(x+4)(x-1)}{(x+3)(x^2-4x-3)}$ **77.** $\dfrac{3(x-6)(x-1)}{(x+2)^2(x+3)}$

79. binomial; 2; 4, 2 **81.** $2x-4=x+2$

Section 7.4 Exercises

1. $3x^2$ **3.** $-2y$ **5.** $12ab$ **7.** $5x^2y^2$

9. $-12axy^2$ **11.** $40x^4y$ **13.** $5(x-2)$

15. $4ax(a+2)(a-2)$ **17.** $4x(b^4-5)$

19. $3s(s-7)$ **21.** $(a+2)(a-4)$ **23.** $5m(m-2)$

25. $9(b-4)$ **27.** $-2x(x+3)$ **29.** $4y(y+8)$

31. $(y-3)^2$ **33.** $(z-4)^2$ **35.** $b+4$

37. $-3(x-4)$ **39.** $(x-1)(3x+1)$

41. $(y+6)(y+2)$ **43.** $(a+3)(a+5)$

45. $(x+9)(x+5)$ **47.** $-7a$

49. $-k-6$ **51.** $\dfrac{5b}{b^3},\dfrac{4}{b^3}$ **53.** $\dfrac{36}{4x^2},\dfrac{x}{4x^2}$

55. $\dfrac{2(x-5)}{(x+5)(x-5)},\dfrac{4(x+5)}{(x+5)(x-5)}$

57. $\dfrac{10(y+8)}{(y+2)(y+8)},\dfrac{y+2}{(y+2)(y+8)}$

59. $\dfrac{-3(b+5)}{b^2(b+5)},\dfrac{b^4}{b^2(b+5)}$

61. $\dfrac{10a^2}{a(a-6)},\dfrac{2}{a(a-6)}$

63. $\dfrac{(x+1)(x-1)}{(x-1)(x+2)(x-3)},\dfrac{(x+4)(x-3)}{(x-1)(x+2)(x-3)}$

65. $\dfrac{-4(b+1)}{(b+1)(b-1)(b+6)},\dfrac{(b+6)^2}{(b+1)(b-1)(b+6)}$

67. $\dfrac{(x+7)(x-7)}{(x+1)(x-3)(x-7)},\dfrac{(x+3)(x-3)}{(x+1)(x-3)(x-7)}$

69. $\dfrac{(x-2)^2}{(x+1)(x-2)(x+6)},\dfrac{2x(x+1)}{(x+1)(x-2)(x+6)}$

71. $\dfrac{2}{x-5},\dfrac{3}{x-5}$ **73.** $\dfrac{-6}{x-2},\dfrac{5}{x-2}$

75. $\dfrac{2m}{m-8},\dfrac{-7}{m-8}$ **77.** $(y-7)(y-3)$ **79.** $y+2$

Section 7.5 Exercises

1. $\dfrac{1}{2}$ **3.** $\dfrac{3}{10}$ **5.** $\dfrac{2}{x}$ **7.** $\dfrac{14y}{5x}$ **9.** $\dfrac{9n}{2m}$

11. $\dfrac{2y+12}{y-6}$ **13.** $\dfrac{4a+11}{a-1}$ **15.** $\dfrac{2x+4}{5x}$

17. $\dfrac{2b+3}{b-3}$ **19.** $\dfrac{8}{b-6}$ **21.** $\dfrac{y+9}{y+8}$

23. $\dfrac{-7x-8}{x+2}$ **25.** $\dfrac{2(x+1)}{3x^2}$ **27.** $\dfrac{4a-1}{10a^3}$

29. $\dfrac{5(x-2)}{(x-6)(x-1)}$ **31.** $\dfrac{2y(4y+13)}{(y+4)(y+3)}$

33. $\dfrac{2x^2-2x+9}{(x-5)(x+2)}$ **35.** $\dfrac{-6y}{(y-1)(y-4)}$

37. $\dfrac{y^2}{(y+1)(y+6)}$ **39.** $\dfrac{-2a^2-4a+9}{(a+4)(a-1)}$

41. $\dfrac{7a-9}{(a+1)(a-3)}$ **43.** $\dfrac{2(x^2+x+4)}{(x+2)(x-2)(x+4)}$

45. $\dfrac{2b^2+3b+29}{(b-3)(b+4)(b+5)}$ **47.** $\dfrac{-x+29}{(x-2)(x+2)(x+7)}$

49. $\dfrac{5x^4-3x^3-34x^2+34x-60}{(x-2)(x+2)(x-3)(x+3)(x+4)}$

51. $\dfrac{(a+5)(a-2)}{(a+2)(a-3)(a-6)}$ **53.** $\dfrac{-a^3-8a^2-18a-1}{a^2(a+3)(a-1)}$

55. $\dfrac{-x^3+2x^2+6x+18}{4x^4(x-2)(x+3)}$

57. $\dfrac{14x^4-9x^3-2x^2+9x-36}{8x^3(x+3)(x-3)}$ **59.** $\dfrac{8x+50}{x+6}$

61. $\dfrac{3x-13}{x-6}$ **63.** $\dfrac{2a+1}{a-1}$ **65.** $\dfrac{3x^2+2x-4}{x+1}$

67. $\dfrac{2x^2+x+2}{x-1}$ **69.** $\dfrac{-5x+31}{x-4}$

71. $\dfrac{-(y^2+y+1)}{y+4}$ **73.** $\dfrac{7a^3+35a^2+85a-1}{(a+7)(a+4)}$

75. $\dfrac{-2m}{m-6}$ **77.** $\dfrac{2y-13}{5y-4}$ **79.** $x^{22}y^{14}z^{32}$

81.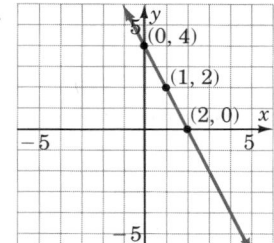

83. $(x+3)(x-2)$

Section 7.6 Exercises

1. $x=6$ **3.** $y=12$ **5.** $x=7$ **7.** $y=-9$

9. $a=15$ **11.** $b=-47$ **13.** $a=-4$

15. $y = -\dfrac{1}{2}$ **17.** $m = -3$ **19.** $x = 2$

21. $a = 6$ **23.** $b = -\dfrac{9}{5}$ **25.** $x = -2$

27. $a = 2$ **29.** No solution;   6 is an excluded value.

31. $y = -12$ **33.** $b = 8$ **35.** no solution

37. No solution; -4 is an excluded value.

39. $y = -6, 1$ **41.** $x = 4, -5$ **43.** $y = 4, -1$

45. $y = 4, -4$ **47.** $y = 6, -6$ **49.** $a = \dfrac{1}{3}, -2$

51. $a = -\dfrac{1}{3}, \dfrac{4}{3}$ **53.** $x = -\dfrac{4}{3}, -2$ **55.** $a = \dfrac{4}{5}, 1$

57. $n = \dfrac{PV}{rt}$ **59.** $W = \dfrac{P-2}{2}$ **61.** $r = \dfrac{A-P}{Pt}$

63. $S_y^2 = \dfrac{S_x^2}{F}$ **65.** $S_2 = \dfrac{2K}{h} - S_1$ or $\dfrac{2K - hS_1}{h}$

67. $h^2 = \dfrac{6V - 3\pi a^2}{\pi}$ **69.** $\dfrac{y^8}{16x^6}$ **71.** steepness

73. $\dfrac{2x^2 - 5x + 1}{(x+1)(x-3)}$

Section 7.7   Exercises

1. The number added is 5.

3. The number added is -2.

5. The number subtracted is -17.   **7.** $x = \dfrac{1}{2}, 6$

9. 2   **11.** $3\dfrac{1}{13}$ hours   **13.** two minutes

15. No. $x = -20$ hours.   **17.** $2\dfrac{8}{11}$ hours

19. First person: 12 hours; second person: 4 hours

21. 16, 17   **23.** width $= 9$ ft;   length $= 21$ ft

25. side $1 = 9$ inches;   side $2 = 12$ inches;   side $3 = 9$ inches

27. (a) $f = \dfrac{15}{4}$ ft   (b) $p = 15$ inches   (c) $q = 8$ cm

29. $1\dfrac{11}{13}$ hours   **31.** 30 hours   **33.** $m = 1$ **35.** 0

Section 7.8   Exercises

1. $\dfrac{5}{3}$ **3.** $\dfrac{y-1}{y+1}$ **5.** $\dfrac{a+c}{a-c}$ **7.** x **9.** $\dfrac{2a+7}{2a-3}$

11. $\dfrac{2m-1}{m}$ **13.** $k-1$ **15.** $\dfrac{3y^2}{(2x-y)^2}$

17. $\dfrac{-3}{x}$ **19.** $\dfrac{y-x}{xy}$ **21.** $\dfrac{y-1}{y+5}$ **23.** $3x - 4$

25. $\dfrac{4}{a^2 + 4}$ **27.** $\dfrac{(x-2)(x+1)}{(x-1)(x+2)}$

29. $\dfrac{c^2(V_1 + V_2)}{c^2 + V_1 V_2}$ **31.** $9x^2 + 24x + 16$ **33.** $x = 7$

Section 7.9   Exercises

1. $3a + 6$ **3.** $-2y + 1$ **5.** $-x(x-2)$ **7.** $3a + 1$

9. $3x^2 + x + 4$ **11.** $ab + 4a + 6b - 10$

13. $-x^2y^2 + 3xy - 4$ **15.** $\dfrac{1}{4}y^2 + \dfrac{1}{2}y + \dfrac{3}{4}$

17. $\dfrac{8x^2 - 10x^3 + 12x^4}{3a^3}$ or $\dfrac{12x^4 - 10x^3 + 8x^2}{3a^3}$

19. $\dfrac{2b^2 - 3b^3c + 4c}{a^2c}$ **21.** $1 + \dfrac{8}{x-2}$

23. $x - 3 + \dfrac{10}{x+2}$ **25.** $x - 2 + \dfrac{5}{x+1}$

27. $x - 1 - \dfrac{1}{x+1}$ **29.** $y - 2 + \dfrac{8}{y+2}$

31. $x^2 - x + 1 - \dfrac{2}{x+1}$ **33.** $x^2 + x + 1$

35. $x^2 + 5x + 11 + \dfrac{20}{x-2}$ **37.** $a^2 + a + 2 + \dfrac{8}{a-1}$

39. $y^2 + y - 2 + \dfrac{8}{y+2}$ **41.** x^2 **43.** $1 + \dfrac{1}{x+1}$

45. $2 + \dfrac{11}{y-4}$ **47.** $x + \dfrac{4}{2x-1}$ **49.** $2x - \dfrac{1}{3x+4}$

51. $2x^2 + 3x - \dfrac{2}{2x-1}$ **53.** $4x^3 + 2x - \dfrac{1}{x-1}$

55. $3 + \dfrac{2}{y^2 + y + 1}$

57. $4z^5 + 4z^4 + 2z^3 + 7z^2 + 12z + 11 + \dfrac{33}{2z-3}$

59. $x^4 - \dfrac{1}{2x+5}$ **61.** $\dfrac{x+4}{2(x-3)}$ **63.** $x = \dfrac{1}{2}$

65. $\dfrac{x-5}{4}$

Chapter 7   Exercise Supplement

1. $x \neq -4$ **3.** $x \neq \dfrac{5}{2}$ **5.** $m \neq 0, 1$

7. $s \neq -8, -\dfrac{7}{4}, 0$ **9.** $y \neq \dfrac{1}{6}, -\dfrac{5}{2}$

11. $-(4 \cdot 5) = -20, -4(5) = -20$

13. $-(7 \cdot 10) = -70, -7(10) = -70$

15. $-6a$ **17.** 9 **19.** $5m + 1$ **21.** $\dfrac{2}{x+4}$

23. $\dfrac{m+5}{m(2m+3)}$ **25.** $\dfrac{3a+4}{a(5a+6)}$ **27.** $\dfrac{5}{y-4}$

29. $\dfrac{3a^2(a^2-2)}{a^2-7}$ **31.** $-3yz^{10}$ **33.** $\dfrac{x+6}{x^2}$

35. $\dfrac{y-4}{y-1}$ **37.** $\dfrac{3r+2}{r-2}$ **39.** $a(a-2)$

41. $\dfrac{6a+1}{5a+2}$ **43.** $\dfrac{1}{6x}$ **45.** $\dfrac{x+7}{x-1}$

47. $20mr(2m^2-1)$ **49.** $\dfrac{-7a}{3a-1}$ **51.** $\dfrac{13y-5}{8y+1}$

53. $\dfrac{2a(6a-13)}{(a+6)(a-8)}$ **55.** $\dfrac{y^2-7y-23}{(y+4)(y+1)}$

57. $\dfrac{2(8a^2-a+1)}{(2a+1)(4a-3)}$ **59.** $\dfrac{3r(3r+7)}{(r-1)(r-2)(r+9)}$

61. $\dfrac{-16a^2-18a-17}{2(4a-1)(4a+1)(2a-1)}$ **63.** $\dfrac{(x+4)(x-4)}{(x-2)(x-6)}$

65. $(r+3)(r+4)$ **67.** $(x-7)(x+1)$

69. $\dfrac{7x^2-20x+1}{(x-4)}$ **71.** $\dfrac{(y+2)}{(y+3)(y+7)}$

73. $\dfrac{5x^3-26x^2-192x-105}{(x^2-2x-35)(x+1)(x+2)}$

75. $\dfrac{13x^3-39x^2+51x-100}{(2x+3)(x-7)(5x-2)}$ **77.** $a = 4$

79. $y = -2$ **81.** $a = 1$

83. No solution; $m = 8$ is excluded.

85. No solution; $b = 7$ is excluded.   **87.** $t = \dfrac{A-P}{Pr}$

89. $t = \dfrac{2mn-Qs}{Q}$ **91.** 2 **93.** No rational solution.

95. $3\dfrac{15}{16}$ hrs   **97.** 24 hrs

99. (a) 176 units of pressure;   **(b)** $\dfrac{5}{4}$ units of length

101. $a + 6$ **103.** $x^2 + 2x + 4$

105. $m^3 + 4m^2 - 1$ **107.** $a^2 + 6a - 2 - \dfrac{8}{a-9}$

109. $y^2 - 3y + 11 - \dfrac{44}{y+4}$ **111.** $y^6 + 3$

113. $x^2 - x + 1$ **115.** $y^5 + 3$

Chapter 7   Proficiency Exam

1. $a \neq -3, 8$ **2.** -3 **3.** $-2x - 5$

4. $5x^5(x-3)^2(x+5)$ **5.** $\dfrac{x+4}{x-5}$ **6.** $\dfrac{4x+3}{2x+7}$

7. $(x-3)(x+2)$ **8.** $(a+2)(a-2)(a+3)(a-4)$

9. $\dfrac{a+5}{a+6}$ **10.** $9abx$ **11.** $\dfrac{(y+3)(y+8)}{(y+1)(y-3)}$

12. $\dfrac{2(y^2-22)}{(y-8)(y-3)(y+6)}$

13. $\dfrac{62x+19}{(2x+7)(6x-1)}$ **14.** $\dfrac{8x^4(x+1)}{3(x-1)}$

15. $\dfrac{5m+1}{2}$ **16.** $\dfrac{-7y^2+15y+63}{(4y-1)(2y+3)(y+6)}$

17. $x = -2$ **18.** No solution; $m = 4$ is excluded.

19. 7 **20.** 12 hours   **21.** 8 ft by 14 ft

22. $\dfrac{4x-3}{4x+3}$ **23.** $\dfrac{x-6}{x+5}$ **24.** $x^2 + 4x - 3$

25. $2x^2 + 4x + 13 + \dfrac{25}{x-2}$

CHAPTER 8

Section 8.1   Exercises

1. two   **3.** secondary   **5.** 9 and -9

7. 11 and -11 **9.** 15 and -15 **11.** $\dfrac{1}{4}$ and $-\dfrac{1}{4}$

13. $\dfrac{5}{6}$ and $-\dfrac{5}{6}$ **15.** 0.2 and -0.2

17. 1.1 and -1.1 **19.** 8 **21.** -10 **23.** $-\dfrac{2}{3}$

25. not a real number   **27.** not a real number

29. 3 **31.** $y \geq -10$ **33.** $a \geq 16$ **35.** $k \geq \dfrac{1}{2}$

37. $x \leq -4$ **39.** m^3 **41.** a^4 **43.** $x^2 y^5$

45. $a^2 b^3$ **47.** $9ab$ **49.** $10m^4 n$ **51.** $6x^{11}y^{22}$

53. $5x^6(y-1)^2$ **55.** $3m^3 n^2(m+n)^9$

57. $(f-2)(g+6)^4$ **59.** $-8r^2 s^{11}$

61. $w + 6$ **63.** $1.1h^2 k^2$ **65.** $-\dfrac{13ab^2c^3}{14x^2y^3z^4}$

67. $\dfrac{x+1}{2x-1}$ **69.** No solution; $x = 2$ is excluded.

71. $x^2 - 2x + 7$

Section 8.2   Exercises

1. $2b\sqrt{2}$ 3. $2x^2\sqrt{6}$ 5. $a^2\sqrt{a}$ 7. $x^5\sqrt{x}$

9. $6n^4\sqrt{n}$ 11. $10x^2y^5\sqrt{xy}$ 13. $20m^3n^3\sqrt{n}$

15. $12x\sqrt{x}$ 17. $2a^2\sqrt{3}$ 19. $4x^3\sqrt{2x}$

21. $5ab^4\sqrt{2ab}$ 23. $12a^2b^8\sqrt{2ab}$ 25. $-20a^2b^3\sqrt{3}$

27. $-b^6$ 29. abc 31. $-3ab\sqrt{b}$

33. $m^3n^4p^6q^{10}$ 35. p 37. $\dfrac{1}{4}$ 39. $\dfrac{3}{7}$

41. $\dfrac{8\sqrt{6}}{3}$ 43. $\dfrac{\sqrt{14}}{7}$ 45. $\dfrac{2\sqrt{3}}{3}$ 47. $-\dfrac{\sqrt{30}}{10}$

49. $\dfrac{2a^2\sqrt{42a}}{7}$ 51. $\dfrac{\sqrt{2a}}{a}$ 53. $\dfrac{\sqrt{6x}}{x^2}$

55. $\dfrac{7xy^2z^4\sqrt{abyz}}{5a^2b^6}$ 57. $(b+2)^2$ 59. $(x+2)^3$

61. $(a-3)^2(a-1)$ 63. $(a-5)$ 65. $(a-1)^4$

67. $a \geq -\dfrac{2}{3}$ 69. positive; negative   71. $11x^2w^3z^4$

Section 8.3   Exercises

1. $2\sqrt{5}$ 3. $2\sqrt{14}$ 5. $12\sqrt{6}$ 7. 5   9. 8

11. 36   13. $\sqrt{5m}$ 15. $\sqrt{6m}$ 17. $2\sqrt{5a}$

19. $5\sqrt{3y}$ 21. a 23. y 25. 3   27. k

29. $m\sqrt{m}$ 31. x^2 33. $y^2\sqrt{y}$ 35. a^4

37. x^6 39. $y^3\sqrt{y}$ 41. $\sqrt{(x+2)(x-3)}$

43. $\sqrt{(y+3)(y-2)}$ 45. $(x+9)\sqrt{x+9}$

47. $3a^2\sqrt{5a}$ 49. $6(p-q)^4$ 51. $10m^5n^2r^4\sqrt{10mr}$

53. y^5 55. $2a^7\sqrt{5}$ 57. y^{3n} 59. $2m^{2n+2}\sqrt{5}$

61. $2(2+\sqrt{3})$ 63. $\sqrt{3x}+\sqrt{6}$ 65. $2\sqrt{2a}-2\sqrt{6a}$

67. $y^2(y+\sqrt{3})$ 69. $6m^2(m^3\sqrt{2}-1)$

71. $(a^2y^2+5w)(a^2y^2-5w)$ 73. $\dfrac{4(x+5)}{(x+3)^2}$

75. $2xy^2z^4\sqrt{3xy}$

Section 8.4   Exercises

1. $\sqrt{14}$ 3. 2   5. 6   7. 3   9. $\dfrac{6\sqrt{35}}{35}$ 11. $\dfrac{7}{15}$

13. $\dfrac{\sqrt{21}}{7}$ 15. $\dfrac{\sqrt{10}}{2}$ 17. $\dfrac{\sqrt{15}}{6}$ 19. $\dfrac{\sqrt{7}}{5}$

21. $\dfrac{5\sqrt{2}}{9}$ 23. $6m^2$ 25. $5y^3$ 27. $5a^4\sqrt{a}$

29. $6x^8\sqrt{x}$ 31. $3ab^3\sqrt{c}$ 33. $p^2q^3\sqrt{6pq}$

35. $\dfrac{\sqrt{10x}}{2}$ 37. $\dfrac{m\sqrt{2m}}{2n}$ 39. 0   41. $\dfrac{\sqrt{s^2-9}}{s-3}$

43. $\sqrt{x-6}$ 45. $\sqrt{x-1}$ 47. $\dfrac{-20+5\sqrt{5}}{11}$

49. $\dfrac{2(1+\sqrt{a})}{1-a}$ 51. $-2(\sqrt{7}-2)$ 53. $\sqrt{6}-\sqrt{2}$

55. $3+\sqrt{6}$ 57. $\dfrac{\sqrt{6y}-3y\sqrt{2}}{1-3y}$ 59. $\dfrac{a-\sqrt{ab}}{a-b}$

61. $\dfrac{\sqrt{35}-\sqrt{7}}{4}$ 63. x^9y^{11}

65.

(3, −2)

67. $a+4$

Section 8.5   Exercises

1. $2\sqrt{5}$ 3. $-15\sqrt{6}$ 5. $5\sqrt{7x}$ 7. $26\sqrt{2}$

9. $2\sqrt{2}$ 11. $12\sqrt{10}+32\sqrt{5}$ 13. $13\sqrt{15}$

15. $3x\sqrt{x}$ 17. $2xy^2\sqrt{2xy}$ 19. $-4\sqrt{6}-8\sqrt{3}$

21. $-\sqrt{5}$ 23. $5\sqrt{14}-6\sqrt{34}$ 25. $13ax\sqrt{3x}$

27. $\sqrt{6}+\sqrt{2}$ 29. $\sqrt{15}-\sqrt{10}$ 31. $2(\sqrt{6}+2)$

33. $-1+\sqrt{3}$ 35. $7(2-\sqrt{2})$ 37. $17+4\sqrt{10}$

39. $54-2\sqrt{15}$ 41. -42 43. $14+6\sqrt{5}$

45. $11-4\sqrt{7}$ 47. $4+4\sqrt{5x}+5x$

49. $64-16\sqrt{6b}+6b$ 51. $9y^2-6y\sqrt{7y}+7y$

53. -1 55. 29   57. 3   59. $x-y$

61. $\dfrac{2(6-\sqrt{2})}{17}$ 63. $\dfrac{4+\sqrt{3}}{13}$ 65. $3+\sqrt{7}$

67. $\dfrac{2\sqrt{3}-\sqrt{2}}{10}$ 69. $\dfrac{21+8\sqrt{5}}{11}$

71. $\dfrac{-16+2\sqrt{3}+24\sqrt{2}-3\sqrt{6}}{14}$ 73. $5-2\sqrt{6}$

75. $2\sqrt{6}-5$ 77. $64x^{14}y^{14}$ 79. $3x^2y^5z\sqrt{3xz}$

Section 8.6   Exercises

1. $x=25$ 3. $a=100$ 5. no solution

7. $x=0$ 9. $x=6$ 11. $a=34$ 13. $y=20$

15. $x=16$ 17. $m=3$ 19. no solution

21. $a=2$ 23. no solution   25. $m=4$

27. no solution   29. no solution

31. (a) $S=175$;   (b) $E=190$

33. (a) $F = 32$ **(b)** $r = 8$ cm

35. (a) $T = 1.57$ sec   **(b)** $L = 95.99$ cm

37. $x^{12}(x + 7)^5$ **39.** positive; negative   **41.** $7 + \sqrt{5}$

Chapter 8   Exercise Supplement

1. $2\sqrt{5}$ **3.** $12\sqrt{5}$ **5.** y **7.** $m + 3$ **9.** $x + 2$

11. $\dfrac{\sqrt{(x+5)(x+2)}}{x+2}$ **13.** 5 **15.** $\sqrt{a+3}$

17. $\sqrt{x-3}$ **19.** $3 + \sqrt{15}$ **21.** $a - \sqrt{abc}$

23. $a^2\sqrt{14} - 2a^3\sqrt{7}$ **25.** $\dfrac{2\sqrt{5}}{5}$ **27.** $8\sqrt{y}$ **29.** 1

31. $16y^2 - 3x$ **33.** $-\dfrac{2(2 - \sqrt{7})}{3}$ **35.** $\dfrac{6(x - \sqrt{y})}{x^2 - y}$

37. $\dfrac{a\sqrt{5} - \sqrt{15}}{a^2 - 3}$ **39.** $\dfrac{8\sqrt{6} - 2\sqrt{3} + 3\sqrt{2} - 16}{2}$

41. $\dfrac{6a^2b^2c^5\sqrt{bcy}}{xy^3}$ **43.** $4x^2(x - 2)\sqrt{2xy(x - 2)}$

45. $\dfrac{\sqrt{3}}{4}$ **47.** $\dfrac{3}{4}$ **49.** $\dfrac{\sqrt{6}}{6}$

51. $(x + 4)^2(x - 1)^2(\sqrt{x - 1})$

53. $(y - 3z)^6(y + 3z)^5(y - 5z)\sqrt{y - 5z}$ **55.** $12\sqrt{11}$

57. $3\sqrt{15}$ **59.** $-73\sqrt{6}$ **61.** $(2 + \sqrt{10})(x + 4)$

63. $(-20ab^3 + 6a^2b^3)\sqrt{14ab}$

65. $\dfrac{-3\sqrt{2} - \sqrt{30} - \sqrt{6} - \sqrt{10}}{2}$ **67.** $a = 64$

69. no solution   **71.** $m = 31$ **73.** $a = 43$

75. $x = 3$ **77.** $b = 1$ **79.** $x = -2$

81. (a) $S = 180$;   **(b)** $E = 258$

83. (a) $F = 15.8125$ **(b)** $r = 12.31$ cm

Chapter 8   Proficiency Exam

1. $2\sqrt{10}$ **2.** $\dfrac{2\sqrt{15}}{3}$ **3.** $\sqrt{n + 4}$ **4.** $2ab^2c^4\sqrt{6ab}$

5. $\dfrac{8x^2y^2z^3\sqrt{acy}}{7a^2bc^5}$ **6.** $(x - 2)(x + 1)^2$ **7.** $a - 4$

8. $\dfrac{8 - 4\sqrt{x}}{4 - x}$ **9.** $\dfrac{\sqrt{15} - \sqrt{6}}{3}$ **10.** $8x\sqrt{3}$

11. $-a^3b\sqrt{ab}$ **12.** $2\sqrt{5}$

13. $6 + 3\sqrt{5} + 2\sqrt{6} + \sqrt{30}$

14. $5\sqrt{2} + 2\sqrt{5} - \sqrt{15} - \sqrt{6}$ **15.** $16 - 8\sqrt{5y} + 5y$

16. $\dfrac{24 - 6\sqrt{2} - 4\sqrt{3} + \sqrt{6}}{14}$ **17.** $-\dfrac{3 + \sqrt{6} + \sqrt{10} + \sqrt{15}}{2}$

18. $x = 8$ **19.** $a = 5$ **20.** no solution

21. no real solution   **22.** $m = 6$ **23.** $a = -1$

24. no solution   **25. (a)** $S = 217$ **(b)** $E = 52$

CHAPTER 9

Section 9.1   Exercises

1. $3, 4, -7$ **3.** $2, -5, 5$ **5.** $-3, 4, -1$

7. $2, 5, 0$ **9.** $8, -2, 0$ **11.** $4, 0, 0$ **13.** $1, 1, 10$

15. $3, 5, -7$ **17.** $1, 6, -1$ **19.** $1, 0, -9$

21. $x = 0$ **23.** $a = 0$ **25.** $k = -7$ **27.** $x = -4$

29. $y = 0, 1$ **31.** $n = 0, -4$ **33.** $a = 4$

35. $x = -7$ or $x = 0$ **37.** $y = 4$ or $y = 8$

39. $x = -4$ or $x = -5$ **41.** $x = \dfrac{6}{5}$ or $x = 3$

43. $m = -\dfrac{5}{6}$ or $m = \dfrac{6}{11}$ **45.** $x = \dfrac{-5}{4}, \dfrac{7}{2}$

47. $a = \dfrac{-6}{7}, \dfrac{6}{7}$ **49.** $x = \dfrac{14}{5}, \dfrac{-10}{3}$ **51.** $y = \dfrac{-5}{2}$

53. $m = \dfrac{6}{5}$

55.

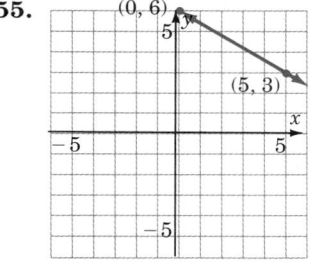

57. $\sqrt{14} + 2\sqrt{7}$

Section 9.2   Exercises

1. $x = -1, \ -3$ **3.** $x = 1, \ 5$ **5.** $x = -2, \ 4$

7. $x = -\dfrac{1}{2}, \ 7$ **9.** $x = -\dfrac{3}{4}, \ \dfrac{2}{3}$ **11.** $x = -\dfrac{5}{6}, \ \dfrac{4}{9}$

13. $x = -4, \ 0$ **15.** $y = 0, \ \dfrac{4}{3}$

17. $x = -4, \ -\dfrac{1}{2}, \ 0$ **19.** $x = 8$ **21.** $b = -7$

23. $x = 0, \ 4$ **25.** $y = 0, \ 7$ **27.** $x = -2, \ 2$

29. no solution   **31.** $a = -10, \ 10$

33. $b = 7, \ -7$ **35.** $a = 5, \ -5$

37. $y = 0, \ 1, \ -1$ **39.** $b = 2, \ -2$

41. $a = 6, \ -6$ **43.** $x = \sqrt{2}, \ -\sqrt{2}$ **45.** $b = 3, \ -3$

47. $b = 4, \ -4$ **49.** $y = -5$ **51.** no solution

53. $a = -2$ **55.** $b = 7$ **57.** $m = 0, \ -1$

59. $a = -3, \ 1$ **61.** $x = -7, \ -2$

63. $b = -9, \ -3$ **65.** $x = 6, \ 7$

67. $a = -\dfrac{5}{3}, \ -\dfrac{1}{2}$ **69.** $a = -\dfrac{1}{4}, \ -1$

71. $a = -1$ **73.** $x = -\dfrac{5}{2}, \ 3$ **75.** no solution

77. $x^{12} y^{14}$ **79.** $\dfrac{x^2 + 1}{(x+1)(x-1)(x-2)}$

81. $x = -4, \ \dfrac{-1}{3}$

Section 9.3   Exercises

1. $x = \pm 6$ **3.** $a = \pm 3$ **5.** $b = \pm 1$ **7.** $x = \pm 5$

9. $a = \pm\sqrt{5}$ **11.** $b = \pm 2\sqrt{3}$ **13.** $y = \pm\sqrt{3}$

15. $a = \pm 2\sqrt{2}$ **17.** $a = \pm\sqrt{5}$ **19.** $x = \pm\sqrt{10}$

21. $x = \pm 3$ **23.** $x = \pm 5$ **25.** $x = \pm 2\sqrt{3}$

27. $x = \pm 3b$ **29.** $k = \pm mn$ **31.** $y = \pm an$

33. $x = \pm z$ **35.** $a = b\sqrt{2}, -b\sqrt{2}$ **37.** $x = 5, -1$

39. $x = 11, -1$ **41.** $a = -8, -10$

43. $a = -4 \pm \sqrt{5}$ **45.** $x = -1 \pm \sqrt{a}$

47. $y = -2 \pm a$ **49.** $x = a \pm b$ **51.** $a = \pm 4.58$

53. $y = \pm 7.30$ **55.** $x = \pm 1.00$ **57.** $x = \dfrac{-11}{9}$

59. $x = 0, 4$

Section 9.4   Exercises

1. $x = -4, 2$ **3.** $a = -3, -4$ **5.** $y = -4, 6$

7. No real number solution.   **9.** $x = -2$

11. $b = \dfrac{-5 \pm \sqrt{37}}{2}$ **13.** $a = -5 \pm \sqrt{34}$ **15.** $x = 10, 0$

17. $a = 6, 0$ **19.** $x = -13, -1$ **21.** $a = \dfrac{-1 \pm \sqrt{3}}{2}$

23. $x = \dfrac{1}{3}, -\dfrac{5}{3}$ **25.** $x = \dfrac{-5 \pm \sqrt{57}}{4}$

27. No real number solution.   **29.** $a = \dfrac{-3 \pm \sqrt{37}}{14}$

31. $y = 0.91, -2.57$ **33.** No real number solution.

35.

(number line from -2 to 5 with points at 3 and 4)

37. $\dfrac{x+1}{x+3}$

Section 9.5   Exercises

1. $x = 3, -1$ **3.** $y = 1, 4$ **5.** $a = -2, -10$

7. $b = -2$ **9.** $x = 3, -\dfrac{1}{2}$ **11.** $x = \dfrac{1 \pm \sqrt{5}}{4}$

13. $a = 1, -\dfrac{3}{5}$ **15.** $x = \dfrac{5 \pm \sqrt{41}}{2}$

17. $a = \dfrac{1 \pm \sqrt{89}}{2}$ **19.** $b = \pm 5$ **21.** $y = \dfrac{-5 \pm \sqrt{41}}{2}$

23. $x = -1$ **25.** $a = -4, 1$ **27.** $b = -1, -2$

29. No real number solution.   **31.** $\dfrac{-2 \pm \sqrt{6}}{2}$

33. $x = 4 \pm \sqrt{3}$ **35.** $b = 6 \pm 2\sqrt{2}$

37. $x = \dfrac{1 \pm \sqrt{34}}{3}$ **39.** No real number solution.

41. $x^8 y^2 z^6$ **43.** $6y^2 + 19y - 7$ **45.** $x = -2, 6$

Section 9.6   Exercises

1. $8 or $14

3. (a) 12,036   (b) 3 and 12 years from now

5. (a) 48 feet   (b) 64 feet   (c) $t = 3$

7. length = 8;   width = 1

9. width = $\sqrt{21}$ length = $\dfrac{2}{3}\sqrt{21}$ **11.** $b = 7$; $h = 4$

13. $-9, -8$ or $8, 9$

15. $-13, -11$ or $11, 13$ **17.** $n = 2, -5$

19. $4 + \sqrt{12.5}$ inches

21. (a) carbon monoxide now 2.2 parts per million   (b) 2.37 years

23. $x = 2$ **25.** $\dfrac{3x + 14}{(x+2)(x-2)}$ **27.** $51\dfrac{3}{7}$

29. $x = 0, \dfrac{3}{4}$

Section 9.7   Exercises

1. $y = x^2$

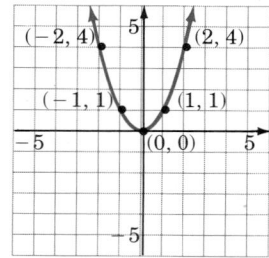

3. $y = (x - 1)^2$

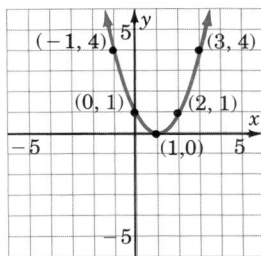

5. $y = (x + 3)^2$

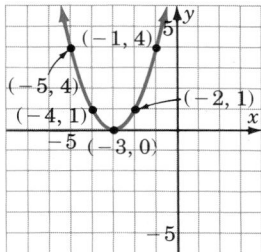

7. $y = x^2 - 3$

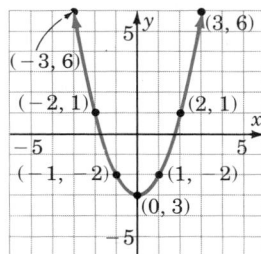

9. $y = x^2 + 2$

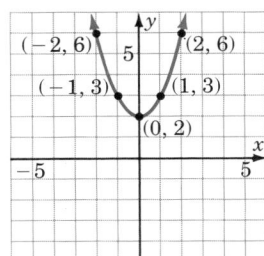

11. $y = x^2 - \dfrac{1}{2}$

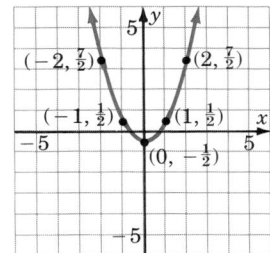

13. $y = -x^2 - 1$

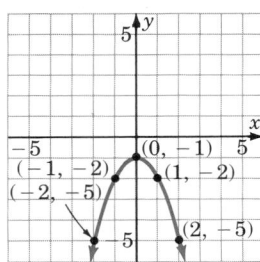

15. $y = (x + 3)^2 + 2$

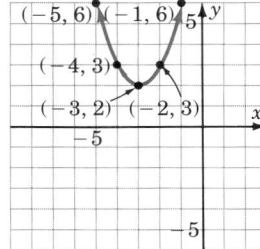

17. $y = -(x + 3)^2$

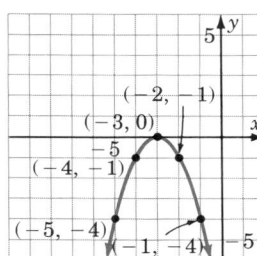

19. $y = 3x^2$

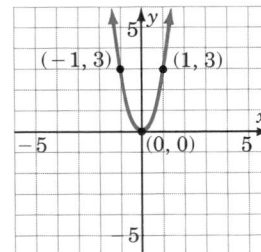

21. $y = \dfrac{1}{3} x^2$

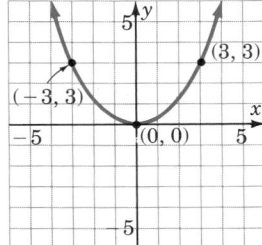

23. $y = (x - 3)^2$ **25.** $y = -(x + 3)^2 + 2$

27. $(y + 6)(y - 7)$ **29.** $\dfrac{8 - 2\sqrt{5}}{11}$

Chapter 9 Exercise Supplement

1. $x = 2, 5$ **3.** $a = -10, 5$ **5.** $m = 8, -1$

7. $x = -2, \dfrac{1}{3}$ **9.** $x = 0, -\dfrac{3}{2}$ **11.** $y = -3$

13. $y = \pm\dfrac{7}{4}$ **15.** $a = -4, -2$ **17.** $s = 1, 8$

19. $y = \dfrac{1}{6}, -2$ **21.** $m = 2$ **23.** $a = 0; \ b = -\dfrac{7}{4}$

25. $m = \pm 9$ **27.** $a = \pm 5$ **29.** $m = 5, 1$

31. $x = -7, 4$ **33.** $y = 4, -1$ **35.** $r = -\dfrac{5}{2}, 1$

37. $a = \pm 11$ **39.** $m = \pm\sqrt{2}$ **41.** $s = \pm\sqrt{10}$

43. $y = \pm 5$ **45.** $m = \pm 4np^2$ **47.** $y = \pm x$

49. $x = 5, -1$ **51.** $a = -11, -9$

53. $r = 8 \pm \sqrt{10}$ **55.** No real number solution.

57. $y = -6 \pm \sqrt{a}$ **59.** $m = \dfrac{11 \pm a\sqrt{2}}{3}$

61. $s = 4, -6$ **63.** $b = \dfrac{-1 \pm \sqrt{33}}{2}$

65. $a = \dfrac{-1 \pm \sqrt{31}}{5}$ **67.** $y = 1 \pm \sqrt{2}$

69. $r = -1 \pm \sqrt{10}$ **71.** $x = 0, \dfrac{-1 \pm \sqrt{221}}{10}$

73. $x = 0, \dfrac{1 \pm \sqrt{181}}{12}$ **75.** $x = \dfrac{1 \pm \sqrt{61}}{2}$

77. No real number solution. **79.** $r = \dfrac{-1 \pm \sqrt{161}}{10}$

81. $a = \dfrac{-3 \pm \sqrt{17}}{4}$ **83.** $r = 8 \pm \sqrt{70}$

85. No solution. **87.** $x \approx 1.29$ feet **89.** $w = 6$

91. $x = -15, -14$, or $14, 15$ **93.** $x = \dfrac{18 + 5\sqrt{3}}{3}$

Chapter 9 Proficiency Exam

1. $a = 2, b = -3, c = 10$ **2.** $a = 10, b = -3, c = 0$

3. $x = -\dfrac{5}{3}, 1$ **4.** $b = 0, \dfrac{1}{2}$ **5.** $a = 8$

6. $x = -2, 2$ **7.** $y = 4, 8$ **8.** $-2, 7$

9. $a = -\dfrac{5}{2}, \dfrac{2}{3}$ **10.** $x = -2, -\dfrac{1}{2}$ **11.** $x = 0, -5, 5$

12. $c = -9, 9$ **13.** $x = -\sqrt{15}, \sqrt{15}$ **14.** $a = -\sqrt{6}, \sqrt{6}$

15. $x = 4, 6$ **16.** $y = -8, -14$ **17.** $y = -5z, 5z$

18. $a = \pm bc\sqrt{3}$ **19.** $x = -2, 8$

20. $y = 1 \pm 2\sqrt{2}$ **21.** $m = -2 \pm \sqrt{5}$

22. $x = -b \pm c$ **23.** $x = \dfrac{-5 \pm \sqrt{33}}{2}$

24. $z = \dfrac{2 \pm \sqrt{19}}{3}$ **25.** $m = 0, \dfrac{5}{2}$

26. $x = -3 \pm \sqrt{17}$ **27.** $x = \dfrac{-7 \pm \sqrt{145}}{4}$

28. 11 and 13 or -11 and -13

29. (a) 3.1 **(b)** 6 years from now

30. width $= \dfrac{-1 + \sqrt{1793}}{4}$;

length $= \dfrac{1 + \sqrt{1793}}{4}$

31.

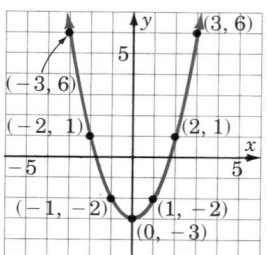

32.

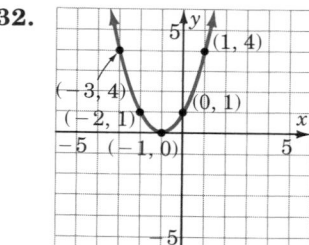

33.

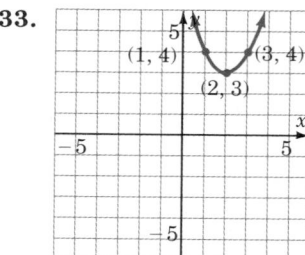

34. $y = (x - 2)^2 + 1$ or $y = x^2 - 4x + 5$

35. $y = -(x + 3)^2 - 2$ or $y = -x^2 - 6x - 11$

CHAPTER 10

Section 10.1   Exercises

1. $(-3, -2)$

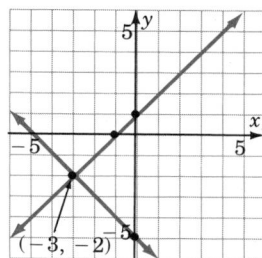

3. $(-1, 2)$

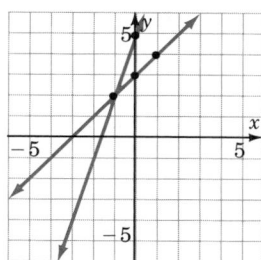

5. $\left(\dfrac{12}{13}, -\dfrac{36}{13}\right)$

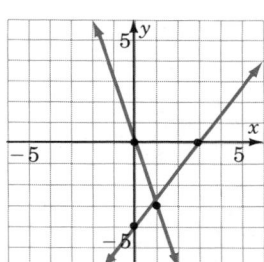

7. These coordinates are hard to estimate. This problem illustrates that the graphical method is not always the most accurate. $(-6, 6)$

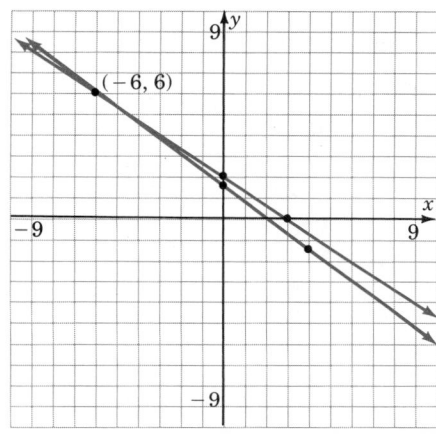

9. inconsistent

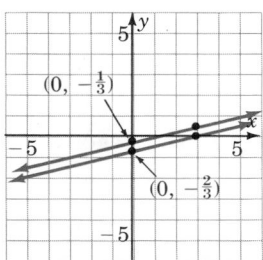

11. dependent

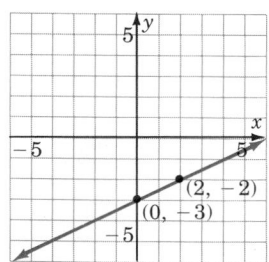

13. 4.26×10^{-4} **15.** slope

17.

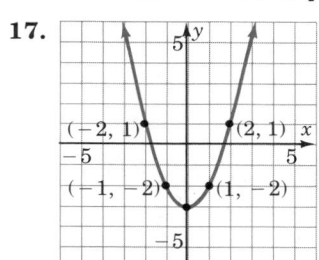

Section 10.2   Exercises

1. $(1, 3)$ **3.** $(-1, 1)$ **5.** $(2, 2)$ **7.** $\left(4, -\dfrac{1}{3}\right)$

9. Dependent (same line)   **11.** $(1, -2)$

13. inconsistent (parallel lines)   **15.** $(-1, -5)$

17. $\left(\dfrac{15}{4}, 4\right)$ **19.** $(-1, -3)$ **21.** $(4, -1)$

23. inconsistent (parallel lines)   **25.** $(1, -1)$

27. $\dfrac{(x-4)^2}{(x-5)^2}$ **29.** $-9x^4y^2z^2\sqrt{y}$

31. $(2, 1)$

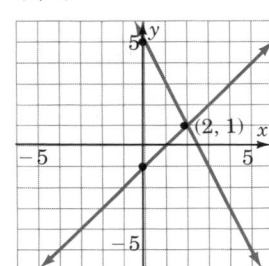

Section 10.3   Exercises

1. $(5, 6)$ **3.** $(2, 2)$ **5.** $(0, -6)$ **7.** $(2, -2)$

9. dependent   **11.** $(1, 1)$ **13.** $(-2, 2)$

15. $(-4, 3)$ **17.** $(-1, 6)$ **19.** dependent

21. $(1, 1)$ **23.** inconsistent   **25.** dependent

27. inconsistent   **29.** $(0, 4)$ **31.** $\left(-\dfrac{258}{7}, \dfrac{519}{7}\right)$

33. dependent   **35.** $17\sqrt{2}$

37. $(1, 3)$

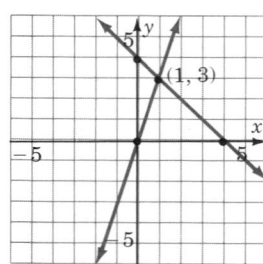

Section 10.4   Exercises

1. The two numbers are 14 and 8.

3. The two numbers are 18 and 6.

5. The numbers are 16 and 12.

7. 6 pounds of Type 1 and 8 pounds of Type 2.

9. 12 grams of Type 1 and 30 grams of Type 2.

11. 37 non-student tickets.

13. 30 ml of 80% solution; 20 ml of 60% solution.

15. **(a)** cannot form 12% solution;
(b) $\dfrac{3}{5}$ liter of 4% solution.

17. 25 ml of pure water; 975 ml of 80% salt solution.

19. 18 nickels; 60 pennies.   **21.** $9x^2 - 30x + 25$

23. $(0, -3)$

Chapter 10   Exercise Supplement

1. $(2, -3)$ **3.** $\left(\dfrac{86}{13}, -\dfrac{45}{13}\right)$ **5.** No solution.

7. $(8, -4)$ **9.** $(-6, -1)$ **11.** $(6, -1)$

13. Dependent (same line)   **15.** $(4, -2)$

17. Inconsistent (parallel lines)   **19.** $\left(\dfrac{2}{7}, \dfrac{9}{7}\right)$

21. The numbers are 14 and 21.

23. 26 pounds at 90¢;   9 pounds at 75¢

25. $28\dfrac{3}{4}$ ml of solution A; $17\dfrac{3}{4}$ ml of solution B.

27. $x = 32$, $y = 4$

29. **(a)** No solution
(b) 1.6 liters (1600 ml) of the 14% solution;
2.4 liters (2400 ml) of the 24% solution.

Chapter 10   Proficiency Exam

1. inconsistent

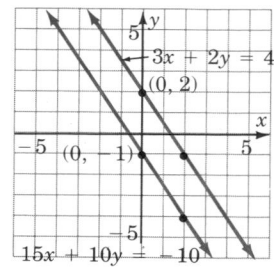

2. $(4, 2)$

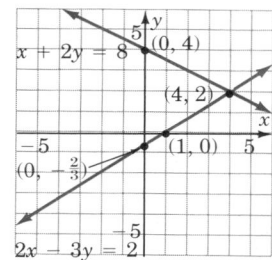

3. $(-1, 3)$ **4.** $(1, -1)$ **5.** inconsistent

6. $\left(\dfrac{4}{3}, 0\right)$ **7.** 18 and 25

8. 32 ml of solution A;   48 ml of solution B

9. 17 nickels and 15 quarters

10. \$11,250 at 8%;   \$3,750 at 12%

Index

A

Absolute value
  algebraic definition of, 48, 82
  geometric definition of, 47, 82
  notation $|a|$, 47, 82
Addition, 2
  coincident lines and, 529, 541
  method of solving system, 526
  of decimals, R-18
  of fractions, R-14
  of numbers with like signs, 50,
    51, 82
  of numbers with unlike signs,
    52, 82
  of rational expressions, 347, 350
  of signed numbers, 82
  parallel lines and, 529, 541
  with zero, 52, 82
Additive identity, 17, 52, 82
Additive inverse, 17
Algebraic expression, 88, 136
  simplifying, 104
Applied problems, 370, 486, 533
  five-step method for, 160–161, 187,
    370, 487, 533
  mathematical translations for, 155
  of quadratic equations, 486
  of rational equations, 370
  of systems of equations, 533
Approximately equal to, 411
Associative properties, 14
Axes, 240

B

Base, 21, 25
Betweenness, 175
Binary operation, 2, 36
Binomials, 100, 119–120
  conjugate of, 429
  squares of, 119, 120
Black holes, 242
Boundary line, 292, 301
  dotted, 293
  solid, 293

C

Cancelling, 317, 401
Closed circles, 11, 243
Closure properties, 13
Coefficient, 90, 126
  exponents and, 91, 126
Coin problems, 535
Combine-divide method, 384
Combining like terms, 104, 105
Common factors, 89, 126, 202
Common multiple, R-7
Commutative properties, 14
Completing the square, 471, 472, 504
Component, 180
  first, 180

second, 180
Composite number, R-4
Compound inequality, 174–175
Conditional equations, 134, 187
Conjugate, 429
Constant, 2, 36
Contradiction, 134, 187
  recognition of, 150, 187
Coordinate, 8
  of point, 246
Coordinate system, 240
  rectangular, 246, 300
Cubic polynomial, 101, 126

D

Decimal fractions, R-17
  addition of, R-18
  division of, R-18
  fraction conversion of, R-20
  multiplication of, R-18
  percent conversion of, R-25
  subtraction of, R-18
Decline, 301
Degree
  classification of polynomial
    equations by, 101
  of polynomial, 101
  of term with one variable, 101
  of term with several variables, 101
Denominator
  rationalizing of, 419, 427, 429, 446
Dependent variable, 124, 127, 180
Descending order, 106, 126
Dictionary
  mathematical, 155
Difference, 3
  of two squares, 210
Distributive property, 16, 106, 110
Division, 2
  by one, 9
  by zero, 9
  equality property of, 139
  of decimals, R-13
  of fractions, R-13
  of numbers with opposite signs, 64
  of numbers with same sign, 63
  of polynomial by monomial, 390
  of polynomial by polynomial, 393
  of rational expressions, 326, 401
  of square roots, 427, 429, 447
  property of square roots, 427, 446
Domain, 124, 127
  of rational expression, 310, 401
Double-negative property, 45, 82

E

Elimination
  by addition, 526, 541
  by substitution, 519, 541
Equality
  symbol of, 3

Equality property
  of fractions, 313, 333
Equation(s), 94, 126
  conditional, 134
  contradiction, 134, 150
  domain of, 124, 127
  equivalent, 134
  graph of, 240, 251, 300
    in slope-intercept form,
      277–278
  identity, 134, 150
  linear, 512, 541. *See also* Systems
    of equations.
    in two variables, 180–181,
      251–253, 259, 300
  literal, 135
  of line, 284–285
  quadratic, 454, 504
    standard form of, 454, 478
  rational, 360, 402
  solutions to, 134. *See also* Solution.
  solving of, 134–137, 139–140,
    145, 150
  square root, 441
  system of, 512, 541. *See also*
    Systems of equations.
  types of, 134
  variables of, 124, 127
Equivalent equations, 134
Equivalent fractions, R-9
Evaluation
  numerical, 94, 126
Expansion
  of quantity, 17
Exponent(s), R-3, 19, 25, 36
  base of, 21
  caution with, 20
  coefficients and, 91
  factors and, 68, 91
  negative, 67–68, 82
  positive, 82
  power rule for powers, 31
  power rule for quotients, 33
  power rules for, 31–33
  product rule for, 25–26
  quotient rule for, 27
  rules of, 25, 27, 31–33
  working with negative, 68
  zero as, 28
Exponential notation, R-2, R-3,
  19, 21
Expression, 88. *See also* Polynomials.
  algebraic, 88, 126
  meaningful, 412
  rational, 310, 401
Extraction of roots, 465
Extraneous solutions, 362, 402, 441

F

Factor(s), R-2, 19, 88, 194, 202
  common, 89, 126, 202
  exponents and, 68, 91

Factor(s) (*Continued*)
 greatest common, 203
Factoring, 194, 233
 by grouping, 207, 233
 collect and discard method of, 228
 difference of two squares, 210
 for perfect square trinomials,
  212, 213
 for quadratic equation, 458, 504
 for trinomials, 217, 219–220, 233
 fundamental rules of, 211, 233
 hints for, 220
 monomial from polynomial, 197,
  202, 233
 special products, 233
 trial and error method for, 233
 with leading coefficient 1, 217
 with leading coefficient other
  than 1, 222
First component, 180
Five-step method, 160, 187, 370,
  487, 533
FOIL, 112
Forms of equations of lines, 301
Formula, 94
Fraction(s), 9, 310. *See also* Rational
  expression.
 addition of, R-14
 algebraic, 310
 clearing equation of, 361, 402
 combine-divide method for, 384
 complex, 383
 decimal, R-17
  addition of, R-18
  division of, R-19
  fraction conversion of, R-20
  multiplication of, R-18
  percent conversion of, R-25
 decimal conversion of, R-21
 division of, R-13
 equality property of, 313, 333
 equivalent, R-9
 LCD-multiply-divide method for,
  386
 multiplication of, R-12
 negative property of, 313, 401
 percent conversion of, R-24
 raised to higher terms, R-10
 reduced to lowest terms, R-9
 simple, 383
From, definition of, 157
Fundamental Principle of Arithmetic,
  R-4
Fundamental rule of factoring, 211

G

Graph(s), 8
 four-dimensional, 242
 lines and, 300
 n-dimensional, 242, 300
 of equation, 240, 251, 300
 of linear equation, 300

 of parabola, 498
 of quadratic equation, 498
 one-dimensional, 241, 300
 three-dimensional, 241, 300
 two-dimensional, 241, 300
 uses of, 300
Graphing
 of compound inequality, 243
 of equation, 240, 300
 of linear equation, 243, 251–253
 of linear inequality, 243, 292–293
 of parabola, 498
 of systems of equations, 512, 514
 using any two points, 256
 using intercept method, 253
Greatest common factor, 203
Grouping
 factoring by, 207
 symbols for, 3

H

Half-planes, 292, 301
Horizontal lines
 equation of, 258, 259, 300
 slope of, 258

I

Identity, 134, 187
 additive, 17, 52, 82
 multiplicative, 17
 recognition of, 150, 187
Identity properties, 17
Inconsistent systems 513, 541
Independent systems, 513
Independent variable, 124, 127, 180
Inequality, 171, 187
 betweenness, 175
 compound, 174–175, 187
 in two variables, 292, 301
 linear, 171, 187
  algebra of, 172
  graphing of, 292–293
 notation for, 171
 symbols of, 3
Integer, 9
Intercept(s), 253, 265, 300
 method of graphing, 253, 300
 x, 253
 y, 253, 267, 268
Inverse
 additive, 17
Inverse property, 17
 additive, 17
 multiplicative, 17
Invert and multiply, R-13
Irrational number, 9, 417, 446

L

LCD-multiply-divide method, 386
Least common denominator (LCD),
  337, 401

 finding of, 338
Least common multiple (LCM),
  R-7
Like terms, 104, 126
 combining of, 104, 105
Line(s)
 boundary, 292, 293, 301
 coincident, 521, 529
 equation of, 184–185
 general form of, 264
 graphs of, 252, 277–278
 horizontal, 258
 parallel, 520, 529
 slanted, 258
 slope formula for, 269–270
 slope-intercept form of, 265, 268
 vertical, 258
Linear equality, 171, 187
Linear equation. *See also* Systems of
  equations.
 graphing of, 512, 514
 in two variables, 180–181
  general form of, 252–253,
   259, 300
  graphing of, 251–253
Linear inequality
 algebra of, 172
 graphing of, 292–293
Linear polynomial, 101, 126
Literal equation, 135
Location of solutions in the plane,
  292, 301

M

Mathematics dictionary, 155
Meaningful expressions, 412
Mixture problems, 535
Monomials, 100
 factors of, 194
 multiplication by polynomial,
  110, 126
Multiple, R-6
 common, R-7
 least common (LCM), R-7
Multiplication, R-2, 2
 equality property of, 139
 of decimals, R-2, R-18
 of fractions, R-12
 of numbers in scientific notation,
  77
 of numbers with like signs, 60
 of numbers with unlike signs, 61
 of polynomial by monomial,
  110, 126
 of polynomial by polynomial, 112,
  113, 127
 of rational expressions,
  323–324, 401
 of signed numbers, 82
 of square root expressions,
  423, 446
 of square roots, 422–423

Multiplicative identity, 17
Multiplicative inverse, 17

N

Natural numbers, 9, 36
Nature of solutions, 466
Negative exponents, 67
  working with, 68
Negative property
  of fractions, 313, 401
Negative real numbers, 8, 44, 45, 82
Notation, 2, 3
Number(s)
  addition of, 50, 51, 52
  composite, R-4
  dissociation from letters, 145
  division of, 62
  integer, 9, 36
  irrational, 9, 36
  multiplication of, 60
  natural, 9, 36
  prime, R-4
  prime factorization of, R-4
  rational, 9, 36
  real. *See also* Real number(s).
    negative, 8, 44, 82
    positive, 8, 44, 82
  scientific form of, 75–77
  signed, 44–45
  standard form of, 75–76
  subtraction of, 56
  whole, 9, 36
    multiples of, R-6–R-7
    prime factorization of, R-5
Number line, 8, 36, 240
  addition using, 52
  closed circles and, 11, 243
  construction of, 8
  open circles and, 11, 243
Number problems, 534
Numerical coefficient, 90
Numerical evaluation, 94, 126

O

Open circles, 11, 243
Operations
  binary, 2, 36
  dictionary of, 155
  order of, 4, 22
  symbols of, 2
Opposites, 45, 82
Order
  descending, 106, 126
  of operations, 4, 22
Ordered pair, 180, 186, 245
Ordering real numbers, 10
Origin, 246

P

Parabola, 498, 504

graphs of, 498
  vertex of, 498, 504
Parallel lines
  slopes of, 273
  substitution and, 520
Parentheses
  removal of, 106
Percent, R-23
  decimal conversion of, R-23
  fraction conversion of, R-26
Perfect square, 210, 416, 446
Perfect square trinomial, 212, 213
Plane, 246
  plotting points in, 247
  solutions in, 292, 301
  xy, 246
Polynomials
  classification of, 100
  cubic, 101, 126
  degree of, 101, 126
    fourth, 101
    *n*th, 100
    0th, 100
  division by monomial, 390
  division by polynomial, 393
  factoring of, 197, 202, 207
  linear, 101, 126
  multiplication by monomial,
    110, 126
  multiplication by polynomial, 112,
    113, 127
  products of, 194
  quadratic, 101, 126
  simplifying of, 104, 105, 106, 111
  subtraction of, 392
Positive real number, 8, 44, 82
Power, 21, 25
  rule for powers, 32
  rule for products, 32
  rule for quotients, 33
Prime factorization, R-4
Prime numbers, R-4
Product, R-2, 3, 88, 202
  factoring of, 194
  of polynomial, 194
  power rule for, 32
  property of square roots, 417, 446
  special binomial, 119
Properties of real numbers, 13, 36
  associative, 14, 36
  closure, 13, 36
  commutative, 14, 36
  distributive, 16, 36
  double-negative, 36, 45
  identity, 17, 36
  inverse, 17, 36
  zero-factor, 311

Q

Quadrant, 246
Quadratic equation, 454, 504
  applications of, 486–487

graphing of, 498
  solution by completing the square,
    471, 472, 504
  solution by extraction of roots, 465
  solution by factoring, 458, 504
  solution by quadratic formula, 479,
    480, 504
  solutions of, 466
  standard form of, 454, 478, 504
  zero-factor property and, 455
Quadratic formula, 479, 504
  derivation of, 480
Quadratic polynomial, 101, 126
Quotient, 3
  power rule for, 33

R

Radical, 411, 446
Radical sign, 411, 446
Radicand, 411, 446
Raised to higher terms, R-10
Rational equation, 360, 402
  applications of, 370
Rational expression, 310, 401
  addition of, 347, 402
  building, 334, 337, 401
  cancelling of, 317, 401
  combine-divide method for, 384
  complex, 383
  division of, 326, 401
  domain of, 310
  LCD-multiply-divide method
    for, 386
  multiplication of, 323–324, 401
  reducing, 318, 401
  simple, 383
  subtraction of, 347, 402
Rational number, 9
  multiplying, 318
Rationalizing the denominator, 419,
    427, 429, 446
Real number(s), 8–10, 36
  associative properties of, 14
  closed circles and, 11
  closure properties of, 13–14
  commutative properties of, 14
  distributive properties of, 15–16
  identity properties of, 17
  integer, 8
  inverse properties of, 17
  irrational, 8
  natural, 8
  negative, 8, 44, 82
  open circles and, 11
  opposites of, 45, 82
  ordering of, 10, 11
  positive, 8, 44, 82
  properties of, 13–19
  whole, 8
  zero-factor property of, 311
Real number line, 8, 36
  construction of, 8

Reciprocal, R-13, 67
Rectangular coordinate system, 246, 300
Reduced to lowest terms, R-9, 319, 401
Reducing rational expression, 318, 401
Rise, 301
Rules of exponents, 37

S

Scientific form, 75–77
  standard form conversion to, 76
Scientific notation, 75–77, 82
  working in, 77
Second component, 180
Simplifying algebraic expression, 104
  containing parentheses, 106
Simplifying square root expression, 412–413, 416–419
  involving fractions, 419, 446
  not involving fractions, 418, 446
Slanted lines
  slope of, 258, 300
Slope, 265, 267, 268, 300
  as change, 267
  as measure of steepness, 268
  formula for, 269–270, 273, 300
  negative, 273
  no slope, 273
  of horizontal lines, 273
  of parallel lines, 273
  of perpendicular lines, 273
  of slanted lines, 273
  of vertical lines, 273
  positive, 273
  zero, 273
Slope-intercept form, 265, 268
  graphing equation in, 277–278
Solution
  extraneous, 362, 402, 441
  location in plane, 292
  nature of, 466
  of equation, 134
  of equation in two variables, 180, 187
  of inequality in two variables, 292
  of system, 512, 541
    by graphing, 512, 514, 541
Special binomial products, 119–120, 127
Square
  completing of, 471, 472, 504
  perfect, 416
    difference of, 210
Square root, 410, 446
  addition of, 433, 446
  division of, 427, 429
  multiplication rule for, 423
  principal, 410, 446
  product property of, 417, 422, 446

quotient property of, 417, 446
  secondary, 410, 446
  simplified form of, 418, 419
  simplifying of, 412–413, 416–419
  subtraction of, 433, 446
Square root equation, 441, 447
Squaring binomial, 119, 120
Standard form of numbers, 75–76
Substitution
  coincident lines and, 521
  method of solving system, 519, 541
  parallel lines and, 520
Subtraction, 2
  definition of, 56, 82
  of decimals, R-18
  of fractions, R-14
  of numbers, 56
  of polynomials, 392
  of rational expressions, 347, 402
  of signed numbers, 56
Sum and difference of two terms, 120
Symbols, 2–4
  graphing, 3
  of equality, 3
  of grouping, 3
  of inequality, 3
  of operation, 2
Systems of equations, 512, 541
  applications of, 533
  dependent, 513, 541
  inconsistent, 513, 541
  independent, 513
  solutions by elimination by addition, 526, 541
  solutions by elimination by substitution, 519, 520, 521, 541
  solutions by graphing, 512, 541

T

Terms, 88, 126
  like, 104, 126
Test point, 293
Trinomials, 100
  factoring of, 217, 219–220
  perfect square, 210
  with leading coefficient, 126, 217

V

Value and rate problems, 535
Variable, 2, 36
  axes for, 240
  dependent, 124, 127, 180
  independent, 124, 127, 180
Verbal expression
  mathematical translation of, 155
Vertex, 498, 504
Vertical lines
  equation of, 258, 259, 300
  slope of, 258, 300

W

Whole numbers, 9, 36
  multiples of, R-6, R-7
  prime factorization of, R-5
Word problems
  five-step method for, 160–161

X

x-intercept, 253
xy-plane, 246

Y

y-intercept, 253

Z

Zero
  addition with, 52
  as exponent, 28
  slope, 273
Zero-factor property, 310, 455, 504
  quadratic equations and, 455